# Dairy Cattle Feeding and Management

# Dairy Cattle Feeding and Management

Sixth Edition

## William M. Etgen
Professor of Dairy Science
Virginia Polytechnic Institute and State University

## Paul M. Reaves
Professor Emeritus of Dairy Science
Virginia Polytechnic Institute and State University

**JOHN WILEY & SONS**
New York • Chichester • Brisbane • Toronto

Library of Congress number-63-20646

Library of Congress Cataloging in Publication Data

Etgen, William M.
Dairy cattle feeding and management.

Previous editions are entered under title:
Includes index.
1. Dairy farming. I. Reaves, Paul Marvin,
joint author. II. Title.
SF239.E83 1978      636.2'1'4      78-17944

ISBN 0-471-71199-3

Printed in the United States of America

10 9 8 7 6 5 4 3

# Preface

The sixth edition of *Dairy Cattle Feeding and Management* endeavors to keep pace with the many rapid changes in our great dairy industry. Advances in the science of dairy production have made it necessary to revise most of the material from the previous editions and to add much new material, including several new topics. Especially, we have added material on business management and labor efficiency.

Research results in practically every branch of the dairy industry have made the work with dairy cattle and milk production a science as well as an art. Improved methods of applying the basic principles of nutrition, genetics, physiology, engineering, and business to the management of the feeding, breeding, and handling of dairy cows and managing the dairy farm business have been developed and are being used to improve the efficiency of the dairy industry.

Most of the chapters of this edition have been completely rewritten and brought up to date, and new chapters have been added. A list of references for further study is given at the end of each chapter. Problems and exercises are included in the appendix. They serve as aids in teaching certain chapters and as a guide to students who desire to pursue further study in selected areas.

Both English and metric systems of weights and measurements are included in the text where appropriate, since the United States is currently in a transitional period with regard to the use of the metric system. Metric conversion tables are also included in the appendix.

The main purpose in preparing the present edition has been to bring together in a compact and teachable form the basic principles of dairy herd management, current knowledge in the many areas of dairy science, and present-day practices of successful dairymen. Although the book is written primarily for students in dairy cattle management and feeding courses, it will be useful to commercial dairymen, breeders of purebred cattle, herdsmen, technicians, and those who work in various areas that provide services to the dairy industry.

The first edition was the work of Carl W. Larson and Fred S. Putney. The second and third editions were revisions by Harry O. Henderson, who with Paul M. Reaves prepared the fourth edition. Reaves prepared the fifth edition, and he and William M. Etgen have written the sixth edition, with the latter taking a major responsibility. Throughout, they have worked closely together with one object in mind, to make this a teachable and useful text.

The fifth edition was translated and published in Spanish for use in Spanish-speaking countries. The Ministry of Education of the Government of India reprinted it in English for use in India and other developing countries and is now having it translated into Hindi.

Appreciation is due to many dairymen, students, former students, colleagues in the Departments of Dairy Science, Agriculture Economics, Agriculture Engineering and Veterinary Science at Virginia Polytechnic Institute and State University and to numerous persons elsewhere for many of the ideas and a great deal of the information included in this book. We also express our thanks to individuals, associations, and publishers for the use of various data and illustrations.

Paul M. Reaves
William M. Etgen

Blacksburg, Virginia
*(February 1978)*

# Contents

# Dairy Cattle Feeding and Management

# The Dairy Industry

Milk is a universal food, and the most nearly perfect. The dairy cow has been called the foster mother of the human race. The dairy cow is the basic unit of production of the dairy industry—an industry that exists because there is consumer demand for milk and dairy products. The dairy industry makes efficient use of resources and offers opportunity for profit for those concerned with the production, processing, and distribution of milk and dairy products.

It is an industry of change. Continuing research has resulted in improved cow performance; increased efficiency of production, processing, and distribution; improved wholesomeness and keeping quality of milk and dairy products; and the development of a wider variety of nutritious dairy foods. These factors, along with the increasing worldwide need for high-quality food, especially protein, form a sound basis for the continued development of the dairy industry.

## CHANGES AND TRENDS
## IN THE U.S. DAIRY INDUSTRY

The overall picture of the U.S. dairy industry in recent years has been characterized by a decreasing number of dairy farms and dairy cows, increased herd size and milk production per cow, and a slight reduction in total milk production (Table 1-1). During the period 1954–1974, total number of milk cows decreased by nearly 50% (from 21.6 to 11.3 million) and production per cow almost doubled (from 5,678 to 10,100 pounds per cow). The number of farms reporting one or more milk cows decreased from 2.17 million in 1954 to 300,000 in 1974. Reduction in dairy farm numbers, however, occurred on farms with less than 30 cows. The average herd size increased from 10.0 to 37.6 (Table 1-1). The percentage of herds with 30 or more cows has increased from 14 to 58% during this period (Table 1-2). These trends toward fewer but larger herds and fewer but higher-producing cows will greatly influence future dairy herd management. Greater management skill than ever before will be required to combine cows, land, labor, capital, and other resources into a profitable dairy enterprise.

Use of the U.S. milk supply has also changed in response to consumer preference, although the percentage of milk utilized as fluid products has remained relatively constant (Table 1-3).

Consumers spend about 20% of their disposable income for milk and dairy products. The average wholesale price of milk to producers was $9.66 per hundredweight in 1976. This amounted to $11.4 billion from the sale of milk and milk products from the farm in 1976. In addition, it is estimated that

**Table 1-1** Trends and Projections for Populations, Milk Production, Per Capita Consumption, Milk Cows, Number of Dairy Herds, and Average Milk Production Per Cow and Cows Per Farm—1954, 1964, and 1974 and Projected for 1984[a]

| Item | Year | | | |
|---|---|---|---|---|
| | 1954 | 1964 | 1974 | 1984 (projected) |
| Population (in thousands) | 161,884 | 191,141 | 212,800 | 238.000 |
| **Milk production (million lb.)** | | | | |
| On total farms reporting milk cows | 122,094 | 126,598 | 115,416 | 120,000 |
| On farms with 10 or more milk cows | 111,105 | 121,534 | 113,450 | 120,000 |
| **Per capita consumption** | | | | |
| Total milk equivalent lb. | 699 | 631 | 543 | 505 |
| **Milk cows (thousands)** | | | | |
| Total cows | 21,581 | 16,061 | 11,280 | 9,400 |
| Cows on farms with 10 or more cows | 15,590 | 14,070 | 11,110 | 9,400 |
| **Number of dairy herds (thousands)** | | | | |
| Total reporting milk cows[b] | 2,167 | 1,134 | 300 | 150 |
| Farms with 10 or more milk cows | 689 | 452 | 230 | 150 |
| **Milk production per cow** | | | | |
| On all farms | 5,678 | 8,099 | 10,100 | 12,800 |
| On farms with 10 or more milk cows | 7,120 | 8,640 | 10,200 | 12,800 |
| **Milk cows per farm** | | | | |
| All farms | 10.0 | 14.2 | 37.6 | 62.7 |
| Farms with 10 or more cows | 22.6 | 26.1 | 48.3 | 62.7 |

[a] Adapted from *Research Report 275* from the Mich. St. Univ. Agr. Exp. Sta. East Lansing, 1975, and *Milk Facts 1977*, Milk Industry Foundation, Washington, D.C.
[b] U.S. Census figures adjusted to whole population of herds for 1954 and 1964.

approximately $2 billion was received from the sale of cows, heifers, veal calves, and so on, from dairy herds, giving a total cash farm income of nearly $12 billion.

## U.S. DAIRY INDUSTRY

Herds of dairy cows are present in all 50 states (Table 1-4); however, in 1976, 65.3% of the U.S. milk supply was produced in the top 10 states in total milk production: Wisconsin 16.9%, California 9.6%, New York 8.5%, Minnesota 7.8%, Pennsylvania 6.3%, Michigan 3.8%, Ohio 3.7%, Iowa 3.4%, Texas 2.8%, and Missouri 2.5%. The second 10 states produced an additional 17.2% while the other 30 states contributed only 17.5% of the total U.S. milk production. The relative economic importance of the cash income from the sale of milk as a percentage of total farm income varied from over 50% in

**Table 1-2**  Percentage Distribution of U.S. Dairy Farms, Milk Cows, and Value of Milk Sold By Size of Herd, Herds of 10 or More Cows 1954–1984[a]

| Herd Size | 1954 | 1964 | Estimated 1974 | 1984 |
|---|---|---|---|---|
| **Percentage of farms with** | | | | |
| 10–29 cows | 86.0% | 66.1% | 42.0% | 22.0% |
| 30–49 cows | 10.8 | 24.3 | 30.5 | 34.0 |
| 50–99 cows | 2.7 | 7.4 | 20.0 | 32.0 |
| 100 or more cows | .5 | 2.2 | 7.5 | 12.0 |
| **Totals** | 100.0 | 100.0 | 100.0 | 100.0 |
| **Percentage of cows in herds of** | | | | |
| 10–29 cows | 66.6 | 39.1 | 17.5 | 6.3 |
| 30–49 cows | 18.7 | 29.2 | 24.0 | 20.4 |
| 50–99 cows | 8.7 | 18.8 | 28.9 | 35.0 |
| 100 or more cows | 6.0 | 12.9 | 29.6 | 37.3 |
| **Totals** | 100.0 | 100.0 | 100.0 | 100.0 |
| **Value of milk sold from herds of** | | | | |
| 10–29 cows | 56.5 | 34.6 | 14.5 | 5.1 |
| 30–49 cows | 21.7 | 30.4 | 23.3 | 19.4 |
| 50–99 cows | 11.3 | 20.0 | 30.5 | 35.7 |
| 100 or more cows | 10.5 | 15.0 | 31.7 | 39.8 |
| **Totals** | 100.0 | 100.0 | 100.0 | 100.0 |

[a] From *Research Report 275* from the Mich. St. Univ. Agr. Exp. Sta., East Lansing, 1975. Based on U.S. Agriculture for 1954 and 1964; 1984 is projected on basis of 1969 Census of Agriculture and recent trends and imputs from knowledgeable dairy oriented research, extension, and market organization personnel.

states such as Vermont, Wisconsin, and New York (states where profitable alternative agricultural enterprises are less available) to less than 5% in Iowa, Nebraska, and Illinois (where land and climate are well suited to other profitable agricultural enterprises).

The major areas of concentration of dairy cows are in the Northeast, the Lake states, the West Coast, and Florida. Tradition, suitability of climate and land for dairying rather than grain production, and the proximity of large population centers are some of the factors responsible for the development of dairying in these areas. In recent years (1960–1972), the biggest percentage increases in milk production have occurred in Florida (43.0%), Louisiana (31.8%), California (29.4%), and Washington (20.7%). These areas are charac-

**Table 1-3** Percentage Utilization of Milk Supply, United States, 1960-1976[a]

| Item | 1960 | 1965 | 1970 | 1971 | 1972 | 1973 | 1974 | 1975 | 1976 |
|---|---|---|---|---|---|---|---|---|---|
| **Total milk supply** | **100.0** | **100.0** | **100.0** | **100.0** | **100.0** | **100.0** | **100.0** | **100.0** | **100.0** |
| Utilization | | | | | | | | | |
| Fluid products | | | | | | | | | |
| Dealer sales | 41.4 | 43.1 | 42.8 | 42.3 | 43.1 | 44.2 | 42.0 | 43.1 | 41.2 |
| Producer sales | 1.7 | 1.5 | 1.4 | 1.3 | 1.3 | 1.2 | 1.3 | 1.3 | 1.3 |
| **Totals** | **43.1** | **44.6** | **44.2** | **43.6** | **44.3** | **45.4** | **43.3** | **44.4** | **42.5** |
| Manufacturing products | | | | | | | | | |
| Butter | 23.9 | 23.0 | 20.4 | 20.1 | 18.9 | 16.3 | 16.7 | 17.0 | 16.6 |
| Cheese | | | | | | | | | |
| American | 7.9 | 9.2 | 12.2 | 12.7 | 13.6 | 14.2 | — | — | — |
| Other | 3.0 | 3.5 | 4.5 | 5.0 | 5.4 | 5.7 | — | — | — |
| **Totals** | **10.9** | **12.7** | **16.7** | **17.7** | **19.0** | **19.9** | **22.3** | **20.8** | **23.8** |
| Ice Cream[b] | 7.7 | 8.5 | 9.4 | 9.3 | 10.8 | 11.3 | 9.7 | 10.4 | 9.5 |
| Cottage cheese | .8 | .8 | 1.1 | 1.1 | 1.1 | 1.1 | — | — | — |
| Canned and bulk | | | | | | | | | |
| condensed milk | 4.4 | 3.7 | 2.7 | 2.7 | 2.6 | 2.4 | 2.3 | 2.2 | 2.2 |
| Other products[c] | .9 | 1.0 | .8 | .9 | 1.0 | 1.2 | 2.8 | 2.4 | 2.8 |
| **Total manufactured[d]** | **48.6** | **49.7** | **51.1** | **51.8** | **51.9** | **50.6** | **53.8** | **52.8** | **54.9** |
| Used on farms | 7.4 | 4.8 | 3.4 | 3.2 | 3.0 | 2.9 | 2.9 | 2.8 | 2.6 |
| Other utilization | .9 | .9 | 1.3 | 1.4 | .8 | 1.1 | — | — | — |

[a] From *Research Report 275, Mich. St. Univ. Agr. Exp. Sta.*, East Lansing, 1975; and *Milk Facts*, 1975, 1976, and 1977 Milk Industry Foundation.

[b] Includes ice milk and other frozen dairy products.

[c] Includes dry milk, dry cream and other miscellaneous products.

[d] Residual. Includes minor uses and corrections for inaccuracies in reporting production and utilization.

**Table 1-4** Milk Cows, Production and Income by States, 1976[a]

| State | Milk cows[b] | Milk per cow | Milk production | Milk to plants and dealers | | | Farm cash receipts from milk and cream | |
|---|---|---|---|---|---|---|---|---|
| | | | | All milk | Percent Grade A[c] | Average price | Value | Percent of total Receipts[d] |
| | Thousands | Pounds | Million pounds | Million pounds | Percent | Dollars per cwt. | 1,000 dollars | Percent |
| Alabama | 88 | 7,750 | 682 | 630 | 98 | 10.80 | 72,226 | 4.7 |
| Alaska | 1.5 | 10,667 | 16.0 | 14.4 | 100 | 17.90 | 2,725 | 37.3 |
| Arizona | 69 | 12,783 | 882 | 840 | 100 | 10.10 | 89,901 | 7.0 |
| Arkansas | 88 | 8,364 | 736 | 685 | 69 | 10.00 | 70,788 | 2.9 |
| California | 811 | 14,273 | 11,575 | 11,140 | 95 | 9.27 | 1,088,678 | 11.7 |
| Colorado | 72 | 11,597 | 835 | 740 | e | 10.50 | 85,119 | 4.4 |
| Connecticut | 53 | 11,623 | 616 | 585 | 100 | 10.70 | 66,660 | 29.8 |
| Delaware | 11.9 | 10,924 | 130 | 126 | 100 | 10.40 | 13,290 | 4.9 |
| Florida | 201 | 10,234 | 2,057 | 2,010 | 100 | 11.90 | 244,315 | 9.2 |
| Georgia | 129 | 9,915 | 1,279 | 1,255 | 100 | 10.70 | 135,196 | 5.9 |
| Hawaii | 13 | 11,212 | 148 | 145.8 | 100 | 15.05 | 21,943 | 9.1 |
| Idaho | 143 | 10,867 | 1,554 | 1,490 | 37 | 9.05 | 136,492 | 10.5 |
| Illinois | 241 | 10,369 | 2,499 | 2,460 | 78 | 9.35 | 230,345 | 3.7 |
| Indiana | 211 | 10,768 | 2,272 | 2,225 | 88 | 9.85 | 221,046 | 6.9 |
| Iowa | 392 | 10,510 | 4,120 | 3,965 | 51 | 9.05 | 361,701 | 5.2 |
| Kansas | 140 | 10,436 | 1,461 | 1,395 | 84 | 9.35 | 133,497 | 3.8 |
| Kentucky | 281 | 8,480 | 2,383 | 2,150 | 81 | 9.45 | 206,710 | 12.5 |
| Louisiana | 137 | 7,942 | 1,088 | 1,045 | 100 | 11.00 | 117,308 | 9.5 |
| Maine | 59 | 10,729 | 633 | 615 | 100 | 10.80 | 67,397 | 15.6 |
| Maryland | 139 | 11,079 | 1,540 | 1,505 | 100 | 10.30 | 156,503 | 23.3 |
| Massachusetts | 54 | 11,074 | 598 | 550 | 100 | 10.70 | 65,882 | 30.2 |
| Michigan | 405 | 11,407 | 4,620 | 4,485 | 95 | 9.85 | 447,099 | 26.6 |
| Minnesota | 878 | 10,523 | 9,239 | 9,010 | 46 | 8.71 | 785,353 | 20.4 |
| Mississippi | 116 | 7,431 | 862 | 830 | 96 | 10.30 | 86,197 | 9.9 |

| | | | | | | | |
|---|---|---|---|---|---|---|---|
| Missouri | 307 | 9,681 | 2,972 | 2,870 | 62 | 9.35 | 272,341 | 5.1 |
| Montana | 26 | 10,808 | 281 | 247 | 96 | 9.45 | 24,293 | 2.5 |
| Nebraska | 142 | 9,923 | 1,409 | 1,320 | 53 | 9.20 | 122,916 | 3.1 |
| Nevada | 14 | 12,786 | 179 | 173 | 100 | 9.50 | 16,598 | 10.7 |
| New Hampshire | 31 | 10,806 | 335 | 319 | 100 | 10.60 | 35,572 | 46.4 |
| New Jersey | 46 | 11,848 | 545 | 515 | 100 | 10.30 | 56,627 | 16.4 |
| New Mexico | 31 | 13,065 | 405 | 374 | 100 | 10.70 | 44,260 | 6.1 |
| New York | 912 | 11,232 | 10,244 | 9,925 | 100 | 9.83 | 992,316 | 58.6 |
| North Carolina | 154 | 10,818 | 1,666 | 1,510 | 98 | 10.60 | 163,502 | 5.9 |
| North Dakota | 117 | 7,786 | 911 | 810 | 36 | 8.55 | 71,602 | 4.6 |
| Ohio | 397 | 11,343 | 4,503 | 4,380 | 91 | 9.90 | 436,332 | 15.4 |
| Oklahoma | 117 | 9,419 | 1,102 | 1,040 | 90 | 9.90 | 106,015 | 5.3 |
| Oregon | 91 | 11,286 | 1,027 | 945 | 84 | 9.90 | 101,495 | 9.8 |
| Pennsylvania | 706 | 10,633 | 7,507 | 7,040 | 98 | 10.40 | 777,522 | 43.3 |
| Rhode Island | 5.5 | 10,909 | 60 | 56 | 100 | 10.70 | 6,420 | 23.2 |
| South Carolina | 56 | 9,339 | 523 | 494 | 100 | 11.30 | 57,496 | 6.8 |
| South Dakota | 163 | 9,515 | 1,551 | 1,445 | 23 | 8.65 | 127,491 | 7.2 |
| Tennessee | 210 | 9,267 | 1,946 | 1,825 | 79 | 9.70 | 179,257 | 13.6 |
| Texas | 320 | 10,341 | 3,309 | 3,220 | 100 | 10.60 | 347,390 | 5.4 |
| Utah | 79 | 11,696 | 924 | 855 | 73 | 9.45 | 87,756 | 24.7 |
| Vermont | 194 | 10,954 | 2,125 | 2,070 | 100 | 10.40 | 216,954 | 83.7 |
| Virginia | 163 | 11,252 | 1,834 | 1,775 | 91 | 10.20 | 182,166 | 17.2 |
| Washington | 180 | 13,489 | 2,428 | 2,240 | 100 | 9.65 | 240,182 | 14.1 |
| West Virginia | 38 | 8,842 | 336 | 298 | e | 10.10 | 31,623 | 21.1 |
| Wisconsin | 1,807 | 11,232 | 20,296 | 19,740 | 65 | 9.16 | 1,810,417 | 59.9 |
| Wyoming | 11.5 | 9,826 | 113 | 100 | e | 9.85 | 10,453 | 2.5 |
| United States | 11,049 | 10,893 | 120,356 | 115,482 | 80.8 | 9.66 | 11,425,367 | 12.1 |

[a] USDA Data from *Milk Facts*, 1977.
[b] Average number on farms during year, including dry cows but excluding heifers not yet fresh.
[c] Percent of milk sold to plants and dealers that is approved by health authorities for fluid use.
[d] Based on preliminary estimates of total cash receipts from all farm products.
[e] Not published to avoid disclosing individual manufacturing operations.

terized by rapid increases in population and are distant from traditional milk producing areas. Largest decreases in milk production occurred in the Corn Belt states of Illinois, Indiana, Iowa, and Missouri, and the Plains states of North Dakota, Nebraska, Kansas, and Oklahoma. In these areas the smaller dairy herds did not survive the competition of alternative enterprises such as cash crops, hogs, and beef cattle.

In addition to these major areas of dairy concentration, there are numerous smaller areas of concentrated dairying, usually near urban centers or in areas suitable for efficient forage production.

## PRODUCTIVE FARMS

Dairying is an effective means of improving the productivity of the farm. The cow is a conserver of soil resources. The extended use of pastures keeps more land in sod and prevents soil erosion. Because forages and other feeds are consumed on the farm, a high percentage of the fertility value of the crop will be returned to the soil in the form of manure. Thus less fertility is sold off the farm.

## DAIRY EFFICIENCY

The dairy cow is a highly efficient animal in converting feed protein and energy (Table 1-5). As a ruminant she can also obtain as much as 70% of her total feed intake from nonhuman food sources such as forages and nonprotein nitrogen. This places the dairy cow in a strong competitive position as a major supplier of high-quality human food now and in the future.

The dairy cow has achieved her unique status as the most efficient ruminant through the efforts of agricultural science, and particularly dairy science. Science is knowledge. Scientists seek knowledge through research. Agricultural and dairy scientists have sought knowledge—basic understanding of the dairy cow—to improve her ability to produce ever-increasing amounts of nature's most perfect food—milk. They have been very successful. They must continue to be so to enable the dairy industry to help meet the world demand for high-quality food.

Improved breeding practices have increased the productivity of the dairy cattle in this country. The six dairy breeds were imported from Europe. Less than 10% of our dairy cattle are purebreds; however, most of the others are high-grade cattle, carrying many top crosses of high-performance-proved registered sires, and often are indistinguishable from purebreds either by characteristics, type, or production. Very few present-day dairy cattle can be classified as scrubs or nondescripts.

**Table 1-5** Efficiency of Various Classes of
Domestic Livestock in Converting Feed Nutrients
to Edible Products[a,b]

| Class of livestock | Efficiency of conversion of indicated nutrient, % | |
| | Crude protein | Energy |
| --- | --- | --- |
| **Nonruminants** | | |
| Broilers | 23 | 11 |
| Turkeys | 22 | 9 |
| Hens (eggs) | 26 | 18 |
| Swine[c] | 14 | 14 |
| **Ruminants** | | |
| Dairy cattle | 25 | 17 |
| Beef cattle[c] | 4 | 3 |
| Lambs[c] | 4 | — |

[a] *Ruminants as Food Producers.* Council for Agricultural Science and Technology (CAST) Special Publication No. 4, 1975.
[b] (Total lifetime production) × (100)/(total lifetime input); e.g., for milk protein the calculation is as follows: (protein in all milk produced + protein in edible cuts of carcass) × (100)/(total crude protein input in the feed).
[c] Includes only protein and energy in edible cuts. Therefore, if tallow or lard (not included here) were also included, the energy efficiency would be higher. Protein efficiency would be increased further if offal were included.

Production testing furnished the first real tool for improvement. It provided records to cull by, to serve as guides for feeding and management practices, and to identify genetically superior cattle that could serve as parents of future herd members.

Programs for sire proving or progeny testing through production testing paved the way for using genetically superior sires more extensively. The development of the techniques of artificial insemination and its extensive use in dairy cattle have made it possible to extend the use of the genetically superior sires to most dairy herds. Identification of the superior bulls through production testing of their daughters and making widespread use of those bulls through artificial insemination are the largest factors in improving the productive inheritance of the mass of dairy cattle. Today a commercial dairyman* can use as good a group of bulls as those available to any breeder, regardless of the number or quality of his cows. About 50% of U.S. dairy cattle are bred artificially.

The words "dairyman" and "dairymen" are used throughout this book without reference to sex.

A high-producing herd must be a healthy herd. Progress in control and eradication of diseases has contributed to increased production and profit. Vaccination against brucellosis, leptospirosis, blackleg, IBR, BVD, PI-3, and other diseases is a means of guarding the health of dairy herds. Nutrition, management, and effective treatments are decreasing the loss from milk fever. More judicious feeding of dry cows and fresh cows is lowering the incidence of acetonemia, or ketosis. The mastitis problem is one of the most difficult to conquer. Programs based on improved milking practices, teat dipping, and dry cow treatment offer the best solution.

Advances in nutrition and the application of better feeding practices furnish the greatest immediate means of increasing production per cow. The greater value of early-cut forages and the high energy content of silages from high-grain hybrid corn enables cows to consume more nutrients from forage in their diet. The change from higher-moisture silages to wilted grass silage, haylage, and more mature corn silage has increased forage dry-matter intake and milk-producing value. Heavier concentrate feeding to high-producing cows is paying off. The use of complete mixed rations is spreading, as this practice is conducive to increased labor efficiency and insuring the consumption of a nutritionally balanced ration. The fears of udder edema, mastitis, and burning out of cows by heavy feeding, especially of concentrates, have not been substantiated by controlled experiments.

Automatic equipment for feeding not only reduces labor and makes the job more pleasant, it can also get the job done better. Bulk handling of pelleted feeds speeds up the job of feeding and enables the cow to eat more grain during her limited time in the milking parlor. Automatic silo unloaders, in conjunction with a silage distributor in a feed bunk, let two motors do the silage-feeding job. Mix wagons with scales enhance complete ration-feeding programs and allow dairymen to provide their cattle with precise nutrient balance in each bite of feed.

Milking parlors, pipeline milkers, automated milking systems, and bulk tanks used in conjunction with free-stall, loose, and corral housing systems increase the labor efficiency. Today many herds ship 500,000 to 1 million pounds or more of milk per man per year.

## MANAGEMENT

Management can be defined as the judicious use of means to achieve goals. Managers must define goals and allocate available resources to achieve those goals. Management can also be defined as the art and science of combining resources and people to market a product profitably. This means that profit is one of the goals of a dairy operation and that the ability to work with people as well as cows, land, capital, and other resources is an important aspect of management.

Management of a dairy operation (Figure 1-1) is especially challenging, as every dairy manager has in his hands a part of "the biggest farm business in America." In addition, management of any dairy farm must be tailor-made to fit the resources and goals of that farm; both amount and type of resources vary tremendously depending on climate, soil type, available buildings and equipment, labor, capital, and so on. The manager of a dairy operation, large or small, purebred or commercial, is the key to that operation. He or she must have the ability to accurately identify problems or weak links, carefully identify and thoroughly evaluate alternatives in view of probable costs and returns, and make and implement plans to strengthen the weak links. The

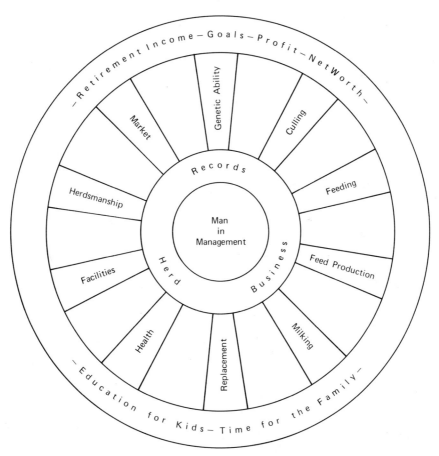

**Fig. 1-1.** The man in management is the hub of the dairy operation.

manager's decisions concerning definition of goals, allocation of resources, planning, implementing, evaluating, and revising will largely determine the success or failure of the operation.

Although dairy management must be tailor made to fit the resources and goals of individual farms, certain basic principles of business and herd management, as well as certain characteristics of successful managers do apply to most dairy operations. Some of the personal characteristics of successful managers are the following.

*1. Attitude.*   They are positive, confident, optimistic, and flexible. They have a "can do" attitude that enables them to look for ways to solve problems rather than reasons why they can't be solved. They are managers who become part of the solution rather than part of the problem. They like people and people like to be associated with them. They know how to work with people. They inspire and motivate people. They are proud to be what they are, proud of their association with the industry and optimistic about their and the industry's future.

*2. Planner.*   They set specific, achievable goals, both financial and nonfinancial, both long and short range. They carefully plan the route to goal achievement.

*3. Worker.*   They welcome work, both physical and mental, for they know that both are essential to success.

*4. Thinker.*   They gather the facts, evaluate them objectively, and consider alternatives before arriving at a decision. They are constantly looking for new ideas, techniques, and methods that will enable them to do things efficiently or more productively.

*5. Evaluator.*   They constantly evaluate records, business, and herd performance, looking for weak links. When weaknesses are identified, successful managers assign priorities and correct them. They evaluate overall progress toward goal achievement, and, when progress is slow or absent, they revise their plans to improve progress.

*6. Foresight.*   They have the ability to foresee problems and avoid them. Thus they avoid forced actions when hasty decisions without careful evaluation of the facts can be costly.

*7. Knowledge.*   They have thorough and up-to-date knowledge in all areas of dairying. They realize that research is constantly providing new knowledge that can be applied to dairying to increase productivity of both cows and labor. They read, listen, and travel to keep up to date.

These personal characteristics enable managers to make the majority of their decisions correctly and this helps ensure the success of their operations. They are often known as "lucky," but most of their luck results from an ability to plan, work, think, evaluate, foresee, and acquire knowledge. These attributes also enable managers to combine the characteristics of a profitable herd and the characteristics of a profitable business, to avoid the pitfalls to profitable dairying, and to harvest the achievement of both financial and nonfinancial goals.

## Characteristics of a Profitable Herd of Any Size

1. A breeding program that results in cattle with the genetic ability for high performance.

2. A rigid culling program that weeds out unprofitable producers.

3. A feeding program that encourages maximum economical production.

4. A feed-production program that maximizes use of available land resources and results in ample quantities of high-quality forage.

5. A milking program that results in maximum let-down of high-quality milk with minimum damage to the udder.

6. A replacement program that results in an adequate supply of healthy, well-grown, high-genetic-potential replacements ready to take their place in the herd at 24 to 26 months of age.

7. Economical, yet durable, labor-efficient buildings and equipment.

8. A preventive health-care program that results in minimal nongenetic culling and high reproductive efficiency.

9. Cowmanship—interest in and concern for cattle by those who work with the cattle.

10. A market with a high Class I usage and in a strong competitive position for the future.

## Characteristics of a Profitable Business

1. It pays all operating expenses.

2. It pays interest on all capital invested.

3. It maintains productivity.

4. It earns a reasonable return for the operator.

## Pitfalls to Profitable Dairying

1. Failure to use accurate business and performance records as the primary basis for most management decisions.

2. Low production per cow.

3. High feed cost per unit of production.

4. Low reproductive efficiency.

5. Low genetic culling and high nongenetic culling.

6. Failure to accept dairying as a business as well as a way of life, which results in unsound business practices such as:

   *a.* Overinvestment per unit of production.
   *b.* Poor investment priority.
   *c.* Too many purchased inputs.
   *d.* Improper financial planning.

7. Poor systems approach.

8. Failure to save time to manage.

Successful dairying, then, involves application of management skills in many areas. These include cattle management, land management, labor management, and business management. Goals of successful dairymen are both financial and nonfinancial. The key to successful goal achievement is the person in management. He or she is the hub of the operation.

## PROFITABLE DAIRYING

Profit is one of the goals of most dairy farms. It may be defined broadly as the difference between gross income and production expenses. Increased profit can be achieved by (1) increasing income, (2) decreasing production expenses, or (3) increasing income and decreasing production expenses concurrently.

The sale of milk accounts for approximately 90% of the gross income in average dairy operations. Therefore income can be increased significantly by (1) increasing milk production per cow, (2) milking more cows, and (3) receiving a higher price per unit of milk sold.

The remaining gross income from most dairy operations results from the sale of bull calves, cull cows, breeding stock, and extra income enterprises. Significant increases in gross income can result from maximizing income from these sources.

Feed expense, including both home-grown and purchased feed, is the largest production expense on most dairy farms. It usually accounts for 45 to 60% of the total. Increasing production per cow by maximizing intake of high-quality forage can lower production expenses significantly. Labor (10–25%), buildings and equipment annual expenses (10–25%), taxes, interest, supplies, livestock expense, utilities, etc., account for the remaining production expenses. Careful management can also decrease these production expenses per unit of milk produced.

## SUMMARY

In this chapter we have discussed some general characteristics of the dairy industry in the United States, dairy herd management, and profitable dairying. Succeeding chapters will be devoted to principles and practices that affect profitable dairying.

## REFERENCES FOR FURTHER STUDY

Council for Agricultural Science and Technology, *Special Publication No. 4.* Dept. of Agronomy, Iowa State University, Ames, Iowa, 1975.

Hoglund, C. R., *Research Report 275,* Michigan State University, East Lansing, 1975.

*Milk Facts,* Milk Industry Foundation, Washington, D.C., 1977 (published annually).

# 2

# Dairy Herd Records

**2**

A dairy herd program cannot be conducted in an efficient, businesslike manner without accurate records. Accurate records provide the information needed to make objective management decisions based on facts. A good herd manager may be defined as one who has the ability to make the majority of his or her decisions good ones. Accurate herd records provide the factual information needed for good herd decisions. They are a valuable management tool.

Records that are accurately maintained should be the basis for most management decisions on dairy farms. Accurate individual cow performance records should be the basis of a selection program. They should be used to: (1) identify cows to remain in the herd and those that should be culled, (2) select herd replacements, and (3) select bulls to use in the herd or to sell for herd sires. Accurate recording of heats, breedings, calvings, vaccinations, disease diagnosis, and treatment are important in sire identification, registration of purebreds, and preventive health care.

One essential of records is that they be simple, though they should contain all the necessary information. Many dairymen become discouraged in the keeping of records because the system is too complicated. The correct record system for any farm is the one that will provide the necessary information, yet will be easily and accurately maintained.

## PRESERVATION OF PERMANENT RECORDS

As many records as possible should be final and not require copying. This is not possible with all forms of records, however.

There are three general methods of preserving records. Individual conditions will determine which one is best suited to each case. Records may be kept in any of the following ways: (1) in books with permanent leaves, (2) in loose-leaf books or files, and (3) in envelopes.

There are advantages in using books with permanent leaves for some records, whereas loose-leaf books, files, or envelopes are better for others. For breeding records, the permanent-leaf book has the advantage that separate pages cannot be lost; but it is cumbersome, as old records must be handled frequently, there is the danger that a whole book may be lost. The loose-leaf records, however, may be divided so that only those records that are in use at a particular time need be kept at hand; records that are used only for reference, such as those of cattle that have gone out of the herd, may be put away for safekeeping. Loose-leaf records also have the advantage that all data relating to an individual animal may be kept together in one book. Files have the same advantages as the loose-leaf book, but are not so easily carried around, nor are they so convenient to handle.

In the envelope system all records pertaining to each cow are kept in one envelope. This system is not so convenient as some of the others, but is often used for the filing of registration papers.

## KINDS OF HERD RECORD

Several kinds of herd record should be kept on a dairy farm. These are (1) individual cow and herd performance (production) records, (2) individual identification and parentage records, and (3) individual breeding and health records. These may be incorporated collectively in the Dairy Herd Improvement (DHI) program, or separate records may be kept for (2) and (3).

Figures 2-1 and 2-2 give a lifetime permanent record of an animal. This is used as a loose-leaf record. The front side (Figure 2-1), in addition to the name, registration number, date of birth, and so on also has spaces for complete breeding, calving, and lactation records for the lifetime of the animal. The reverse side (Figure 2-2) has space for three generations of ancestors, health record, progeny information, and solids-not-fat information. Many purebred breeders find that a record of this type is convenient in recording information needed to help sales of breeding stock. This form can be modified to include more or less information as desired by the individual.

For large herds a herdsman's daily or weekly report is needed. Figure 2-3 has proved to be a satisfactory type. Modifications may be incorporated into this form to meet individual needs. It may be bound in pads with alternate sheets perforated and carbon sheets to make duplicate copies. One copy can then be kept at the barn and the other taken to the house or office for transfer to permanent records.

## PERFORMANCE-TESTING PROGRAMS

Individual cow and herd performance records may be under either one or both of two types of testing. These are the Dairy Herd Improvement program (DHI) and the Dairy Herd Improvement Registry (DHIR). The DHI program includes four types of production testing: (1) Standard DHI, (2) Owner-Sampler, (3) AM-PM Program, and (4) Weigh-A-Day-A-Month. The DHIR is the official testing program of the Breed Registry Associations.

20

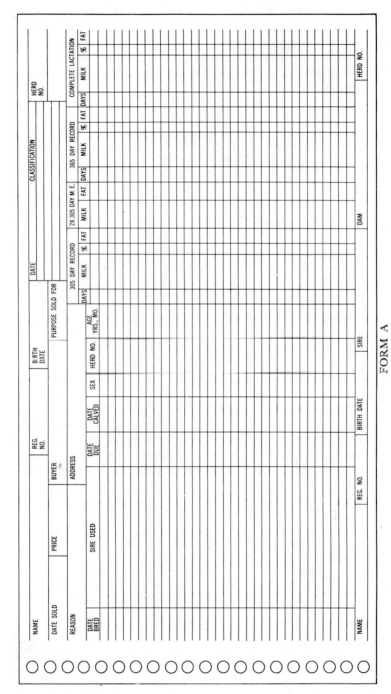

**Fig. 2-1.** Form A. Lifetime record.

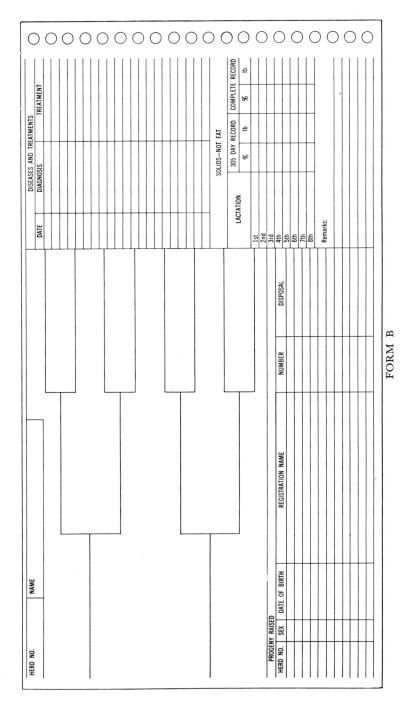

**Fig. 2-2.** Form B. Lifetime record, continued.

21

Date_____, 19 __

Gallons of milk shipped_____

| FEED GROUND AND MIXED | | | MILK FED TO CALVES | |
|---|---|---|---|---|
| | | | | Number Calves |
| Mixture | Amount | Lb. | | Fed |
| _____ | _____ | Whole Milk_____ | | _____ |
| _____ | _____ | Skim Milk_____ | | _____ |
| _____ | _____ | | | |

COWS IN HEAT

_____  _____  _____
_____  _____  _____

BREEDING RECORD

| Name | No. | Sire Used | Artificial |
|---|---|---|---|
| _____ | _____ | _____ | _____ |
| _____ | _____ | _____ | _____ |
| _____ | _____ | _____ | _____ |
| _____ | _____ | _____ | _____ |

CALVING RECORD

| Name of Cow | Sex of Calf | Weight of Calf | Herd Number | Normal Calving |
|---|---|---|---|---|
| _____ | _____ | _____ | _____ | _____ |
| _____ | _____ | _____ | _____ | _____ |
| _____ | _____ | _____ | _____ | _____ |

ANIMALS TREATED

| Name | Diagnosis | Treatment |
|---|---|---|
| _____ | _____ | _____ |
| _____ | _____ | _____ |

ANIMALS SOLD

| Name | Purchaser |
|---|---|
| _____ | _____ |
| _____ | _____ |

Remarks: _____

_____

Signed: _____
Herdsman

**Fig. 2-3.** Form C. Herdsman report.

## Dairy Herd Improvement (DHI) Program

The DHI is a performance testing program sponsored and supervised by the USDA and the State Extension Service in cooperation with the participating dairyman (Figure 2-4). In a Memorandum of Understanding between the Cooperative Extension Service of the State and Animal Husbandry Research Division, Agricultural Research Service of the USDA, adopted by the coordinating groups for National DHI programs, the following purposes and objectives are set forth.

A. Purposes
   1. To improve the producing ability of dairy cattle by providing guides for breeding, feeding, and management practices.
   2. To provide information by which the breeding value and transmitting ability of individual cows and sires can be measured reliably and the superior ones selected and used for breeding purposes.
   3. To improve the efficiency and financial position of all dairymen through improved herd management.
   4. To provide data and results for dairy extension workers and others in developing and conducting effective educational and demonstrational programs.
   5. To provide data to state experiment stations and to the Division for Research.

B. Objectives
   1. To maintain a uniform system of record-keeping to be used in guiding cooperating dairymen in their herd improvement program and to supply reliable records and herd improvement information to the Division, Agency, Service and other cooperating groups.
   2. To establish sources of superior dairy inheritance as a means of improving the producing ability of all dairy cattle.
   3. To maintain uniformity and a high standard of integrity in the Program that will insure reliability to scientific studies and educational demonstrations.
   4. To summarize and analyze information from DHI herds and make these summaries available to dairymen, research, educational, and extension workers, and to others as appropriate. No copyright shall exist in the materials published pursuant hereto.
   5. To demonstrate the fundamental benefits of the dairy herd improvement program to dairy farmers.

The information recorded and summarized in the DHI program has been a primary basis for improved dairying in herds enrolled on the program, as well as improving the U.S. dairy industry. Few businesses in America have record systems that compare in usefulness and efficiency at such reasonable cost as the Cooperative National DHI Program. Building on a background of

24

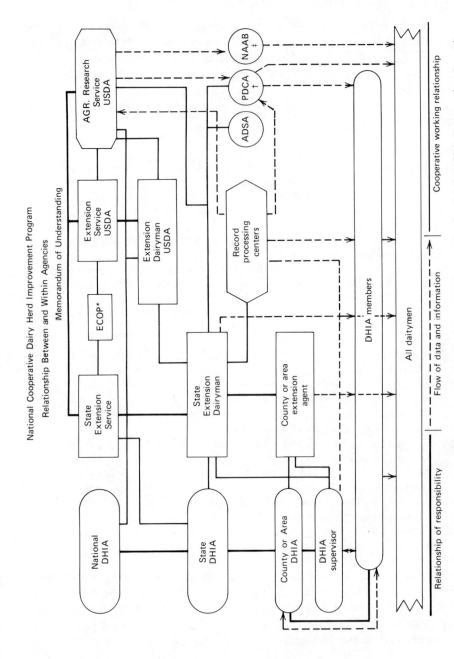

**Fig. 2-4.** Organizational structure of the National D.H.I. Program [*USDA Agr. Handbook* 248 (1973)]. (* Extension Committee on Organization and Policy. † Includes breed associations. ‡ Includes bull studs.)

70 years of organized dairy record keeping, the program has evolved and continues to improve as new data-processing techniques become available. The system always has a backlog of data needing manipulation for more effective use in the daily decision-making process. This program, when properly used, has proved over and over again to result in efficient dairy production. The organizational setup is such that data collected from thousands of farms nationwide are sent to a central point, summarized, and analyzed; this information is available to the entire industry.

In DHI testing programs all cows in the herd, both purebreds and grades, are included. Yearly herd averages and individual cow records of production, feed consumption, feed costs, and returns are determined and are available from the dairyman's herd book and the State Extension office. Lactation records are reported on the first 305 days of each lactation.

**Standard DHI.** A DHI supervisor visits the farm periodically (usually monthly) to weigh and sample milk from each cow for a 24-hour period. The supervisor may test the sample for butterfat (required) and leucocyte count (optional in most states), or send the sample to a central testing laboratory for butterfat and leucocyte testing. Various herd and individual cow information such as milk price, feed intake and cost, calving information, dry dates, breeding information, etc., is recorded on the DHI Barn Sheet, form DHI-201, as shown in Figure 2-5. Forms in Figures 2-5 to 2-14 are reproduced from *Dairyman's DHI Manual* of the Southern Regional Processing Center at Raleigh, N.C. These forms may vary somewhat from those used by other DHI processing centers. These data are processed and summarized at a central processing center using high-speed electronic data-processing machines. The summary information on an individual cow basis (Figure 2-6) and herd basis (Figures 2-7 and 2-8) is returned to the dairyfarmer and a copy is sent to the State Extension office. A copy of all 305 lactations from cows whose sire is identified is forwarded to the USDA for use in sire summaries.

Careful analysis of the monthly report (Figure 2-6) reveals that it contains, in addition to herd identification and administrative information (items (1) through (3), cow identification (4), individual cow performance data concerning milk and butterfat production and percentage of butterfat for the test day (5), and current lactation on an actual and projected Mature Equivalent basis (6), the value of milk produced (6), income over feed cost (6), and other management factors such as breeding record, action code, etc., (7). These are the facts that can enable managers to identify profitable and unprofitable cows and to cull the unprofitable ones. These facts also provide a basis for management decisions in areas such as feeding and breeding.

The monthly herd summary and management report, DHI Form 202, as shown in Figures 2-7 and 2-8, contains average per cow test day information,

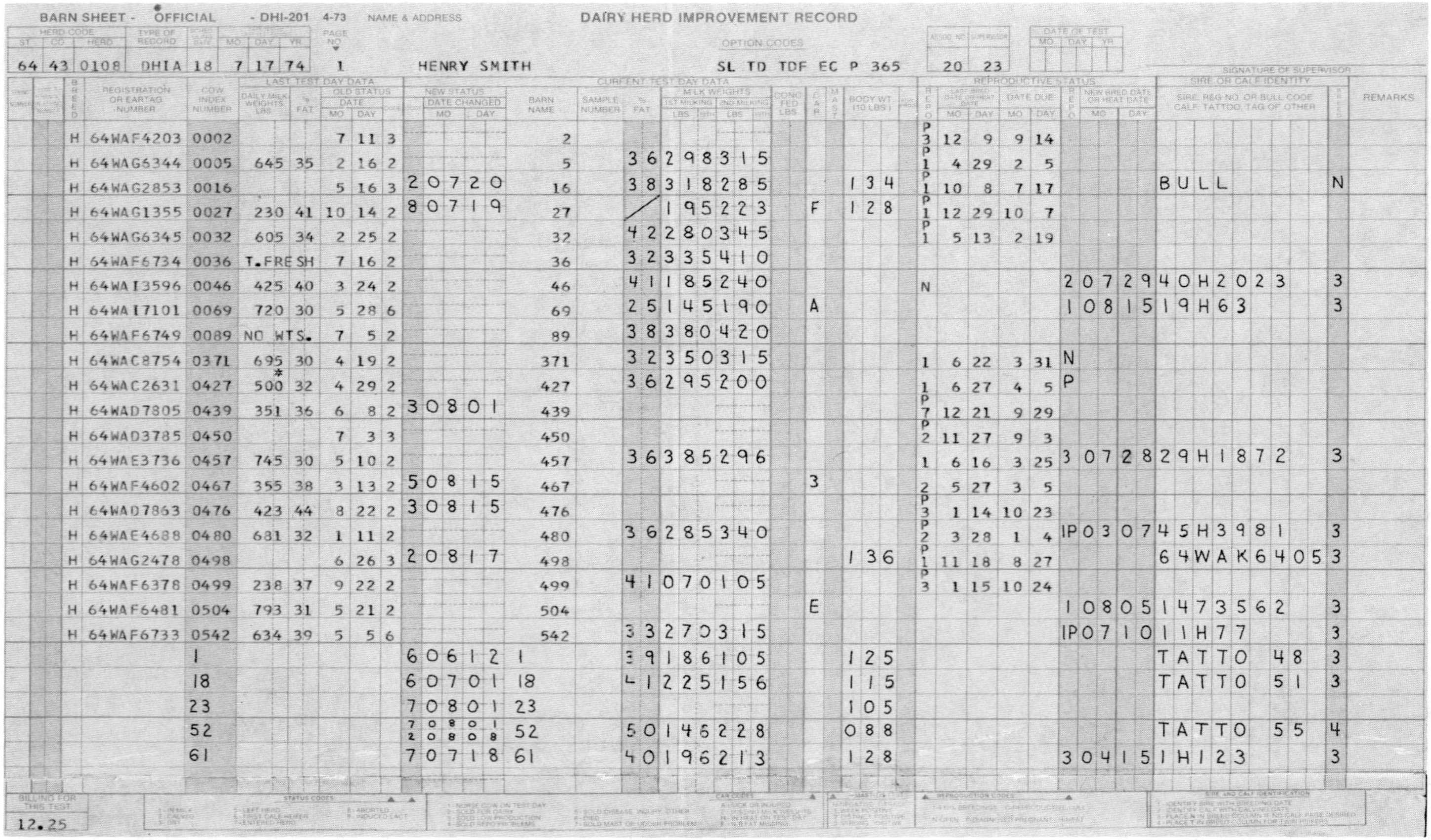

BARN SHEET - OFFICIAL - DHI-201 4-73 NAME & ADDRESS

DAIRY HERD IMPROVEMENT RECORD

| Herd code (St, Co, Herd) | Type of record | | Mo Day Yr | Page no. | Name & address | Option codes | | |
|---|---|---|---|---|---|---|---|---|
| 64 43 0108 | DHIA | 18 | 7 17 74 | 1 | HENRY SMITH | SL TD TDF EC P 365 | 20 | 23 |

| Breed | Registration or eartag number | Cow index number | Last test day milk weights lbs. | % fat | Old status Mo | Day | Code | New status date changed | Barn name | Current test day data (% fat, milk weights) | Car | Body wt. (10 lbs.) | Repro | Last bred date Mo Day | Date due Mo Day | New bred date or heat date | Sire reg. no. or bull code, calf tattoo, tag of other | Code |
|---|---|---|---|---|---|---|---|---|---|---|---|---|---|---|---|---|---|---|
| H | 64WAF4203 | 0002 | | | 7 | 11 | 3 | | 2 | | | | P3 | 12 9 | 9 14 | | | |
| H | 64WAG6344 | 0005 | 645 | 35 | 2 | 16 | 2 | | 5 | 36298315 | | | P1 | 4 29 | 2 5 | | | |
| H | 64WAG2853 | 0016 | | | 5 | 16 | 3 | 20720 | 16 | 38318285 | | 134 | P1 | 10 8 | 7 17 | | BULL | N |
| H | 64WAG1355 | 0027 | 230 | 41 | 10 | 14 | 2 | 80719 | 27 | / 195223 | F | 128 | P1 | 12 29 | 10 7 | | | |
| H | 64WAG6345 | 0032 | 605 | 34 | 2 | 25 | 2 | | 32 | 42280345 | | | P1 | 5 13 | 2 19 | | | |
| H | 64WAF6734 | 0036 | T.FRESH | | 7 | 16 | 2 | | 36 | 32335410 | | | | | | | | |
| H | 64WAI3596 | 0046 | 425 | 40 | 3 | 24 | 2 | | 46 | 41185240 | | | N | | | 20729 | 40H2023 | 3 |
| H | 64WAI7101 | 0069 | 720 | 30 | 5 | 28 | 6 | | 69 | 25145190 | A | | | | | 10815 | 19H63 | 3 |
| H | 64WAF6749 | 0089 | NO WTS. | | 7 | 5 | 2 | | 89 | 38380420 | | | | | | | | |
| H | 64WAC8754 | 0371 | 695* | 30 | 4 | 19 | 2 | | 371 | 32350315 | | | 1 | 6 22 | 3 31 | N | | |
| H | 64WAC2631 | 0427 | 500 | 32 | 4 | 29 | 2 | | 427 | 36295200 | | | 1 | 6 27 | 4 5 | P | | |
| H | 64WAD7805 | 0439 | 351 | 36 | 6 | 8 | 2 | 30801 | 439 | | | | P7 | 12 21 | 9 29 | | | |
| H | 64WAD3785 | 0450 | | | 7 | 3 | 3 | | 450 | | | | P2 | 11 27 | 9 3 | | | |
| H | 64WAE3736 | 0457 | 745 | 30 | 5 | 10 | 2 | | 457 | 36385296 | | | 1 | 6 16 | 3 25 | 30728 | 29H1872 | 3 |
| H | 64WAF4602 | 0467 | 355 | 38 | 3 | 13 | 2 | 50815 | 467 | | 3 | | 2 | 5 27 | 3 5 | | | |
| H | 64WAD7863 | 0476 | 423 | 44 | 8 | 22 | 2 | 30815 | 476 | | | | P3 | 1 14 | 10 23 | | | |
| H | 64WAE4688 | 0480 | 681 | 32 | 1 | 11 | 2 | | 480 | 36285340 | | | P2 | 3 28 | 1 4 | IP0307 | 45H3981 | 3 |
| H | 64WAG2478 | 0498 | | | 6 | 26 | 3 | 20817 | 498 | | | 136 | P1 | 11 18 | 8 27 | | 64WAK6405 | 3 |
| H | 64WAF6378 | 0499 | 238 | 37 | 9 | 22 | 2 | | 499 | 41070105 | | | P3 | 1 15 | 10 24 | | | |
| H | 64WAF6481 | 0504 | 793 | 31 | 5 | 21 | 2 | | 504 | | E | | | | | 10805 | 1473562 | 3 |
| H | 64WAF6733 | 0542 | 634 | 39 | 5 | 5 | 6 | | 542 | 33270315 | | | | | | IP0710 | 11H77 | 3 |
| | | 1 | | | | | | 606121 | 1 | 39186105 | | 125 | | | | | TATTO 48 | 3 |
| | | 18 | | | | | | 60701 | 18 | 41225156 | | 115 | | | | | TATTO 51 | 3 |
| | | 23 | | | | | | 70801 | 23 | | | 105 | | | | | | |
| | | 52 | | | | | | 70801 / 20808 | 52 | 50146228 | | 088 | | | | | TATTO 55 | 4 |
| | | 61 | | | | | | 70718 | 61 | 40196213 | | 128 | | | | 30415 | 1H123 | 3 |

BILLING FOR THIS TEST 12.25

**Fig. 2-5.** D.H.I. barn sheet.

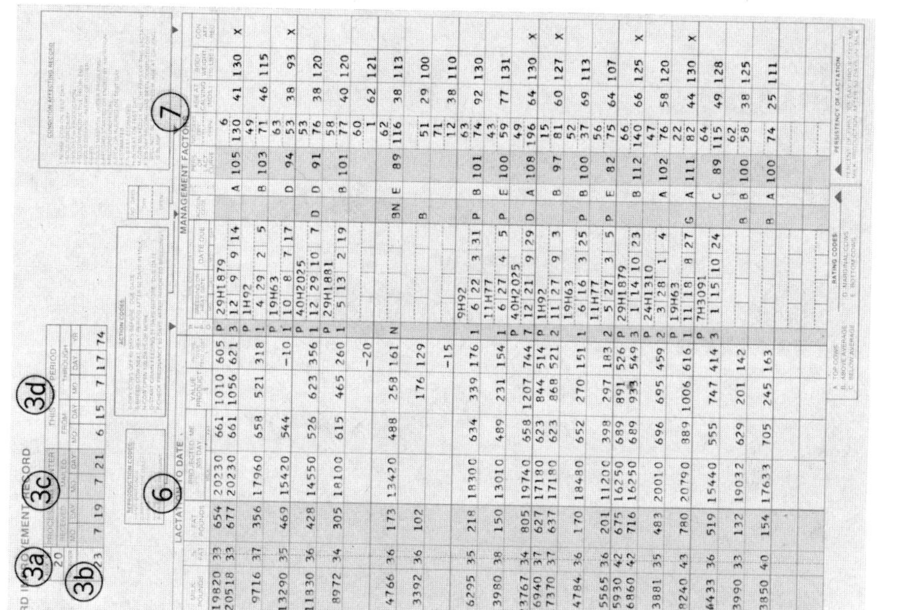

**Fig. 2-6.** D.H.I. monthly report.

## ③ PRODUCTION, INCOME, AND FEED COST SUMMARY

| DESCRIPTION | DAILY AVERAGE PER COW ON TEST DAY | | YEARLY HERD TOTALS | ROLLING YEARLY HERD AVERAGES | |
|---|---|---|---|---|---|
| TOTAL COWS | 79 | | 27,476 | 75.2 | |
| % IN-MILK | 92 | | ///// | 87 | |
| MILK LBS.   (ALL COWS) | 41.6 | | 1,116,239 | 14,828 | |
| B'FAT LBS.   (ALL COWS) | 1.41 | | 41,720 | 554 | |
| B'FAT PERCENT | 3.4 | | ///// | 3.7 | |
| MILK LBS.   (MILKING COWS) | 45.2 | | ///// | ///// | |
| B'FAT LBS.   (MILKING COWS) | 1.54 | | ///// | ///// | |
| | LBS. CONSUMED | ENE VALUE | LBS. CONSUMED | LBS. CONSUMED | % ENE |
| SILAGE | 55 | 19 | 1,442,535 | 19,183 | 35 |
| | LBS. CONSUMED | ENE VALUE | LBS. CONSUMED | LBS. CONSUMED | % ENE |
| OTHER SUCCULENTS | | | 95,224 | 1,266 | 3 |
| | LBS. CONSUMED | ENE VALUE | LBS. CONSUMED | LBS. CONSUMED | % ENE |
| DRY FORAGE | 5 | 38 | 257,380 | 3,423 | 13 |
| | LBS. CONSUMED | ENE VALUE | LBS. CONSUMED | LBS. CONSUMED | % ENE |
| OTHER FEEDS | | | 30,843 | 410 | 1 |
| | QUALITY | | TOTAL DAYS | DAYS | % ENE |
| PASTURE | | | 9,550 | 127 | 3 |
| | LBS. CONSUMED | ENE VALUE | LBS. CONSUMED | LBS. CONSUMED | % ENE |
| CONCENTRATES | 16 | 72 | 480,950 | 6,396 | 45 |
| VALUE OF PRODUCT $ | 3.82 | | 96,635 | 1,285 | |
| COST OF CONCENTRATES $ | 1.19 | | 24,418 | 324 | |
| TOTAL FEED COST $ | 1.68 | | 37,978 | 505 | |
| INCOME OVER FEED COST $ | 2.14 | | 58,657 | 780 | |
| FEED COST PER CWT. MILK $ | 4.04 | | ///// | 3.40 | |
| | PER CWT | % FAT | | PER CWT | % FAT |
| MILK BLEND PRICE | 9.79 | 3.5 | ///// | 8.66 | 3.8 |

④ BILLING FOR PERIOD FOLLOWING DATE OF TEST

12.26

**Fig. 2.7.** D.H.I. herd summary (1).

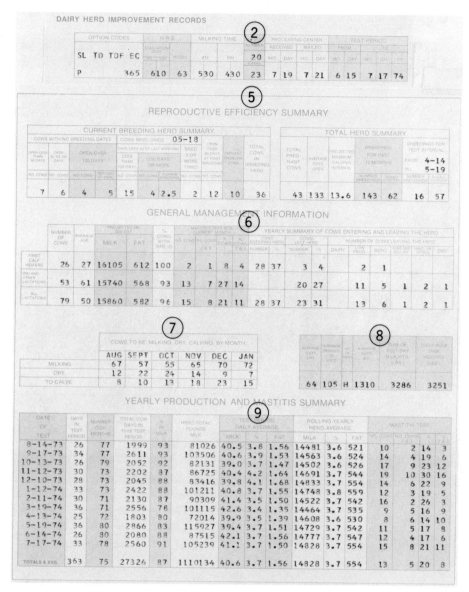

**Fig. 2-8.** D.H.I. herd summary (2).

yearly herd totals, rolling herd averages (3), and summaries of other herd management factors (5), (6), (7), (8) and (9) that affect herd performance and profitability. Information contained in this report is especially useful in identifying weak and strong points in total herd performance.

The individual cow record, DHI Form 203, as shown in Figure 2-9,

**INDIVIDUAL COW RECORD OFFICIAL** — **DAIRY HERD IMPROVEMENT RECORD**

DHI-203  673

| HERD CODE NO | | | NAME & ADDRESS | (1a) | DATE MAILED MO DAY YR | BARN NAME | INDEX NUMBER |
|---|---|---|---|---|---|---|---|
| ST. | CO. | HERD | HENRY SMITH, SMITH DAIRY FARMS, RFD 3, AUBURN ALA. 36830 | | 7 21 74 | 450 | 0450 |
| 64 | 43 | 0108 | | | | | |

| BARN NAME | REGISTRATION OR EARTAG NUMBER | B R | | REGISTRATION OR EARTAG NUMBER | B R | | REGISTRATION OR EARTAG NUMBER | B R | DATE OF BIRTH MO DAY YR |
|---|---|---|---|---|---|---|---|---|---|
| 371 | 64WAC8754 | H | | 1092490 | H | | 64WAD3785 | H | 7 24 68 |
| DAM | | | | SIRE | | | COW | | |

**LACTATION PRODUCTION SUMMARY (1b)**

| CALVING DATE MO DAY YR | AGE AT CALVING (MOS.) | DAYS DRY BEFORE CALVING | CON. AFF REC | 3X | 305 DAY LACTATION MILK | FAT | FAT | COMPLETE LACTATION DAYS IN MILK | MILK | FAT | VALUE PRODUCT | INCOME OVER FEED COST | LIFETIME PRODUCTION MILK | FAT | TYPE OF RECORD | REMARKS |
|---|---|---|---|---|---|---|---|---|---|---|---|---|---|---|---|---|
| 8 01 70 | 25 | | C | | 12260 | 35 | 426 | 305 | 12260 | 426 | 631 | 316 | 12260 | 426 | DHIA | |
| 8 07 71 | 37 | 67 | C | | 13830 | 34 | 472 | 323 | 14070 | 482 | 722 | 373 | 26330 | 908 | DHIA | |
| 8 24 72 | 49 | 59 | B | | 16240 | 36 | 585 | 278 | 16240 | 585 | 812 | 421 | 42570 | 1493 | DHIA | |
| 7 28 73 | 60 | 60 | B | | 16840 | 37 | 617 | 321 | 17370 | 637 | 896 | 453 | 59940 | 2130 | DHIA | |

| STATUS CHANGE (1c) | MO DAY YR | LAST TEST DAY MILK WT %FAT | | ESTIMATED RELATIVE PRODUCING ABILITY (1d) MILK | FAT |
|---|---|---|---|---|---|
| TURNED DRY | 6 17 74 | | | +680 | +9 |

**CALF AND BREEDING RECORD (1e)**

| CALVING DATE MO DAY YR | BODY WT (10 LBS) | DAYS OPEN | NO. BR | SUCCESSFUL BREEDING DATE MO DAY YR | SIRE IDENTITY REG. EARTAG OR OTHER NUMBER | CALF IDENTITY EARTAG OR OTHER | ME LACTATION MILK | FAT | ADJUSTED HERDMATE AVERAGE MILK | FAT | HERDMATE DEVIATION MILK | FAT |
|---|---|---|---|---|---|---|---|---|---|---|---|---|
| 8 01 70 | 91 | 90 | | | 1238123 H | DIED | 15570 | 528 | 15210 | 545 | +360 | -17 |
| 8 07 71 | 107 | 101 | 1 | 10 30 70 | 1244009 H | NEK-CH-60 | 16040 | 533 | 15452 | 556 | +588 | -23 |
| 8 24 72 | 123 | 62 | 2 | 11 22 71 | 1291498 H | BULL | 17380 | 614 | 15606 | 562 | +1774 | +52 |
| 7 28 73 | 127 | 103 | 1 | 10 25 72 | 29H1842 H | TATTCO-87 | 17180 | 623 | | | | |

**HEALTH AND VETERINARY RECORD (1f)**

| DATE | DISEASE, AILMENT, VACCINATION OR TEST | TREATMENT AND REMARKS |
|---|---|---|
| | | |

Fig. 2-9. D.H.I. individual cow record.

contains the animal's identification (1a), lifetime production (1b), status change (1c), estimated relative producing ability (ERPA) (1d), calving and breeding record (1e), and space for recording health and veterinary information (1f). This form, containing the most recent lactation-production and other information, is produced whenever a cow is turned dry and when she leaves the herd. This record summarizes the individual cow's lifetime performance to date.

Individual heifer calf pages, DHI Form 204, as shown in Figure 2-10, are automatically printed for heifer calves born into the herd unless the supervi-

**INDIVIDUAL HEIFER CALF PAGE**

DHI-204
3-72

| HERD CODE | | | NAME | HENRY SMITH |
|---|---|---|---|---|
| ST | CO | HERD | AND | SMITH DAIRY FARMS |
| | | | ADDRESS | RFD 3 |
| 64 | 43 | 0108 | | AUBURN ALA.    36830 |

NAME_____

| DATE OF BIRTH | | | B R | EARTAG OR OTHER IDENTIFICATION |
|---|---|---|---|---|
| MO. | DAY | YR. | | |
| 4 | 19 | 74 | H | TATTO-101 |

CALF

NAME_____

| REGISTRATION OR EARTAG NUMBER | B R |
|---|---|
| 29H1842 | H |

SIRE

NAME_____

| BARN NAME | INDEX NUMBER | REGISTRATION OR EARTAG NUMBER | B R | DATE OF BIRTH | | | SUCCESSFUL BREEDING DATE | | | GESTATION LENGTH |
|---|---|---|---|---|---|---|---|---|---|---|
| | | | | MO. | DAY | YR. | MO. | DAY | YR | |
| 371 | 0371 | 64WAC8754 | H | 8 | 04 | 66 | 7 | 14 | 73 | 279 |

DAM

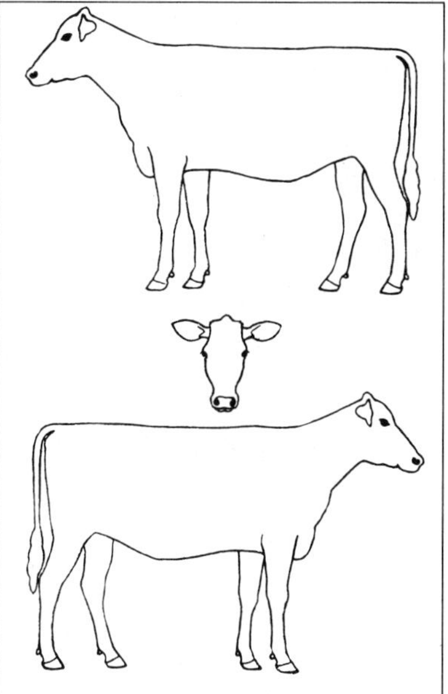

USE THIS SPACE TO SKETCH CALF

**BREEDING RECORD**

| 1st BREEDING | | 2nd BREEDING | | 3rd BREEDING | | 4th BREEDING | | 5th BREEDING | | REMARKS |
|---|---|---|---|---|---|---|---|---|---|---|
| DATE | BULL | DATE | BULL | DATE | BULL | DATE | BULL | DATE | BULL | |
| | | | | | | | | | | |

DATE BRUCELLOSIS VACCINATED_____ TAG AND TATTOO NO._____

OTHER VACCINATION_____ DATE_____

OTHER INFORMATION_____

**Fig. 2-10.** D.H.I. individual heifer calf page.

sor instructs the computer otherwise on the barn sheet. This form provides a convenient way to record permanent identification concerning the heifer until she freshens.

For purebred herds the owner has the option of having the application for

NYDHI 321 7/71

| ST. | CO. | TTT | HERD NO. |
|-----|-----|-----|----------|
| 64 | 43 | 0 | 108 |

PROC. CENTER
SOUTHERN
REGIONAL

HENRY SMITH
SMITH DAIRY FARMS
RFD 3
AUBURN AL 36830

**DHIA - HEIFER CALF IDENTIFICATION**

| Date Processed | Control Number |
|----------------|----------------|
| 5-10-72 | |

CALF

| Registration Name | | | | Registration Number |
|---|---|---|---|---|

| Breed | Date of Birth | Color | Eartag Number - Barn Name or No. | LE | Tattoo | RE |
|-------|---------------|-------|----------------------------------|----|--------|-----|
| HOLSTEIN | 04-09-72 | | TATTO-101 | | | |

Twin: No____ Yes____ Sex of other twin: Male____ Female____ Check if naturally polled_____
Date and reason left herd_____

SIRE

| Registration Name | | Registration Number |
|---|---|---|

| Breed | A.I. Code No. | A.I. Code Name |
|-------|---------------|----------------|
| HOLSTEIN | ABS-1842 | |

DAM

| Registration Name | | Registration Number |
|---|---|---|
| | | 6489143 |

| Breed | Index No. | Barn Name or Chain Number | Date Last Bred | Days in Gestation Period | Eartag Number |
|-------|-----------|---------------------------|----------------|--------------------------|---------------|
| H | 0371 | 371 | 04-04-71 | 280 | |

DAIRY BREEDS UNIFIED APPLICATION FOR REGISTRY _____

Registry Fees Must Accompany Application _____ (Breed Association)

If sire and dam are registered, these additional items are necessary to apply for registry. Check instructions and procedures for the breed involved

____ Check here if the dam was artificially bred, and attach the breeding receipt.

In making this application, I hereby subject myself to all the provisions of the Constitution and Bylaws of the breed association as they now exist or may from time to time be amended, knowledge of which I now have or will immediately acquire. I guarantee that all matters stated herein are true and that careful comparison of the diagram, tattoo or the photographs of the color markings has been made with the animal and found to be correct. In addition, I agree that all records on the animals in my herd, whether maintained by me or by others, including production records, may be obtained and used by the breed association in its several programs.

Applicant's Signature _____
(owner of dam at time of calving)

Address _____ I.D. or Owner's No. _____

Breeding Name _____
(owner of dam when bred)

Address _____ I.D. or Owner's No. _____

Provide tattoo for Brown Swiss or Jersey in the spaces at the top of this application: tattoo or sketch optional for Ayrshire and Guernsey. For Ayrshire and Guernsey color markings, use sketch below. Trace with dark pencil or black ink. OR attach two sets of photos (one set for Guernsey) in which head, feet and tail show clearly and distinctly. Indicate sex, birth date and dam's number on back of each print. Clip photos to the application - do not paste them.

SKETCH HOLSTEINS ON THE BACK. THIS OUTLINE FOR AYRSHIRE AND GUERNSEY ONLY.

**Fig. 2-11.** D.H.I. application for registry in breed registry association.

Registry form, NYDHI Form 321, as shown in Figure 2-11, partially completed by the computer.

An annual heifer calf listing, DHI Form 209, as shown in Figure 2-12, is also printed. This can provide an annual inventory of heifer calves and assist dairymen in identifying animals that are old enough to breed, identify sire groups, and so on.

ANNUAL HEIFER CALF LISTING

| DHI-209 5-74 | | | | | | | | | | | | | CALVES BORN SINCE | | | DATE PRINTED | | |
|---|---|---|---|---|---|---|---|---|---|---|---|---|---|---|---|---|---|---|
| HERD CODE | | | DATE OF TEST | | | P A G E 0 1 | NAME & ADDRESS | | | | | | | MO | DAY | YEAR | MO | DAY | YR |
| ST | CO | HERD | MO | DAY | YEAR | | HENRY SMITH | | | | | | | 07 | 20 | 73 | 07 | 21 | 74 |
| 64 | 43 | C108 | 07 | 17 | 74 | | SMITH DAIRY FARMS RFD3 AUBURN   AL   36830 | | | | | | LISTED IN ORDER BY: | CALF"S BIRTH DATE | | | | | |

| CALF IDENTITY | | | SIRE IDENTITY | | | | DAM IDENTITY | | | | | CALF'S BIRTH DATE | | | VACCINATION, BREEDING, |
|---|---|---|---|---|---|---|---|---|---|---|---|---|---|---|---|
| REGISTRATION OR EARTAG NUMBER | B R | BARN NAME OR NUMBER | REGISTRATION OR EARTAG NUMBER | B R | CODE NUMBER OR CODE NAME | | REGISTRATION OR EARTAG NUMBER | B R | BARN NAME | INDEX NUMBER | | MO | DAY | YEAR | OTHER INFORMATION |
| 64WAK7130 | H | | 1516004 | H | | | 64WAG2853 | H | 16 | 0016 | | 08 | 19 | 73 | |
| 64WAK7129 | H | | 1525496 | H | | | 64WAF4203 | H | 2 | 0002 | | 08 | 22 | 73 | |
| 64WAK7131 | H | | 1489802 | H | | | 64WAD3785 | H | 450 | 0450 | | 08 | 28 | 73 | |
| | H | TAG-405 | 1447893 | H | | | 6481109 | H | 25 | 0025 | | 08 | 31 | 73 | |
| 64WAK8182 | H | | 1527804 | H | | | 64WAF6378 | H | 499 | 0499 | | 09 | 22 | 73 | |
| 64WAK8184 | H | | 1525496 | H | | | 64WAG2478 | H | 498 | 0498 | | 09 | 30 | 73 | |
| 64WAK8181 | H | | 1489802 | H | | | 64WAC8181 | H | 10 | 0010 | | 10 | 05 | 73 | |
| 64WAK8183 | H | | 1516004 | H | | | 64WAG1355 | H | 27 | 0027 | | 10 | 14 | 73 | |
| | H | TAG-401 | 1516004 | H | | | 6785382 | H | 318 | 0318 | | 11 | 12 | 73 | |
| | H | TAG-404 | 1525496 | H | | | 6444482 | H | 30 | 0030 | | 11 | 24 | 73 | |
| | H | TAG-411 | 1525496 | H | | | 7181892 | H | 492 | 0492 | | 12 | 15 | 73 | |
| 64WAK8315 | H | | | H | | 7H676 | 64WAF4688 | H | 480 | 0480 | | 01 | 11 | 74 | |
| | H | TAG-412 | | H | | 40H2025 | 6811121 | H | 301 | 0301 | | 02 | 27 | 74 | |
| | H | TATTO-101 | | H | | 29H1842 | 64WAC8754 | H | 371 | 0371 | | 04 | 19 | 74 | |
| | H | TAG-416 | | H | | 29H1881 | 6998121 | H | 563 | 0563 | | 04 | 26 | 74 | |
| | H | TATTO-102 | | H | | 11H105 | 64WAF6481 | H | 504 | 0504 | | 05 | 21 | 74 | |
| | H | TATTO-103 | | H | | 1H159 | 64WAF6734 | H | 36 | 0036 | | 07 | 16 | 74 | |

**************** CALVES WILL ALSO BE LISTED IN ORDER BY SIRE GROUPS ****************

**Fig. 2-12.** D.H.I. annual heifer calf listing.

Other special management lists, DHI Form 205, as shown in Figure 2-13, and ERPA lists, DHI Form 206, as shown in Figure 2-14, are also available through the DHI program. These can be valuable management tools in maintaining overall herd efficiency.

**Official AM-PM (Not Accepted for DHIR).** This program differs from Standard DHI in that: (1) milk is weighed and sampled alternately at morning and evening milkings in consecutive test periods, (2) the entire herd must be milked 2 times daily and (3) at least one component of the milking system must be equipped with an *approved* continuous monitoring device that provides an authentic record of the milking intervals. On test day the supervisor will determine and record the reference time at the beginning and ending of the test milking and the two previous milkings.

**Owner-Sampler.** This program is similar to Standard DHI, except that the owner rather than a supervisor weighs and samples the milk and records other feeding and breeding, etc., information. The supervisor leaves sample bottles and record sheets, tests the samples for butterfat, and reports the information to the computing center or sends samples to a central testing laboratory. Similar herd and cow information (Figures 2-2 to 2-8) is generated and available for herd management use. Production information collected from Owner-Sampler herds is not considered authentic and is not included in USDA sire summaries.

**Weigh-A-Day-A-Month.** In this program the dairyman weighs the milk from each cow monthly. No milk samples are tested for butterfat. Simple feed records may be kept.

The necessary forms and instructions can be obtained from the county agricultural agents. These forms are sent to a central computing office for calculation. A monthly report for each cow and the herd is mailed to the dairyman. These are not authenticated records and are not included in USDA sire summaries.

### Dairy Herd Improvement Registry (DHIR)

The DHIR is a program sponsored by the various national purebred dairy cattle associations and coordinated with the DHI program. The major differences between DHIR and Standard DHI are that in the DHIR program only registered cows are included and the completed records (both 305 and 365 days) are reported to the breed association as well as to the USDA.

DHI-205    1-75

## HENRY SMITH ①     SPECIAL MANAGEMENT LIST

HERD CODE NUMBER | DATE TESTED
ST | CO | HERD | MO | DAY

| 64 | 43 | 0108 | 7 | 17 | COWS OPEN | | COWS BRED IN LAST 60 DAYS |

### COWS OPEN

| BARN NAME | DATE FRESH | 60TH DAY | (a) |
|---|---|---|---|
| ***COWS OPEN | OVER | 100 DAYS*** | |
| 843 | 2-26 | 4-26 | |
| 985 | 3-10 | 5-08 | |
| 1043 | 3-18 | 5-16 | |
| 46 | 3-24 | 5-22 | |
| **COWS OPEN 60 TO 100 DAYS*** | | | |
| 86 | 4-12 | 6-10 | |
| 1099 | 4-17 | 6-15 | |
| 1184 | 4-30 | 6-28 | |
| 542 | 5-05 | 7-04 | |
| 504 | 5-21 | 7-20 | |
| *COWS OPEN LESS THAN 60 DAYS* | | | |
| 69 | 5-28 | 7-27 | |
| 97 | 6-04 | 8-02 | |
| 877 | 6-08 | 8-06 | |
| 1104 | 6-11 | 8-09 | |
| 738 | 6-12 | 8-10 | |
| 1263 | 6-23 | 8-21 | |
| 1401 | 7-05 | 9-02 | |
| 933 | 7-16 | 9-13 | |

### COWS BRED IN LAST 60 DAYS

| BARN NAME | NO. OF BR | DATE BRED | (b) |
|---|---|---|---|
| 427 | 1 | 6-27 | |
| 271 | 1 | 6-22 | |
| 457 | 3 | 6-16 | |
| 467 | 2 | 5-27 | |
| 539 | 1 | 5-20 | |

PAGE 1      PAGE 2

```
* * * * * * * * *            * * * * * * * * * * *
*COWS DUE TO CALVE*  (c)     *COWS TO TURN DRY *  (d)
* * * * * * * * *            * * * * * * * * * * *
```

| BARN NAME | DATE DUE | SERVICE SIRE | BARN NAME | DATE DUE | DATE TO DRY | MILK PER DAY |
|---|---|---|---|---|---|---|
| 16 | 7-17 | 19H63 | 439 | 9-28 | 7-31 | 35.1 |
| 498 | 8-27 | 19H63 | 27 | 10-06 | 8-08 | 37.5 |
| 450 | 9-03 | 1H92 | 476 | 10-22 | 8-24 | 28.2 |
| 618 | 9-06 | 7H3991 | 499 | 10-23 | 8-25 | 41.4 |
| 814 | 9-10 | 11H77 | 1429 | 10-26 | 8-28 | 36.3 |

---

DHI-205    1-75

## HENRY SMITH ②     SPECIAL MANAGEMENT LIST

HERD CODE NUMBER | DATE TESTED
ST | CO | HERD | MO | DAY

| 64 | 43 | 0108 | 7 | 17 | COWS BY TEST DAY MILK | | COWS BY TEST DAY MILK |

| BARN NAME | DAILY LBS. MILK | DAILY % FAT | PROJECTED 305 DAY-ME MILK   FAT | DAYS IN MILK | DAILY $ VALUE | RAT-ING | DATE DUE | SERVICE SIRE |
|---|---|---|---|---|---|---|---|---|
| 738 | 95.3 | 3.4 | | 29 | 7.93 | | OPEN | |
| 1104 | 89.8 | 3.0 | | 35 | 7.29 | | OPEN | |
| 97 | 85.2 | 3.4 | | 45 | 7.41 | | OPEN | |
| 504 | 79.3 | 3.1 | 19032   629 | 55 | 6.83 | B | OPEN | |
| 69 | 78.3 | 3.6 | | 49 | 7.05 | | OPEN | |

---

DHI-205    1-75

## HENRY SMITH ④     SPECIAL MANAGEMENT LIST

HERD CODE NUMBER | DATE TESTED
ST | CO | HERD | MO | DAY

| 64 | 43 | 0108 | 7 | 17 | CULLING GUIDE | CULL VALUE PROD. LEVEL = $2.32 |

| BARN NAME | DAILY $ VALUE | DATE DUE | PROJECTED 305 DAY-ME MILK   FAT | DAILY LBS. MILK | DAYS IN MILK | AGE IN MONTH | DAYS OPEN | RAT-ING | PRP |
|---|---|---|---|---|---|---|---|---|---|
| 1 | 2.59 | 3-11 | 7917   277 | 27.1 | 261 | 27 | 245 | E | 297 |
| 152 | 2.42 | 4-09 | 10862   389 | 23.5 | 285 | 49 | 282 | E | 464 |
| 209 | DRY | 2-17 | 10903   351 | DRY | 331 | 43 | 284 | E | 486 |
| 206 | 2.39 | 3-26 | 11973   363 | 24.8 | 285 | 68 | 272 | D | 508 |
| 56 | 2.49 | OPEN | 12424   370 | 25.5 | 322 | 25 | 322 | D | 520 |

**Fig. 2-13.** D.H.I. special management lists.

DAIRY HERD IMPROVEMENT RECORDS
DHI-206  3-74

| TYPE OF RECORD | DHIA | COWS CALVING BEFORE | PAGE NO |
|---|---|---|---|
| HERD CODE | | MO YR | |
| ST CO HERD | 64 43-0108 | 11 73 | 1 |

NAME AND ADDRESS
HENRY SMITH
SMITH DAIRY FARM
RFD 3
AUBURN    ALA    36830

UNITED STATES DEPARTMENT OF AGRICULTURE
AGRICULTURAL RESEARCH SERVICE AND
STATE AGRICULTURAL EXTENSION SERVICE COOPERATING

ESTIMATED RELATIVE
PRODUCING ABILITY LIST (ERPA) 1

| BREED | REGISTRATION OR EARTAG NUMBER | COW INDEX NUMBER | BARN NAME | LAST LACTATION DATE FRESH MO | YR | AGE | M.E. PRODUCTION MILK | FAT | R | No. LACTATIONS | AVERAGE OF ALL 305 DAY M.E. RECORDS MILK | FAT | AVERAGE OF ALL HERDMATE RECORDS MILK | FAT | ESTIMATED RELATIVE PRODUCING ABILITY MILK | FAT |
|---|---|---|---|---|---|---|---|---|---|---|---|---|---|---|---|---|
| H | 64WAG2478 | 0498 | 498 | 8 | 72 | 44 | 20790 | 899 | | 2 | 20404 | 805 | 15489 | 573 | 3293 | 155 |
| H | 64WAE3736 | 0457 | 457 | 3 | 72 | 55 | 17140 | 592 | | 3 | 19620 | 726 | 16282 | 602 | 2504 | 93 |
| H | 7281802 | 0439 | 439 | 6 | 72 | 64 | 19740 | 658 | | 4 | 17743 | 621 | 15401 | 570 | 1874 | 41 |
| H | 64WAF4203 | 0002 | 2 | 8 | 72 | 41 | 20230 | 661 | | 2 | 18235 | 602 | 15502 | 574 | 1831 | 19 |
| H | 64WAC8754 | 0371 | 371 | 4 | 72 | 80 | 18420 | 643 | | 5 | 17378 | 626 | 15591 | 577 | 1483 | 41 |
| H | 64WAF6378 | 0499 | 499 | 9 | 72 | 49 | 15505 | 559 | | 3 | 16896 | 591 | 15241 | 564 | 1241 | 20 |
| H | 64WAD3785 | 0450 | 450 | 7 | 72 | 60 | 17180 | 623 | | 4 | 16004 | 608 | 14881 | 551 | 898 | 46 |
| H | 64WAG2853 | 0016 | 16 | 8 | 72 | 38 | 15420 | 544 | | 2 | 16655 | 583 | 15532 | 575 | 752 | 5 |
| H | 64WAC8146 | 0008 | 8 | 3 | 72 | 83 | 16040 | 564 | 4 | 5 | 16320 | 555 | 15485 | 573 | 693 | -15 |
| H | 64WAF4688 | 0480 | 480 | 11 | 70 | 44 | 17720 | 590 | | 2 | 16740 | 569 | 15781 | 584 | 643 | -10 |
| H | 64WAD7863 | 0476 | 476 | 8 | 72 | 54 | 16250 | 689 | | 3 | 15920 | 610 | 15201 | 562 | 539 | 36 |
| H | 64WAF6481 | 0504 | 504 | 4 | 72 | 27 | 17620 | 561 | | 1 | 17620 | 561 | 16598 | 614 | 511 | -27 |
| H | 64WAC2631 | 0427 | 427 | 2 | 72 | 63 | 17140 | 634 | | 4 | 16680 | 632 | 16110 | 601 | 456 | 25 |
| H | 64WAG6344 | 0005 | 5 | 2 | 72 | 34 | 16420 | 621 | | 2 | 16680 | 634 | 16520 | 611 | 107 | 15 |
| H | 64WAI3556 | 0046 | 46 | 3 | 72 | 28 | 16300 | 580 | | 1 | 16300 | 580 | 16621 | 615 | -161 | -18 |
| H | 64WAG6345 | 0032 | 32 | 1 | 72 | 27 | 15983 | 592 | | 1 | 15983 | 592 | 16644 | 616 | -330 | -12 |
| H | 64WAF6623 | 0034 | 34 | 7 | 72 | 26 | 13204 | 487 | 3 | 1 | 13204 | 487 | 15545 | 575 | -1171 | -44 |
| H | 64WAG6482 | 0003 | 3 | 9 | 72 | 39 | 13890 | 505 | 3 | 2 | 13660 | 557 | 15543 | 575 | -1262 | -12 |
| H | 64WAF2122 | 0061 | 61 | 8 | 72 | 28 | 12482 | 453 | 3 | 1 | 12482 | 453 | 15574 | 576 | -1546 | -62 |
| H | 64WAG1355 | 0027 | 27 | 10 | 72 | 38 | 14505 | 525 | | 2 | 12985 | 519 | 15556 | 576 | -1723 | -38 |
| H | 64WAF4602 | 0467 | 467 | 2 | 72 | 51 | 13140 | 503 | | 3 | 11421 | 423 | 16196 | 599 | -3581 | -132 |

| TOTALS AND AVERAGES | NUMBER OF LACTATIONS | 305-2X-M.E. LACTATION AVERAGE MILK | FAT | LEFT HERD - LOW PRODUCTION NUMBER | % | AVERAGE E R P A MILK | FAT | LEFT HERD - ALL OTHER REASONS NUMBER | % | AVERAGE E R P A MILK | FAT |
|---|---|---|---|---|---|---|---|---|---|---|---|
| HERD | 86 | 15696 | 578 | 9 | 10 | -1141 | -35 | 17 | 20 | 200 | 6 |
| STATE | 109 | 12012 | 427 | 15 | 14 | -860 | -32 | 10 | 9 | 24 | 1 |

Fig. 2-14.  D.H.I. estimated relative producing ability list (ERPA).

36

## VALUE OF TESTING PROGRAMS

Authenticated performance-testing programs (Standard DHI and DHIR) provide the U.S. dairy industry with information needed for sire progeny testing (USDA sire summaries) and for research and educational programs. Enrollment in any of the four programs or the Owner-Sampler program provides individual cow and herd performance information that can be used as a primary basis by the owner or manager for many management decisions. Enrollment in any of the programs except Weigh-A-Day-A-Month can, in addition, provide a convenient method of maintaining individual identification, parentage, breeding, and health records. These records provide additional information needed for herd management decisions. Additional tools are constantly being developed (special management lists) and additional information is being recorded (mastitis tests, etc.) to make the management information more complete.

The success of the performance-testing program and the value of the information generated are primarily dependent on the cooperation of individual dairymen whose herds are enrolled, and on the local supervisor. Accurate reporting of feeding, breeding, identification, and other such information by the dairyman to the supervisor is a "must." Some dairymen complain that the information reported on the monthly and annual summaries is not accurate. When this is true, it is because the dairyman did not report accurate input information concerning these items to the supervisor. Simply stated, if accurate information is recorded by dairymen and supervisors, accurate summary information is generated. If inaccurate information is recorded, the summary information will be inaccurate. Accurate reporting of information by the dairyman, accurate recording of milk weights and butterfat testing by the supervisor, and accurate and prompt reporting by the supervisor are all essential.

The value of the program is evidenced by the fact that in the 1975–76 test year, cows enrolled on official (DHI and DHIR) performance testing programs produced about 25% more milk than all other cows not enrolled on official test (an average of 13,326 versus 10,619 lbs). In addition, information obtained from the program has been the basis for research that has led to the development of sire progeny testing programs (USDA sire summaries) that are primarily responsible for the rapid improvement in genetic ability for milk production in U.S. dairy cattle and for development of improved feeding and management practices. The development, use, and continued improvement in national performance-testing programs has probably been the most significant single factor in improving dairying in the United States.

## SOLIDS-NOT-FAT, OR PROTEIN, TESTING

In addition to milk and butterfat records there is also interest in having protein, or solids-not-fat (SNF), tests on cows. One of these may well become a part of the basis for pricing milk. In that event dairymen will want this information for each cow as an additional basis for selection and breeding.

Experiment stations are collecting data on many cows, and a few dairy herd-improvement associations have added the SNF test to their program. Some central DHI processing centers and the breed associations added this additional information to their record keeping for those dairymen who may desire it. The Holstein-Friesian Association of America initiated a program for solids-not-fat testing and recording for registered cows. This is an addition to their conventional production-testing program.

The SNF testing programs are being labeled with various names, such as TNT (total nutrient testing) and PLM (protein, lactose, mineral) program.

## METHODS OF
## CALCULATING HERD AVERAGES

Three methods of calculating the herd average for production are now in use. There is need for a standardized method so that whenever a herd average is given it will always mean the same thing.

One method is on the *cow-year basis*. At the close of the year the number of days each cow is in the herd during the yearly testing period is determined and totaled. The total cow days is then divided by 365 (366 if leap year). This gives the number of cow years. The total production of milk and butterfat is then divided by the number of cow years, which gives the average production per cow for the year. This is the Standard DHI method, and uses actual records.

A second method is to count in the herd average only such cows as have been in the herd for 300 days or more. The production of these cows is totaled and averaged. This plan is used by some of the breed associations for calculating DHIR averages, and uses actual records.

A third method is to calculate each lactation record completed during the testing year to a twice daily milking, 305 day lactation, mature equivalent basis (2×, 305–day, ME). The standardized records are then averaged to determine the herd average.

This method is used by some other breed associations. It is estimated that the 2×, 305–day, ME record is approximately 9% greater than when actual records are averaged. This method does place the herd average on the

same basis as sire proofs and other programs of some of the breed associations.

## SUMMARY

Herd performance records that are accurately maintained can be used as the
basis for management decisions. They are valuable assets to management.

Records need to be complete enough to contain the necessary information, yet not so complicated that dairymen become discouraged in maintaining
them.

The National DHI Program is an excellent herd record system and can be
used to record herd and individual cow performance information, to record
parentage, and to maintain health records in addition to production, feed, and
management information.

## REFERENCES FOR FURTHER STUDY

Crandall, B. H., Using DHI-EDP Information in Herd Management, *J. Dairy
Sci.,* **58**:230 (1975).

Patterson, W. N., Wall, O. G., Carpenter, M., Gramling, G. E., Griffin, C.
D., Webb, D. W., and Wright, E. E., *Dairyman's DHI Manual,* Southern
Regional Dairy Records Processing Center, North Carolina State University,
Raleigh, 1976.

# The Business of Dairying

**3**

Dairying is a business as well as a way of life. Increased herd size, mechanization, and costs of resources require capital investment in significant amounts. This has created a need for increased skill in business and financial management as well as herd and land management if success is to be achieved in the business of dairying.

Herd performance goals, personal goals, and financial goals are not antagonistic, but rather are interrelated and complementary to each other. Achieving a high level of milk production increases gross income from the sale of milk. This, when accompanied by production cost control, increases the opportunity for a reasonable profit margin. Earning a reasonable profit enhances the opportunity to achieve personal goals.

Some of the factors that are consistantly associated with successful dairy businesses are: (1) adequate herd size, (2) high production per cow, (3) high production of quality feed per acre of land, (4) investment cost control, (5) production cost control, and (6) accurate and complete financial records that can be analyzed on an enterprise basis.

The first two factors largely determine the gross income of a dairy operation, as about 90% of the gross income on most dairy farms is from the sale of milk. Adequate herd size cannot be specifically defined for all dairymen. It must, however, be large enough to generate sufficient income to offer the opportunity for a reasonable level of profit after operating and fixed expenses have been paid. Herds of less than 25 to 30 cows have decreased rapidly while herds with more than 30 cows have increased in number in recent years. Therefore it appears that a minimum herd size of 25 to 30 cows is necessary. High production per cow yields more dollar income per producing unit. The effect of this on profit potential is well documented.

The third factor, high yield of quality feed per acre, is important to those dairymen whose operation includes the use of land resources to produce a part of the herd feed supply. Both yield and quality of homegrown feed have a significant effect on milk production and feed cost per unit of milk production, feed cost is the largest single production cost on most dairy farms. The effects of these factors will be discussed in detail in Chapters 4 through 8.

The fourth factor, investment cost control, is critical if a major portion of the investment is financed by borrowed capital. Debt-service costs, both interest and principal payments, must be paid from operating profits; thus debt-service costs are proportionate to the size of the debt load on a per-cow and total basis. Excessive investment costs paid for with borrowed capital, especially in convenience, low productive, and highly depreciable assets, can significantly increase production costs and decrease profit. Alternative investments should be evaluated in view of three basic criteria. These are:

*1. The benefit-cost ratio of the investment.* In financial terms, will the invest-ment generate more after-tax income than its cost? This is a measure of net return on investment. For example, if $900 is invested in a high-producing cow (15,000 lb. of 3.5 milk valued at $10 per cwt.), the value of her annual production is $1,600 ($1,500 for milk and $100 for ½ a heifer calf), and the total annual production cost is $1,200, the before-tax annual return on this purchase is $400. If we assume a 25% charge for income taxes, the net after-tax return on this purchase is $300, or a 33¹/₃% annual return on the investment. If the cost of the capital is 10%, this investment did show a desirable benefit-cost ratio.

*2. The timing of the benefits of the investment.* Will the benefits be realized early, for example, in the form of increased annual net dollar profit, or later, for example, in the form of appreciated value of land or other long-term assets? Early benefits are desirable because the net dollars generated by the investment can be used to service debt, pay operating expenses, and to reinvest. Long-term appreciation benefits are also desirable, but these bene-fits cannot be used to service debt, meet expenses, or reinvest unless the investment is sold or additional capital is borrowed on the basis of the collateral value of the investment. The timing of benefits, therefore, should be evaluated in terms of individual needs and goals. When cash flow needs are high and asset-liability ratio is low, as is usually the case with beginning dairymen, early realization or benefits to meet expenses and improve capital position is highly desirable. However, when net operating profit and asset-liability ratio are high, as is often the case with older, well-established dairymen, investments that will return benefits later and can be used as a source of retirement income may be desirable.

*3. The risk involved with the investment.* Comparing the safety of risk (risk of loss of the investment as well as the probability that the investment will generate a certain level of benefits) of alternative investments is often a difficult task. However, the degree of risk can be predicted with some confidence by knowledge of the probable productivity of the investment, supply and demand, useful life, flexibility of use, and probable loss of the investment. If one evaluates the usual dairy investments in view of these factors, productive land can be described as a relatively safe investment because it is in great demand and the supply is limited, it has an indefinite useful life when properly managed, it can produce a high yield of a wide variety of useful products annually, and it rarely is destroyed by fire, wind, disease, etc. Milking dairy cows and general farming equipment might be classified as intermediate in risk, and specialized dairy facilities and equipment might be classified as higher-risk items because of their adequate supply and limited demand and their limited flexibility of usage.

In summary, both the amount and type of investment can profoundly affect the business of dairying. The total investment must be within the income-generating capacity of the investment. Types of investment must be evaluated in view of the needs of each situation in regard to benefit-cost ratio, timing of benefits, and risk.

The fifth factor, production cost control, involves controlling the day-to-day costs of producing milk. These costs include feed, labor, building and equipment annual costs, interest, fuel, utilities, taxes, etc: If one or several of these cost areas becomes excessive, the cost of producing a unit of milk becomes excessive and the profit margin is reduced. In severe cases production costs may exceed the value of the milk.

## BUSINESS RECORDS

An accurate and complete business record program can provide the dairy business manager with the information needed for good financial management. It can provide data for tax reporting and management, total and enterprise business analysis, budgeting, planning, and other management decisions that are essential to maintaining the business efficiency of the dairy business operation. There is no nationwide dairy business record program such as the national DHI herd performance record program; however, many individual states, local banks, and business services do have such programs. Some of these are computerized and provide annual and monthly financial analysis. One may wish to enroll in one of these programs or to personally maintain the necessary business records.

The essentials of a good business record system are: (1) that they provide the necessary information, and (2) that they are simple to keep. Necessary information includes an itemized record of receipts and expenses and an inventory of assets and liabilities. From this information one can calculate taxes, a profit-and-loss or operating statement, a net-worth statement, and a cash-flow statement. These, in turn, can be used to prepare budgets, support loan applications, for business analysis, and as the basis for other management decisions.

A record of income and expenses may be kept in a calendar pad, a diary, a multicolumn record book (Fig. 3-1) or some type of farm record book. Information should be recorded regularly (preferably at the time of the business transaction) and can be summarized monthly, quarterly, or annually. If one also notes the enterprise (dairy herd, corn for silage, etc.) involved in

INCOME AND EXPENSE RECORD　　Month *Jan*　Year *1977*

| MO. | DAY | DESCRIPTION | Check No. | Receipts | Expenses |
|---|---|---|---|---|---|
| | | Totals brought forward | | | |
| Jan. | 4 | 10 tons 16% dairy feed - Pure Feed Co. | 419 | | 1320 00 |
| Jan. | 4 | Salary - owner | 420 | | 600 00 |
| Jan. | 6 | Fertilizer -10 tons (15-15-15) - corn Pure Fertilizer Co. | 421 | | 1500 00 |
| Jan. | 8 | Milk sold (gross) 30,000 lbs. @ $9.60/cwt. | | 4 880 00 | |
| Jan. | 8 | Milk hauling -30,000 lbs. @ $.40/cwt. | | | 120 00 |
| Jan. | 8 | Milk marketing charges | | | 30 00 |
| Jan. | 10 | Interest on tractor loan - P.C.A. | 422 | | 136 80 |
| Jan. | 10 | Principal payment on tractor loan - P.C.A. | 423 | | 500 00 |
| Jan. | 15 | Treatment of #484 - Dr. Jones | 424 | | 26 00 |
| Jan. | 18 | Supplies for the milk house - N.C. Coop. | 425 | | 94 00 |
| Jan. | 20 | 1 cow for beef (#618) - Acme Packing | | 246 20 | |
| Jan. | 20 | 3 bull calves - Acme Packing | | 31 10 | |
| Jan. | 31 | Salary - John Smith for month of January | 426 | | 600 00 |
| Jan. | 31 | Repair parts for corn forage harvester Acme Implement Co. | 427 | | 196 20 |
| Jan. | 31 | Fuel for dairy - N.C. Coop. | 428 | | 84 70 |
| | | Totals carried forward | | 5157 30 | 5207 70 |

**Fig. 3-1.** Example of a Dairy farm income and expense ledger.

each transaction, enterprise profit-and-loss statements can be calculated and are useful in identifying weak or unprofitable enterprises.

An inventory record should include an itemized list, with values, of the assets and liabilities of the dairy farm business at a given point in time. The inventory should contain separate listings of nondepreciable and depreciable assets by type of asset (machinery, cattle, etc). Maintaining depreciation schedules on depreciable assets facilitates tax preparation. Farm inventory record books are available from the Extension Service and other sources. The inventory should include:

REAL ESTATE

| | |
|---|---|
| Cropland | Valued at normal market value |
| Pasture, woods, etc. | Valued at normal market value |
| Personal residence | Valued at normal market value |
| Farm buildings | Valued at cost less depreciation |
| Fences | Valued at cost less depreciation |
| Growing crops | Valued at present market value |

LIVESTOCK

| | |
|---|---|
| Kept for production | Valued at normal market value |
| To be sold | Valued at present market value |

EQUIPMENT

| | |
|---|---|
| Farm equipment | Valued at cost less depreciation |
| Dairy equipment | Valued at cost less depreciation |

FEED AND SUPPLIES

| | |
|---|---|
| Homegrown | Valued at present market value |
| Purchased | Valued at present market value |

Inventories are usually recorded annually at the end of the business year. For most, this date is December 31. This is a convenient time to complete the inventory as it coincides with the tax year and few growing crops need to be inventoried.

## Successful Business Criteria

As mentioned in Chapter 1, a dairy business must meet the same criteria as other successful businesses if it is to remain in business for a long period of time. These criteria are as follows.

1. It must generate enough cash income to pay all cash operating costs, meet current debt commitments, and cover family living expenses. It must maintain a positive cash position. Continued failure to meet this requirement results in loss of ability to acquire needed supplies or resources to continue operations, and may result in legal actions by creditors.

2. It should pay interest on owned as well as borrowed capital invested in the business. Interest on owned capital invested in the business should be considered a business expense and should be credited as income to the owner of the capital. The interest rate charged on owned capital

should be 1 to 1½% more than current interest rates paid on short-term savings accounts. If the business is not generating enough income to pay interest on owned as well as borrowed capital, this indicates that capital resources are not being used profitably. In this case careful evaluation of current practices to identify weak links, with consequent strengthening of these weak links, is advised. Consideration of alternative use of capital resources may also be considered.

3. It must maintain productivity. On a temporary basis, a business may meet criteria (1) and (2) by selling assets such as heifers, cows, land, etc. or by using these assets as collateral to obtain capital to meet cash expenses. However, this practice can reduce future productivity of the business or increase debt-service needs and further decrease future income. Continual sacrifice of productive assets to meet operating expenses business is not a sound business practice.

4. It must earn a reasonable return for the operator. Personal needs and goals determine what is reasonable.

## MEASURES OF PROFIT

Profit can be defined and is determined in various ways. Some of the usual ones are described in this section.

### Farm Net Cash Income

Farm net cash income is the amount of cash available for current debt principal repayment, family living expenses, savings, payment of income taxes, and new debt service. It is the difference between total cash farm receipts and total cash farm operating expenses.

**Example 1.** **Farm Net Cash Income**

|  |  |
|---|---|
| Cash farm receipts | $85,500 |
| (sale of milk, cows, crops, etc.) | |
| Cash farm operating expenses | $71,000 |
| (feed, labor, interest on farm loans, etc.) | |
| | |
| **Farm net cash income** | **$14,500** |

Farm net cash income measures the ability to pay operating expenses for a given period of time. It does not provide information needed to determine if the farm business is generating enough capital to pay interest on all invested capital, to maintain productivity, and to earn a reasonable return for the operator. Farm net cash income may be high because of sacrificed inventory, which will decrease future productivity, because members of the family worked on the farm without receiving wages, or because of a large owned investment and few or no debts. Family net cash income can be determined by including current debt principal obligations and income from nonfarm sources such as wages of family members. If family living expenses are then deducted from the family net cash income, the cash position of the family can be determined. This figure is useful in determining the money available to service new debt. For example,

|  |  |  |
|---|---|---|
| **Farm net cash income** | $14,500 |  |
| Family income from non-farm sources | $ 5,500 |  |
| **Total** |  | **$20,000** |
| Current debt principal payment | $10,000 |  |
| Family living expenses | $ 7,000 |  |
| **Total** |  | **$17,000** |
| Available to service new debt |  | $ 3,000 |

## Farm Income

Farm income is the amount available to pay for the dairyman's time and for the use of invested capital. It is a commonly used measure of farm net profit. It is the farm net cash income plus or minus the net change in inventory and depreciation minus the cash farm operating expenses and unpaid family labor.

| **Example 2.** | **Farm Income** |  |  |
|---|---|---|---|
|  | Cash farm receipts | $85,500 |  |
|  | Net change in farm inventory less machinery and building depreciation | 9,000 |  |
|  | **Total** |  | **$94,500** |
|  | Cash farm operating expenses | $71,000 |  |
|  | Unpaid family labor | 6,500 |  |
|  | **Total** |  | **$77,500** |
|  | **Farm income** |  | $17,000 |

Farm income measures dairy farm profitability more accurately than does farm net cash income, as it includes adjustments for inventory and depreciation. It can also be used to evaluate the continued productivity of the business.

## Labor Income

Labor income is the amount left after paying all farm business expenses and interest on all invested capital. It is the farm income plus farm interest minus an interest charge on all invested capital. The labor income of a dairyman is comparable to the cash wages of a hired hand who is furnished with a house and farm products in addition to cash wages. It can be calculated as follows.

**Example 3.** **Labor Income**

| | |
|---|---:|
| Farm income | $17,000 |
| Interest on owned capital ($112,500 × 8%) | 9,000 |
| **Labor income** | **$8,000** |

In this example, interest on owned capital is included at 8%. Interest on borrowed capital is included as a cash operating expense and is accounted for in the calculation of farm income. An alternate method of calculating labor income is to exclude interest on borrowed capital from the cash farm operating expenses and charge the same rate for all invested capital, owned and borrowed. Labor income is a widely used measure of profitability and is useful in evaluating all four criteria of a successful business.

## Labor Earnings

Labor earnings is the labor income plus the value of the farm products used in the operator's household and the value of one year's use of the operator's house.

**Example 4.** Labor Earnings

| | |
|---|---:|
| Labor income | $8,000 |
| Value of house rent | 1,800 |
| Value of milk, meat, etc. | 700 |
| **Labor earnings** | **$10,500** |

Labor earnings is a more accurate measure of farm profitability than labor income. It is not so widely used because of the difficulty in obtaining accurate information on the amount and value of farm products used in the household and in estimating the rental value of the dwelling.

## Percent Return on Capital

Percent return on capital (investment) is a ratio of earnings to investment. It measures the productivity of the business in relationship to the investment.

$$\text{Percent return on capital} = \frac{\text{Farm income} + \text{interest} - \text{Value of operators time}}{\text{Average capital investment}} \times 100$$

**Example 5.**   **Percent Return on Capital**

| | |
|---|---:|
| Farm income + interest (on farm loans) | $20,000 |
| Less value of operator's labor and management | 7,500 |
| | ——— |
| Return on capital | $ 12,500 |
| Average capital invested | 150,000 |
| **Percent return on capital** | **8.33%** |

Percent return on capital can be used to evaluate productivity of invested capital. Comparisons can be made with alternative uses of that capital. Theoretically, if the return on capital is less than interest rates on savings accounts, one might consider alternative investments. However, other factors such as providing steady employment for the operator's family, land appreciation, etc., should also be considered.

## Net Worth

Net worth is total assets minus total liabilities. It indicates, on a specific date, the value of the assets that would be left if all outside claims against the business were paid. It is an indicator of the financial progress of a business over time if it is calculated at regular intervals.

**Example 6.**   Net Worth

| | |
|---|---:|
| Current assets (feed, supplies, and market livestock) | $34,500 |
| Intermediate assets (dairy and breeding livestock and machinery) | 43,500 |
| Long-term assets (land and buildings) | 72,000 |
| | ——— |
| **Total assets** | **$150,000** |

| | |
|---|---:|
| Current liabilities | $13,500 |
| (payable within next 12 months) | |
| Intermediate liabilities | 12,500 |
| (payable in over 1 but less than 10 years) | |
| Long-term liabilities | 14,000 |
| (mortgage on real estate less portion included in current liabilities) | |
| | ——— |
| **Total liabilities** | **$ 40,000** |
| | |
| **Net worth** | **$110,000** |

All six of the methods discussed here are used as measures of profitability by various dairymen. The question of which is best is a difficult one to answer and is dependent on personal goals and needs. For example, an elderly dairyman nearing retirement with a net worth in excess of $500,000 may not be interested primarily in maximizing percent return on capital or increasing net worth, but rather in farm net cash income. A young dairyman with assets of $200,000, liabilities of $180,000, and net worth of $20,000 needs to be very concerned about increasing net worth and maximizing percent return on capital as well as farm and family net cash income to meet debt service and family living expenses and increase the asset-liability ratio.

A combination of farm or family net cash income, farm income, labor income or labor earnings, percent return on capital, and net worth are valuable measures of profitability for most dairymen. Farm net cash income is important in determining ability to pay cash costs, family living expenses, and to estimate debt-paying ability. Farm income indicates return for labor and invested capital, and labor income or earnings enable dairy managers to evaluate the productivity of their labor and management skills. Percent return on capital is a good tool to use to evaluate the efficiency of the business's use of resources, and periodic (annual) net-worth analysis is a reliable measure of the financial progress and of the solvency of the business.

## FINANCIAL TOOLS

Three tools useful in determining if the criteria for a successful business are being met and in financial planning and decision making are: (1) the profit-and-loss or operating statement, (2) the cash-flow statement, and (3) the net-worth statement. These tools, compiled from accurate financial records, can provide the information needed for many of the financial decisions faced by dairy farm managers. "These tools, along with associated estimates, calculations, ratios, etc., can provide an effective means of analyzing financial problems and situations, controlling financial matters, and forestalling possible financial disaster."*

## Profit-and-Loss Statement

The profit-and-loss statement, or operating statement, shows the earnings of a business for a specific period of time, usually the business year. It lists the source and amount of cash receipts and the nature and amount of cash expenses and reveals the influence of inventory changes and depreciation on profitability. It also shows the net cash income available for family living, debt-principal repayment, paying old bills, or servicing new debt. From the profit-and-loss statement one can evaluate the business in view of the criteria for a successful business mentioned previously. Is the business paying operating expenses (farm net cash income), maintaining productivity (net change or difference between inventory increase and depreciation), and earning a reasonable return for the operator (labor income or earnings)? If the total investment is known, the interest on investment can be calculated. This can be achieved by dividing the profit (farm income − value of operator's time) by the total investment.

If one also maintains enterprise records (cost and returns for each part of the total farm operation, i.e., corn raised for silage, hay raised for feed, the dairy herd, etc.), the profitability or lack of profitability for each enterprise, as well that of the total operation, can be determined. This *enterprise analysis* can be a valuable method of determining weak links in the total operation. Often profitability of the total operation can be increased by identifying and correcting weak or unprofitable enterprises. If one computes a profit-and-loss statement for the total farm only, profitable enterprises can mask the identity of the unprofitable ones.

Profit-and-loss statements are also useful in income-tax computations; in loan applications, as an indicator of management skill (ability to combine available resources to market a product profitably) and ability to repay a loan; in budgeting for future expansion or new purchases; and in examining alternative enterprises or uses of capital resources.

An example of an abbreviated profit-and-loss statement is shown in Table 3-1. From the information in this statement, one can see that this business can pay operating expenses (farm net cash income = $29,000) and is earning a reasonable return for the operator (farm income = $30,000). If the value of the

---

* Reynolds, R. K. and Chamblis, R. L. Managing Farm Finances, Va. Agric. Econ. Newsletter, Dec. 1973.

**Table 3-1**  Abbreviated Profit-and-Loss Statement for a 100-Cow Dairy Farm, January 1, 1976 through December 31, 1976

| | | |
|---|---:|---:|
| **Cash Receipts** | | |
| Net milk sales (hauling and service charges deducted) | $126,000 | |
| Dairy calves, steers, cull cows | 10,000 | |
| Breeding livestock | 2,500 | |
| Other miscellaneous receipts | 500 | |
| TOTAL CASH RECEIPTS | | $139,000 |
| **Cash Operating Expenses** | | |
| Purchased feed | $ 37,400 | |
| Hired labor | 16,500 | |
| Crop expenses | 20,350 | |
| Repairs, fuel, supplies (dairy) | 11,200 | |
| Livestock expenses (breeding, veterinarian, etc.) | 3,950 | |
| Interest (at 9% on $105,000) | 9,450 | |
| Other farm and dairy expenses | 6,000 | |
| Insurance and taxes | 5,150 | |
| TOTAL CASH OPERATING EXPENSES | | $110,000 |
| **Farm Net Cash Income** | | **$ 29,000** |
| Net change in inventory—including cattle and feed | $ 14,700 | |
| Machinery and building depreciation | 13,700 | |
| NET CHANGE (difference between inventory increase and depreciation) | | +$  1,000 |
| NET RETURN TO MANAGEMENT, CAPITAL AND UNPAID OPERATOR'S LABOR (Farm income) | | $ 30,000 |
| Interest on average owned capital during the year ($181,000 × 8%) | $ 14,480 | |
| LABOR INCOME | | $ 15,520 |

operator's time is estimated and the total value of the assets of the business are known, one can determine whether this business is paying a reasonable rate of return on all capital invested. For example, if the value of the operator's time was estimated at $7,600 and the average investment in this business during 1976 was $306,000, the return on investment would be cash receipts ($139,000), minus cash operating expenses except interest ($100,550) and value of operator's time ($7,600), divided by average investment ($306,000), or

$$\frac{\$139,000 - \$108,150}{\$306,000} = 10.1\%.$$

This abbreviated profit-and-loss statement, however, cannot identify weak links in the individual enterprises that contribute to the total dairy business. Examples of enterprise profit-and-loss statements are given in Tables 3-2, 3-3, and 3-4. Note that in Table 3-2 cash and noncash receipt items are accounted for as is interest on borrowed capital (a cash expense) and interest on owned investment (a noncash expense). Homegrown feeds are charged as a noncash expense to the dairy herd at opportunity cost, and manure value is credited as a noncash receipt. This dairy herd operation meets the criteria for a successful business as operating expenses were paid, reasonable interest on all capital (a cash expense on borrowed capital and a noncash expense on owned capital investment) was earned, productivity was maintained, as indicated by a net increase in inventory-depreciation, and there was a reasonable return to the operator (labor income = $9,640).

Table 3-3 is the profit-and-loss statement for the alfalfa hay enterprise on the farm. Analysis of this information identifies the alfalfa hay enterprise as an unprofitable one and indicates alternatives should be investigated.

Table 3-4 is the profit-and-loss statement for the corn silage enterprise on the farm. Analysis of this statement indicates that this enterprise is a very profitable one and that the use of land resources to produce this feed for dairy cattle on this farm is a wise management decision.

## The Net-Worth Statement

The net-worth statement, or balance sheet compares business assets with business liabilities. The amount by which assets exceed liabilities is the owner's net worth. It should be calculated periodically (annually) to evaluate financial progress and solvency.

Three categories of assets and liabilities—current, intermediate, and long term—must be included in the balance sheet to make it most useful in financial management analysis (Table 3-5).

Current assets include the value of feed, supplies, and market livestock on hand. Current liabilities include all accounts payable within the next 12 months, including principal payments on intermediate and long-term notes. This means that the business in Table 3-5 must generate $20,000 in cash after paying cash operating costs and family living costs during the 1977 calendar year to meet existing debt commitments. Failure to do so, even though the business is sound in the long run, can lead to financial problems. If sufficient cash is not generated to meet current liabilities, alternatives such as selling assets (current, intermediate, or long term) or refinancing should be investigated. Selling assets can lead to decreased productivity; refinancing can be expensive and increases intermediate and/or long-term liabilities.

The ratio of current assets to current liabilities can be calculated from the

**Table 3-2** Dairy Herd Profit-and-Loss Statement for a 100-Cow Dairy Herd, January 1, 1976 through December 31, 1976

| | | |
|---|---:|---:|
| **Cash receipts** | | |
| Milk sales (14,000 cwt. × $9.50/cwt.) | $133,000 | |
| Cull cows and heifers (38) | 9,160 | |
| Bull calves (56) | 840 | |
| Breeding livestock (3 heifers, 1 cow) | 2,500 | |
| Miscellaneous | 500 | |
| TOTAL CASH RECEIPTS | | $146,000 |
| **Noncash receipts** | | |
| Manure credit | $ 5,000 | |
| Net increase in cattle inventory | 10,000 | |
| Net increase in stored feed inventory | 4,700 | |
| TOTAL NONCASH RECEIPTS | | $ 19,700 |
| TOTAL RECEIPTS | | $165,700 |
| **Cash operating expenses** | | |
| Purchased feed | $ 37,400 | |
| Hired labor | 15,500 | |
| Fuel and repairs for dairy equipment | 4,650 | |
| Building maintenance and repairs | 4,350 | |
| Dairy supplies | 2,200 | |
| Breeding fees | 1,000 | |
| Veterinarian and medicine | 2,500 | |
| Utilities (heat, light, and telephone) | 2,500 | |
| Insurance and taxes (cattle, dairy buildings, and equipment) | 3,450 | |
| Milk hauling (14,000 cwt. × $.40/cwt.) | 5,600 | |
| Marketing charges (14,000 cwt. × $.10/cwt.) | 1,400 | |
| Livestock hauling | 450 | |
| Interest on borrowed capital ($54,000 @ 9%) | 4,860 | |
| Miscellaneous | 2,600 | |
| TOTAL CASH OPERATING EXPENSES | | $ 88,460 |
| **Noncash operating expenses** | | |
| Homegrown feed: 2000 tons of corn silage @ $20/ton | $ 40,000 | |
| 150 tons of alfalfa hay @ $60/ton | 9,000 | |
| Depreciation: dairy buildings, machinery, & equipment | 8,200 | |
| Interest on owned capital ($130,000 @ 8%) | 10,400 | |
| TOTAL NONCASH OPERATING EXPENSES | | $ 67,600 |
| TOTAL OPERATING EXPENSES | | 156,060 |
| LABOR INCOME FOR THE DAIRY ENTERPRISE | | $ 9,640 |

**Table 3-3** Profit-and-Loss Statement for 50 Acres of Alfalfa Hay

| | | |
|---|---:|---:|
| **Cash receipts** | 0 | |
| **Noncash receipts:** 150 tons @ $60/ton | $ 9,000 | |
| TOTAL RECEIPTS | | $ 9,000 |
| **Cash operating expenses** | | |
| Fertilizer | $ 1,500 | |
| Custom baling | 1,500 | |
| Hired labor | 1,000 | |
| Seed (reseeded ¹/₃) | 400 | |
| Lime | 900 | |
| Interest on borrowed capital ($15,000 @ 9%) | 1,350 | |
| Maintenance, fuel, and repairs on haying equipment | 1,700 | |
| Taxes and insurance | 500 | |
| Miscellaneous | 300 | |
| TOTAL CASH OPERATING EXPENSES | | $ 9,150 |
| **Noncash operating expenses** | | |
| Depreciation (equipment and storage) | $ 1,500 | |
| Interest on owned capital ($15,000 @ 8%) | 1,200 | |
| TOTAL NONCASH OPERATING EXPENSES | | $ 2,700 |
| TOTAL OPERATING EXPENSES | | $11,850 |
| LABOR INCOME (DEFICIT) FOR THE ALFALFA HAY ENTERPRISE | | ($ 2,850) |

net-worth statement and can be applied to a dairy business. This ratio measures the ability of the business to pay its current debts from current assets. To ensure a reasonable margin of safety, the ratio should not fall below 1.7 to 1 ($1.70 current assets per $1.00 current liabilities). Analysis of the example in Table 3-5 reveals a current-assets-to-current-liability ratio of 2:1 ($40,000 current assets and $20,000 current liabilities), an acceptable business ratio. Analysis of Table 3-1 (profit-and-loss statement for the previous year) also indicates a healthy business situation as the farm net cash income was comfortably above $20,000 for the previous year.

Perhaps of equal or greater importance is the ratio of liquid assets to current liabilities. Liquid assets may be defined as those that can be sold for cash without interfering with the normal operation of the business. Examples of liquid assets are checking and savings accounts, market livestock held for sale, cash value of life insurance policies or bonds, etc.

The ratio of intermediate assets to intermediate liabilities in Table 3-5 is also very favorable ($120,000 to $20,000), indicating very little intermediate debt on breeding cattle and machinery.

**Table 3-4**  Profit-and-Loss Statement for 120 Acres of Corn for Silage

| | | |
|---|---:|---:|
| **Cash receipts** | 0 | |
| **Noncash receipts** (2,000 tons corn silage @ $20/ton) | $40,000 | |
| | | |
| TOTAL RECEIPTS | | $40,000 |
| **Cash operating expenses** | | |
| Fertilizer and lime | $ 7,200 | |
| Seed and herbicide | 3,600 | |
| Harvesting (custom) | 2,400 | |
| Fuel and repairs (plowing, planting, etc.) | 1,150 | |
| Interest on borrowed capital ($36,000 × 9%) | 3,240 | |
| Taxes and insurance | 1,200 | |
| Miscellaneous | 600 | |
| | | |
| TOTAL CASH OPERATING EXPENSES | | $19,390 |
| **Noncash operating expenses** | | |
| Manure credit from dairy | $ 5,000 | |
| Depreciation (equipment and storage) | 4,000 | |
| Interest on owned capital ($36,000 × 8%) | 2,880 | |
| | | |
| TOTAL NONCASH OPERATING EXPENSES | | $11,880 |
| | | |
| TOTAL OPERATING EXPENSES | | $31,270 |
| LABOR INCOME FOR THE CORN SILAGE ENTERPRISE | | $ 8,730 |

**Table 3-5**  Abbreviated Net-Worth Statement (Balance Sheet) for a 100-Cow Dairy Business, January 1, 1977

| Assets | | Liabilities | |
|---|---:|---|---:|
| Current assets (value of feed, supplies, and market livestock) | $ 40,000 | Current liabilities (accounts and notes payable within the next 12 months, including the part of the intermediate and long-term debt due within 12 months) | $ 20,000 |
| Intermediate assets (value of breeding livestock and machinery) | 120,000 | Intermediate liabilities (accounts and notes payable in over 1 but less than 10 years, less amount included in current liabilities) | 20,000 |
| Long-term (fixed) assets (value of land and buildings) | 146,000 | Long-term liabilities (mortgage on real estate, less amount included in current liabilities) | 85,000 |
| | | | |
| TOTAL ASSETS | $306,000 | TOTAL LIABILITIES | $125,000 |
| NET WORTH (total assets − total liabilities) | | | $181,000 |

The ratio of long-term assets to long-term liabilities in our example ($146,000 to $85,000) indicates a rather heavy debt load on the real estate and buildings. However, annual liability for this debt load is accounted for in current liabilities without applying undue pressure to the business. This illustrates an advantage of maximum use of borrowed capital for long-term investment items. The repayment of the loan can be distributed over a longer time period with lower annual repayments, thus exerting less financial pressure on the business. Because of recent land appreciation values, additional noncash income from land appreciation is reflected in substantial gains in net worth.

The ratio of total liabilities to net worth is a good method of evaluating the solvency of the business. The owner's invested funds serve as guarantee of his ability to repay the total liabilities if liquidation should occur. If the total liabilities exceed the net worth, then creditors have more at stake in the business than the owner. In our example (Table 3-5), this ratio is $125,000 to $181,000, or .69:1, and indicates that the owner does own the majority of the assets. This enhances ability to borrow additional capital should that be necessary. Ratios of 1:1, 2:1, or higher are not uncommon in dairy businesses, especially for young persons just starting in the business. The financial importance of these higher ratios varies, however, with the proportion of the liabilities that are current, intermediate, and long term. If most of the liabilities are current, serious financial problems can result, as management will have little time (1 year) to correct the situation; however, if most of the liabilities are intermediate or long term, management has a longer period of time to correct the situation. For these reasons it is usually a good policy to use borrowed capital for intermediate and long-term assets, especially those assets that are likely to generate income (breeding cattle) and appreciate in value (land), and to request the use of the capital for as long a term as possible (20 to 30 years on land). This reduces financial pressure because of lower annual repayment schedules and helps avoid short-term financial pressures that can lead to refinancing or liquidation of assets to meet current financial obligations.

The net-worth statement can also be used to evaluate financial progress over time if it is calculated on a regular basis. To illustrate this point, let us assume that the dairyman in our example (Table 3-5) is making annual principal payments of $4,000 on intermediate liabilities and $6,000 on long-term liabilities. He is then increasing net worth $10,000 per year, assuming no land appreciation and no increase or decrease in intermediate or current assets. If the farm appreciates 10% in market value during the year, this increases net worth an additional $14,600. If intermediate assets increase (inventory or breeding cattle) more than depreciation of equipment, this also increases net worth. If the feed and supply inventory increases (current assets), this also increases net worth. Steady progress in net worth is an indication of good financial and herd management.

These are some of the ways the net-worth statement can be a valuable tool in financial management. It is important for every dairy manager to prepare this statement regularly (annually) and use it to analyze financial progress, evaluate business solvency, and identify potential problem areas so that adjustments can be made before financial problems occur.

## The Cash-Flow Statement

A cash-flow statement is a calendar of receipts and expenses. It may be calculated for a variety of time periods but usually on a monthly or quarterly basis on dairy farms. An example of a quarterly cash flow for our 100-cow dairy herd is given in Table 3-6. Careful analysis of the figures in Table 3-6 reveals several financial aspects rather typical of dairy farm operations. Both receipts and expenses are subject to variation when examined on a quarterly basis. Receipts vary with milk production, which is somewhat seasonally variable on most farms, and with wholesale milk price, which also varies seasonally. Cash expenses are normally heaviest during the cropping season (April to September) and unfortunately milk prices are usually lowest during this part of the year. Variations such as these can cause financial problems if they are not anticipated. If adequate financial reserves are not available (as in the example in Table 3-6), loans can be arranged to meet financial obligations during those periods when cash expenses exceeds cash income. Even though the dairyman in our example had $3,000 cash on hand on January 1, this was reduced to a balance of $2,600 during the second quarter. If all of the $3,000 had been spent for a new car or other personal or capital items during the first quarter, this would have necessitated borrowing additional capital for use during the second and third quarters. This, in turn, would create an additional expense (interest on the additional capital) and would decrease annual profitability.

A cash-flow statement for the past year is an excellent basis for projecting future cash flow, especially when some change in the business is anticipated (expansion, major improvements, large capital purchases, etc.). It enables management to anticipate the times additional capital will be needed and appropriate times for scheduling repayment, thus avoiding scheduling of loan repayments in times of cash shortages. Projected cash-flow statements provide valuable information to lenders in evaluating an applicant's ability to repay the loan as well as the timing of interest and principal payments.

**Table 3-6** Cash-Flow Statement (Quarterly), January 1, 1976–December 31, 1976

| | Total | January–March | April–June | July–September | October–December |
|---|---|---|---|---|---|
| **Operating receipts** | | | | | |
| Milk (net) | $126,000 | $33,000 | $31,000 | $26,000 | $36,000 |
| Other | 0 | 0 | 0 | 0 | 0 |
| Total | **$126,000** | **$33,000** | **$31,000** | **$26,000** | **$36,000** |
| **Capital receipts** | | | | | |
| Livestock | $ 12,500 | $ 3,500 | $ 3,000 | $ 4,000 | $ 2,000 |
| Other | 500 | 500 | 0 | 0 | 0 |
| Total | **$ 13,000** | **$ 4,000** | **$ 3,000** | **$ 4,000** | **$ 2,000** |
| **Operating expenses** | | | | | |
| Hired labor | $ 16,500 | $ 3,900 | $ 4,300 | $ 4,400 | $ 3,900 |
| Feed purchases | 37,400 | 10,000 | 9,000 | 8,000 | 10,400 |
| Crop expenses | 20,350 | 5,000 | 10,600 | 3,600 | 1,150 |
| Livestock, other | 26,300 | 7,352 | 5,610 | 5,666 | 7,672 |
| Total | **$100,550** | **$26,252** | **$29,510** | **$21,666** | **$23,122** |
| **Capital expenditures** | | | | | |
| Cows | $ 1,000 | $ 0 | $ 1,000 | $ 0 | $ 0 |
| Machinery | 1,000 | 0 | 1,000 | 0 | 0 |
| Total | **$ 2,000** | **$ 0** | **$ 2,000** | **$ 0** | **$ 0** |
| Family living expenses | $ 7,450 | $ 2,000 | $ 1,800 | $ 1,600 | $ 2,050 |
| Debt principal payments | $ 10,000 | $ 2,500 | $ 2,500 | $ 2,500 | $ 2,500 |
| Debt interest payments | $ 9,450 | $ 2,448 | $ 2,390 | $ 2,334 | $ 2,278 |
| **Total receipts** | **$139,000** | **$37,000** | **$34,000** | **$30,000** | **$38,000** |
| **Total expenses** | **$129,450** | **$33,200** | **$38,200** | **$28,100** | **$29,950** |
| Cash surplus (deficit) | $ 9,550 | $ 3,800 | ($ 4,200) | $ 1,900 | $ 8,050 |
| Cash start of period | $ 3,000 | $ 3,000 | $ 6,800 | $ 2,600 | $ 4,500 |
| Cash end of period | $ 12,550 | $ 6,800 | $ 2,600 | $ 4,500 | $12,550 |

## SUMMARY

In recent years capital investment and debt load per dairy farm have increased rapidly. Output value as well as purchased inputs have also increased. Investments of $3,000 to 4,000 per cow and annual income of $1,500 or more per cow are not unusual. These developments have made good financial management a must for dairymen. It is a larger and more critical component of dairy herd management than ever before. The dairy manager must be knowledgeable about the tools of the professional business manager so that he or she can use them to make financial decisions based on accurate financial data. Operating, or profit-and-loss, statements are needed to determine the profitability of the business. Past operating statements, along with past cash-flow sheets, are an excellent basis for budgeting. This is especially true when major capital outlays for remodeling, expansion, etc., are being considered. Financial progress and business solvency can be monitored from net-worth statements. All three of these tools provide valuable information to lenders when additional borrowed capital is needed. They can provide insight for future financial planning and realistic repayment scheduling, and help avoid many financial problems. Current and past financial records can help evaluate the use of resources and reveal whether the criteria for a successful business are being met. Enterprise operating statements can identify weak links in various enterprises that comprise the total dairy operation. Accurate business records can also help identify many existing conditions and projected actions other than those previously discussed that can lead to business failure. Some of these are discussed below.

### Overinvestment per Unit of Production

In the example in Tables 3-1 to 3-5, total investment was $306,000 for a 100-cow operation with replacements, or an investment of $3,060 per cow. What would be the effect on profit (farm income) if the investment were $4,060 per cow and the additional $1,000 per cow were rented capital at 9% interest? This would increase cash expenses by $90 per cow or $9,000 per year for the business. This added expense would reduce farm net cash income from $29,000 to $20,000 and labor income from $15,520 to $6,520.

### Poor Investment Priority

In Table 3-5, let us assume that $80,000 of the intermediate assets were accounted for in breeding livestock and $40,000 in equipment. This $40,000 equipment investment plus $146,000 for land and buildings totals $186,000 invested in land, buildings, and equipment for a 100-cow dairy herd, or $1,860 per cow. What would happen to profitability of this operation if the owner decided to build or install new milking and feeding equipment at a cost of $120,000 (investment A, Table 3-7) to reduce hired labor expense by $8,000 per year? If the purchase were financed at 9% interest, the interest expense alone would increase expenses $2,800 more than the reduction in labor expense. Compare this investment with that of purchasing 40 additional cows (@$600 each) at a cost of $24,000, adding 40 freestalls to the barn (@$200 each) at a cost of $8,000, increasing feed storage and feeding area at a cost of $12,000, or a total additional cost of $44,000 (investment B, Table 3-7). Based on current business records (Table 3-2), one could expect this investment to generate $53,200 additional cash income from the sale of milk ($1,330 per cow × 40 cows) and create additional cash expenses of $48,360 ($1,110 per cow × 40 cows plus the interest on $44,000 at 9%). Investment B increases farm net cash income by $4,840, a sum that can be used for principal payment. This is a much sounder investment than investment A, assuming availability of the additional resources, because the additional capital is used to purchase items that generate more income than their cost.

## Too Many Purchased Inputs

In our example, if all feed were purchased at a cash expense of $80,000 (currently $31,000 for purchased feed plus a $49,000 value for the homegrown feed) instead of the current expense of $69,120 ($31,000 for purchased feed plus $11,850 for the alfalfa hay and $26,270 for the corn silage), this would

**Table 3-7**  Comparison of Investments

|  | Current | Investment A[a] | Investment B[b] |
|---|---|---|---|
| Total investment in land, buildings and equipment | $186,000 | $306,000 | $206,000 |
| Number of cows | 100 | 100 | 140 |
| Investment per cow (land, bldgs. and equip.) | $1,860 | $3,060 | $1,471 |
| Additional cash receipts | 0 | 0 | 53,200[c] |
| Additional cash expenses | 0 | 10,800[d] | 48,360[e] |
| Reduced cash expenses | 0 | 8,000[f] | 0 |
| Farm net cash income | 29,000 | 26,200 | 33,840 |

[a] Added investment of $120,000 in new milking and feeding equipment.
[b] Added investment of $20,000 in feed storage and feeding equipment.
[c] Sale of milk from the 40 added cows.
[d] Interest on $120,000 @ 9%.
[e] Interest on $44,000 @ 9%, plus additional feed, labor, etc., expense for 40 cows.
[f] Reduced labor expense.

**Table 3-8** Comparison of Homegrown and Purchased Feed Costs

|  | Homegrown | Purchased |
| --- | --- | --- |
| Purchased feed expense | $31,000 | $80,000 |
| Expense of producing alfalfa hay | 11,850 | — |
| Expense of producing corn silage | 26,270 | — |
| Total feed expense | $69,120 | $80,000 |
| Farm net cash income | 29,000 | 18,120 |
| Labor income | 15,520 | 4,640 |

decrease his farm net cash income by $10,880 per year (from $29,000 to $18,120) and his labor income from $15,520 to $4,640 (Table 3-8).

Good financial management is one of the skills that today's dairymen must possess if they are to avoid financial problems. It is a powerful ally, but not a substitute for good cow, land, machinery and equipment, and labor management. Collectively, skills in all of these areas increase the probability for achievement of herd, financial, and personal goals.

## SELECTED REFERENCES

Brown, L. H., and Speicher, J. A., Business Analysis for Dairy Farms, *Ext. Bull.* E-685, Michigan State University, East Lansing, 1970.

James, S. C. and Stoneberg, E. Farm Accounting Business Analysis. Iowa State University Press, Ames, Iowa. 1974.

McGilliard, M. L., Is That Your Best Investment? *The Virginia Dairyman*, Harrisonburg, Va. October 1975.

Pardue, D. C., Using Business Records for Dairy Management, *Guernsey Breeders' Journal*, Peterborough, N.H. July 1971.

Reynolds, R. K., The Skill of Management, *The Virginia Dairyman*, Harrisonburg, Va. October 1975.

Reynolds, R. K., Your Business Balance Sheet. *The Virginia Dairyman*, Harrisonburg, Va. August 1975.

Reynolds, R. K., and Chambliss, R. L., Managing Farm Finances, *VAE Newsletter*. Blacksburg, Va., December 1973.

Schneeberger, K. C. and Osburn, D. D. Financial Planning in Agriculture. Interstate Printers and Publishers, Danville, Ill. 1977.

Wells, R. C., and Pardue, D. C., Dairy Farm Business Summary, *N. C. Ext. Serv. Circ.* 532 (1971).

# Feeding Dairy Cattle

## FEEDING GOALS

The goal of the feeding program is to offer each animal a ration that will encourage maximum economical production—to provide each animal with a ration that:

*1. Meets the animal's nutrient requirements.* The ration must provide adequate amounts of each nutrient needed for maintenance, growth (if immature), reproduction (if pregnant), and production at maximum or most economical levels.

*2. Is palatable.* Ingredients of the ration must be in a form and condition that have appetite appeal to the cow.

*3. Is economical.* Feed costs constitute about 50% of the total costs of producing milk on most dairy farms. An economical feeding program, on a feed-cost-per-unit-of-production basis, is critical in controlling production costs.

*4. Is conducive to the health of the animal and the production of milk of normal composition.* Feeding practices that are likely to result in an increase in incidence of metabolic disorders such as milk fever, ketosis and displaced abomasum, or in severe milk fat depression should be avoided.

Assuming equal genetic ability, the quality of the feeding program is the single most important factor in determining level of production. Deficiencies in total nutrient intake, deficiencies in specific nutrients, or lack of a balanced ration will result in less than optimum production. Generally, feed costs per unit of production decreases as level of production increases. Analysis of the Virginia DHI summary data (Table 4-1) clearly illustrates the relationship between level of production, feed intake, total feed cost per hundredweight of milk, and income over feed cost.

Many alternatives are available to the manager of a dairy farm and many decisions must be made in regard to the feeding program. Decisions must be made concerning the use of available land resources for feed production, in feed purchases, in developing a satisfactory ration, and in the feeding system. New discoveries and developments are continually being made in all these areas. The challenge to management is to utilize new information and ideas as they become known and to utilize available resources to the maximum to develop a feeding program that meets the basic principles that are not likely to change; that is, to feed cattle a ration that meets nutrient requirements, is palatable, economical, and conducive to good health.

To achieve this goal, the manager should have some understanding of the basic principles of ruminant nutrition and thorough knowledge of the following.

1. *Nutrient requirements of dairy cattle.* This information will assist in estimating the nutrient needs of cattle and in identifying deficiencies.

2. *The nutrient content, palatability, and cost,* on a per-unit-of-nutrient basis, of various feedstuffs. This knowledge will assist in decisions regarding making maximum use of land resources for feed production, and for purchasing feeds.

3. *Ration formulation,* so that the nutrient intake approximates the nutrient requirements and enhances the opportunity for each animal to receive a nutritionally balanced ration.

4. *The various systems of feeding dairy cattle,* in order to develop an efficient method of feeding his cattle.

**Table 4-1**   The Relationship of Level of Production and Income over Feed Cost (1974–1975 Virginia DHI Summary)

| Grouping by lb. fat | 601–710 | 551–598 | 501–550 | 451–500 | 401–450 | 246–400 |
|---|---|---|---|---|---|---|
| Milk per cow (lb.) | 16,800 | 15,495 | 14,178 | 12,771 | 11,553 | 9,812 |
| Fat per cow (lb.) | 630 | 570 | 524 | 478 | 427 | 365 |
| Concentrates per cow (lb.) | 5,500 | 5,700 | 5,100 | 4,600 | 4,400 | 4,100 |
| Rate of forage feeding[a] (lb.) | 2.6 | 2.4 | 2.4 | 2.4 | 2.3 | 2.3 |
| Total feed cost per cow ($) | 657 | 635 | 585 | 542 | 496 | 473 |
| Feed cost/cwt. milk ($) | 3.91 | 4.10 | 4.13 | 4.24 | 4.30 | 4.82 |
| Income over feed cost per cow ($) | 949 | 818 | 743 | 673 | 578 | 452 |

[a] The amount of good hay equivalent fed per 100 lb. of body weight per day.

## NUTRIENTS AND THEIR USE IN THE ANIMAL BODY

Nutrients are used by animals for maintenance, repair, growth, production, and reproduction. A *nutrient* can be defined as a specific element or compound derived from ingested food and used to support the physiological processes of life.

Nutrients are classified as either essential or nonessential. An *essential* nutrient is one that cannot be synthesized by the body and therefore must be supplied in the ration. *Nonessential* nutrients are those that can be synthesized by an animal's body or, in the case of the ruminant, by the rumen microorganisms. Plants and animals are composed of six classes of compounds. These are (1) water, (2) ash, or minerals, (3) protein, (4) carbohydrates, (5) fats, and (6) vitamins.

## Water

Water, composed of hydrogen and oxygen, is a necessary compound of plants and animals. Growing plants contain as much as 70 to 80% water, and even the bodies of animals are 70 to 90% water. Water has several important functions in the animal body. It gives elasticity and rigidity to the supportive tissue of the animal, helps to dissolve the food, acts as a carrier of food and waste, helps to maintain the osmotic pressure of the body, and prevents extensive changes of temperature. Any interference with the normal amount of water, in either plants or animals, produces disastrous results. Water, therefore, should always be supplied to livestock in large amounts.

## Dry Matter

If a substance is heated to a temperature at or above that of boiling water until it ceases to lose weight, the remaining residue is known as dry matter. The loss of weight represents moisture, or water. Dry matter is divided into organic matter and ash, or inorganic matter.

### Ash, or Mineral Matter.

When dry matter is burned the organic matter is burned out, leaving what is known as ash, or mineral matter. There is considerable ash in all the common feeding stuffs. In the animal the dry matter of the bones consists largely of ash, whereas the dry matter of the rest of the body contains, on the average, about 7% ash.

The functions of the minerals in the animal body are numerous. They furnish material for the formation of new tissues, especially that of the skeleton and of the mineral part of milk. Minerals are especially important for young, growing animals, and for lactating cows.

Minerals help to maintain osmotic pressure. The cells of the various body tissues draw their nourishment from the lymph, from which they are separated by cell walls. These walls have the nature of a semipermeable membrane. In order to maintain normal conditions in the protoplasm of the cells, the osmotic pressure of the lymph, and therefore that of the blood from which it is derived, must be maintained approximately constant. The constant osmotic pressure is maintained largely by minerals contained in solution.

Minerals help maintain homeostasis of the body. The body is continually producing acids, especially carbonic and phosphoric acid, which tend to increase the acidity of the blood. These are in part neutralized by salts in the blood serum, especially sodium phosphate and bicarbonate, which play an important part in maintaining its neutrality. Salts of short-chain fatty acids are present in blood in considerable quantities and influence the acid-base balance.

Minerals aid in respiration. Iron is an essential part of the hemoglobin, by means of which the oxygen is distributed through the body. Iodine is an essential constituent of thyroxine, an important hormone secreted by the thyroid gland, which influences the rate of metabolism of all body cells.

Minerals are also necessary in putting certain materials into solution. Certain proteins, for example, are soluble only in dilute salt solutions. Some minerals also aid in digestion, especially of fats and proteins, and others are useful in protein and carbohydrate metabolism.

**Organic Matter.** Chemists determine the amount of organic matter by measuring the difference between the total dry matter and the ash. Organic matter is divided into three groups: protein, fat, and carbohydrates.

PROTEIN. The protein of feeding stuffs is not determined directly by the chemist. The usual method of analysis is to determine the amount of nitrogen and then to multiply this amount by the factor 6.25 as most proteins contain about 16% nitrogen.

COMPOSITIONS OF PROTEINS. All proteins contain carbon, hydrogen, oxygen, and nitrogen; many contain sulfur, and a few contain phosphorus or iron.

STRUCTURE OF PROTEINS. Proteins are made up of amino acids linked together. The number of amino acids contained in a protein molecule varies in different proteins. No two proteins are alike in this regard. Some proteins contain none of the more important amino acids. This fact is very important in the study of nutrition, since aminals with single stomachs cannot produce these amino acids, many of which are essential for proper nutrition. However, in ruminants the microorganisms in the rumen are able to synthesize proteins from nitrogenous compounds. The dairy cow is able to digest the bacteria and thus secure amino acids, even though they are not present in the feed. For this reason it is not so necessary to feed high-quality proteins to dairy cows as it is to pigs or chickens.

FUNCTION OF PROTEINS. The function of proteins in the animal body is to repair and replace living tissue, to provide amino acids for hormone synthesis, and, in cows, to supply the protein content of the milk. Protein can also be used as a source of energy.

NONPROTEIN NITROGEN-CONTAINING SUBSTANCES. Feeding stuffs contain a great variety of nitrogen-containing substances that are not proteins but have a very much less complex molecular structure. The most important of these compounds are amides, amines, amino acids, nucleic acids, and urea. Because plants build up proteins through the utilization of some of these compounds, they are found in more abundance in young, growing plants. Because they are ruminants, dairy cattle can utilize many of these nonprotein nitrogen-containing substances for protein synthesis.

CARBOHYDRATES. Carbohydrates are found only in small amounts in animals but are especially characteristic of plants, in which they form about 75% of the entire dry matter. They are the chief source of energy for dairy cattle. The carbohydrates of feeding stuffs are divided into fiber and nitrogen-free extract.

*Fiber* is the more insoluble portion of the carbohydrates and consists of hemicelluloses, celluloses, and pentosans. These compounds usually contain a considerable amount of lignin. Lignin is the more fibrous part of a plant and is even less digestible than the cellulose. It is deposited around cellulose fibers. The microorganisms in the rumen can convert much of the cellulose to soluble compounds for their own use and for absorption. However, as a plant matures and the amount of lignin covering the cellulose increases, the rumen microorganisms have increasing difficulty in digesting the cellulose; even these microorganisms are unable to digest the lignin. Thus every plant stem that has become mature and encrusted with lignin is less digestible.

*The nitrogen-free extract* is the more soluble portion of the carbohydrates. It includes sugars, starches, and the more soluble portions of the cellulose and pentosans. It also includes organic acids such as lactic, butyric, and acetic acid. It is easily digested and has a high nutritive value.

COMPOSITION OF CARBOHYDRATES. Carbohydrates are composed of carbon, hydrogen, and oxygen. They are unstable, easily oxidized, and easily reduced. Carbohydrates can be broken down into organic acids or changed into sugars. In ruminants much of the carbohydrate in the diet is fermented by rumen microorganisms; the end products of this fermentation are primarily short-chain fatty acids. These are absorbed from the rumen and provide energy. By alternative metabolic routes, they may be reduced to form fats. The short-chain volatile fatty acids, especially acetic acid, are primary precursors of milk fat. Any glucose that is absorbed from the digestive tract or arises from metabolism may also be oxidized for energy or reduced to form fats. In this manner the carbohydrates can serve as a source of either immediate energy or stored energy.

FUNCTION OF CARBOHYDRATES. Carbohydrates are the dairy animal's major source of energy. They are not stored in the body in large amounts in their original form, but are changed into fat, the form best suited for storing energy.

FATS. Fats supply energy and serve as a source of the fat-soluble vitamins. They are made and stored in large quantities by farm animals. It appears that a limited amount of fat in the feed of the dairy cow is essential.

STRUCTURE OF FATS. Fats contain the same elements as carbohydrates, but the proportion of oxygen is much smaller and that of carbon and hydrogen much greater.

FUNCTION OF FATS. The fats in the animal body are a concentrated form of energy. They contain much more energy per unit than other nutrients and thus are well adapted for storing the body's reserve energy. They also have, to a limited degree, certain structural functions.

## ANATOMY AND FUNCTION
## OF THE RUMINANT DIGESTIVE SYSTEM

The animal body secures all the nutrients for its growth and function from solutions of the food, in much the same way that plants secure nutrients for their growth and function from water solutions in the soil. The process of preparing food ingredients for passage into the bloodstream is *digestion*. Digestion involves mechanical breaking of large particles into smaller particles (by chewing), chemical action of breaking large molecules into smaller ones (by gastric digestion), and, in the ruminant, microbial action of breaking large molecules into smaller ones (cellulose to short-chain volatile fatty acids) and microbial synthesis of elements and small molecules into amino acids and vitamins.

Digestion takes place in the alimentary canal (Figure 4-1), which in ruminants such as the dairy cow is much more complex than that of other animals. In cows it includes the mouth, the esophagus, the four compartments of the stomach, the small intestine, and the large intestine. These together form a long, winding canal approximately 180 feet long in the average mature dairy animal. The functions of the total digestive system are to (1) ingest feed, (2) store it for a short period of time, (3) digest and absorb the digestible portion, and (4) reject the undigested portion.

The mouth is an organ of prehension, mastication, insalivation, and rumination. The initial process of food gathering is called prehension, which is seizing and conveying food into the mouth. The strong, mobile, rough tongue is the main organ of prehension. The tongue readily curves around forages and other feed, drawing it between the eight lower incisor teeth and the dental pad above, and cutting it off.

In pasturing, the width of the jaw, about 2½ inches, limits the width of the swath. The position of the teeth in relation to the dental pad makes it impossible for the cow to graze closer than about 2½ inches from the ground. The head of the cow moves from side to side as she moves forward, the neck flexing within an arc of 90 degrees. Cattle have been observed to graze continuously for as long as 40 minutes without raising their heads, although such a long time is unusual.

The second process that takes place in the mouth is mastication, which is simply the chewing of the feed preliminary to swallowing. This occurs between the molar teeth in the back part of the mouth. In this process several

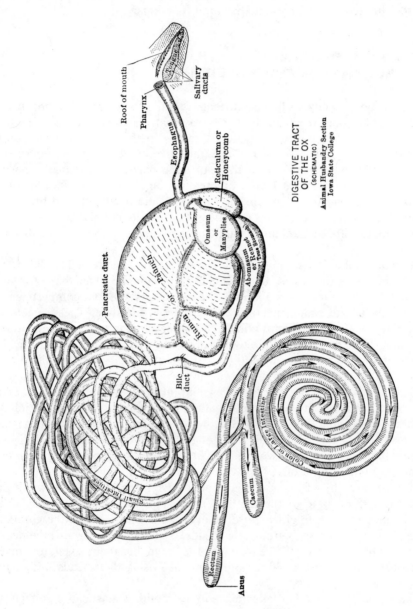

Roof of mouth

Tongue & lips

Salivary ducts

Pharynx

Esophagus

Reticulum or Honeycomb

Omasum or Manyplies

Abomasum Rennet or True Stomach

Rumen or Paunch

Rumen

Pancreatic duct.

Bile duct

Caecum

Small Intestine

Colon or Large Intestine

Rectum

Anus

DIGESTIVE TRACT
OF THE OX
(SCHEMATIC)
Animal Husbandry Section
Iowa State College

72

Fig. 4-1.   The digestive tract of the ox (schematic).

objectives are accomplished: the coarse roughage is broken down into smaller particles, some of the whole grains are crushed, and the feed is mixed with saliva, a process known as insalivation. The process of mastication stimulates the salivary glands, causing them to secrete a large amount of saliva, which readily mixes with the food. The amount of saliva secreted in a day's time is enormous. It is estimated that a cow secretes as much as 112 pounds of saliva in one day. This amount is increased if the food is unusually dry. Cows eating hay secrete 50 pounds of saliva for each 10 pounds of hay eaten, but only 13 pounds for the same amount of concentrates. Saliva is alkaline, with a pH of about 8.2, because of its high bicarbonate and phosphate content.

Saliva assists mastication and swallowing by acting as a lubricant, stimulates the taste nerves, and provides a buffering action in the rumen because of its alkalinity. Many feeds, such as corn silage, are quite acid. Organic acids are also produced by rumen microorganisms. These are neutralized by saliva, maintaining rumen pH between 6.5 and 7.5. This provides a good growth medium for bacteria and reduces frothing, which can lead to bloat.

In the process of mastication and insalivation, a mass of feed, called a bolus, gradually forms in the rear of the mouth. At intervals, this bolus is swallowed, passing down the esophagus into the rumen. It has been noted that when a cow ate whole corn, $1^{1}/_{3}$ boluses were formed per minute; with whole oats, $2^{1}/_{3}$ boluses were formed per minute; and with ground feed, $3^{1}/_{4}$ boluses were formed per minute.

The fourth process that takes place in the mouth is rumination. After the feed has been stored in the rumen and the cow has completed her feeding, she will begin to chew her cud, the coarse material forced back to the mouth by the process known as regurgitation. The bolus, or cud, weighs 3 to 4 ounces and requires about 3 seconds to ascend and $1\frac{1}{2}$ seconds to descend after complete mastication. The chewing of the cud occupies about 50 seconds at the rate of about one chew per second, but continues until all coarse particles have been thoroughly rechewed. The time between swallowing one cud and regurgitating the next is 5 to 10 seconds. Rumination is, therefore, a slow process; it occupies a cow's time for about 8 hours out of 24. If a cow is alarmed or disturbed she immediately ceases to ruminate. One of the first signs of ill health is the suspension of rumination.

After the food has been thoroughly masticated and mixed with saliva, the act of deglutition, or swallowing, takes place. This is brought about by the muscular action of the throat and tongue, which forces the food into the esophagus, or gullet, the tubelike passage extending to the stomach. The esophagus of the cow is easily stretched; as a result, animals are sometimes choked by food which has passed into the esophagus without proper mastication, such as an apple swallowed in large pieces.

## Digestion in the Stomach

**The Rumen.** As has been pointed out, the food passes from the mouth down the esophagus into a very important part of the stomach called the rumen, or paunch. It has a capacity of about 50 gallons, or as much as 300 pounds of material. The rumen is divided into four compartments by constrictions in the wall produced by large, muscular bands. The interior of the organ is lined with a well-developed muscular membrane covered with many pointed papillae.

The first function of the rumen is to act as a storage place: it holds the feed that the cow gathers while feeding. Later, when she has finished feeding, she rests and regurgitates the larger particles back to the mouth, to grind them more completely.

The rumen also refines the coarser pieces of food so that the bacteria and later the digestive juices can have a large surface upon which to act. Liquid fills the lower half of the rumen, and by means of a churning motion, caused by the muscular bands, the feed is driven down and thoroughly soaked in this warm liquid. The material also comes in contact with the rough pointed papillae that line the rumen. The coarse particles that are not thus broken down are returned to the mouth for further mastication. Thus, by churning, soaking, and rechewing, the particles are greatly reduced in size.

The third function of the rumen is to provide a place for fermentation. As already indicated, the saliva of the dairy cow does not contain digestive juices; nor are there any secreted into the rumen. Yet important changes take place in the feed during the time it remains in the rumen. These changes take place as a result of the action of bacteria and yeasts, both of which are plants, and one-celled animals called protozoa, all of which grow in great numbers in the rumen content. The temperature, food, and moisture provide ideal conditions for their growth and multiplication. The types of bacteria in the rumen vary according to the kind of feed. The type changes when an animal is fed largely on roughage or high fiber feeds and then is changed to largely a concentrate feed, or vice versa. These organisms in the rumen have three main functions.

HELP DIGEST THE FIBER. Enzymes secreted by these organisms cause the softening and disintegration of the particles of roughage and bring about the breakdown of the starches and cellulose. This action on the cellulose, the fiber of the feed, is of the greatest importance, because only the organisms of the rumen, or enzymes secreted by them, are able to break this cellulose down into organic acids, the final aim in digestion. However, even these enzymes are unable to break down the material when it is too mature and mixed with lignin. The organic acids produced in the rumen are mainly acetic, propionic, and butyric acids. They are absorbed through the rumen wall into the bloodstream.

High-roughage rations favor the production of acetic acid, which is involved in the manufacture of milk fat. Low-roughage, high-concentrate rations are more conducive to the production of propionic acid. In this situation the butterfat content of milk may be lowered. When the roughage is entirely finely ground hay or is very young, succulent grass, a decrease in fat test of the milk is often encountered.

SYNTHESIZE PROTEINS. Bacteria and yeasts have the capacity to synthesize proteins from nonprotein nitrogen-containing substances into their own bodies. As these organisms die, they can in turn be digested by the cow and used in her body. Much of the dietary protein is broken down, assimilated, and resynthesized into bacterial protein. Compounds such as urea can be utilized to a certain degree by dairy cows.

MANUFACTURE THE B-COMPLEX VITAMINS. A third function of the rumen organisms is the production of a considerable amount of various B-complex vitamins and vitamin K. Because the organisms can manufacture them, diary cows need not be fed these vitamins that are so essential to single-stomach animals. Cows are able to absorb these vitamins from the bacteria into the bloodstream, and they then become available for the functions of the body or for inclusion in the milk. Not all the vitamins present are absorbed from the digestive tract; some of them pass out in the feces. It has been found that dried cows' feces are rich in these vitamins.

The fermentation process produces large amounts of carbon dioxide, methane, and ammonia; smaller amounts of hydrogen, hydrogen sulfide, and carbon monoxide; and probably trace amounts of other gases. Normally, these gases are passed off by the reflex action of belching. Sometimes, however, the cow is unable to get rid of this gas, and bloating results.

**The Reticulum, or Honeycomb.** The reticulum lies directly in front of the rumen. Since they are not completely separated, food particles pass freely from one to the other. The capacity of the reticulum in the cow is about 13 quarts. Its interior is lined like a honeycomb, hence the popular name. The esophagus passes the bolus of food through the esophageal groove into the reticulo-rumen area.

Since the cow does not chew her food thoroughly, she sometimes swallows nails, stones, and various foreign objects along with the bolus of food. The churning movement of the rumen causes these heavy objects to be driven to the front portion of the reticulum. The constant churning action at times forces sharp metal objects into the reticulum wall. This condition causes the cow to go off feed. If an object is forced through the wall and penetrates the heart cavity, it causes death.

The contents of the reticulum are fluid. There is no secretion from its walls. Its functions are to assist the passing of the bolus up the esophagus and to regulate the passage of the food from the rumen to the omasum and from the rumen to the esophagus, in both cases through the esophageal groove.

**The Omasum, or Manyplies.** After the food has been thoroughly masticated and broken down, it goes to the omasum, which has a capacity of about 20 quarts. Some of the food goes directly on to the abomasum. The omasum, like the reticulum and rumen, possesses no secretive powers; it consists of powerful, muscular leaves that squeeze the water out of the food that it receives. Most of the water and some organic acids are absorbed by the organ. The solid portion remains in the omasum to be further acted upon by the leaves. The movement of the leaves is not simultaneous, but successive, in such a way that the rasping of the food is continuous. This movement also gives a pumping action, passing the food into the abomasum. When illness occurs, rumination ceases, cutting off the chief supply of fluid to the omasum. The content then becomes dry and sometimes cakes, resulting in a condition in which it is practically impossible to pass anything through the animal. This is called impaction.

**The Calf's Stomach.** The digestion in the young calf is more like that of a simple-stomach animal than that of a ruminant. Milk consumed normally by the calf bypasses the rumen and reticulum and goes directly to the abomasum for digestion. Milk and a few other substances have a sensory-stimulating effect, which causes the esophageal groove to close and prevent entrance into the rumen and reticulum. When milk is drunk in too large a swallow, it will force the groove open and allow milk to enter the rumen and reticulum. Digestive disturbances may result, since the calf is not functioning as a ruminant. As the calf begins to eat grain and hay, it begins gradually to function as a ruminant.

**The Abomasum, or True Stomach.** This organ, which has a capacity of about 20 quarts, is the true digestive stomach of the cow. The walls of this stomach secrete the gastric juices, which contain hydrochloric acid, and the two enzymes pepsin and rennin. Pepsin can act only in an acid medium; hence it is the function of the hydrochloric acid to change the alkaline condition, which the food has maintained up to this point, to an acid one. Pepsin acts on the proteins and breaks them down into simpler compounds, mainly peptides, which are short chains of amino acids, but does not break them down into amino acids.

Rennin is an enzyme that curdles milk, and is therefore very important in young calves that are fed milk. If it were not for the action of the rennin, the milk might pass through the digestive tract without being acted on by the other digestive enzymes.

### Digestion in the Intestines

The intestines are composed of two well-defined parts, the small and the large intestine. The small intestine is a long, folded tube into which the stomach empties. Its length in the cow is about 135 feet, and it has a capacity of about 40 quarts. The walls of the intestines are covered with very small, fingerlike projections called villi, which have a lashing movement, helping to mix the content of the intestines. Chyle, the partially digested material, is carried along in the intestines by a peristaltic movement. This is a wave of constriction followed by a wave of relaxation. The chyle moves very slowly, and the digestive juices have ample time to do their work. The upper part of the intestinal tract is specialized for secretion, and the lower part for absorption. In the intestines the chyle comes into contact with three digestive juices: pancreatic juice, bile, and intestinal juice.

**The Large Intestine.** When the content of the small intestine reaches the large intestine, it still contains undigested food. The food remains in the large intestine a relatively long time, permitting the digestive processes started in the small intestine to continue and also permitting more complete absorption of digested food. In the large intestine the food undergoes a great deal of bacterial action. Putrefaction takes place, causing the offensive odor of feces and often setting free large quantities of poisonous products. No digestive fluid is secreted in the large intestine, but many catabolic products are there returned to the digestive tract. Often the food remains in the large intestine for some time and becomes more solid, much of the water being absorbed. It is finally passed out through the anus as feces. The feces consist of the undigested residue of the feed, the remains of the digestive secretions, waste material resulting from wear and tear on the digestive tract, certain excretory products, and the bacterial flora.

## DIGESTIBILITY OF FEEDSTUFFS

The word digestion includes all the processes necessary for conversion of feed into the soluble forms in which it is assimilable. However, not all feed can be converted into soluble forms so that it can be absorbed. Feeds, especially forages, vary considerable in their nutritive content as well as their digestibility. To balance a ration properly, it is necessary to evaluate the nutritive content of the feed and the digestibility of the nutrients. Nutrient content can be evaluated by chemical procedures, and digestibility by the use of digestion trials. For digestion trials one determines the nutrient content of the feeds consumed and the feces excreted; the difference between what is eaten and what is excreted is said to be digestible feed. The coefficient of digestion is the percentage of feed that is digested.

## Evaluation of Nutrient Content

Most of the existing information concerning the nutrient content of feeds is based on a laboratory procedure known as *Weende proximate analysis*. This analysis includes the determination of moisture or dry matter (DM), crude or total protein (TP), ether extract (EE) or fat, crude fiber (CF), ash or mineral and nitrogen-free extract (NFE). Analysis of the DM compounds may be expressed as a percentage of the feed on an as-fed basis (weight of the nutrient divided by weight of the feed × 100) or on a DM basis (weight of the nutrient divided by the DM weight of the feed × 100). The general scheme of proximate analysis is given in Figure 4-2. The parenthetical numbers in the figure refer to the steps in the list that follows, in which definitions and computations are given.

*1. Dry Matter.* That material remaining after drying a feed sample at 100°C for a given period of time.

$$\% \text{ DM} = \frac{\text{Dry weight}}{\text{Wet weight}} \times 100$$

$$\% \text{ Moisture} = \frac{\text{Wet weight} - \text{dry weight}}{\text{Wet weight}} \times 100$$

*2. Total Protein.* The percentage of nitrogen (N) of a sample of feed is multiplied by the factor 6.25. The factor 6.25 is used because average protein contains 16% nitrogen.

$$100 \text{ units of protein} \div 16 \text{ units of nitrogen} = 6.25$$

$$\% \text{ TP (as-fed basis)} = \frac{6.25 \times \text{units of N}}{\text{As-fed weight of sample}} \times 100$$

$$\% \text{ TP (DM basis)} = \frac{6.25 \times \text{units of N}}{\text{Dry weight of sample}} \times 100$$

*3. Ether Extract or Fat.* The fat and other ether-soluble substances are

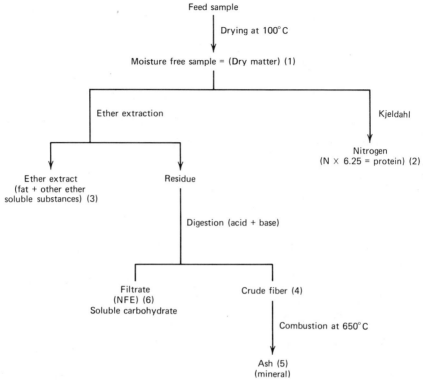

**Fig. 4-2.** Diagram of proximate analysis.

determined by subjecting a known amount of the dry matter of a feedstuff to an ether extraction. The ether is then evaporated and the extract weighed.

$$\% \text{ EE (as-fed basis)} = \frac{\text{Weight EE}}{\text{As-fed weight of sample}} \times 100$$

$$\% \text{ EE (DM basis)} = \frac{\text{Weight EE}}{\text{Dry weight of sample}} \times 100$$

Some feeds, especially coarse roughages, contain small amounts of gums, resins, and waxes that are soluble in ether and are included in the ether extract.

*4. Crude Fiber.* In the laboratory crude fiber is measured by refluxing a dry sample in acid and then in base. The residue is filtered out of the solution, dried, and weighed. Crude fiber represents the majority of the cellulose and lignin in the feed.

$$\% \text{ CF (as-fed basis)} = \frac{\text{Weight of fiber residue}}{\text{As-fed weight of sample}} \times 100$$

$$\% \text{ CF (DM basis)} = \frac{\text{Weight of fiber residue}}{\text{Dry weight of sample}} \times 100$$

**5. Ash.**   This represents the mineral components of a feed. A dry sample is placed in a crucible and completely combusted in a furnace at 650°C. The residue is the ash.

$$\% \text{ ash (as-fed basis)} = \frac{\text{Weight of ash}}{\text{As-fed weight of sample}} \times 100$$

$$\% \text{ ash (DM basis)} = \frac{\text{Weight of ash}}{\text{Dry weight of sample}} \times 100$$

**6. Nitrogen-Free Extract.**   This represents the more soluble carbohydrates such as starches and sugars. The nitrogen-free extract is determined mathematically by difference and not by actual analysis.

$$\% \text{ NFE} = 100 - (\% \text{ H}_2\text{O} + \% \text{ ash} + \% \text{ EE} + \% \text{ TP} + \% \text{ CF})$$

The coefficient of digestibility for any of the nutrients may be calculated as follows.

$$\frac{\text{Weight of the nutrient in the feed} - \text{Weight of the nutrient in the feces}}{\text{Weight of the nutrient in the feed}}$$

The Weende proximate analysis system for analyzing feed does not define precisely the nutrient composition of feeds. It is, however, a useful analysis because it provides an index of the nutrient value of feeds, the procedure is simple, the fractions that it isolates are associated with some of the properties of feed that have nutritional significance, and a general distinction can be made between the more digestible carbohydrates (NFE) and those that are less digestible (CF).

An important limitation of this procedure is that certain portions of the fiber components are lost during the procedure. Part of the lignin is removed by the alkali, and the nitrogen-free extract contains certain polysaccharides that are not completely digestible. This results in an incomplete recovery of the fiber and an underestimation of the fiber content of some feeds, especially high-fiber forages. The nutritional significance of this is not clear, as crude fiber was intended to provide a measure of the undigestible part of the feed, but part of it is digested by ruminants.

In many laboratories feed-analysis data obtained from Weende proximate analysis are supplemented with more detailed analysis by new and improved techniques.

In a procedure developed by Van Soest and associates at the USDA's research laboratory, feed dry matter is separated into two fractions, one highly digestible and the other of low digestibility. In this procedure detergents are used to isolate the cell contents (neutral detergent solubles, NDS) from the cell wall material (neutral detergent fiber, NDF). The NDS consists of sugars, starches, proteins, lipids, soluble carbohydrates, and other water-soluble materials. They are all high in digestibility, averaging about 98%. The NDF fraction is composed of the cell wall contents, including essentially all of the lignin and hemicellulose; variable amounts of these two components are lost from the crude fiber fraction to the nitrogen-free extract fraction in the Weende proximate analysis. Because of this, the NDF corresponds more closely to the total fiber portion of the feed. The nutritive availability of the NDF is influenced to a major degree by the amount of lignin present. In procedures developed by Van Soest and associates, the air dry feed material is boiled in an acid detergent solution, and the residue or insoluble part is known as acid detergent fiber (ADF). It consists primarily of cellulose, lignin and variable amounts of silica. The difference between NDF and ADF is an estimate of the amount of lignin in the feed. The amount of lignin can be more precisely determined by subjecting the ADF to an acid digestion and then combusting the residue to determine the ash content. This is referred to as the acid detergent lignin (ADL) fraction.

The crude fiber (CF) of forages as determined by the Weende proximate analysis system varies from 75 to 95% of the acid detergent fiber (ADF) of forages as determined by the Van Soest procedures.

**Determining Digestability of a Feed**   To illustrate how the digestibility of a feed is determined, let us assume that during a digestion experiment a cow consumed 3,000 grams of timothy hay, analyzing at 7% protein, 34% fiber, 50% nitrogen-free extract, and 1.8% ether extract. During the trial the cow excreted 1,300 grams of feces, which analyzed at 10% crude protein, 35% crude fiber, 46% nitrogen-free extract, and 2% ether extract.

From these figures it can be calculated that the amount of the feed digested and the coefficient of digestibility of the dry matter of timothy are as shown in Table 4-2.

It is assumed that the feces contain only undigested food. This is not strictly true, since many metabolic substances are added from the blood and

**Table 4-2**  Amount of Nutrients Fed, Excreted, and Digested, and Coefficient of Digestibility of Dry Matter in Timothy Hay

|  | Crude protein, grams | Crude fiber, grams | Nitrogen-free extract, grams | Ether extract, grams |
|---|---|---|---|---|
| Timothy hay | 210 | 1020 | 1500 | 54 |
| Feces | 130 | 455 | 598 | 26 |
| Digested | 80 | 565 | 902 | 28 |
| Coefficient of digestibility | .381 | .554 | .601 | .518 |

from the excretions that enter the digestive tract, and there is no method of determining these metabolic products; hence that which is really apparent digestibility is called digestibility. Because some of the minerals, especially calcium, phosphorus, iron, and part of magnesium, are excreted mainly through the intestines, this method cannot be used to determine their digestibility or absorption.

**The Indicator Method.**  To eliminate many of the laborious features of collection in the digestion trial method, the indicator method has been developed. This is an indirect method and involves the use of an inert substance which is not digested or absorbed. It may be a natural constituent of the feed, such as lignin or chromogens, or an added substance such as chromic oxide. The ratio of the amount of the indicator to each of the nutrients in the feed is determined. Samples, not complete collections, of the feces are analyzed for the ratio of the indicator to the nutrients in the feces. From these two ratios the portion that was digested and absorbed is calculated.

**Total Digestible Nutrients.**  To find the percentage of digestible nutrients in a feeding stuff, the percentage of each nutrient is multiplied by the coefficient of digestibility for that nutrient. As an example, the analysis of timothy hay and the coefficients given in Table 4-2 will be used. The results are given in Table 4-3.

The total digestible nutrients of any feeding stuff are determined by taking the sum of the digestible crude protein, the digestible carbohydrates (crude fiber and nitrogen-free extract), and 2¼ times the digestible fat (ether extract). In the foregoing example it would be

$$2.67 + 18.84 + 30.05 + 2¼(0.93) = 53.65$$

There would therefore be 53.65 pounds of total digestible nutrients in 100 pounds of the timothy hay.

The total digestible nutrients (TDN) simply represents the nutrients of the feeding stuff converted into carbohydrate equivalents, and is a measure of the energy value of a feed. On an average, fat contains about 2¼ times as much energy as carbohydrates, and protein contains about the same amount as carbohydrates.

Since protein has certain functions that cannot be performed by the other nutrients, the percentage of digestible protein is given in Appendix Table E, along with that of the total digestible nutrients; however, it should be borne in mind that the digestible protein is included in the total digestible nutrients.

## Factors Influencing Digestibility

Several factors influence digestibility. Animals of different species vary widely in the percentage of food digested, but animals of the same species are nearly the same in this respect. There is very little difference between the various breeds of dairy cows. The main differences are individual and relate to faulty teeth, diseased digestive organs, intestinal worms, etc., but these differences rarely exceed 3 or 4%. In the young calf the first three parts of the stomach are not well developed, and until the calf is old enough to eat roughage it cannot ruminate and properly develop the first three compartments. After the stomach is fully developed, the age of the animal seems to have little influence on the percentage of feed digested. Heavy feeding seems to decrease the digestibility of the feed, probably because of the greater bulk, the relatively rapid passage through the digestive tract, and consequently the less time for bacterial fermentation. Roughages, especially when overripe, are as a rule less digestible than concentrates, because large amounts of fiber and lignin in such feeds tend to protect them from the action of the digestive juices. It has also been observed that when cattle are fed very poor rations

**Table 4-3** The Composition, Coefficient of Digestibility, and Digestible Nutrients of the Dry Matter of Timothy Hay

|  | Crude protein (%) | Crude fiber (%) | Nitrogen-free extract (%) | Ether extract (%) |
|---|---|---|---|---|
| Analysis of timothy | 7.0 | 34 | 50 | 1.8 |
| Coefficient of digestibility | .381 | .554 | .601 | .518 |
| Digestible nutrients | 2.67 | 18.84 | 30.05 | 0.93 |

consisting of straw, cornstalks, or other feeds of like nature, the number of microorganisms in the rumen becomes very low; in other words, these bacteria, yeasts, and protozoa must be fed suitable nutrients if they are to be helpful in digesting the feed. Palatability may have some effect on digestibility, as it has been shown that palatability influences the secretion of the digestive juices. Grinding does increase digestibility of hard seeds, which otherwise would go through the digestive tract unbroken. Grain should therefore be ground when fed to dairy cows. Fine grinding of roughages decreases their digestibility, apparently by causing or allowing their quicker passage through the rumen with less bacterial action on them. The chopping of roughages or the soaking of feeds apparently does not affect digestibility.

## REFERENCES FOR FURTHER STUDY

Church, D. C. and Pond, W. G., *Basic Animal Nutrition and Feeding,* D. C. Church, Corvallis, Ore., 1975.

Crampton, E. W. and Harris, L. E. *Applied Animal Nutrition,* 2nd edition, W. H. Freeman Co., San Francisco, 1969.

Cullison, A. E. *Feeds and Feeding,* Reston Publishing Company, Reston, Va., 1975.

Ensminger, M. E. and Olentine, Jr. C. G. *Feeds and Feeding.* Ensminger Publishing Co. Clovis, Calif. 1977.

Hillman, D., et al. Basic Dairy Cattle Nutrition. *Ext. Bull.* E-702, Michigan State University, East Lansing, 1973.

Maynard L. A., and Loosli, J. K. *Animal Nutrition,* McGraw-Hill Co., New York, 1969.

*Nutrient Requirements of Dairy Cattle,* 4th rev. ed., National Research Council Publication, 1971.

# 5

# Nutrient Requirements of Dairy Cattle

Dairy cattle require nutrients for maintenance, for growth if immature, for reproduction when pregnant, and for production. Minimum requirements of each of the nutrients needed for these functions and to prevent deficiency symptoms have been studied for many years. Dairy nutrition research published in the nineteenth century compared the feeding value of various feedstuffs and found that it required more pounds of some feeds than of others to maintain similar production. Later in the nineteenth century standards based on requirements for digestible carbohydrates, digestible protein, and digestible fats were proposed. Further refinements, such as separating requirements for maintenance and milk production and listing requirements for various minerals and vitamins, were made during the first half of the twentieth century.

National Research Council (N.R.C.) standards were first published in 1945. These established a recommended nutrient allowance for dairy cattle and were prepared by a subcommittee of the Committee on Animal Nutrition of the National Research Council. It was the first attempt in the history of animal nutrition to develop nutrient standards for dairy cattle by nutritionists who had made a specialty of dairy cattle nutrition. The original publication has been revised periodically as additional information has been discovered, with the latest revision in 1971.

The most recent N.R.C. recommended nutrient requirements for dairy cattle are contained in Appendix Tables A and B. The levels recommended are adequate to prevent deficiencies, and they provide for acceptable rates of growth, reproduction, and milk production with feeds of at least average composition and digestibility. Requirements are given for dry feed, protein (total and digestible) and energy [net energy for maintenance ($NE_m$) and net energy for gain ($NE_g$) for nonlactating animals, and net energy for lactating cows (NE lactating cows), which includes energy requirements for maintenance, reproduction, and milk production]. Energy requirements expressed in terms of digestible energy (DE), metabolizable energy (ME) and total digestible nutrients (TDN) are also included. Calcium, phosphorus, carotene, vitamin A, and vitamin D requirements are listed.

All dairy animals have nutrient requirements for maintenance. These can be defined as the nutrients needed to maintain the animal body at homeostasis, that is, neither gaining nor losing weight, not pregnant, and not producing (Appendix Tables A and B). These requirements are in proportion to the body weight of the animal.

Immature animals need additional nutrients for growth, the laying down of new muscle and bone. Growth requirements for dairy cattle continue until they reach five to six years of age (Appendix Tables A and B).

Pregnant animals need additional nutrients for reproduction or growth of the fetus. The quantity of nutrients needed for this function is very small until the last two months of gestation, when most fetal growth takes place (Appendix Table B).

Lactating cows also require additional nutrients for the production of milk. The amount required for this function is proportionate to the amount and composition (% butterfat) of the milk.

## ESTIMATING NUTRIENT REQUIREMENTS

If the weight and age of the animal, the stage of lactation, the daily milk production, and the butterfat percentage are known, the animal's daily nutrient requirements can be estimated for energy, protein, calcium, phosphorus, carotene, vitamin A and vitamin D from the information in Appendix Tables A and B.

The following is an illustration of estimating nutrient requirements for a lactating dairy cow (Betsy).

**Betsy:** Weight = 600 kg.
Age = Second lactation
Production = 20 kg./day of milk that contains 4% butterfat
Reproductive status = Open

Determination of Nutrient Requirements:

**Nutrients**

| | TP,[a] (g) | NE,[a] (Mcal.) | Calcium, (g.) | Phos- phorus, (g.) | Caro- tene, (mg.) | Vita- min A, (1,000 I.U.) |
|---|---|---|---|---|---|---|
| Maintenance | 734 | 10.3 | 22.0 | 17.0 | 64.0 | 26.0 |
| [b]Growth | 73 | 1.0 | 2.2 | 1.7 | 6.4 | 2.6 |
| [c]Production | 1560 | 14.8 | 54.0 | 40.0 | — | — |
| Reproduc- tion | — | — | — | — | — | — |
| **Total** | **2367 g. or 2.367 kg.** | **26.1** | **78.2** | **58.7** | **70.4** | **28.6** |

[a] Currently, total protein (TP) and net energy (NE) are widely accepted methods of stating protein and energy requirements for dairy cattle.
[b] 10% of maintenance requirements for growth during the second lactation.
[c] The value in Appendix Table B multiplied by the daily production (kg.).

Estimating nutrient requirements should be looked upon as a guide or base point in feeding dairy cattle. The N.R.C. requirements are based on research data accumulated from many animals under a variety of environmental conditions. They are the best current estimates available concerning nutrient requirements of dairy cattle. They are, however, still estimates, and may change as new information becomes available. The current N.R.C. committee is working on the fifth revised edition.

## NUTRIENTS REQUIRED FOR DAIRY CATTLE

Dairy cattle require five classes of nutrients—water, energy, protein, minerals, and vitamins—to meet their needs for maintenance, growth, reproduction, and production. Identification of the specific nutrients needed and the amount of each that is needed to prevent deficiency symptoms and provide for acceptable rates of growth, reproduction, and production have been researched since the early nineteenth century. Many outstanding agricultural scientists have made valuable contributions to the dairy industry through their discoveries of nutritional requirements of dairy cattle and their development of the various feeding standards. This intense and continuing effort on the part of many dedicated scientists has resulted in our current knowledge of this subject.

### Water

Large amounts of clean, fresh drinking water should be available to cattle at all times. Lack of sufficient amounts of water or water of poor drinking quality (stagnant; contaminated by feces, urine, or spoiled feed; or otherwise made unpalatable) can seriously restrict milk production. Cattle usually consume three to four units of water for each unit of dry feed. This amounts to 60 to 100 liters (15 to 25 gallons) or more per day for high-producing cows.

Water is also the cheapest nutrient on a cost-per-unit basis; therefore, it should be provided to all cattle in unlimited amounts.

### Energy

A simple definition of energy is "the ability to do work." Energy is required to maintain normal functions of the animal body such as respiration, digestion, metabolism, growth, and production. Large amounts of energy are required by dairy cattle. With the exception of water, energy is quantitatively the major nutrient required by dairy cattle, and normally comprises 70 to 80% of the nonwater nutrient intake. Daily requirements of 20 to 40 megacalories (Mcal.) are common for lactating dairy cattle.

In mature, lactating cows deficiency of energy results in decreased production and loss of body weight. In young animals energy deficiency results in retarded growth and delayed puberty. Insufficient intake of energy to support body maintenance plus the production of large amounts of milk is a serious problem on many U.S. dairy farms.

Symptoms of energy deficiency in young cattle include slow growth rates, a thin or emaciated appearance, the head disproportionately large when compared to the rest of the animal body, and delayed puberty. In prolonged, severe cases of energy deficiency the onset of puberty may not occur until the animal reaches 18 to 24 months of age. In mature, lactating cattle energy deficiency symptoms include rapid loss of weight after freshening, low peak production levels, lack of persistency of lactation (rapid decrease in rate of daily milk production), short lactations and, in severe cases, irregular reproductive patterns. This is particularly a problem with high-producing cows, which are usually in negative energy balance. These cows may fail to exhibit symptoms of heat (estrus) and/or poor conception rates may result. For best reproductive efficiency, cows should be in positive energy balance and gaining weight.

Energy can be provided from carbohydrates, fats, and protein. Most of the energy for dairy cattle is normally supplied from carbohydrates, as they are usually the most economical source of energy for dairy cattle. Protein is usually 5 to 10 times as expensive per unit as carbohydrates as an energy source for dairy cattle. Fat, although a more concentrated source of energy, cannot be utilized in large amounts by dairy cattle.

The energy requirements of dairy cattle and the energy content of feedstuffs may be expressed or measured in several ways. TDN is not a widely used measure of energy content of feeds or of energy requirements currently because it does not consider the quality of the end products of rumen digestion (proportions of various volatile fatty acids produced, their energy value and physiological function) or the large amount of energy used in the digestion of fibrous feeds. It is thought by many nutritionists that the TDN system overestimates the energy value of forages in the ration.

Other measures of energy—digestible energy (DE) metabolizable energy (ME), and net energy (NE) are based on the caloric value of feeds when different body functions or losses have been subtracted. In order to determine these measures of energy, it is necessary to measure the energy content of the feed, feces, urine, gases, and heat produced by the cow.

DE = caloric content of the feed (gross energy) − caloric content of the feces (fecal energy)

ME = DE − caloric content of the urine and gases

NE = ME − heat increment (calories lost as result of microbial fermentation and metabolism of ingested feed)

Net energy is the energy available to the animal for maintenance (NEm), for gain (NEg), for reproduction, and for production.

The caloric energy system may be shown as follows.

|  | Mega-calories | % of total | % remaining |
|---|---|---|---|
| Gross energy | 40 | = 100 | 100 |
| Minus fecal energy | −12 | = 30 | |
| Digestible energy | 28 | = | 70 |
| Minus urinary and gaseous energy | −4 | = 10 | |
| Metabolizable energy | 24 | = | 60 |
| Minus heat increment | −4 | = 10 | |
| Net energy for maintenance, gain, production, and reproduction | 20 | = | 50 |
| Minus net energy for maintenance | −8 | = 20 | |
| Net energy for production, gain, and reproduction | 12 | = | 30 |

The net energy system is accepted by most as a more accurate and precise system of expressing energy requirements of animals and energy content of feeds. It is currently the more commonly used system.

## Protein

Protein is required in animal rations to provide a supply of amino acids needed for tissue repair and synthesis, hormone synthesis, milk synthesis, and many other physiological functions. In dairy cattle rations it comprises 10 to 18% of the dry matter nutrient intake. Daily requirements for lactating cows vary from 2 to 4 or more kg. per day of total protein, depending on the body size and level of milk production.

Protein deficiency in immature dairy animals results in decreased rate of growth and maturation. In lactating cows protein-deficient diets cause lowered production, loss of body protein (negative nitrogen balance), and depressed appetite. In severe cases, the solids-not-fat content of the milk is also depressed. Overfeeding of protein does not cause harm to the animal, but because of the high cost per unit of protein, it is usually uneconomical to overfeed protein.

Body reserves of protein are very limited. Therefore the ration should contain adequate protein to meet the animal's protein needs on a daily basis to avoid protein deficiency symptoms and ensure adequate growth and maximum production.

All protein required by dairy cattle need not be supplied from dietary protein, as ruminants have the ability to synthesize amino acids from nonprotein nitrogen (NPN) and energy sources. Rumen microorganisms convert nitrogen from NPN sources into amino acids for their use and in turn are digested in the small intestine; the microbial amino acids are utilized by the cow to provide part of her protein requirement. Additional energy is required, however, to supply the carbon, hydrogen, and oxygen needed for microbial protein synthesis.

The usual source of NPN in dairy cattle rations is feed-grade urea. Urea usually contains 42 to 45% nitrogen or 262.5 to 281% protein equivalent (42 or 45% nitrogen × 6.25). Use of urea in dairy cattle rations is limited by its lack of palatability and the animal's ability to utilize it for protein synthesis. Its utilization is affected by the way it is fed (in large amounts fed once or twice a day or distributed throughout the ration), the availability of a source of the carbon compounds (CHO) needed for protein synthesis, and the level of protein in the total ration. Starches are a very effective source of carbon compounds for ruminal amino acid synthesis. Cellulose is less effective, as it is degraded too slowly, and simple sugars are degraded too rapidly to be most effective. The feeding of urea uniformly mixed in an ample supply of starches increases the amount that can be fed and the efficiency of its utilization by rumen microorganisms. NPN is not very effectively utilized by high-producing cows being fed relatively high levels of total ration protein (14 to 16% of the DM). It can, however, be more effectively utilized by lower-producing cows being fed lower levels of total ration protein (up to 12 to 13% of the DM).

When economic conditions are such that protein from natural sources is very expensive and that from NPN is reasonably priced, providing 10 to 20% of the daily protein requirement from NPN for lower-producing cows may be justified. Its use in higher-protein-content rations for high-producing cows is questionable as it may depress appetite and lower production. Urea levels in dairy rations should be raised gradually; about three weeks are required for the rumen microflora to adapt to urea containing rations.

Nonprotein nitrogen is not effectively utilized by young dairy calves that are not ruminating or receiving significant quantities of dry concentrate feeds. It should not be fed to calves until they reach five to six months of age.

Protein requirements and protein content of feedstuffs may be expressed either as total or crude protein (TP or CP) or as digestible protein (DP). However, not all nitrogen in a feedstuff or in animal tissue is in the form of protein. Some is contained in other nitrogenous compounds, so multiplying the nitrogen content of a substance by 6.25 can be slightly erroneous in estimating protein content. The use of total protein in expressing protein requirements of ruminants and protein content of feeds for ruminants is acceptable, however, because of the ruminants' ability to utilize the nonprotein as well as the protein nitrogen.

Protein is one of the most expensive nutrients needed by dairy cattle because of the relatively large amount needed and the high cost per unit. Protein-deficient rations severely and quickly restrict milk production because of meager body reserves. Feeding excessive protein results in the extra protein being utilized by the animal as energy, an uneconomical use of an expensive ingredient. Therefore it is recommended that adequate, but not excessive, amounts of protein be included in dairy rations.

## Minerals

The mineral elements required by dairy cattle are sodium, chlorine, calcium, phosphorus, iodine, copper, iron, cobalt, magnesium, potassium, manganese, zinc, sulphur, fluorine, molybdenum, and selenium. Other minerals are found in the animal body but the need for them is not established. Calcium, phosphorus, potassium, sodium, and chlorine are required in large amounts, but the so-called trace minerals are also critical to maintaining proper body function.

Animals appear to suffer more from a deficiency of minerals than they did in the early days. This is true for two reasons. (1) The amount of minerals in the natural feed, especially the roughage, is often less than it was formerly. This because the mineral content of the soil has decreased in the older farming sections, and more especially the trace minerals, which are not added in fertilizer, and because the use of heavy fertilizer with the resulting higher yields takes much more of these elements out of the soil than when such large crops were not grown. (2) The production of the dairy cow has increased greatly, which increases the requirements for these minerals. Milk is rich in minerals, and if a cow produces more milk, more minerals must be supplied in the diet.

**Reasons Feeds are Deficient in Minerals.**   When average-producing cows are fed legume roughages grown on limed and fertilized soils and a grain mixture containing some of the high-protein ingredients, there will usually not be a need for additional minerals except sodium and chlorine, which are usually supplied as sodium chloride (salt). Nonlegumes and low-quality legume feeds are much lower in minerals.

Legume forages are the best sources of calcium; the high-protein oil meals and wheat bran are the best sources of phosphorus.

High-roughage, low-grain rations may be low in phosphorus; high-grain, low-roughage rations are likely to be low in calcium.

Cows at high levels of production have proportionally greater needs for minerals.

Animals suffer at times from an inadequate supply of minerals in relation to their needs. This is brought about under several conditions, such as the following.

*1. Lack of Minerals in the Soil.* Some soils never did have sufficient minerals, whereas other soils have been depleted of their minerals by heavy and continuous cropping. Whatever the cause, crops grown on such soil give serious trouble to the dairyman, as cattle fed crops produced on such soils may be affected by a lack of sufficient minerals.

*2. Lack of Minerals in Ration.* Insufficient feed or rations low in mineral content may result in mineral deficiency, which shows up in the dairy herd. Cattle that subsist on a very low level of feed intake often do not obtain sufficient protein, energy, or minerals, and as a result become very unthrifty. Sometimes, however, the ration may seem adequate as far as protein and energy are concerned and still lack in mineral content. This is because the crops are grown on soil that lacks minerals or the cows are fed a forage(s) that is naturally deficient in one or more minerals. It will be necessary in that case either to fertilize the soil or to feed the animals additional minerals.

*3. Lack of Vitamin D in the Ration.* Even though the ration may contain sufficient protein, energy, and minerals, a deficiency disease may nevertheless develop. The calcium and phosphorus of a ration cannot be properly assimilated unless vitamin D is present. If this vitamin is lacking in the ration for several months, stiffness in legs and joints, an arched back, and general unthriftiness may occur. This vitamin is found most abundantly in field-cured hay and is manufactured in the body when animals are exposed to the ultraviolet rays of the sun.

**Salt (NaCl).** Salt is required in dairy rations at the rate of 4 to 6 g. per 100 kg. body weight per day for maintenance, plus 1.5 to 1.8 g. per kilogram of daily milk production. This requirement is sometimes stated as 0.45% salt in the ration dry matter. Daily salt requirements for lactating dairy cows vary from 50 to 100 or more grams per day, depending on body size and level of production. Expressed another way, dry cows should be fed salt at the rate of about 1 ounce per day and lactating cows 2 to 3 ounces per day.

Early salt deficiency symptoms are usually the craving for salt, shown in licking clothing or other objects and an appetite for urine from other cows and soil. Later symptoms, usually after several months of a deficient diet, include loss of appetite, decreased milk flow, and unthrifty or haggard appearance.

Ordinary feeds for dairy cattle do not contain sufficient salt to meet their requirements. Providing salt blocks is usually adequate for dry cows and older heifers. Lactating cows should be fed salt as part of their daily ration.

**Calcium.** Calcium has been long recognized as a critical nutrient requirement of dairy cattle. Large amounts of this element are secreted in the milk (milk and dairy products provide over 75% of the calcium requirement of the U.S. human population) and it is a major constituent of bone and other body tissues and fluids. Precise calcium requirements are not available, however, for all phases of growth, reproduction, and production of high-producing cows. Results of various researchers studying calcium requirements have been inconsistent with each other and with feeding practices. This situation is not so critical as it might be, however, as body calcium reserves are quite large, and some natural feeds for dairy cattle such as legume roughages are excellent sources of calcium.

Current recommendations for lactating cows are to feed 15 to 25 g. of calcium per cow per day to meet maintenance requirements plus 2.5 to 3 g. of calcium for each kilogram of milk produced. Daily calcium requirements, then, vary from 50 to 125 g. or 2 to 5 oz. per day. This requirement is sometimes stated as 0.5 to 0.6% calcium in the total ration dry matter.

In growing calves and heifers calcium deficiency prevents normal bone growth, low bone content of calcium and phosphorus, and depressed normal growth and maturation. In mature lactating cows calcium-deficient rations over prolonged periods cause depletion of bone stores of calcium and phosphorus, which results in fragile, easily broken bones. Eventually reduced milk production also results. There is, however, no reduction in the calcium content of the milk.

A large cow, if she mobilized 10% of the calcium in her skeletal reserves, would have sufficient calcium for 2,000 pounds of milk. Therefore calcium-deficiency symptoms are not likely to appear quickly when cows are fed calcium-deficient diets. If cows are required to draw on skeletal reserves of calcium, for example, during peak production periods, it is important that they be given the opportunity to replace these reserves during late lactation.

Legume forages are excellent sources of calcium, and dairy cattle being fed ample amounts of these forages need little, if any, additional calcium. When ration forages are primarily grasses or corn silage, however, calcium supplied by these natural ration ingredients is not sufficient to meet requirements and supplemental calcium must be fed.

**Phosphorus.** Dairy cattle require relatively large amounts of phosphorus, as it is a major component of both bone and milk (milk and dairy products supply about 35% of the phosphorus requirements of the U.S. human population). Phosphorus requirements for dairy cattle are somewhat less than calcium requirements; a ratio of about 1.0 parts phosphorus to 1.3 to 1.4 parts calcium is acceptable. Daily requirements for lactating cows include 11 to 21 g. of phosphorus per day for maintenance plus 1.7 to 2.4 g. per kilogram of milk produced. Daily phosphorus requirements vary from 40 to 80 g., or 1.5 to 3 oz., per day, depending on body size and level of production. This requirement is sometimes stated as 0.3 to 0.4% phosphorus of the total ration dry matter.

Symptoms of phosphorus deficiency are most likely to occur during growth and during heavy lactation. During growth the symptoms are swollen, stiff and aching joints, arched back, beaded and deformed ribs, and retarded growth. This condition is called rickets. In mature animals the symptoms may not be so pronounced, as the animals will draw on their reserve supply for some weeks before they show any outward sign. Cattle suffering from such a deficiency often have a depraved appetite and exhibit a craving for wood, bark, bones, hair, and other foreign material; irregular heats, anestrus, and lowered conception rates may also occur. Internally, the bones lose part of their calcium and phosphorus and become more fragile and easily broken. Phosphorus content of the blood is also lower.

Most roughages and grains are only fair to poor sources of phosphorus. If the roughages are grown on soils low in phosphorus, they may not contain sufficient phosphorus and deficiencies may develop, especially if the animals are being fed largely upon roughages without some protein supplement. The cereal grains are richer in phosphorus than in calcium, but contain only a moderate amount. However, most of the protein-rich by-products of plant origin are much higher in phosphorus than are the cereals and roughages. Wheat bran is especially rich in phosphorus; standard wheat middlings, cottonseed meal, and linseed meal are medium-rich in phosphorus. When these feeds are fed in large amounts there will probably be no lack of phosphorus in the ration for medium-producing cows.

**Calcium-Phosphorus Ratio.** Not only should dairy animals be fed sufficient calcium and phosphorus, but a proper ratio should be maintained between the two minerals. If there is a great excess of one or the other of these minerals, bad effects may result even though both are being fed in sufficient amounts. Less vitamin D is required when the proper calcium-phosphorus ratio is maintained. A 1 : 1 or 2 : 1 ratio seems to be satisfactory to dairy animals, though if sufficient vitamin D is fed a wider ratio may be fed without harm, provided that there is an adequate supply of both minerals.

**Iodine.** Iodine is a component of the hormone thyroxine, which controls metabolic rate. It is present in small quantities in milk. The precise amount required by cattle has not been established.

Iodine deficiency results in enlargement of the thyroid gland, commonly called goiter, in calves at birth. In more severe deficiencies calves may be borne with thin hair or hairless, or be stillborn.

Iodine deficiency is usually a problem only in certain parts of the country (the Northwest, Florida, and the Great Lakes region), where feeds and water are low in iodine. The feeding of iodized salt provides ample iodine to prevent deficiencies, except when large amounts of goitrogenic substances such as soybean meal are fed. Then a minimum of 8 to 12 mg. of iodine should be included in the ration.

**Copper and Iron.** Copper and iron are needed in the body for the formation of hemoglobin in the blood, and when they are deficient in the feed an animal may suffer from nutritional anemia. Ordinarily there is no lack of these elements in the ration of the dairy cow. However, in certain areas, notably in the sandy region of Florida,* the forage is so low in these elements that cattle do suffer from anemia. It is known also that calves fed on rations consisting of milk, as an exclusive diet or with grain, will suffer from anemia. When roughage is fed such trouble will be avoided.

**Cobalt.** If the soil is deficient in cobalt, ruminants fed on the feed produced on this soil may show a disease because of the lack of cobalt. Animals so affected will lose their appetite, become unthrifty and emaciated, and may exhibit depraved appetites. Young animals may fail to grow normally and have delayed sexual development.

Cobalt deficiency was first noted in New Zealand and Australia. Often called "salt sickness," it has been found in various places in the United States; the trouble has been reported in Florida, North Carolina, New York, New Hampshire, Massachusetts, Michigan, and Wisconsin, sometimes combined with a deficiency of copper and iron.

Cobalt is essential in the production of vitamin $B_{12}$ by rumen bacteria. A cobalt salt must be added to the ration when there is a deficiency, rather than injected into the animal.

**Magnesium.** The animal body contains about 0.05% magnesium, of which about 60% is stored in the bones. This bone reserve, however, is not easily mobilized in mature animals. Consequently, dietary intake should correspond to usage. Abrupt changes to magnesium-deficient rations can result in magnesium deficiency symptoms in 2 to 18 days in mature cattle. Minimum daily requirements are approximately 0.9–1.3 g. per day per 100 kg. body weight of calves. For cows, 2.0 to 2.5 g. per day for maintenance plus 0.12 g.

* Fla. Exp. Sta. Bul. 231.

per kilogram of milk produced is required. Inclusion of 0.06% magnesium in calf diets and 0.20% in cow diets is adequate to meet minimum requirements.

Grass tetany, sometimes called hypomagnesemic tetany results from magnesium deficiency. Symptoms include skin twitching, unsteady gait, falling on the animal's side with legs alternately extended and contracted, frothing at the mouth, and profuse salivation. It commonly occurs when cattle graze lush, succulent pasture or are fed similar material as green chop.

**Potassium.** Potassium functions to maintain acid-base and osmotic pressure in intracellular fluids, in electrolyte balance, and in helping to control muscle and nerve excitability in the animal body. It is also secreted in the milk. The minimum requirement for potassium is not well defined, but is believed to be approximately 0.7 to 0.8% of the total ration dry matter for calves and growing animals and 1.0% for heavily lactating cows.

Deficiency symptoms include general muscle weakness; loss of muscle tone, especially intestinal and cardiac muscle; decreased hide pliability; and loss of hair glossiness. Milk and blood plasma levels of potassium are also lowered.

Forages normally contain high levels of potassium, thus cattle on high forage diets rarely encounter deficiencies. However, cattle fed high-concentrate rations or forages grown on potassium-deficient soils can suffer deficiency symptoms if supplemental potassium is not added to the ration. These deficiencies can be corrected by including 0.5 to 1% potassium chloride (KCl) (muriate of potash) in the grain mixture.

**Trace Minerals.** The other minerals are required in very small amounts. The normal feed of the dairy cow usually contains sufficient amounts of these elements to satisfy the needs of the animal. However, as milk production levels increase and composition of feeds change, more deficiencies are being observed. A trace-mineral mixture may be added to the total ration or a trace-mineral salt may be fed on a free-choice basis to help avoid such deficiencies. A brief summary of the requirements, as far as is known, and deficiency symptoms are listed in Table 5-1.

**Vitamins**

Vitamins play an important role in animal nutrition. These organic compounds profoundly affect dairy farming by increasing the efficiency of production and by preventing nutritional diseases. Dairy cattle require the same vitamins as other animals; however, under normal conditions their feeds furnish the vitamins or they can be synthesized in the animal body. The B vitamins and

**Table 5-1** Nutrient Requirements and Deficiency Symptoms[a]

| Nutrient | Daily requirement | Deficiency symptoms |
|---|---|---|
| ENERGY (glucose, fats, fatty acids) | Variable with body size, rate of growth, milk production, and % milk fat (see Appendix Tables A–B) | Low milk production; slow growth rate; poor body condition; silent estrus (heat); lowered protein content of milk<br><br>*Excess energy relative to requirement*; fattening; high blood-fat levels; fatty liver; tendency for depressed appetite and ketosis post calving; unsaturated fats in tissue and fat deposits; tendency toward low resistance to infectious diseases, retained placenta and metritis; sudden increase causes lactic acidosis, death |
| PROTEIN | 10–18% of ration dry matter depending on age and level of production. | Emaciation (poor body condition); retarded growth; low milk production; reduced digestion of feed; poor conversion of feed to growth, fat, or milk; lower blood protein and possibly immune fractions; underdeveloped reproductive organs possibly result of retarded growth |
| FIBER | Minimum 14–15% of ration dry matter for lactating cows; higher with finely chopped feeds. | Rumenitis; founder; rumen stasis; tendency toward displaced abomasum postcalving; low milk-fat test; higher unsaturated fats in tissue fats; may contribute to poor muscle contractility |
| SALT (Na) sodium and (Cl) chloride | 5–9 g./kg./day; 0.18% sodium or 0.45% salt (NaCl) of ration dry matter | Lack of appetite; unthrifty, low production; craving for salt; appetite for soil, clothing; licking objects; drinking urine from other cows during urination |
| CALCIUM (Ca) | Maintenance 15–25 g. + 2.5–3.0 g./kg. milk; 0.5–0.6% of ration dry matter for lactating cows | Bones and teeth easily broken; low calcium content in bones |
| PHOSPHORUS (P) | Maintenance 10–20 g. + 1.7–2.4 g./kg. milk; 0.3–0.4% of dry matter | Lack of appetite; irregular estrus (heat periods); depraved appetite for bones, wood, bark, etc. |
| IODINE (I) | 0.1 ppm. of ration dry matter; for nonlactating, 2 mg./head: 0.8 ppm. or 8–12 mg. for pregnancy; more may be required when soybean meal or other goitrogenic feeds are fed heavily | Goiterous (big neck) calves frequently dead or hairless; failure to show estrus; high incidence of retained placentas in mature cows |

| Mineral | Requirement | Deficiency / Toxicity Symptoms |
|---|---|---|
| COPPER (Cu) | 5 ppm. of ration dry matter, 6–8 ppm. suggested; increases with high molybdenum intake | "Coast disease" or "salt sick" (in Florida); anemia, stillbirth of young; incoordination of hind legs; sudden death in cows due to heart degeneration |
| IRON (Fe) | 330 mg./kg. for growth, 100 ppm. in adult ration (2 g.) | Anemia, particularly in calves maintained on milk; seldom in adult cattle |
| COBALT (Co) | 0.1 ppm. in dry ration (2 mg.) | Loss of appetite; anemia, emaciation; low appetite for grain; calves unthrifty, poor appetite, first to exhibit symptoms because low vitamin $B_{12}$ content of milk |
| MAGNESIUM (Mg) | Calves: 12–16 mg./kg; Cows: 2.0–2.5 g. for maintenance + 0.12 g./kg. milk or 0.15–0.20% of ration dry matter. | Grass tetany (or grass staggers); twitching of the skin; staggering or unsteady on feet; down; common with cattle grazing rapidly growing, succulent pasture, or similar green chop and occasionally stored feeds; may be aggravated by high N and K levels in feeds |
| POTASSIUM (K) | 0.7–0.8% of the ration dry matter | Overall muscle weakness; loss of appetite; poor intestinal tone with intestinal distension; cardiac and respiratory muscle weakness and failure |
| MANGANESE (Mn) | Calves: 7–10 ppm. of ration dry matter; Cows: 20 ppm. of ration dry matter | Newborn: deformed bones, enlarged joints, stiffness, twisted legs, shorter humerus (foreleg), general physical weakness; deficiency could occur in cattle fed high-grain, low-roughage rations; symptom could be ataxia (uncoordinated movements) |
| ZINC (Zn) | Calves: 8–10 ppm. of ration dry matter; 40 ppm. of ration dry matter for lactating cows | Itch, hair slicking, stiff gait, swelling of hocks and knees, soft swelling above rear feet, rough and thickened skin, dermatitis between rear legs and behind elbows; under-size testicles in bull calves and low fertility in cows |
| SULFUR (S) | 0.2% in dry ration, a ratio of 10 parts N to 1 part S in high NPN rations | Lowered production; poor nitrogen utilization; cellulose digestion and conversion of lactate to propionate |
| FLUORINE (F) | Small amount appears to prevent dental cavities | Toxic above 10 ppm.: deformed teeth and bones |
| SELENIUM (Se) | 0.05 ppm. minimum: recommend 0.10 ppm. of ration dry matter for lactating cows | Nutritional muscular dystrophy; high calf mortality; retained placentas increased; toxic above 3 ppm. |

**Table 5-1** (Continued)

| Nutrient | Daily requirement | Deficiency symptoms |
|---|---|---|
| VITAMIN A | 4,000–5,000 I.U. vitamin A/100 kg. body weight, or 10–11 mg. carotene/100 kg. body weight. Double this amount for pregnant cows | Night blindness; bulging and watery eyes; muscle incoordination; bronchitis and coughing may progress to pneumonia; chronic symptoms; roughened haircoat; emaciated, hairless or blind calves if dam deficient; edema or swelling of the brisket and forelegs (anasarca); abortions preterm; young calves, weakness at birth; susceptible pneumonia and digestive infections; watering of the eyes, cloudiness of the cornea, protrusion or "bulging" of the eye followed by permanent blindness and death |
| VITAMIN D | 800–1,000 I.U./100 kg. body weight; $D_2$ or $D_3$, 5,000–10,000 I.U./head | Rickets; enlarged joints; wobbly gait; lack of appetite; stiff legs; arched back; swelling of pasterns; lameness; calves deficient in Ca, P, or vitamin D |
| VITAMIN E (alpha tocopherol) | Calves: less than 40 mg./day; Adults: not established; 1–2 g. (1,000–2,000 I.U.) fed to cattle prevents oxidized flavor milk having high copper content | Nutritional muscular dystrophy; white muscle disease in calves; sudden death from heart muscle degeneration; heart, diaphram, and intercostal muscles show light streaking |
| B VITAMINS | Synthesized by organisms in normal-functioning rumen to meet requirement; calves to weaning; contained in milk | Deficiency symptoms produced only with abnormal restricted diets; may occur in calves with prolonged severe scours or fed artificial milk diets |
| VITAMIN C | Synthesized in tissue of calves and adult bovines | Other species—scurvy; loosening of teeth; subepithelial hemorrhage |

[a] Adapted from *Nutrient Requirements of Dairy Cattle*, National Research Council, 1971; and *Ext. Bul.* E-702, Michigan State University, East Lansing, 1973.

vitamin K are synthesized in the rumen. Vitamin C is synthesized in body tissues. Only vitamins A, D, and E must be provided in the ration.

**Vitamin A and Carotene.** Carotene is the precursor of vitamin A. Plants do not have vitamin A in their composition, but they do contain carotene. The animal can change the yellow-colored carotene to vitamin A in the intestinal wall and/or liver. Vitamin A is necessary for proper reproduction; it influences synthesis of certain hormones, has a role in protein synthesis, and is a part of the visual pigment of the eye.

Requirements for vitamin A are expressed as requirements for carotene or for vitamin A. These are expressed as milligrams (mg.) of carotene or as International Units (I.U.) of vitamin A. One milligram of carotene is approximately equal to 300 to 400 I.U. of vitamin A. Calves require a minimum of 10 to 11 mg. of carotene or 4,000 to 5,000 I.U. of vitamin A per 100 kg. of body weight per day. For cows, 10 to 11 mg. of carotene or 4,000 to 5,000 I.U. of vitamin A per 100 kg. body weight is required for maintenance. Doubling that amount meets minimum requirements for maintenance, gestation, and lactation. A large, pregnant cow would need 100 to 150 mg. of carotene or 40,000 to 60,000 I.U. of vitamin A per day.

Vitamin A deficiency symptoms include night blindness, degeneration of the mucosa of the respiratory tract, mouth, eyes, intestinal tract, urethra, and vagina. Calves become susceptible to diarrhea, colds, pneumonia, and other infections. Calves are often weak at birth. In cows, shortened gestation periods and high incidence of retained placentas are also common.

Carotene is widely distributed in plants but is more abundant in those of green, yellow, or orange color. The feeds increase in their carotene content as the color intensifies. The dairy cow must be given carotene or vitamin A in her feed, as she cannot synthesize vitamin A or carotene in her body, and it is necessary for life and well-being. The liver acts as the storehouse of carotene and also regulates the amount of vitamin A to be transferred to other parts of the body. In milk are found both vitamin A and carotene. Since vitamin A is colorless, milk that is yellow in color contains a higher percentage of unchanged yellow carotene.

CAROTENE IN FEEDS. Feeds with green, yellow, or orange color are rich in carotene. The animal turns part of this into vitamin A in its own body, and part is used as carotene. Hence, to assure an adequate supply, feeds rich in carotene must be fed. The amount of vitamin A or carotene in milk depends to a large extent upon the amount in the feed. Cows fed rations lacking in carotene will produce milk lacking in vitamin A and carotene.

Probably the best source of carotene is good, green pasture. Often the milk from cows on pasture will contain more than twice as much vitamin A and carotene as when they are barn fed. However, even pasture, when it becomes dry and weathered, loses much of its carotene content and cannot be counted on as a good source of carotene.

In the field-curing of hay or other dry roughages a considerable portion of the carotene is destroyed, especially if it is bleached by either the sun or rain. Hay retaining a green color is a good source of carotene, but much of the carotene may be lost during storage for a long period of time.

Corn silage varies considerably in its carotene content, depending upon its greenness when ensiled; if ensiled when very green, it may have as much carotene as well-cured alfalfa hay. Hay-crop silages ensiled when the crop is green are good sources of this vitamin. Corn stover, straws, and similar feed lacking in green color are extremely low in carotene.

The concentrates, with the exception of yellow corn and by-products made from it, are practically free of carotene. Even yellow corn is not a good source of carotene, since it contains only about one-tenth to one-quarter as much carotene as does well-cured alfalfa hay, and the amount fed is so much less.

Vitamin A palmitate and other synthetic sources are now the best and most economical means of supplying supplemental vitamin A.

Cod-liver oil and other fish-liver oils are rich sources of vitamin A. These oils are often fed to dairy calves; cod-liver oil, however, is not suitable for dairy cows because of the depressing influence it has on the fat percentage of the cow's milk.

VITAMIN A AND NITRATE. The heavy nitrogen fertilization of forages has produced crops higher in nitrogen content. This is usually expressed as higher protein. However, in some cases some of the nitrogen is in the form of nitrates and other nonprotein compounds. Nitrogen dioxide is associated with silo-fillers disease. Nitrate and nitrite poisoning results in some instances.

There are indications that high nitrates in the feed interfere with the conversion of carotene to vitamin A. There may be a need for the addition of greater amounts of vitamin A or carotene in the ration when such forages are fed, if they are suspected of having an abnormal nitrate content.

**Vitamin D.** Green forage and silages are low in vitamin D. The green or respiring portion of forage plants is essentially devoid of vitamin D activity. Sunlight or other lights that contain ultraviolet rays have the power of changing the provitamin D, ergosterol, and related compounds, into vitamin D. Small amounts of these provitamins are found in most of the common feeds, which, when exposed to the sunshine, change into vitamin D. Hays cured away from the sun are usually low in this vitamin, as are hay-crop silages. Corn silage may have a moderate amount. High-quality field-cured hay is the best forage for supplying vitamin D. Cattle exposed to the sun's rays usually will not be lacking in this vitamin, for the ultraviolet rays from the sun, acting upon these provitamins in their bodies, produce sufficient vitamin D for their needs. During the season of the year when cows are housed, or when there is little or no sunshine, cattle may be lacking in this vitamin. The trend toward heavier silage and grain feeding accompanied by less hay and pasture fed to dairy cows means less vitamin D in their rations.

Vitamin D deficiency results in low levels of blood calcium and phosphorus, reduced growth, deformed and weak bones, and depressed appetite. The urine may contain high levels of amino acids and phosphorus.

Requirements for vitamin D are usually expressed in International Units (I.U.). Mature cows have a daily requirement of about 1000 I.U. per 100 kg. of body weight daily, or about 6,000 I.U. per day for a 600-kg. cow. Growing calves have a daily requirement of about 660 I.U. per 100 kg. of body weight. This is about 250 to 300 I.U. daily for baby calves of the larger breeds and 150 to 200 I.U. daily for baby calves of the smaller breeds. These amounts are sufficient to prevent rickets if sufficient calcium and phosphorus are also present. However, large daily amounts of about 600 to 800 I.U. per kilogram are usually recommended for calves. Excessive vitamin D is toxic. Feeding 5 to 100 times the daily requirement has produced excess calcification.

Most silages and field-cured hays contain over 200 I.U. of vitamin D per kilogram. Cows fed liberal amounts of these forages and given exposure to sunshine should obtain sufficient vitamin D. However, vitamin D supplementation may be necessary under the following conditions: high-corn-silage or all-silage feeding; high levels of milk production; dry-lot feeding; high-grain and limited-forage feeding; and increased use of high-carotene feeds as silage and barn cured hays. Under these conditions, the addition of 4,000 to 6,000 I.U. daily per cow to the total ration may be advisable.

**Vitamin B Complex.** Vitamin B complex is a term now applied to a group of vitamins that was formerly considered a single factor, vitamin B. There seem to be at least 12 or more of these vitamins, and perhaps others that have not been isolated as yet.

However, it is unlikely that these vitamins will be lacking in rations of dairy cows for two reasons. (1) Green forage, such as pastures, are rich in these vitamins. Well-cured legume hay and silages are good sources. Good mixed hay also helps supply them. (2) The dairy cow, through the action of the organisms in her rumen, has the ability to synthesize these vitamins. An ample supply is therefore assured to the cow, even though the feed may be lacking in them.

A dairy cow will seldom have a deficiency in these important vitamins. No benefits have been obtained by feeding calves vitamin B supplements, even before the normal rumen digestion starts, after which they can manufacture their own in the rumen. The milk that is fed to young calves will supply their needs.

**Vitamin C (Ascorbic Acid).** The lack of ascorbic acid in the ration of certain animals (man, monkey, and guinea pig) leads to a disease known as scurvy. With dairy cattle, however, such a disease is not known. The cow has a requirement for the vitamin, but is able to synthesize sufficient amounts for her needs in her body tissues. The amount of vitamin C in the milk is not greatly affected by the amount that is consumed in the feed. Experiments have shown that dairy cows may be fed large amounts of ascorbic acid in their rations or may be fed rations practically devoid of it, and, in both cases, the amount in their milk is normal.

Green forage and sprouted seeds are rich in vitamin C, and the silages are good sources. Dried roughages, grains, and their by-products are lacking in this vitamin. Vitamin C is easily oxidized, so it is almost useless to feed it to cows, because it would be oxidized in their rumen and they would receive little benefit from it.

**Vitamin E.** Vitamin E is necessary for successful reproduction in certain species of animals, but with the dairy cow there seems to be no deficiency. This is because the common feeds contain an ample supply of this vitamin. The requirement for cattle is not known, but calves require less than 40 mg. per day*.

Gullickson** fed dairy cattle on rations deficient in vitamin E. They reproduced normally. Several of the experimental animals died suddenly during the trials, apparently from cardiac failure, probably as a result of vitamin E deficiency. Symptoms of muscular dystrophy (white muscle disease) and heart muscle lesions have been produced in calves fed vitamin-E-deficient diets. Feeding 1,000 to 2,000 I.U.s of vitamin E daily has also been effective in preventing oxidized flavor in milk having a high copper content.

## REFERENCES FOR FURTHER STUDY

Annison, E. F. and Lewis, D., Metabolism in the Rumen. J. Wiley & Sons, New York, 1959.

Crampton, E. W. and Harris, L. E., Applied Animal Nutrition, 2nd ed. W. H. Freeman and Co., San Francisco. 1969.

---

* Basic Dairy Cattle Nutrition, *Mich. St. Ext. Bul.* E-702, 1973.
** *J. Dairy Sci.*, **32**:495 (1949).

Church, D. C., and W. G. Pond, *Basic Animal Nutrition and Feeding,* D. C. Church, Corvallis, Ore., 1975.

Cullison, A. E., *Feeds and Feeding.* Reston Publishing Company, Reston, Va. 1975.

Hillman, D., et al., Basic Dairy Cattle Nutrition, *Ext. Bul.* E-702. Michigan State University, East Lansing. 1973.

Maynard, L. A., and Loosli, J. K., *Animal Nutrition.* McGraw-Hill, New York, 1969.

Minerals, *J. of Dairy Sci.* **54:** 655 (1971); **55:** 768 (1972); **56:** 1276 (1973); **58:** 1538 (1975); **58:** 1561 (1975); **59:** 324 (1976).

Morrison, F. B., *Feeds and Feeding,* 22nd ed. The Morrison Publishing Company, Ithaca, N.Y., 1959.

*Nutrient Requirements of Dairy Cattle,* revised ed. National Research Council. 1971.

Proteins, *J. of Dairy Sci.,* **54:** 1629 (1971); **56:** 390 (1973); **58:** 1219 (1975).

Vitamins, *J. of Dairy Sci.,* **57:** 985 (1974).

# Forages for Dairy Cattle: Pasture, Hay, Silage

**6** Forages may be defined as vegetable feed for domestic animals. This term usually refers to feeds that contain the entire plant, such as pasturage, green chopped feeds, silage, haylage, and hay. They contain a relatively high percentage of fiber and a relatively low percentage of energy, in contrast to the concentrate feeds (grains, seeds, and by-products), which have a relatively low-fiber, high-energy content.

Forages are a primary constituent of most dairy rations for physiological and economic reasons. They can make up 60 to 70% of the total dry matter intake for dairy cattle. Dairy cattle must consume adequate amounts of fiber (a minimum of 15% of DM), usually supplied by forages, to ensure adequate rumen function. Consumption of rations too low in fiber may result in decreased milk-fat percentage and decreased ration digestibility. Because forages are relatively useless as a nutrient source for humans and other monogastric animals, and the entire plant rather than just the seed is consumed, the cost per unit of nutrient from forages is usually much lower than the cost per unit of nutrient from concentrate feeds. However, when economic conditions prevail that make the cost per unit of nutrient from concentrates and forages competitive, the cow can adapt to the situation and consume relatively larger quantities of grain.

Decisions regarding the production, harvesting, storing, and feeding of forages are complex and critical to the total profitability of the dairy farm. On most dairy farms an ample supply of high-quality forage is the key to high levels of production at minimal feed cost per unit of production. If the forage is of high quality, dairy cattle will consume larger amounts per unit of body weight, the forage will contain a higher percentage of nutrients, and the nutrients will be more highly digestible. These three factors combine to ensure a high percentage of total digestible nutrient intake from forages and minimize the need for supplemental concentrate feeds, which are usually a more expensive source of nutrients.

Dairy cattle will consume and can utilize a wide variety of crops as forage, including grasses, legumes, corn, small grains, and other crops. They can utilize these as hay, silage, or haylage, or as pasture or green chop. The objective of the forage program on a dairy farm should be to utilize available land resources to provide an adequate supply of uniformly high-quality forage at the lowest cost per unit of nutrient, which when balanced with other feeds will give the lowest total feed cost per unit of milk produced. The three basic alternatives available to most dairymen are: (1) pasture or green chop, (2) silage or haylage, and (3) hay. Various combinations of these three basic alternatives are also practical on many dairy farms. Decisions as to which forage(s) to grow and feed depend on the following.

1. Climatic conditions: rainfall, length of growing season (in some areas double cropping may be advisable), and temperature.

2. Land resources: amount of land (when land is limited, maximum yield of nutrients per acre becomes more important), topography, and type of soil.

3. Labor resources: is adequate labor available to harvest the forage and transport it to the cow or should the cow harvest the forage?

4. Feed storage resources: amount and type of silos and hay storage facilities.

5. Harvesting and feeding equipment: availability and cost.

6. Supplemental feeds: availability and cost of feeds to be fed with the forage to balance the ration.

The forage program for any given dairy farm should meet the stated objective while choosing between the alternatives in forage producing, harvesting, storing, and feeding systems. Many combinations can be quite satisfactory depending on the resources of a given dairy farm; a thorough analysis of the alternatives should be made before deciding. Analysis should begin with thorough study of climatic conditions and inventory of the available resources. This step alone should narrow the alternatives considerably. This should then be followed by an economic analysis: preparation and comparison of enterprise budgets for feasible alternative crops, harvesting, storage, and feeding systems. An example of a corn-silage crop budget is presented in Table 6-1. A comparison of corn silage, alfalfa-orchardgrass hay, and corn for grain is presented in Table 6-2. In this example it is easily seen that maximum energy per acre, at least cost per unit of energy, is attained from the corn silage. It can also be seen that alfalfa-orchardgrass hay provides maximum protein per acre at least cost per unit of protein. However, when one considers the ratio of energy to protein needed to balance a lactating dairy cow ration (7:1), the large difference in energy yield between the forages in the example (5,183 Mcal./acre), and the small difference in protein yield (55 kg.), the choice of forage to provide maximum nutrient yield per acre is in favor of corn silage. Another factor that must be considered is the availability and relative cost of other feedstuffs needed to balance dairy rations.

In summary, dairy cattle are well adapted to utilizing large amounts of forage. They can provide up to 60 to 65 percent of the total dietary intake of high-producing cows and even greater percentages of the diet of low-producing cows, dry cows, and heifers over 10 to 12 months of age. Forage

**Table 6-1** Corn Silage Enterprise Budget (Per-Acre Basis) (Prices Will Vary)

|  | Quantity | Price per unit | Total | |
|---|---|---|---|---|
| RECEIPTS |  |  |  |  |
| Corn silage (opportunity value) | 15 tons | $20.00 | $300.00 | $300.00 |
| EXPENSES |  |  |  |  |
| Variable |  |  |  |  |
| Seed | ¼ bu. | $48.00 | $ 12.00 |  |
| Fertilizer  N | 150 lb. | 0.20 | 30.00 |  |
| P₂O₅ | 60 lb. | 0.20 | 12.00 |  |
| K₂O | 80 lb. | 0.15 | 12.00 |  |
| Lime | 0.5 ton | 16.00 | 8.00 |  |
| Herbicides (including application) |  | 10.00 | 10.00 |  |
| Interest on operating capital |  | 9.0% | 3.78 |  |
| Machinery | 5 hrs. | 9.00 | 45.00 |  |
| **Total variable expenses** |  |  |  | **$132.78** |
| Fixed |  |  |  |  |
| Labor | 5 hrs. | $ 4.00 | $ 20.00 |  |
| Land (including taxes and maintenance) |  |  | 70.00 |  |
| Storage | 15 tons | 2.00 | 30.00 |  |
| **Total fixed expenses** |  |  |  | **$120.00** |
| **Total expenses** |  |  |  | **$252.78** |
| Profit from corn enterprise |  |  |  | $ 47.22 |

COST PER Mcal. OF NET ENERGY (EXPENSE BASIS)

15-ton yield × 600 Mcal./ton = 9,000 Mcal.NE

$252.78 expense ÷ 9,000 Mcal. = $0.0281/Mcal.NE

can be effectively utilized as hay, silage, haylage, green chop, or pasture. Of major importance in determining which crop to grow and how to harvest, store, and feed it on any given dairy farm are the following.

1. Type, cost, and availability of land resources.

2. Climatic conditions.

3. Yield of nutrients per acre.

4. Cost per unit of nutrients.

5. Labor availability.

6. Machinery costs.

7. Type of storage and feeding facilities available.

Of paramount importance for any type of forage fed to dairy cattle is forage quality. High quality increases palatability, nutrient content, and digestibility. Forage quality is determined primarily by stage of maturity at harvest. It is also affected by harvest and storage conditions.

## SECTION I. PASTURE

Pasture refers to all vegetation on which animals graze. It may consist of perennial grasses such as bluegrass, timothy, and orchardgrass; legumes such as alfalfa and ladino clover; or combinations of grasses and legumes (Figure 6-1) or annual grasses and cereal grain crops. Fresh pasture is palatable and nutritious. It is rich in protein, vitamins, and minerals (See Appendix Table E for nutrient composition of various pastures).

**Table 6-2** Comparative Cost per Kilogram of Protein and Per Megacalorie of NE from Corn Silage, Alfalfa-Orchardgrass Hay, and Corn for Grain (Per-Acre Basis) (Prices Will Vary)

| Items | Corn silage | Alfalfa-Orchardgrass Hay | Corn for grain |
|---|---|---|---|
| EXPENSES | | | |
| Seed | $ 12.00 | $ 8.00 | $ 12.00 |
| Fertilizer | 54.00 | 30.00 | 54.00 |
| Lime | 8.00 | 8.00 | 8.00 |
| Herbicide or insecticide | 10.00 | 2.00 | 10.00 |
| Interest on Operating Capital @ 9% | 3.78 | 2.16 | 3.78 |
| Machinery | 45.00 | 30.00 | 36.00 |
| Twine | | 2.00 | |
| Labor | 20.00 | 24.00 | 17.00 |
| Land | 70.00 | 70.00 | 70.00 |
| Storage | 30.00 | 8.00 | 3.00 |
| **Total** | **$252.78** | **$184.16** | **$213.78** |
| Yield/acre | 15T | 4T | 2.8T |
| % DM | 40% | 90% | 90% |
| DM/acre (tons) | 6 | 3.6 | 2.52 |
| DM/acre (kg.) | 5,454 | 3,272 | 2,291 |
| % protein (DM basis) | 8% | 15% | 10% |
| Protein/acre (lb.) | 959 | 1,080 | 504 |
| Protein/acre (kg.) | 436 | 491 | 229 |
| Mcal./kg. DM | 1.65 | 1.25 | 2.42 |
| Mcal./acre | 9,000 | 4,090 | 5,544 |
| Cost/kg. protein | $0.580 | $0.375 | $0.934 |
| Cost/Mcal. energy | 0.0281 | 0.045 | 0.039 |

**Fig. 6-1.** Ladino clover and orchardgrass pasture (V.P.I. & S.U.).

The major advantages of pasture as a forage for dairy cattle are:

1. It is very labor efficient. The cows harvest the forage so there is little labor involved in harvesting or feeding pasture.

2. Harvesting, storage, and feeding equipment costs are minimal.

3. Land that is too steep or rocky to accommodate tillage and harvesting equipment can be utilized.

4. Its use is especially adapted to small herds.

Following are some disadvantages of pasture that have led to declining use in much of the country.

1. Pasture quality is variable and subject to rapid changes in digestibility and palatability resulting primarily from weather conditions. It is difficult to provide a uniformly high-quality forage for extended periods from pasture.

2. Yield of nutrients per acre is considerably reduced by trampling and by urine and manure contamination.

3. Maximum utilization of available nutrients is not realized because irregular, inefficient grazing leaves considerable waste.

4. Pasture is difficult to handle in most of today's modern feeding programs because pasture quality is generally less consistent than stored forages and an accurate analysis of composition cannot be determined.

Pasture will probably continue to be a major source of nutrients for heifers and nonproducing cows where consistent forage quality is of less importance and under farm conditions of steep land and/or limited labor.

## Desirable Characteristics of Pasture

A pasture, to be of the greatest benefits to the dairyman, must possess the following desirable characteristics.

*1. Young and Growing.* The best pasture is a young pasture, and in order for it to remain good it must be kept actively growing. Young, actively growing pastures are high in protein, often reaching 15% or more on the dry-matter basis. They are soft and tender, containing less crude fiber and lignin than more mature pastures, and hence are more digestible. They are rich in all of the vitamins or their precursors. They are rich in phosphorus and in calcium. The vitamins and minerals decrease as the plant grows older.

*2. Dense and Abundant.* To be good a pasture should have a dense sod, as this determines the amount of feed that is available for the cattle. The job of harvesting the pasture crop by the cow is a big task. An average dairy cow will consume about 150 to 200 lb. of pasture daily, containing 10 to 30% dry matter, when it is young, tender grass. Such a pile of grass would measure 6 ft. in diameter and 3 ft. high at the center. To gather such a quantity of feed is quite a task, even when the pasture is in optimum condition.

*3. The Proper Height.* Cows seem to be able to harvest the pasture best when it is about 6 in. tall. When it is shorter they cannot get as much at one bite. When it is taller the grazing is done by biting into the sward to the depth of 4 or 5 in. As the blades and stems are more widely separated, the amount of grass gathered per bite is considerably less.

*4. Palatable and Digestible.* The palatability and digestibility of the forage will determine the amount of feed that the cows will consume and how much they will be able to convert into milk. Plants change in their composition somewhat differently. For example, orchardgrass, Reed's canary grass, and some of the other grasses become woody and lose their palatability much sooner than some others. Most legumes, with the exception of the perennial lespedeza, brome grass, and Sudan grass, will retain their palatability and nutritive value for a relatively long period of time. Most of the grasses, however, change in their digestibility and palatability as the season advances. It is important that there be a mixture of grasses or legumes in the pasture that will remain palatable and nutritive throughout the entire season.

**5. Even Distribution Throughout the Season.** To have a good pasture it is necessary that there be some actively growing pasture plant from early spring to late fall. Different grasses show different growth characteristics during the season. Bluegrass pasture, for example, grows well in the spring and fall but fails to grow during the hot part of summer. It is necessary then to have some pasture plant that will do well in hot weather to supplement the bluegrass at that period. Unless there is abundant pasture at all seasons, cows cannot keep their milk production uniform. A study* at the Kansas Station showed that cows on good pasture spend about 50% of the time grazing; on fair pasture they spend 55% of the time grazing; and on poor pasture 62%.

**6. Convenient to the Barn.** Dairy cows should be provided pasture that is not too distant from the barn, as they should not be required to walk great distances to and from the pasture. When improving a pasture, a dairyman should start with that nearest the barn. Considerable energy is required for a cow to walk far to and from pasture.

**7. Well Watered.** The importance of water in the pasture during the hot, dry months cannot be overemphasized. Cows require a large amount of water at all times, and in hot weather this requirement is increased greatly. If there is no stream in the pasture, fresh water should be provided by other means.

**8. Fenced Properly.** A good pasture should have a good fence surrounding it, but good pasture management requires a flexible fencing system. Rotational grazing calls for an inexpensive and temporary fence that can be moved as occasion warrants. The electric fence with an automatic control and a one- or two-wire line is quite satisfactory. Such a fence can be changed as often as needed with a minimum of work.

**9. Shade.** Shade must be provided in warm climates. This may be provided by trees or some type of constructed shade.

## Kinds of Pasture

Pastures may be classified by various methods. They may be classified according to the nature of the plant, as perennials or as annuals, or according to the time of year that the plant makes the most growth.

The more usual designation is on the basis of use: permanent, special, temporary, and winter pasture.

* *J. Dairy Sci.*, **25**:779 (1942).

## Permanent Pasture

ECONOMY OF PERMANENT PASTURE. Permanent pasture is one of the most economical feeds available for dairy cattle. This is because it requires very little labor and care, as compared to most other crops. Permanent pastures will last for many years if given proper care, and there is no labor expense in fitting the soil each year or in harvesting, since the cows harvest the crop themselves. Furthermore, though pastures must be fertilized at intervals, they do not require so much fertilizer as land in cultivated crops. Pastures also tend to prevent erosion, because the ground is covered the entire year.

YIELDS OF PASTURES. There is a prevailing opinion that lands put to permanent pasture will not yield a return in proportion to the value of the land or to the returns secured from other crops. Even though this is undoubtedly true of much of the pastureland in the United States as it is now handled, it need not be so. If care is taken that the cattle are not placed on it too early, that it is not overpastured so that the plants are destroyed, and that the sod is as carefully and systematically manured and fertilized as for any other crop, it can produce a high yield of nutrients.

Bluegrass pasture has a wide range of carrying capacity, depending on the soil, the amount of fertilizer applied, its management, and the amount of rainfall. At the Virginia Station the number of cow days of grazing per acre per year ranged from 63 to 218. This represented 3 acres per cow for the poorer pasture and less than 1 acre per cow for the better pasture. The length of the grazing season was about 180 days. The total digestible nutrients produced per acre varied from 1,020 pounds to 3,266 per year. Milk production had a range of 974 pounds to 3,504 pounds per acre for the grazing season. In most areas permanent pastures will furnish grass more cheaply than any other type of pasture; however, they do have certain limitations.

LIMITATIONS OF PERMANENT PASTURE. The main disadvantage of a permanent pasture is that during the latter part of the season it often does not produce as much feed as is needed. As the grasses become mature, they are less digestible and less palatable, and the protein, vitamin, and mineral content decreases. During this period it is usually necessary to have some supplementary feed such as temporary pasture, soiling crop, or summer silage. The so-called dormant period of bluegrass may result not so much from the nature of the plant as from the lack of moisture. Experiments carried on by the Department of Agriculture* at 12 different stations in this country indicate that there is no practical way of producing a uniform growth of grasses throughout the entire growing season. The total yield of dry matter varies greatly not only with the different pastures and localities but also from year to year on the same pasture.

* USDA Tech. Bul. 465.

The curves in Figure 6-2 show the variations in the production of feed nutrients by pastures during different months. It will be seen that the peak months of production are different at the three locations. In order to have sufficient grass during most of the grazing season, there will be surplus grass at some periods. This can be used to advantage as hay or as hay-crop silage.

PERMANENT PASTURE CROPS.   Permanent pastures consist of perennial plants and can be used under favorable conditions for many years in succession. They have the advantage that they do not have to be plowed every few years, and furthermore they are adapted to conditions under which plowing is not possible. A large part of the pastureland in the United States is permanent, and rightfully so, for there are many acres on hillsides and lowlands that should never be broken up but should return a profitable acre income when in pasture and cared for properly.

The best permanent-pasture grasses and legumes vary in different sec-

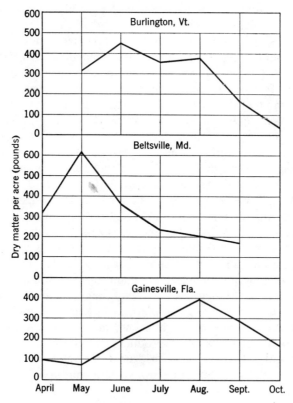

**Fig. 6-2.** Monthly yields of pasture in different areas. (*USDA Misc. Pub.* 194).

tions of the country. Kentucky bluegrass is probably the most universal crop. The many others that are used include Canadian bluegrass, redtop, brome grass, timothy, orchardgrass, carpet grass, Bermuda grass, Dallis grass, rye grass, fescue, white or Dutch clover, alsike clover, bur clover, hop clover, lespedeza, bird's-foot trefoil, and ladino clover.

A mixture consisting of both grasses and legumes is desirable. Also, a mixture containing early-growing and late-growing crops will help to spread out the feed supply more evenly over the entire grazing period. The crops should be selected from the ones best adapted to the specific locality.

MANAGEMENT OF PERMANENT PASTURE. The amount of pasture that can be secured from a given area of permanent pasture depends largely upon the management that it receives. The pasturelands in most sections of the United States have been greatly depleted, and it is necessary that they be built up again if the greatest returns are to be derived from them.

PROPER GRAZING METHODS. Overgrazing is one of the chief causes of the depletion of pastures. The effect of close grazing depends on the crops used for pasture. The legumes and tall grasses cannot stand as close grazing as the shorter, more spreading crops. The reduced leaf area lowers the ability of the plant to grow and to store nutrients in the roots. Under these conditions the yield is lessened and the stand or covering becomes thinner because the plants die out. The effect is greatest in periods of low rainfall or in arid regions.

Undergrazing also has some undesirable effects. With well-established bluegrass pastures, comparatively closely grazed plots give a higher yield than lightly grazed plots. At the end of three years, it was found that the sod on the closely grazed plots was in a much better condition, with fewer weeds than the sod in the lightly grazed plots. When pastures are not kept grazed down, many heads form, and the pasture loses in palatability. Bluegrass should be pastured close enough to keep it from heading.

Cattle should not be turned on pasture until after it has made a good start in the spring, and they should not be pastured too late in the fall.

When dairy cows are kept in the same pasture throughout the entire pasture season, it is often difficult to get the pasture grazed uniformly, as the cows consume the pasture near the barn and leave that away from the barn until it has become too ripe. In order to secure more uniform grazing and to give the pasture a chance to grow, a system of dividing the pasture into two or more separate fields and pasturing the animals in succession has been advocated. By this system the cows are turned on the pasture when it is 4 or 5 in. tall and still palatable. As soon as the cows have eaten the pasture down, they are turned into the second pasture. If they cannot keep all the pastures eaten down, one or more fields may be mowed for hay or silage, to prevent the bad effects of undergrazing.

PASTURE FERTILIZATION. Pastures usually respond remarkably well to fertilization. They cannot be expected to maintain their yield year after year unless some plant food is added to replace that which is removed by the growing crop. If fertilizers of some kind are not added, the yield of pasture gradually decreases until the pasture becomes unprofitable. One should always realize that a pasture should be treated like any other crop, and should not be expected to produce a crop year after year without any fertilizer.

The fertilizer treatment for a pasture varies greatly, depending upon the type of soil and its previous treatment. If the soil is acid, a certain amount of lime must be added before other fertilizer treatments will be of much benefit. It is advisable to have a soil analysis made to be used as a guide to liming and fertilization.

In an experiment with milking cows at the Virginia Station, covering a period of 10 years, permanent bluegrass pasture was limed and fertilized with phosphorus, phosphorus and nitrogen, and phosphorus and potash. A comparison of the average yields is given in Table 6-3.

It can be seen that the addition of fertilizer had a very great effect upon the yield.

A similar experiment of 10 years' duration was conducted in West Virginia, the results of which are given in Table 6-4. All plots were limed.

**Specialized Pastures.** Some dairymen in practically every dairy section are growing specialized pasture crops and treating them more like their main crops. Yields are obtained that compare favorably with any forage crop. Pastures used in these intensive programs must receive heavy applications of fertilizers and be managed differently than the usual permanent pasture.

CROPS TO USE IN INTENSIVE PASTURES PROGRAM. The crops to use in intensive pasture programs must be heavy feeders and produce rank growth quickly. Some of the more popular ones are ladino clover, alfalfa, orchardgrass, brome grass, and Coastal Bermuda. These plants have different adaptations and all

**Table 6-3** Average Pounds of Total Digestible Nutrients Produced per Acre When No Fertilizer Was Applied and When Fertilized with Three Types of Fertilizer

|  | Total digestible nutrients per Acre | Percent of check lot |
|---|---|---|
| Check | 1,245 | 100 |
| Phosphorus | 1,759 | 141 |
| Phosphorus and potash | 2,047 | 164 |
| Phosphorus and nitrogen | 2,166 | 174 |

**Table 6-4** Average Standard Cow Days per Acre and Equivalent Carrying
Capacity in Acres per Cow under Various Fertilizer Treatments

|  | Standard cow days per acre | Carrying capacity, acres per cow | TDN produced per acre, pounds |
|---|---|---|---|
| Check | 43.3 | 4.15 | 693 |
| Phosphorus | 65.9 | 2.75 | 1,055 |
| Phosphorus and potash | 83.9 | 2.15 | 1,393 |
| Phosphorus, potash, and nitrogen | 94.5 | 1.90 | 1,513 |

do not do their best at any one location. Also, there are others that respond
well at certain locations, including alsike clover and bird's-foot trefoil.

*Ladino clover* is a giant strain of white clover. It requires heavy
fertilization and considerable moisture. It is probably the most productive
pasture crop for areas of heavy rainfall and irrigated sections. Ladino cannot
be grazed continuously but is used in a rotational grazing system. The pasture
yield is greater when ladino is grown with one or more of the grasses. When
grown alone there is danger of bloat. The greatest value of ladino is for
pasture. It is seldom made into hay because it is so difficult to cure. Excess
growth is often used for grass silage.

*Alfalfa* has generally been grown for hay, silage, and haylage. It will,
however, furnish an abundance of pasture. Like ladino, it should be grown in
a mixture with grasses. These two crops are two of the foremost milk-
producing pasture crops. Alfalfa pasture must be rotated, as it will not stand
continuous grazing. It will not maintain its stand as long as ladino.

*Orchardgrass* is productive and grows early in the spring and late in the
fall. It responds well to heavy fertilization and if seeded in combination with
alfalfa or ladino has a tendency to crowd them out. It can be kept in the
desired proportion, however, by clipping after the rotational grazing period.
Improved strains of orchardgrass are available.

*Brome grass* is used extensively in the northern and central part of the
country. It grows well with alfalfa for either pasture, hay, or grass silage.
Brome yields well but is not well adapted to the southern area.

*Coastal Bermuda grass** is a tall-growing, palatable hybrid that is well
adapted to climatic and soil conditions in the southern states. With heavy
nitrogen fertilization the protein percentage is similar to that of alfalfa. It can
be used for pasture, silage, or hay. Its heavy yield and feeding value have
been responsible for its increased production in recent years.

*Fescue* is a grass of many varieties and uses. Kentucky 31 fescue gives

* *Ga. Agr. Exp. Sta. Cinc. N.S.* 10 (1957); and *S. C. Agr. Exp. Bul.* 490 (1961).

an abundant growth and holds on well into the fall and winter. It is not so palatable and has some limitations when used in mixtures.

A mixture of one or both of the legumes with one or two of the grasses is used. The specific mixture will depend upon the adaption of the legumes and grasses. Local agronomic recommendations concerning varieties should be followed.

ROTATIONAL GRAZING.  Pastures consisting of one of the grasses with a legume should be grazed in a rotation system, as they will not stand continuous grazing because of the nature of their growth. By this system, the cows are turned on a small area of pasture when it is 4 to 6 in. tall and still palatable. As soon as they have eaten the pasture down, they are turned into the second lot, and so on. Electric fences are valuable for dividing pastures into smaller lots. If the cows cannot keep all the pasture eaten down, one or more of the lots may be mowed for hay or silage to prevent the bad effects of undergrazing. The pastures should be kept mowed in order to keep down the weeds. In addition, a light harrowing will scatter the droppings.

Some dairymen have gone a step farther and practice rationed grazing. This system is based on a change of pasture for the herd every night and morning after milking, grazing a very small acreage completely from one milking to the next.

**Temporary Pastures.**  A temporary pasture lasts for only one year and is designed to carry stock only for such a period. On account of the large amount of work necessary to prepare the ground and to sow the seed, this type of pasture has not been used extensively in the past in many parts of the United States. In some sections, however, temporary pastures are now being widely used. Their advantage is that the yield is greater than that of a permanent pasture, and that they may furnish pasture during the so-called dormant period of bluegrass and other permanent pastures.

*Sudan grass* is one of the most popular temporary pastures, as it can be planted so as to fit into the time when most needed. Because it is drought-resistant and is a warm-weather plant, it should not be planted until the ground has warmed up well in the spring. It is a nonlegume, and is often grown with soybeans. Sudan grass does not cause bloat, but occasionally, especially when the growth has been stopped by drought or by frost, prussic acid poisoning results from the pasturing of Sudan grass. It is recommended that grazing of Sudan be delayed until it reaches a height of 18 in. to reduce the danger. Young Sudan grass and new growth are highest in prussic acid. Frosted Sudan should not be grazed. It is usually safe to use frosted Sudan for silage or hay if cut immediately following frost. It is important to use certified varieties to minimize the hazard of poisoning. Sweet varieties of Sudan have been gaining in popularity; they are very palatable and are not subject to some of the diseases of the common varieties.

*Millet* is about equal to Sudan grass as a temporary pasture crop for green chopping. Each can be seeded so as to be ready to graze or to cut at the time it is most likely to be needed to supplement other pastures. Either should be ready for grazing six to eight weeks after seeding. Millet is not quite so palatable as Sudan; however, cows eat it well when there is no choice. Pearl and star millets are the two most generally used. Millet has the advantage that it does not contain prussic acid, but there are indications that it may have a depressing effect on butterfat test.

Each of these crops should be grazed rotationally. Graze limited areas down to 4 to 8 in. in height and allow them to recover to about 18 in. before turning back on them.

Excess forage not needed for grazing or green chopping may be harvested for silage or hay.

Other crops that are used in various parts of the country for temporary pastures are oats, field peas, and lespedeza. All these have been used successfully. The aftermath of red clover or clover and timothy hay crops may also be used for pasture during late summer months.

**Winter Pastures.** Winter pastures will include those crops adapted to late fall and early spring grazing. In every area of the United States the pasture season can be lengthened both in the spring and in the fall by fertilization and good management practices. This may amount to one to two weeks on each end of the season.

If the number of days of real grazing are to be increased materially, special crops must be used. They must be crops that make their major growth when the usual grasses are not growing. In the deep South, grazing may be had the entire year. However, crops used during the winter months must be different from those used during the summer.

WINTER PASTURE CROPS. Rye grass and crimson clover are used more than other winter crops in the Southeast, Table 6-5. Other crops used for winter grazing in various sections of the country are Austrian winter peas, hairy vetch, and small grains such as rye, winter oats, barley, and wheat. It is preferable to seed a mixture of two or more of these grains together and to sow them very heavily. With a thick stand over the land, there will be fewer days of grazing lost because of wet and soft land.

VALUE. Most of the winter pasture comes from annuals. The cost of preparing the land, seeding, and fertilizing makes it more expensive than summer pastures. The value is determined not by comparing it with summer pasture, but by the value of the barn feeding that it replaces.

**Table 6-5** Percentage of Total Digestible Nutrients Obtained from Rye Grass and Crimson Clover Pasture Each Year, and the Average for Three Years, by 28-Day Periods[a]

| Dates | 28-day periods | 1946–1947 | 1947–1948 | 1948–1949 | Three-year average |
|---|---|---|---|---|---|
| Dec. 2–29 | 1 | 5.3 | 10.2 | 11.1 | 9.6 |
| Dec. 30–Jan. 26 | 2 | 10.6 | 7.5 | 7.9 | 8.3 |
| Jan. 27–Feb. 23 | 3 | 14.7 | 5.9 | 8.0 | 8.6 |
| Feb. 24–Mar. 23 | 4 | 7.3 | 18.9 | 11.7 | 13.5 |
| Mar. 24–Apr. 20 | 5 | 25.1 | 30.2 | 22.3 | 25.8 |
| Apr. 21–May 18 | 6 | 37.0[b] | 24.3 | 27.9 | 28.4 |
| May 19—— | 7 | — | 3.0[b] | 11.1[b] | 5.8[b] |
| Total | | 100.0 | 100.0 | 100.0 | 100.0 |

[a] *S. C. Agri. Exp. Sta. Bul.* 380, 1951.
[b] The last entry in each year is for less than a full 28-day period.

## Use of Surplus Pasture

Whenever sufficient grass is produced for the greater part of the pasture season, there will be a surplus during the heaviest growing period, mainly in the spring and early summer. If this surplus grass is allowed to mature, the pasture will lose much of its milk-producing value. The amount of pasture that is not needed for grazing at this period should be fenced off, and can be cut for either hay or grass silage. This method gives added acreage to graze at other times of the year.

## Green Chop Feeding

The practice of cutting green crops and feeding them in that form is called by many names: soiling crops, green crop feeding, green chop feeding, and zero grazing.

Any forage crop that is in a palatable stage and has sufficient growth can be used. Green chop feeding can be used as a supplement to grazing at times when pastures are inadequate, or it can be used for dry-lot feeding without pasture.

Compared to rotational grazing, green chop feeding is about equal in maintaining milk production. The cut chop requires only about two-thirds as much acreage because there is no trampling of the forage and the crop is cut when it yields more. Labor and equipment costs are higher, however, because of the daily chore of cutting and feeding.

Some dairymen who have used green chop feeding have changed to silage feeding in order to do away with the day-to-day job of cutting and hauling the feed; automation will handle the silage. Silage also gives a uniform feed; green chop feeding is subject to changes due to differences in maturity, crops used, and weather.

The decision on whether to use supplementary pasture, green chop, or silage should be based on the system that best suits the individual operation and will keep the cows well fed and working on an efficient level of production.

## Irrigating Pastures

The growth of pasture grasses is more dependent upon moisture than upon any other single factor. The process of irrigating pastures from water supplies such as ponds, creeks, and rivers is receiving much attention. When the value of the extra feed is calculated on the basis of the cost of barn feeding that it replaces, it appears that the practice is profitable under some conditions. Irrigation is just one means of increasing summer feed supply. Its cost and that of alternatives should be considered.

## Feeding Cows on Pasture

Pasture is not a concentrate, and it is impossible for heavy-milking cows to consume sufficient pasture to furnish the necessary nutrients for their milk production. As a result, such cows will decline rapidly in their milk production unless they are given some feed along with their pasture. Even the best pastures will furnish only enough nutrients for cows producing from 15 to 20 kg., or 30 to 40 lb., of milk per day. It is necessary to recognize this fact and to supply the cows with additional concentrate feed before they drop too rapidly in milk production.

The amount and protein content of the grain mixture to feed will vary with the kind of pasture, the amount of milk that is being produced, and the test of the milk.

The amount of protein in the grain mixture for cows on pasture need not be high when the pasture is good. Young pasture grasses and legumes are rich in protein. Hence a grain mixture consisting of homegrown grains and containing not more than 10 to 16% protein can be fed with good results. As the late summer approaches, the pasture grasses contain considerably less protein, and at that time the protein content of the grain mixture should be raised to 14 to 20%. It is seldom necessary to go over 20% unless the pasture is very poor.

## Summary

The use of pasture as a forage crop for dairy cattle in the United States has decreased steadily in recent years. This can be attributed to increased herd size, more intensive systems of dairying and the difficulty in providing a steady supply of high-quality forage nutrients because of climatic conditions. It has, however, continued to be a major source of forage nutrients for many dairymen, especially for young stock and dry cows in situations where land is steep or rocky and on farms where labor and storage facilities are limited.

# SECTION II. HAY

Hay refers to grasses or legumes that are harvested, dried, and stored at 85 to 90% dry matter. High-quality hay is green in color, leafy, soft and pliable, and free from mustiness. When harvested in the proper physiological stage of growth and well cured to 20% or less moisture at the time of baling, hay can be an excellent feed for dairy cattle.

The use of hay for dairy cattle has declined in recent years in much of the United States. The major reasons for this are:

1. The risk of weather damage during field drying.

2. The amount of labor involved in harvesting and feeding hay.

3. Increasing amounts of insect damage to hay crops, especially alfalfa weevil damage.

## Kinds of Hay

Hays are divided into two groups: legume and nonlegume. Although legume hays are generally superior to nonlegumes in the amount of protein, in the mineral and vitamin content, and in palatability, this difference may not always exist, because a great deal depends upon the soil on which the hays are grown. A grass hay, for example, grown on a soil that is rich in nitrogen, lime, and phosphorus, cut early, and well cured may be superior in many respects to a legume hay not so well grown and cured. The fact that a hay is legume is not always a sign that it is a good hay. Other factors must be considered to determine the difference between good hay and poor hay, such as time of cutting, curing and storing conditions, and the kind of soil on which it is grown.

**Legume Hays.**  Good legume hay has many characteristics that make it of special value for dairy cattle. Some of them are as follows.

1.  With the exception of corn silage, legume hay will produce more digestible nutrients on a given area than any other crop. This is especially true of alfalfa, which will yield two or more crops during the season.

2.  More digestible protein can be grown on an acre with legume hay than with other common crops. Furthermore, the protein of legumes is of excellent quality.

3.  Well-cured legume hays are higher in the vitamins necessary in the nutrition of dairy cattle than any other of the common feeds. They are particularly rich in carotene, the precursor of vitamin A, and may contain considerable vitamin D. They are also a rich source of vitamin E. Nonlegume hays, when well cured, may also contain considerable amounts of these vitamins.

4.  The legume hays are especially rich in calcium. They are only moderately rich in phosphorus, the amount depending upon the amount in the soil on which they are grown. They are usually considered a fair source of this element.

5.  Good legume hay is palatable, adds the necessary fiber to the ration, and has a good physiological effect upon the digestive system of the cow. With the possible exception of silage, no winter roughage is better liked than good legume hay.

6.  Legume hay keeps the ground covered, thus preventing erosion better than the cultivated crops. Alfalfa hay continues for several years with one seeding. Sometimes there is difficulty in getting a good stand of alfalfa, as the soil must be specially prepared for such a crop, well drained, and not too acid. Soybeans and some of the other legumes will grow on soil where the acidity is higher than is required for growing alfalfa.

The principal legume hays grown in the United States are alfalfa, clover (red, mammoth red, alsike, crimson, ladino, and sweet), soybean, cowpea, peanut, lespedeza, and Canada field peas.

ALFALFA.  Alfalfa hay is one of the very best roughages for dairy cattle. It is very palatable, and it has a good effect upon the digestive system, as it is slightly laxative. It is high in protein of excellent quality, and is the highest of all common feeds in calcium. When properly cured, it is rich in carotene, the precursor of vitamin A.

CANADA FIELD PEAS.  Field peas are sometimes grown, either by themselves or more commonly with oats, for hay in the northern states. When cut at the proper time, they make a very nutritious feed, somewhat higher in protein than red clover, but not so palatable.

CLOVERS. Clover hay has the same advantages as alfalfa hay, except that it is a little lower in protein and is slightly less palatable. Five kinds of clover are grown extensively in the United States; alsike, crimson, red, ladino, and sweet. All have about the same feeding value. Ladino clover is primarily a pasture plant, but mixtures of it and some grasses are sometimes used for hay. Sweet clover is hard to cure properly, and should be cut when the first blossoms appear, as the stems rapidly grow woody after this stage is reached. Although it has about the same analysis as the other clovers, a much larger portion is usually refused by dairy cattle. It has been noted that when sweet clover exclusively is fed to cattle, it may cause a loss of the blood's clotting power, because of the chemical dicoumarol, which it often contains when it has not cured properly. Animals fed heavily on sweet-clover hay, especially when it is moldy, will sometimes bleed to death from internal hemorrhages or from outside wounds such as dehorning or castration.

COWPEAS. Cowpea hay, when properly cured, provides a roughage that is even higher in protein than clover or alfalfa. It makes a good roughage for the southern part of the country, and is adapted to warm climates.

LESPEDEZA. Lespedeza has been grown principally in the South and is adapted to land too acid or too poor to grow alfalfa or clover. Certain varieties have been developed that will grow farther north than the common lespedeza. Korean lespedeza, for example, matures earlier and is larger than the common variety and will grow farther north. Lespedeza hay is well liked by dairy cattle and is fairly high in protein and total digestible nutrients, but as a rule the yield is not high. The perennial species, *Lespedeza sericea*, is not as palatable as the annual, and as it is inclined to be stemmy there is greater waste.

SOYBEANS. Soybean hay, when properly cured, makes a good roughage for dairy cattle. It is slightly higher in protein and total digestible nutrients than alfalfa, but it is usually coarser, and for this reason more is wasted. It is very palatable when cut at the proper time but, when allowed to get too ripe, has coarse, woody stems that the cows will refuse. Its chief disadvantages are, first, that it is hard to cure properly, and, second, that the cost of growing is usually higher than that of alfalfa, as it must be seeded each year. However, it can be grown in some places where alfalfa cannot. It is often seeded with Sudan grass, sorghums, or millet, in order to secure a higher yield and to aid in curing.

**Nonlegume Hays.** Grass hays, such as timothy, redtop, bluegrass, orchardgrass, brome grass, Bermuda grass, Sudan grass, and sorghum, are usually not as good feeds for dairy cattle as the legumes. They are, as a rule, less palatable, and contain less protein, mineral matter, and vitamins than the legume hays. However, if they are harvested early and are properly cured, they may be equal to legume hay in palatability, and their protein content may be not far under that of ordinary legume hay. If the soil on which they are grown has been fertilized with a high-nitrogen fertilizer, the protein content will be greatly increased.

Grass hays have been popular in most of the United States, even though legume hays have many advantages over them. The reason for this is the ease with which they can be grown and cured.

TIMOTHY.    Timothy hay as ordinarily harvested is low in protein, minerals, and vitamins. It is not very palatable and is constipating in its effects. When timothy is cut early, however, and from fields that have been nitrogen fertilized, it is much higher in protein and energy and much more palatable than the ordinary timothy hay. It is often grown with clover or alfalfa, the resulting hay being known as mixed hay.

SUDAN GRASS.    When cut early, Sudan grass makes a fairly good hay. It yields heavily and is fairly palatable. It is slightly laxative. It compares favorably with other nonlegume hays in feeding value. When allowed to become too ripe, it is woody and not so palatable or nutritious as when cut early. Often two crops can be harvested per year. It is used for hay, principally as an emergency crop.

COASTAL BERMUDA.    Coastal Bermuda grass has become one of the highest-yielding hay crops in the southern states. It is especially adapted to this area and is quite high in nutritive value. In South Carolina* experiments, Coastal Bermuda was cut every four or five weeks for a total of four or five cuttings a year, when heavily fertilized. Hay yields over a three-year period averaged over 4 tons per acre. The protein content averaged over 14%. It is not so palatable as alfalfa and some other hays.

MILLET.    Millet ranks about the same as timothy in feeding value. It is less palatable, however, especially if allowed to become too ripe. It should be cut when in bloom, but at best is not a desirable feed for dairy cattle. It is used principally as an emergency crop.

OTHER GRASSES.    Many other grasses, such as orchardgrass, Reed's Canary grass, Johnson grass, Italian rye grass, brome grass, fescues, bluegrass, and redtop, are sometimes used for hay. When cut early and well harvested, they make hays very similar in palatability and feeding value to similarly cured timothy hay.

CEREAL HAY.    All the cereals—oats, wheat, barley, and rye—are sometimes used as hay crops. When cut early and well cured, they are fairly palatable and resemble timothy hay in composition. Oats and barley make

* S. C. Agr. Ext. Sta. Bul. 490 (1961).

more palatable hays than wheat or rye. They are adapted more for an emergency hay crop than for a regular hay crop.

MIXED HAYS. Many farmers grow a legume hay and a nonlegume hay in combination, calling it mixed hay. Some of the most common mixed hays are clover and timothy, alfalfa and orchardgrass or brome grass, soybeans and Sudan grass, oats and peas, and oats and vetch. The composition of such hay will depend upon the proportion of each kind of hay that it contains. It is usually cut early, and hence the nonlegumes are higher in protein than if cut at the ordinary time. The practice of growing mixed hay is to be recommended when it is difficult to get a stand of legume hay.

## Effects of Curing on Quality of Hay

Hay should be cured in such a way as to preserve the leaves and the green color and drive out sufficient moisture so that it will keep well in storage.

**Importance of Leaves.** The leaves on some of the hay crops, such as alfalfa, soybeans, and other legumes, are very important. In the alfalfa plant the leaves make up about 47% of the crop, but their protein content is 141% higher than that of the stems. Twenty-eight pounds of alfalfa leaves contain as much protein as 100 pounds of stems.

The leaves of alfalfa are also much higher in calcium and in vitamins than the stems, and they are much more palatable. From Table 6-6 it can be seen that more than half the soybean plant is in the leaves and that they contain very much more than half the food value, being higher in protein, ether extract, nitrogen-free extract, and minerals. The stems are much higher in crude fiber.

Unless care is taken in curing these crops, much shattering will result and

**Table 6-6**  Chemical Composition of Two Samples of Soybean Hay

|  | Leaves | | Stems | |
|---|---|---|---|---|
|  | Sample 1, (%) | Sample 2, (%) | Sample 1, (%) | Sample 2, (%) |
| Part of plant | 55.39 | 64.02 | 44.61 | 35.98 |
| Crude protein | 19.37 | 21.07 | 5.18 | 7.33 |
| Ether extract | 3.48 | 2.38 | 0.87 | 0.76 |
| N-free extract | 35.34 | 35.37 | 28.22 | 26.51 |
| Crude fiber | 22.11 | 21.38 | 49.64 | 49.18 |
| Ash | 9.47 | 9.80 | 5.46 | 5.88 |
| Water | 10.00 | 10.00 | 10.00 | 10.00 |

many of the leaves will be broken off. In this way a considerable part of the protein, ether extract, nitrogen-free extract, and ash will be lost, and the resulting hay will be considerably lessened in food value.

**Importance of Greenness.** The green color of the leaves indicates the amount of carotene, the precursor of vitamin A, that is present. The importance of this element, not only for the nutrition of the animal but also for its value in the milk produced by dairy cows, has already been mentioned.

**Vitamin D Content of Hay.** Hay when it is first cut contains very little, if any, vitamin D, the antirachitic vitamin. However, when the hay is cured in the sun, the ergosterol in such hay changes into vitamin D. Sun-cured hay, with the exception of direct irradiation of the animal herself with the ultraviolet light of the sun's rays, is about the only natural source of this vitamin available to the cow. The antirachitic potency of dried roughage depends upon the intensity of the sunlight, the length of exposure, and the amount of ergosterol in the plant.

Because the vitamin-D potency of hay depends upon its exposure to sunlight, and because sunlight seems to accelerate the destruction of the carotene in hay, it can be seen that it is a difficult process to cure and handle hay in such a way as to obtain the maximum amount of both these factors. Long exposure to sun has other deleterious results, such as loss of leaves. Thus the quality and carotene content of the hay is usually given first consideration in the curing of hay.

### Effect of Soil on Quality of Hay

The soil has an effect upon the hay grown on it. In fact, some soils are so deficient in certain minerals, such as calcium and phosphorus, that the herbage grown upon it will not support normal well-being in dairy cattle. Alfalfa grown on soils low in phosphorus will be low in that element. Grass hays will be affected, also, by the amount of calcium and phosphorus in the soil. The amount of protein in hays can be materially increased by adding a nitrogen-containing fertilizer to the soil. High-quality hay can be produced, then, only upon soils that are rich in available minerals. The fertility of the soil determines the crops that can be grown on it.

### Time of Cutting

One of the common mistakes in making hay is to let the hay become too ripe before cutting. Hay cut early is higher in protein, is lower in crude fiber, contains more of the vitamins, is more palatable, and will shatter less than that allowed to become ripe (Table 6-7).

It is usually recommended that alfalfa hay be cut between the bud and one-tenth bloom stage. After this the protein decreases and the crude fiber increases. Clover hay should be cut when it is one-fourth in bloom. The total yield may increase after this, but the protein content decreases, as do the digestibility and palatability.

Soybeans are best cut for hay when the pods begin to fill. If the crop is cut earlier the percentage of protein is higher, but the total yield is not so large and the difficulty of curing is much greater. If cutting is delayed beyond this point, the stems become hard and woody, and many of the leaves are lost.

Experiments have shown that the protein of timothy decreases rapidly after it starts to bloom, as do its palatability, digestibility, and vitamin content. Timothy should be cut before it comes in bloom, and should never be allowed to ripen.

All grass hays make much better feed for dairy cattle if they are cut before bloom or when they first begin to come into bloom. The small grains, when cut for hay, should be cut when the grain is in the early milk stage and before they begin to harden.

The importance of early cutting cannot be overemphasized. The percentage of protein, the digestibility, the amount of minerals and vitamins, and the palatability decrease as the crop increases in maturity (Table 6-8).

**Curing the Hay**

In curing hay it is necessary to keep it from becoming bleached by the sun and rain, to preserve the leaves from shattering, and at the same time to drive out sufficient moisture so that it will keep in the barn without heating and spoiling.

**Table 6-7**  Stage of Maturity and Nutrient Content

|  | Legume | | Nonlegume | |
|---|---|---|---|---|
|  | Digestible protein (%) | TDN (%) | Digestible protein (%) | TDN (%) |
| Vegetative (pasture) | 19% | 70% | 14% | 70% |
| Bud | 14 | 63 | 10 | 63 |
| Bloom | 10 | 56 | 7 | 56 |
| Mature | 6 | 49 | 4 | 49 |

**Table 6-8**   Forage Values by Cutting Dates (Ohio)[a]

| | Date of cutting | Percentage of digestible dry matter | Relative intake | Percentage of crude protein |
|---|---|---|---|---|
| Alfalfa prebud (early) | May 18 | 66 | 105 | 24.8 |
| Alfalfa prebud (late) | June 3 | 60 | 97 | 21.9 |
| Alfalfa (bloom) | June 15 | 56 | 96 | 17.5 |
| Orchardgrass (boot) | May 13 | 64 | 85 | 17.5 |
| Orchardgrass (bloom) | June 1 | 58 | 68 | 13.8 |
| S. bromegrass (boot) | May 20 | 68 | 89 | 19.9 |
| S. bromegrass (bloom) | June 17 | 52 | 66 | 10.8 |
| Timothy (boot) | May 25 | 70 | 89 | 16.8 |
| Timothy (bloom) | June 22 | 56 | 71 | 9.0 |

[a] From Myers, D. K. Reproduced by permission from May 10, 1972, issue of *Hoard's Dairyman*. Copyright © 1972 by W. D. Hoard and Sons Company, Fort Atkinson, Wisc. 53538.

There is perhaps no best time of day to cut hay, although it should not be cut until the dew is gone. It should then be allowed to lie in the swath until it is thoroughly wilted. When mowing, one may use the hay crusher, which crushes the stems of the cut plants. These crushed stems will dry much more quickly than will the uncrushed, and the stems are softer and the resulting hay more palatable. After wilting, and before the leaves become dry and brittle, the hay should be raked into windrows. Care should be taken not to rake the hay when it is so dry that the leaves will shatter. The length of time hay should remain in the windrow depends upon the weather conditions, the crop, and the method of harvesting and storing. Although the grass hays are easier to cure than the legumes, as the leaves are not so easily broken off, care must be taken with all hays in order to preserve them properly.

When hay is ready to be baled it should not contain more than 20% moisture. If there is more moisture than this, the hay will heat and fermentation will take place, causing loss of nutrients.

## Harvesting the Hay Crop

There is a wide variety of hay-handling equipment and systems available. Included are balers that produce 40 to 60-lb. string-tied bales, round bales as small as 40 lb. or as large as 1,500 lb.; stackers that pick up and stack hay in 2,500-lb. stacks; wafering and pelleting machines. The decision regarding the system of hay handling and the equipment needed should be based on the topography of the land, available feed storage and feeding equipment and volume of hay handled. It is not economically feasible to own some of the more expensive types of equipment if a small volume of hay is handled annually.

In any baling or stacking system the hay should be quite dry, containing 20% or less moisture before it is baled or stacked. It should not contain over 10% moisture if a wafering machine is to be used.

Since it is often difficult to dry the hay completely in the field, several alternative methods are available. The barn hay drier can be used to complete the curing of partially cured hay. The drier is located in the hay mow and a fan is used to drive air through a duct system and the hay. The drier may also utilize a source of heat, usually an oil burner, to heat the air before it is driven through the hay. Driers may be used on loose, chopped, or baled hay.

### Storage of Field-Cured Hay

Field-cured hay is usually stored in sheds, mows, or stacks. Some farmers chop the hay with a forage harvester or with a hay cutter and blow it into the mow. Such a method has the advantage that considerably more hay can be stored in a given space and it is easier to remove, as it is not bound together like loose hay. Care must be taken, however, that the hay be well dried, or it is likely to heat. There is some danger of spontaneous combustion when hay is stored in this way unless it is more thoroughly dried than when stored loose.

For storing hay in bales, the hay needs to be a little drier than for storing loose. Unless quite dry, the bales should be set on edge, with room for air circulation between them, and stacked so as to give a maximum surface area exposed for drying.

Air-dried hay at time of feeding usually contains 10 to 12% moisture. If hay is stored with 20% moisture, which is a safe storage content, nearly 200 lb. of water must evaporate from each ton of hay stored.

**Danger from Spontaneous Combustion.** If hay is put in the mow or stored when it contains too much moisture, it may ferment very rapidly, releasing a large amount of heat. If it is allowed to continue for a month or six weeks, the temperature may rise to 200°F or higher, at which point spontaneous combustion may occur and the hay burst into flame. This may be prevented by taking care not to store hay with an excessive amount of moisture. Leaks in the roof, unknown to the owner, sometimes cause this trouble.

### Method of Feeding Hay

The method of feeding hay is receiving much consideration from the labor-saving viewpoint. The feeding of baled hay has eliminated much of the harder work.

Some feed barns are arranged with the hay stored on the same level as that on which the cows are housed, with movable hay managers arranged between the cows and the hay. This plan reduces the handling of hay to a minimum.

An upright storage with a self-feeder system around the circumference at the bottom has been developed for feeding chopped hay. It is called a hay keeper. It is made of perforated metal or wood slats. A cone is built in the bottom of the storage to cause the hay to slide to the outside, where the cows can reach it. It is wider at the bottom than at the top to keep the hay from wedging. As the cows eat, the weight of the hay above causes more to slide down. This storage can be constructed with an air duct in the center, extending from the ground level to the top, which can be equipped with a hay drier.

**Hay Pellets and Wafers.** There has been interest in pelleted hay for dairy cows. It could be handled and fed by automation, and consumption could be increased by its use; however, the cost at present is very high. When the hay is finely ground before pelleting, the digestibility is lowered; if it is fed as the only roughage, the fat test of the milk is lowered.

The hay wafer, made from chopped hay and pressed into cubes 2 to 3 in. in diameter, is a move toward reducing the difficulties. The wafer can be made with stationary equipment or with field machines. Portable wafering machines, which pick the hay up from the windrow, are heavy and cumbersome. The preferable moisture content is 14 to 15%. The hay is chopped, pressed, and loaded on a trailer or truck.

The cost of wafering is high, but wafers do not reduce the fat test of milk as pelleted hay may. They can be handled with automatic equipment, but there is some problem of their breaking apart. One of the greatest possibilities for their use is in hay-deficit areas. Transportation costs would be much lower per ton because of their compactness. They weigh about 25 lb. per cubic foot, double the weight of baled hay. Their popularity is increasing in the Pacific and southwestern states.

### Storage Space Requirements

The space required to store hay depends on its looseness, the depth of storage, its moisture content when put in storage, and the kind of hay. Table F-2 in the Appendix gives the weight per cubic foot and the cubic feet required to store a ton of hay and various other materials.

Hay that is chopped packs closer than loose hay. The shorter it is cut, the less space it requires for storage. Hay baled in the field with a pickup baler is usually baled loosely and will use less than half as much space as loose-stored hay. Hay that is baled when it is completely dry and pressed very tight will require only about one-fifth as much storage space as long hay in loose storage.

## Summary

Hay quality is primarily dependent on stage of maturity at the time of cutting and weather conditions during curing. Risk of weather damage is high in much of the United States and relatively low in other parts of the country. It is both difficult and expensive to completely mechanize hay-handling systems. For these reasons the use of hay as a forage for dairy cattle has declined in most of the United States; exceptions are the West and Southwest. High-quality hay, however, is an excellent forage for dairy cattle and is fed, at least to a limited extent, on most dairy farms.

# SECTION III. SILAGE AND HAYLAGE

Storing and feeding forage as silage or haylage for dairy cattle is not a new practice (they have been used extensively in the United States since about 1900), but it is a practice that has increased rapidly in recent years. The use of the silo to preserve corn, grasses, legumes, small grains, and other crops for dairy cattle is a common and highly recommended practice.

## Advantages of Silage

Silage is fed on most dairy farms in the United States. This widespread use is the result of the following advantages.

1.  The silo offers the one means of taking the entire forage plant from the field and preserving it in a succulent form. The crop can be harvested and stored at the time in its development when it has the greatest milk-producing value.

2.  More feed nutrients can be grown on an acre of crops used for silage than on an acre used for most other crops.

3.  Less waste results when crops are put into the silo than when they are pastured or handled in the dry state. When the crop is ensiled, harvest losses are minimal. There are losses in the silo from fermentation, but they are much lower than those that occur when the crop is exposed in the field. Figure 6-3 and Table 6-9 illustrate a major reason why wilted silage and haylage are gaining so rapidly in popularity.

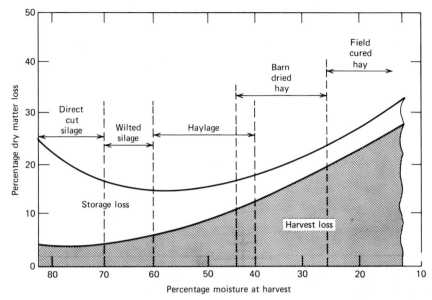

**Fig. 6-3.** Harvest and storage losses at various moisture contents with different forms of storage. (Reproduced from April 25, 1973 issue of Hoard's Dairyman. Copyright 1973 by W. D. Hoard and Sons Company, Ft. Atkinson, Wisconsin 53538.)

4. It is easier and more economical to mechanize the harvesting and feeding of silage than hay, thus resulting in a higher degree of labor efficiency in the forage feeding operation.

5. Forage nutrients can be preserved in a palatable and nutritious state for extended periods of time.

6. Silage is well suited to mixing with other feeds in complete or blended rations that are rapidly gaining in popularity.

**Table 6-9** Percentage of Dry Matter, Protein, and Carotene Preserved, and Relative Amount of Milk Produced from Alfalfa Stored by Three Methods

|  | Dry matter preserved (%) | Protein preserved (%) | Carotene preserved (%) | Milk yield per acre of forage (%) |
|---|---|---|---|---|
| Crop as cut | 100 | 100 | 100 | |
| Wilted silage | 84 | 84 | 28 | 112 |
| Barn-cured hay | 81 | 76 | 7.5 | 108 |
| Field-cured hay, no rain | 75 | 69 | 3.0 | 100 |

**Source.** USDA.

7. Silage requires less space for storage. A ton of silage requires 40 to 50 cubic feet of space, whereas a ton of loose hay will require 8 to 10 times this space, and baled hay, 3 to 4 times as much space. One ton of hay has a feeding value about equal to 3 tons of silage.

8. Feed stored as silage is not so subject to fire hazards as is hay.

9. Practically any green forage crop is suitable for ensiling.

10. In many areas of the United States it is difficult to cure hay satisfactorily because of the weather conditions. The use of the silo makes the saving of hay crops possible, even under unfavorable hay-making weather. It also reduces the risk of severe weather damage or complete crop loss due to inclement weather.

11. Silage provides a palatable succulent feed available at all times of the year.

One might summarize the major advantages of ensiling as (1) the reduction in harvesting losses, (2) the ability to mechanize the handling of the crop from field to cow, and (3) the ability to harvest the forage at the time of maximum digestible nutrient yield and to preserve these nutrients in as good or better state than when harvested.

## Chemical Changes in Silage

Silage traditionally has been defined as the end result of preservation of a crop under conditions suitable for the development of sufficient acid to maintain the mass in a state of preservation. This means that:

1. The crop is stored under anaerobic (without oxygen) conditions.

2. The crop contains or is supplied with adequate carbohydrates such as starches or sugars.

3. It is maintained at an optimum temperature of 80 to 100° F.

These conditions will promote optimum production of acids by bacteria to prevent decomposition. The bacteria ferment the available carbohydrates to acids. Eventually the acids will kill the bacteria and preserve the silage as long as air is excluded from the silo.

When the green, chopped forage is first stored in a compact mass in a silo, the living plant cells continue to respire, thus rapidly using up the oxygen in the trapped air and giving off carbon dioxide. In about 4 or 5 hours the free oxygen is all used up, but the carbon dioxide increases rapidly for about 48 hours, when it comprises from 60 to 70% of the silo gases. Subsequent to this time it begins to decrease. After the oxygen is used up molds do not develop, as they are unable to grow in the absence of oxygen.

### Dangerous Silo Gases

**Nitrogen Dioxide.** When crops are produced by use of heavy nitrogen fertilization, they are high in nitrogen content, especially in nitrates. Under some conditions at the time of filling the silo, nitrates change to nitric oxide and nitrogen dioxide. These products are yellow in color; they are extremely poisonous, even fatal. When they are breathed, they injure the lungs and produce silo-fillers' disease. After material has been put in, the silo should not be entered until after the blower has again been in operation for some time. The production of these gases occurs during the first two days after ensiling, then ceases gradually.

**Danger of Carbon Dioxide.** During the fermentation process carbon dioxide is given off. This gas is dangerous. As this is a heavy gas, it does not pass out if the doors are closed for some distance above the silage. Therefore care should be taken during the filling period, when the machine has not been running for some time, not to enter the silo until the blower has driven off the dangerous gas.

**Temperature Increases.** If air is excluded, the increase in temperature is not great; it will be about 80° to 85°F near the bottom and about 100°F or a little more near the top. The temperature continues to increase for about 15 days and then gradually decreases. If air gets into the silage, the temperature may rise to 130°F, but when the silage is properly packed this increase will occur only near the surface or where the air can reach it. The heat is caused by bacterial fermentation.

**Bacteria Increase.** The plant forage carries with it a large number of bacteria. The conditions of growth found in silage (proper temperature, food, and moisture) are excellent, especially for the lactic-acid bacteria, and their numbers increase very rapidly, frequently to hundreds or thousand of millions per milliliter of the juice. These bacteria, or enzymes produced by them, and enzymes from the cut plant material, attack the sugars and other food material, breaking them down into organic acids, principally lactic, with some acetic and small amounts of other acids, and also some ethyl alcohol. It has been found that besides the sugars, 25% of the pentosans and 25% of the starch contained in the forage are changed as a result of four months' ensiling. Much of it is changed to organic acids, and some of it is used for food for the bacteria and is built up into compounds in their bodies. In this process, some of the proteins of the green forage are also broken down or digested and some are used by the bacteria for their growth, much as occurs in the paunch of the cow.

ACIDITY STOPS BACTERIAL GROWTH.   When the acid in the silage has increased to a certain degree, pH of approximately 4.0, bacteria cease to multiply; hence the action of the enzymes stops, with the result that no more acid is developed, putrefaction ceases, and, if air does not gain entrance, the silage will keep for long periods with but very little change. The amount of acid that will develop depends largely on the kind of crop, especially on its sugar content. If the forage does not contain enough sugar, sufficient acid may not be produced to prevent spoiling of the silage, which often occurs when legumes and certain other crops are made into silage, unless special methods are used. The lactic-acid content of the nonvolatile acid at various stages of fermentation is shown in Table 6-10.

**Causes of Poor Silage.**   Good silage should have a mild, pleasant aroma, an acid taste, and a slightly greenish color. It should be free from sliminess and mold and have sufficient acid to prevent further action of microorganisms. Although ordinary good conditions will produce good silage, sometimes, for various reasons, the silage may not keep satisfactorily, and as a result is lost. Hay-crop silage is more difficult to preserve than corn or sorghum silage. The causes of poor silages can be listed as follows.

*1. Not Enough Acid.*   When the forage that is ensiled does not develop sufficient acid to stop the fermentation, undesirable bacteria cause putrefac-

**Table 6-10**  Lactic Acid Content of the Nonvolatile Acid at Various Stages of Fermentation (calculated for 100 gm. of dry silage)[a]

| Age of silage, days | Total nonvolatile acid, grams | Lactic acid, grams | Nonlactic acid, grams |
| --- | --- | --- | --- |
| 0 | 2.025 | 0.199 | 1.826 |
| 1 | 2.195 | 0.514 | 1.681 |
| 3 | 3.579 | 1.868 | 1.711 |
| 30 | 6.818 | 5.290 | 1.528 |
| 132 | 7.986 | 6.117 | 1.869 |

[a] *Wisc. Res. Bull.* 61, 1925.

tion or rotting. Such bacteria produce enzymes that break down some of the protein, causing an off-flavor and slimy silage.

*2. Too Much Acid.* When forage crops with an exceptionally high sugar content such as immature corn or sorghum are used, the acid may be so high that a sour, unpalatable silage may result. Such silage is not only unpalatable, but when fed in large quantity causes cattle to scour.

*3. Not Enough Moisture.* When not enough moisture is in the forage, the silage will not pack well and more air will be left in the silage. This will result in a moldy silage. Sometimes, when the forage is slightly too dry, air pockets are left through the silage, which will result in occasional batches of mold throughout the silo. Occasionally air will be left in forage with hollow stems, causing spoilage of the silage. Such crop should be packed with special care so that all air will be pressed out.

*4. Too Much Moisture.* When the forage contains too much moisture, the silage is likely to be too sour, and often at the bottom of the silo will be several feet of sour, soggy, unpalatable silage that the cattle will not eat. High moisture causes undesirable fermentation to take place. Hay-crop silages, especially those made of legumes, may not keep satisfactorily if the moisture is too high. Wilting the silage to reduce the moisture content often will result in good silage.

**Methods Used to Ensure Good Hay-Crop Silages.** Little trouble is experienced in the preservation of corn, small grain, or sorghum silage when they are ensiled at the proper time, as they contain an ample supply of sugar, which is quickly converted into lactic acid. Such is not true of the grasses and legumes, which, when used for silage, are usually referred to as hay-crop silages. The chief reasons why special methods are used to ensure proper keeping of these crops are as follows.

1. They may not contain enough sugar to produce sufficient acid to keep it from spoiling or from having a strong, undesirable odor.

2. The legumes are more alkaline than corn or sorghum and neutralize some of the acid that is produced.

3. The forage often is high in moisture content.

4. Such silages, especially legumes, are rather high in protein.

Three primary methods have evolved to help ensure proper keeping of hay-crop silages. These are: (1) ensiling the material after wilting to 60 to 70% moisture (wilted silage); (2) ensiling the crop after drying to 45 to 60%

moisture (haylage); or (3) adding material to the silage at the time of ensiling (silage additives).

WILTING THE CROP. The wilting of the forage for two or three hours in sunshine before ensiling will usually result in a better silage than that which is not wilted. The wilting will reduce the water content so that the sugar content per pound of forage is increased. The amount of moisture in fresh forage cut at the proper time for silage is usually high, around 75% or more. This should be reduced to 60 to 70%. If below 60%, the moisture is not sufficient, and the silage may mold; if the moisture is over 70%, the silage will not go through the normal fermentation. Usually, if the moisture is over 70%, a considerable amount of juice will run out, carrying with it some nutrients and causing a foul odor around the silo. The juice from the silage has a destructive action on the walls of silos. The Dairy Division, USDA,* reports results of various combinations of acidity and moisture; for example, high moisture along with low acidity favors the production of butyric acid and the breakdown of proteins. This gives an off-flavored silage with an objectionable odor. On the other hand, high moisture with high acidity, low moisture and low acidity, or low moisture with high acidity in silages may not cause objectionable odors or flavors.

HAYLAGE. Haylage refers to crops wilted or dried to about 50% moisture before ensiling. It is important to not allow the material to become too dry before ensiling. If it contains less than 50% moisture, field harvest losses increase, it is more difficult to pack tightly enough to exclude air, and the drier silage is more susceptible to heating in the silo, which causes a reaction (browning reaction) that lowers the digestible protein content. This reaction causes some of the protein to be unavailable to the animal (it can vary from a small loss to almost a total loss of the protein). Because of the difficulty in packing haylage, it is usually stored in tower silos rather than horizontal silos. Haylage made from material harvested at the proper stage of growth for optimum nutrient yield and preserved under anaerobic conditions is an excellent feed for dairy cattle.

SILAGE ADDITIVES. Silage additives basically fall into two types.

1.  Strong acids or salts, which are converted to acids in the silo. These reduce the silage pH to 4.0 or below and prevent the occurrence of undesirable fermentation that produces spoilage and strong undesirable odors. Materials used for this include acids such as formic acid, dilute hydrochloric or sulfuric acids, and sodium bisulfite, which is a salt that is converted to an acid in the silo. Formic acid and sodium bisulfite have been shown to be effective in preserving high-moisture silages. Cost and

* *Farmers' Bul. No.* 578 (1941).

inconvenience of adding the material are the major problems. In addition, formic acid can damage the skin and eyes, so exposure to volatile vapors must be avoided when applying the material to crop to be ensiled. There is no problem when feeding the silage. There is considerable current research on various other types of silage additives but results are inconclusive.

2. Starches or sugars. These supply the bacteria with ample food so that fermentation will proceed normally. Commonly used materials include blackstrap, corn, beet, and citrus molasses; ground grains such as corn, barley, wheat; and various by-product feeds such as beet pulp and citrus pulp. Such additions are effective in helping the proper fermentation process. They also increase the palatability of the silage. Much of the feed value of these materials is retained. The major disadvantages are cost and inconvenience.

Another way to ensure the keeping of hay-crop silage is to mix the crop with corn or sorghum forage. These contain sugar sufficient to induce the normal fermentation. Such crops as soybeans or alfalfa can be mixed with corn or sorghum. Equal parts of corn forage and the legume will be satisfactory, but sorghum that contains more sugar than corn can be used in the proportion of three parts of legume to one of sorghum.

## Crops for Silage

A great many different crops are being used for silage. Practically any forage crop that is desirable for pasture or hay will make a desirable silage. In fact, some crops that are not desirable as pasture or cured feed have been reported used for silage with satisfactory results. Some plants that are bitter or produce off-flavor milk if fed green or cured often appear to lose these qualities during the ensiling process. Most weed seeds are destroyed by ensiling. To make good silage, however, good crops cut at the proper stage and put into the silo without losing a part of their feeding value by leaching or by loss of leaves must be used.

**Corn.** Corn is a standard and popular crop for silage in most dairy sections, as its yield in tons of forage per acre is high. Because the corn plant at time of ensiling is high in available sugars, so that normal fermentation takes place without the addition of any preservatives, the making of silage from corn is easier.

Making high-quality corn silage involves (1) selecting a high yielding grain variety of corn, (2) using tillage, weed control, liming, and fertilizing practices that are suited to the type of soil used, (3) harvesting at the stage of maturity that produces the maximum yield of digestible nutrients per acre, and (4) preserving it under anaerobic conditions.

VARIETIES. The best grain varieties of corn for a given area should be planted for silage. In the best grain varieties roughly one-half of the weight of the dry matter is in the ear, which accounts for about two-thirds of the digestible nutrients in the silage.

TILLAGE PRACTICES. The best tillage practices vary considerably in different areas of the United States. In recent years a method known as "no-tillage" or "sod planting" has become increasingly popular in much of the eastern and southeastern parts of the United States. It is especially beneficial when used on drouthy and rolling land, as it is conducive to soil water retention and prevents erosion. In any area it is best to use tillage, weed control, liming and fertilizing practices best suited to local soil and climatic conditions. These recommendations are available through local Extension Offices.

TIME TO HARVEST. Extensive research has been conducted in efforts to determine the stage of maturity to harvest the corn plant to maximize yield of digestible nutrients while minimizing harvesting and storage losses. If corn is harvested too early, a higher proportion of the plant's nutrients are in the vegetative part of the plant rather than the ear and are less digestible. The loss from silo seepage is also significant. If it is harvested too late, many of the lower leaves are lost at harvest and it is more difficult to pack the material to prevent storage losses. The silage is also less palatable. The optimum time to harvest corn silage is in the hard dough or dent stage. At this time the silage will usually contain 32 to 42% dry matter. It may be somewhat lower than this in wet, rainy seasons or somewhat higher in very dry seasons, as the moisture content of the corn plant varies considerably depending on the weather. At maturity, most of the leaves below the ear, as well as the shuck, have turned brown. Nearly all of the kernels will be dented. A black line will appear at the base of the kernel. At this time the maximum nutrients have been deposited in the grain and this is the optimum stage of maturity to harvest the crop. The plant is still wet enough to avoid most harvest losses, dry enough to minimize storage losses, the optimum yield of digestible nutrients can be attained, and the subsequent silage is very palatable to cattle. The increased consumption of more highly digestible material with less fiber results in higher nutrient intake from forage, thus reducing concentrate needed. This, in turn, usually reduces feed costs and increases income over feed cost. Results of research conducted at the Middleburg, Virginia, Forage Research Station, as well as many other research stations, substantiate this recommendation.

**Sunflowers.** Sunflowers have been used successfully for silage in regions of short growing seasons and cool nights where corn cannot be grown so successfully. In such regions their yield is a little higher than that of corn, but they are not quite so high in nutritive value, nor are they so palatable.

**Sorghum.** Sorghum makes good silage when put in the silo at the proper time. The sweet sorghums are better for silage than the grain sorghum, although the taller grain sorghums, especially kafir, are used successfully. Sorghum yields better than corn in the deep South and in the drier parts of the Midwest. Sorghum silage is not so efficient as corn silage for the production of milk. It should be cut in the dough stage. If it is cut too early the silage will be sour and unpalatable; on the other hand, if it is cut too late many of the matured grains will pass through the cow undigested.

**Small Grains.** Small-grain crops can make excellent silage. In many areas they can be planted in the fall, harvested in the spring, and followed with corn, soybeans, or other suitable crop. Barley and wheat are preferred to rye and oats as small-grain silage crops because the yield and energy content are higher. Barley, wheat, and oats should be harvested in the soft dough stage to maximize energy and protein yields per acre. At this stage of maturity these nutrients are highly digestible, the silage can be harvested by direct-cut methods, and the energy, protein, and fiber contents are comparable to corn silage (Table 6-11). Rye should be harvested when in the boot stage and wilted prior to ensiling. If left until the dough stage, it decreases in feeding value and palatability.

**Grasses and Legumes.** Grasses and legumes will be considered together here under the term hay-crop silage, which is used today to include silage made from grasses, legumes, or any combination of these crops. Methods used in the proper ensiling of these crops were discussed in the earlier part of this chapter.

**Table 6-11** Composition of Corn, Barley, and Wheat Silage[a]

| Silage type | Stage of maturity | Dry matter (%) | Dry matter basis, (%) | | |
|---|---|---|---|---|---|
| | | | CP | CF | TDN |
| Corn | Hard dough | 37 | 8 | 25 | 66 |
| Barley | Soft dough | 41 | 9 | 27 | 65 |
| Wheat | Soft dough | 51 | 10 | 25 | 66 |

[a] Average composition of samples analyzed in Virginia Polytechnic Institute and State University Forage Testing Laboratory, 1968–69.

**Hay Crops.** It would be impossible to list all the crops that are suitable for making into hay-crop silage that may be grown within the United States. The most important ones are: all grasses, alfalfa, red clover, ladino clover, sweet clover, crimson clover, lespedeza, soybeans, cowpeas, peas, pea vine, vetch, Coastal Bermuda, millet, excess pasture, and a mixture of any grasses and legumes. Many other crops have been successfully used.

TIME TO HARVEST. The crop material is at its best for ensiling at the same period as that during which it should be cut for hay. The best time for cutting grasses is just before heading. The legumes are ready when in one-tenth bloom and soybeans when the pods have formed and begin to fill. At these stages they contain the maximum milk-producing nutrients, are lower in undigestible fiber, and are most palatable.

COMPOSITION AND FEEDING VALUE. The amount of dry matter in hay-crop silages depends greatly upon weather, how much the crop has been wilted, and also upon the maturity of the crop at time of ensiling. On the average, silage will contain about one-third as much dry matter as does hay; in other words, 3 pounds of silage will on the average equal 1 pound of hay.

Hay-crop silage is lower in total digestible nutrients and net energy than is good corn silage with the same amount of dry matter, but is higher in protein, minerals, and carotene. Because of its higher protein content, less protein is needed in the grain mixture. The higher carotene content is valuable in supplying the needed vitamin A for growth, reproduction, and health, and to ensure a milk rich in vitamin A and carotene. Hay-crop silage does not contain as much vitamin D as does field-cured hay. Green-forage plants do not contain vitamin D but do contain its precursor, ergosterol, which may be changed to vitamin D during field curing, when the hay is exposed to the sun.

Hay-crop silage as a rule is not so palatable as corn silage. Cows eat it readily after they become accustomed to it, but they do not usually relish it as much as corn silage. However, it is an excellent feed, and can support high levels of milk production when properly supplemented with concentrates, hay, or corn silage.

YIELDS. The yield per acre of hay-crop silage is not usually so great as that of corn, but if the soil is well fertilized excellent yields may be obtained. Table 6-12 gives the composition and other nutrient data of various feeds. The yields of silage for the different crops can be estimated by multiplying the expected yield of hay by 3. If an alfalfa field would produce 4 tons of hay per acre during the year, it should furnish about 12 tons of silage.

**Miscellaneous Feeds for Silage.** Numerous feed materials are occasionally used for silage with satisfactory results. Some of these feeds are apple pomace, cull apples, potatoes, beet pulp, and cannery refuse. If too high in moisture, like cull apples and potatoes, for example, they should be put into

**Table 6-12** Composition of Some Forages (Dry Matter Basis)[a]

| Forage | DM (%) | NElactation (mcal./kg.) | Protein Total (%) | Dig. (%) | Fiber (%) | Ca | P | Carotene, (mg./kg.) |
|---|---|---|---|---|---|---|---|---|
| Alfalfa, 1/10 bloom, hay | 90.0 | 1.25 | 18.4 | 12.7 | 29.8 | 1.25 | 0.23 | 127.2 |
| Alfalfa, ½ bloom, hay | 89.2 | 1.21 | 17.1 | 12.1 | 30.9 | 1.35 | 0.22 | 33.3 |
| Alfalfa-orchardgrass silage | 40.0 | 1.14 | 17.2 | 10.0 | 31.4 | | | |
| Clover, red hay | 87.7 | 1.31 | 14.9 | 8.9 | 30.1 | 1.61 | 0.22 | 36.8 |
| Brome, hay | 89.7 | 1.14 | 11.8 | 5.0 | 32.0 | | | |
| Timothy, mid-bloom, hay | 88.4 | 1.28 | 8.5 | 4.6 | 33.5 | 0.41 | 0.19 | |
| Oats, hay | 88.2 | 1.39 | 9.2 | 4.4 | 31.0 | 0.26 | 0.24 | 101.0 |
| Corn silage, well eared | 40.0 | 1.70 | 8.1 | 4.7 | 24.4 | 0.27 | 0.20 | |
| Oats, silage | 31.7 | 1.31 | 9.7 | 5.5 | 31.6 | 0.37 | 0.30 | 119.5 |
| Sorghum silage, grain variety | 29.4 | 1.24 | 7.3 | 2.0 | 26.3 | 0.25 | 0.18 | |

[a] From Table 4, *Nutrient Requirements of Dairy Cattle,* 4th ed., 1971. Reproduced with the permission of the National Academy of Sciences, Washington, D.C.

the silo with dry hay or some material that will absorb the excess moisture. Any of these by-products or surplus crops must be free of spray residues, especially when the silage is to be fed to milking cows.

**Harvesting Silage**

The last several years have brought about many changes in the equipment for handling the silage crop and in the amount of hand labor required. Present-day silage-making machinery not only requires much less labor but also can eliminate entirely the handling of the crop by hand.

**The Field-Chopper Method.** The field-chopper or forage harvester will chop the crop either when it is standing or when it is in the window. Corn is usually cut from the row, but hay-crop silage crops can be cut either from the standing crop or from the windrow. The chopper blows the cut material into a truck or wagon. A dump truck may be used, or the truck or wagon can be equipped with an automatic unloading device, such as an endless belt of canvas or a false front or bottom. Either of these will unload by the use of power at the silo. The trucks or trailers are unloaded directly into the hopper of the silage blower or forage elevator, which, in turn, blows or elevates the material into the silo. The crop should be cut into small pieces not over ½ in. long, and ¼ to 3/8 in. is even better for hay crops, as they will pack better at the shorter length.

**Adding Preservatives.** Preservatives for grasses and legumes may be added in various ways. Many silage blowers have a suction attachment that draws molasses out of the barrel. A small pump may be rigged with a long hose and pump the molasses, slightly diluted, into the silo. The barrel can be placed on a platform to let the molasses run on to the forage at the blower. Acid should be pumped so as to keep it from contact with the silage equipment.

Corn-and-cob meal or other ground grains can be added to the green material at the elevator or blower. This can be done by hand or with a hopper. The amounts of preservatives to use are given in Table 6-13.

**Wilting.** To ensure the making of good-quality hay-crop silage without the use of preservatives, the crop may be wilted to reduce the moisture content. The length of time required for sufficient wilting depends on the maturity of the crop, the moisture in the soil underneath the crop, the humidity, and the sunshine. It may vary from less than an hour to several hours. On good drying days most crops will wilt sufficiently in 2 to 3 hours, unless the material is especially heavy and green.

**Sealing the Top Layer of Silage.** One of the greatest losses of nutrients in silage is in the spoilage on the top of the silo. Various means have been used to reduce this loss. The last few loads may be of less valuable material; for example, if they are corn, the ears can be removed. A few loads of very wet material helps to pack it into a tight mass and exclude most of the air. A covering of heavy paper with sawdust or other material blown on top and kept wet will reduce spoilage. A plastic silo cap or sheet that covers the top has proved very successful in sealing horizontal as well as tower silos. It will need material on top to hold it in place. Old tires are often used for this purpose on horizontal silos.

**Table 6-13** Amount of Preservatives Recommended for Use on Various Crops, per Ton of Forage

| Type of preservative | Grasses and cereals (lb.) | Mixed grasses and legumes (lb.) | Alfalfa or clover (lb.) | Soybeans (lb.) |
|---|---|---|---|---|
| Molasses, per ton | 40 | 60 | 80 | 100 |
| 75% phosphoric acid, per ton | 8 | 10 | 14 | 16 |
| Ground grains, per ton | 75 | 100 | 125 | 150 |
| Sodium bisulfite, per ton | 8 | 8 | 8–10 | 8–10 |

**Types of Silo**

Many types of silo and silo building materials are available, most of which will prove satisfactory as far as keeping the silage is concerned. There is considerable variation, however, in the initial and annual costs, durability, and flexibility of the various kinds. Some are well suited to certain types of forage production and feeding systems whereas others are better suited to other systems.

There are basically three types of silos with additional variations in type of material used in the construction and system of unloading. The three basic types of silo are: (1) upright or conventional tower silos, (2) gas-tight tower silos, and (3) horizontal silos.

**Upright Silos.** These may be constructed from wooden staves, monolithic concrete (poured), concrete staves, tile, brick, or metal. Concrete staves and poured concrete are the two most widely used materials. They vary in size from 10 ft. in diameter and 30 ft. high to 30 ft. in diameter and 80 ft. high or more. They may be used to store various types of silage, haylage, or high-moisture corn. They are moderate in cost and have a 20-year or longer life expectancy. A relatively small surface area is exposed to air, hence storage losses are minimal, especially if the exposed surface is covered after filling. Many upright silos are equipped with mechanical unloaders. These unloaders coupled with silage bunk distributors of various types enable dairymen to automate silage feeding completely (unless the power fails or there is a mechanical breakdown). In some feeding systems the silage is delivered to a mixing area where it is blended with other feedstuffs and then delivered to feed bunks as a blended or complete ration.

**Horizontal Silos.** These may be of the bunker or trench variety. A trench is excavated into a hillside. The floor and walls may be earth or may be finished with concrete. The bunker is constructed above ground using a concrete floor and wooden or concrete sides. These are the most economical to construct; however, a large silage surface area is exposed and high spoilage losses can occur if the silage is not well packed and carefully covered. With proper management, silage losses are not appreciably greater than with conventional upright silos. It is more difficult to pack drier materials such as haylage, high-moisture corn, and very dry corn silage tightly enough to eliminate air and avoid high spoilage losses.

Silage may be self-fed from horizontal silos, it may be fed with a tractor and scoop delivered directly to the feed bunk, or it may be loaded into self-unloading wagons for delivery to feed bunks. More recently, many dairymen are loading the silage into mixing wagons where the silage is mixed with other ration ingredients to make a complete or blended ration before delivery to the feed bunk. Silage can be stacked on the ground or on a hard surface such as concrete or blacktop. This can be a satisfactory temporary silo if the silage is well packed and covered securely.

**Gas-Tight Silos.**   These structures are sealed after filling. Fermentation uses up the oxygen inside and replaces it with carbon dioxide; no oxygen is left for spoilage. The unloading mechanism delivers the ensiled material from the bottom, thus providing a continuous flow of materials through the silo. Older material is removed from the bottom and new material is added at the top. On the average, storage losses are the lowest in these types of structure; however, initial and annual costs are the highest of any of the various types of silos.

A comparison of storage losses is presented in Table 6-14. The choice of silo structure used is dependent on economic factors, the forage production and feeding program, and the size of the herd. Initial investment costs for comparable capacity silos are about twice as much for concrete tower as for concrete bunker silos and about three times as much for gas-tight as for concrete tower silos. When annual costs, investment costs, and spoilage losses are considered it appears that gas-tight structures become economically competitive only when filled more than once per year. The costs are prohibitive when they are used to store one-crop silage. They are more economically competitive when used for repeated fillings with haylage or high-moisture corn.

Horizontal silos appear most suited for large herd operations using one-crop forage systems, whereas conventional upright silos are well suited to smaller herds. Thorough examination of the alternatives in terms of cost, storage losses, convenience, and adaptability to the forage production and feeding system should be made before a decision concerning type of silo is reached.

### Size of Silos

The capacity of a silo is dependent not only upon its diameter and·height, but also upon the kind of forage that is put in, especially the moisture content. The weight per cubic foot will vary from 30 to 40 lb. for a horizontal silo and from 45 to 50 lb. for a tall upright silo. Appendix Table F lists the approximate dry-matter capacity of silos of various diameters and heights. Appendix Table F-1 gives the capacities for horizontal silos.

It is important that the diameter of the silo not be more than will permit at least 2 in. of silage to be fed off per day in winter or 3 in. per day in warm weather; otherwise considerable spoilage will result. The diameter of the silo

**Table 6-14** Estimate of Minimum Dry Matter Losses in Forage Stored as Silage at Different Moisture Levels[a] (Beltsville Data)

| Kind of silo | Moisture content of forage as stored | Dry matter losses | | | | | From cutting of crop to feeding |
| | | Surface spoilage[b] | Fermen- tation[c] | Seepage | Total silo losses | Field losses | |
|---|---|---|---|---|---|---|---|
| Conventional tower silos: | Percent | Percent | Percent | Percent | Percent | Percent | Percent |
| | 85 | 3 | 10 | 10 | 23 | 2 | 25 |
| | 80 | 3 | 9 | 7 | 19 | 2 | 21 |
| | 75 | 3 | 8 | 3 | 14 | 2 | 16 |
| | 70 | 4 | 7 | 1 | 12 | 2 | 14 |
| | 65 | 4 | 8 | 0 | 12 | 4 | 16 |
| | 60 | 4 | 9 | 0 | 13 | 6 | 19 |
| Gastight tower silos: | Percent | | | | | | |
| | 85 | 0 | 10 | 10 | 20 | 2 | 22 |
| | 80 | 0 | 9 | 7 | 16 | 2 | 18 |
| | 75 | 0 | 8 | 3 | 11 | 2 | 13 |
| | 70 | 0 | 7 | 1 | 8 | 2 | 10 |
| | 65 | 0 | 6 | 0 | 6 | 4 | 10 |
| | 60 | 0 | 5 | 0 | 5 | 6 | 11 |
| | 50 | 0 | 4 | 0 | 4 | 10 | 14 |
| | 40 | 0 | 4 | 0 | 4 | 13 | 17 |
| Horizontal silos: | Percent | | | | | | |
| | 85 | 6 | 11 | 10 | 27 | 2 | 29 |
| | 80 | 6 | 10 | 7 | 23 | 2 | 25 |
| | 75 | 8 | 9 | 3 | 18 | 2 | 20 |
| | 70 | 10 | 10 | 1 | 21 | 2 | 23 |
| Stack silos: | Percent | | | | | | |
| | 85 | 12 | 11 | 7 | 30 | 2 | 32 |
| | 75 | 16 | 11 | 3 | 30 | 2 | 32 |
| | 70 | 20 | 12 | 1 | 33 | 2 | 35 |

[a] Conservative estimates for careful filling methods and good drainage based on six months of storage. Plastic caps or other good covers will reduce top spoilage. Poor compacting and sealing of the silage and excessive rainfall or melting snow on uncovered trenches and stacks will increase losses.

[b] Includes side and end spoilage in trenches and stacks.

[c] Allowance made for some heating and flake mold at the lower moisture levels.

would therefore depend upon the height of the silo, and the number of cows to be fed from it, and the length of the feeding period.

**Precautions in Feeding Silage.** When cows are first fed silage after not receiving it for a while, some of the flavor may be carried over in the milk. Silage should not be fed just before or during milking. Silage fed after milking will seldom produce any off flavor in the milk.

Spoiled silage should not be fed. Although cows are not as sensitive to moldy and other types of spoiled feed as are some other animals, there is danger of causing digestive disturbances. Feeding silage before the fermentation process is complete will sometimes cause cows to go off feed. Frozen silage should be thawed before being fed. Good silage is one of our best milk-producing feeds. Use it wisely.

### Summary of Silage and Haylage

The use of silage and/or haylage as the major forage for dairy cattle has increased rapidly in recent years. The major reasons for this are reduction of harvest losses, ease of mechanizing the handling of the forage from field to cow, and ease of providing a uniform supply of high-quality forage nutrients on a year-round basis.

## BUDGETING FORAGE
## NEEDS AND STORAGE CAPACITY

Errors is estimating forage needs and storage capacity needed are quite common and can be very costly, as it is usually more expensive to build additional storage facilities, especially additional silos, than it would have been to build larger ones initially. Unanticipated shortages of forage, which can necessitate costly purchases and/or adversely affect production because of diet changes, can severely affect profitability. Proper calculation of forage storage needs and facilities can usually avoid these problems. The factors that determine the forage storage capacity needed are the following.

1. Size of cows.
2. Dry matter intake from forage.
3. Level of production.
4. Storage losses and wastage.
5. Number of cows.
6. Number of replacements.

It is best to budget forage storage needs on a dry matter basis as it is relatively easy to convert from dry matter to the actual moisture level of the stored material.

## Size of Cows

Larger cows consume more feed than smaller cows. The ratio of feed consumption (dry-matter basis) to total body weight, however, is relatively constant. Breed-average-producing cows consume an average of about 3.0% of their body weight in dry matter per day for the year. It will be higher than this in early lactation and lower in late lactation and during the dry period. From 60 to 70% of this, or 1.8 to 2.1% of their body weight, can be from forage.

## Dry Matter Intake from Forage

This will vary from 30 to 70% of the total DM intake depending on the type of feeding program as well as the quality of the forage. If one assumes an average of two-thirds of the total intake from forage and one-third from grain, then the dry matter intake from forage is about 2.0% of body weight. If more than one forage is fed, the amount of each should be proportioned on a dry matter basis.

## Level of Production

Higher-producing cows consume more feed per unit of body weight than do lower-producing cows. In budgeting feed needs one should increase or decrease total daily dry matter needs by 0.1% and forage daily dry matter needs by 0.06 to 0.07% for each 1,000 lb. increase or decrease in annual milk production. Thus a herd producing at a level of 3,000 lb. over breed average should be budgeted for total feed needs at 3.3% DM and forage needs at 2.0 to 2.3% DM of body weight daily.

## Storage Losses and Wastage

It is wise to allow 10 to 20% for losses and wastage. This includes spoilage, coarse material that is refused by the cattle, and spillage at the manger and during handling. This wastage figure can vary considerable for forages, depending on the type of forage and the storage conditions.

## Number of Cows

When the other calculations have been completed, multiply the per-cow figure times the number of cows in the herd.

## Number of Replacements

If the heifers are on a forage-feeding program similar to the cows, each heifer will need to be allotted about one-half as much forage dry matter as each cow in the herd.

## Calculating Silo Storage Needs

An example of budgeting silage and silo storage needs is presented in Table 6-15. To estimate the size of silo(s) needed, refer to Appendix Tables F and F-1. The figures listed in the tables are for depth of settled silage. It is impossible to fill tower silos to their maximum capacity; therefore, it is wise to allow an extra 10 ft. of silo height to allow for settling, spoilage, and room for the unloader. In our example, two 30 by 70 tower silos with a settled silage depth of 60 ft. would have a capacity of 357 tons each or a total of 714 tons of silage dry matter.

It is often wise to build somewhat larger than needed. If, in our example, production were to increase 2,000 lb. per cow, it is likely that an additional 0.8 to 0.9 kg., or 1.8 to 2.0 lb., of silage dry matter including wastage per cow would be required daily. This would increase the annual silage dry matter needed by the herd about 35 to 40 tons or an additional 100 tons of 35 to 40% DM silage. If the silage were of excellent quality and the consumption of silage dry matter increased from 1.66% to 1.9% of body weight this would create the need for about 1.6 kg., or 3.5 lb., of additional silage dry matter per cow per day. This would amount to an increased annual herd need of 60 to 65 tons dry matter from silage or an additional 150 to 170 tons of silage. This could easily have been accommodated if we had originally constructed two 30 by 74 silos instead of two 30 by 70 silos. The additional 4 ft. in height would have increased our settled silage storage capacity by 68 tons of dry matter at very little additional cost.

Considering all factors such as wastage, the part of the silo that cannot be filled with settled silage, variations in forage quality, and so on, it is wise to budget forage storage dry matter capacity at 2.4 to 2.5% of body weight if one plans on cow intake of 2%. This allows an overrun of 20 to 25% of calculated consumption. In our example we calculated on the basis of 10 kg., or 21.9 lb., of silage dry matter intake for 600 kg., or 1,320 lb., cows, or 1.66% of body weight. Adding 20 to 25% to this for spoilage and wastage would total 2.0 to 2.1% of body weight per day. This would equal about 500 tons of silo capacity for the 100 cows plus 200 tons for the 80 heifers and a total of 700 tons of dry matter silo capacity needed. Referring again to Appendix Table F, we can determine that two 30-by-74 silos with a settled silage depth of 64 ft. would handle our needs (total DM capacity of settled silage = 782 tons).

**Table 6-15** Budgeting Silage and Silo Storage Needs

| | Kilograms | Pounds |
|---|---|---|
| Size of cows = | 600 | 1,320 |
| Level of production = breed average | | |
| Daily dry matter intake from forage = 2% of body weight | 12.0 | 26.4 |
| Daily dry matter intake from hay = 0.34% of body weight | 2.0 | 4.5 |
| Wastage allowance = +20% | | |
| Number of cows = 100 | | |
| Number of replacements = 80 | | |
| Pasture days = 0 | | |
| Silage storage needs = | | |
| Total daily DM intake per cow (3% of body weight) = | 18.0 | 39.6 |
| Daily DM intake per cow from concen. (1% of body weight) = | 6.0 | 13.2 |
| Daily DM intake per cow from forage (2% of body weight) = | 12.0 | 26.4 |
| Daily DM intake per cow from hay (.34% of body weight) = | 2.0 | 4.5 |
| Daily DM intake per cow from silage (1.66% of body weight) = | 10.0 | 21.9 |
| Silage wastage allowance (+20%) = | 2.0 | 4.4 |
| Total daily silage DM needed per cow per day = | 12.0 | 26.3 |
| Total annual silage DM needed per cow (365 days) = | 4,380 | 9,600 |
| Total annual silage DM needed for 100 cows = | 438,000 | 960,000 |
| Total annual silage DM needed for 80 heifers | | |
| (.5 × 80 × 4380) = kg. | 175,200 | |
| (.5 × 80 × 9600) = lb. | | 384,000 |
| **Total annual silage DM needed (cows and heifers)** | **613,200** | **1,344,000** |
| Total tons silage DM needed (1,344,000 ÷ 2000) | | 672 tons |
| Total tons silage needed (35% DM) | | |
| $\dfrac{\text{Tons of silage DM needed (672)}}{\text{DM \% of silage (35)}} \times 100 =$ | | 1,920 tons |

153

## BUDGETING LAND NEEDS
## FOR FORAGE PRODUCTION

Based on the preceeding discussion, one can continue along a similar line of reasoning and budget the use of land resources to provide forage needs.

If the total forage dry matter needs were 2% of body weight per day and one needed to harvest 2.5% of body weight per day to allow for storage losses and wastage, then the total forage dry matter needed in our example herd (Table 6-15) would be about 850 tons of forage dry matter (2.5% of 1,320 lb. = 33.0 lb./day for each of 100 cows and 16.5 lb./day for each of 80 heifers). Of this, 83.0% was from silage and 17.0% from hay. One would need to produce 705 tons of silage dry matter (850 × .83) and 145 tons of hay dry matter (850 × .17). If average yields were 17 tons per acre of 35% dry matter silage and 4 tons of 85% dry matter hay, one would need to plant 118 acres of corn (705 tons ÷ 5.95 tons of DM/acre) and 43 acres of hay (145 tons ÷ 3.4 tons of DM/acre).

This type of budgeting procedure can be useful in determining land use or making maximum use of land resources. In our example it was determined that 161 acres, or about 1.6 acres per cow, were needed to provide an adequate supply of forage for the cows and heifers. What would be the effect of a change to a one-half corn silage and one-half hay program on the amount of this type of land needed for the herd? This can be determined by dividing the 850 tons of forage dry matter by two. The needs would be for 425 tons of corn silage dry matter, or 71 acres of corn (425 ÷ 5.95 TDN/acre) and 425 tons of hay dry matter or 125 acres of hay (425 ÷ 3.4 TDN/acre). This totals 196 acres, or about 2 acres per cow. This example demonstrates the importance of growing forage crops that produce maximum amounts of nutrients per acre when land resources are limited.

## SUMMARY

An ample supply of high-quality forages is a major factor in determining the profitability of most dairy farms. If forage quality is high, cows will consume more and obtain more nutrients per unit, and the digestibility is higher, thus lessening the need for more expensive concentrate feeds. Forages are fed primarily for their energy and protein content, as only small amounts of vitamins and minerals are needed and these can be provided in the concentrate part of the ration. Many suitable alternatives are available concerning crop(s) grown for forage, tillage practices, harvesting, storage, and feeding systems. There is no one system that is best for all dairy farms. The best one for any one farm is the one that utilizes the resources of that farm to "get the job done''; that is, encourages high milk production and low feed cost per unit of milk produced. Identifying this system for any farm will take careful evaluation of the many alternatives in terms of economic and performance effects. The reward for wise management decisions in this area will be a high income over feed cost per unit of milk production.


## REFERENCES FOR FURTHER STUDY

Beene, E. J., The Why and How of Silage Making, *Hoard's Dairyman,* **116:**14:824 (1971).

Browning, C. B., Four-Year Results, Concrete Stave and Gas-Tight Silos. *Silo News,* National Silo Assoc. Winter 1967.

Coppock, C. E., and Stone, J. B., Corn Silage in the Ration of Dairy Cattle. A Review, *Cornell Misc. Bul.* **89**(1968).

Golden, W. I., Economic Considerations of Producing and Storing Forage Crops, *S.C. Ext. Mimeo,* 1971.

Gordon, C. H., et al., Nutrient Looses, Quality and Feeding Value of Wilted and Direct-Cut Orchardgrass Stored in Bunker and Tower Silos. *J. Dairy Sci.,* **42:**1703 (1959).

Gordon, C. H., et al., Preservation and Feeding Value of Alfalfa Stored as Hay, Haylage, and Direct-Cut Silage. *Journal Dairy Sci.,* **44:**1299 (1961).

Hodgson, R. E., et al., Comparative Efficiency of Siloing, Barn Curing and Field Curing Forage Crops, USDA *Burea of Dairy Industry Mimeo* **43**(1946) and *Bureau of Dairy Industry Information,* 117 (1951).

Hoglund, C. R., Comparison of Storage Losses and Feeding Value of Silages. *Silo News,* National Silo Assoc. Summer, 1964.

McCullough, M. E., *Optimum Feeding of Dairy Cattle for Meat and Milk,* 2nd edition, University of Georgia Press, Athens, Ga. 1973.

McCullough, M. E., The Old and New of Silage. *Silo News.* National Silo Assoc. Winter. 1967.

*Proceedings of The Second International Silage Research Conference.* National Silo Assoc. 1975.

# 7

# Concentrates for Dairy Cattle

**7**

Some milk can be produced without feeding concentrates, but the level of production on roughages alone is limited. The terms concentrates, grain, grain mixture, and dairy feed are used interchangeably, meaning grains, seeds, and their by-product feeds. They are low in fiber and high in energy compared to forages. Also, as a rule, concentrates are higher in phosphorus and lower in calcium than forages.

Dairy cows of today are developed with a milk-producing ability so great that they cannot consume and digest enough nutrients from roughages to reach this ability.

Concentrates are fed to make it possible for the cow to use more feed nutrients and therefore work more nearly up to her capacity of milk production. With an increased plane of nutrition, roughage consumption is reduced 0.25 to 0.8 lb. for each additional pound of grain consumed.* This is apparently a space factor.

The economy of various levels of grain feeding is dependent upon many factors, including the relative cost of roughage and grain, the selling price of milk, the potential of the cow, and costs other than feed costs.

## GRAINS

### Barley

Barley is an excellent feed for dairy cattle. It can be substituted for corn, pound for pound, with equally good results. It contains a little more protein than corn but is slightly lower in energy. It is used extensively in countries and localities where corn is not available. It should be ground to medium fineness for dairy cattle. If ground too fine it may become pasty in the mouth.

### Buckwheat

The whole grain of buckwheat is not so desirable as its middlings because of the large amount of crude fiber or hulls that it contains. These, however, give bulk to a ration and may be used when other grains in the ration are heavy. Buckwheat is not a very palatable feed, nor is it equal to corn as a dairy feed. Dairymen believe that buckwheat produces a hard, white butterfat.

* Emery, *Feed Age,* **12**:8:107 (1962).

## Corn

Corn is grown on most dairy farms and should usually be included in the dairy ration. Not only is it highly palatable; it also supplies a large amount of energy economically. It is low in protein and minerals; therefore some high-protein feed must be used in order to supply these deficiencies. Yellow corn has considerable carotene, the precursor of vitamin A, but white corn is low in carotene. It is usual to feed corn ground as a meal, although calves prefer cracked corn.

Sometimes whole ears are ground, making corn-and-cob meal. This is used in dairy rations in place of ground corn and may be valuable in adding fiber when other high-fiber feeds are not available. The cob is high in pentosans and therefore can be utilized, to a certain extent, by the digestive system of the dairy cow. It has been found that when fed to steers, about 2.5 lb. of cob were equal to 1 lb. of ground corn.

When corn is purchased on the market it should be purchased on federal grade, which depends upon the amount of water and damaged grains that it contains. According to the grade, the moisture in corn must not exceed 14% in No. 1; 15.5% in No. 2; 17.5% in No. 3; 20% in No. 4; and 23% in No. 5. Corn with more moisture than No. 2 will not keep well in storage.

High-moisture corn containing 25 to 35% moisture can be stored in upright silos. It can be stored as ground or whole shelled corn or as ground ear corn (corn-and-cob meal). Its nutritive value is based on the dry matter content.

## Oats

Oats are an excellent feed for dairy cattle and when not too high in price should be used in the dairy ration. They are considerably higher in food value than wheat bran. They are bulky, palatable, and higher in protein and mineral matter than corn; if they are homegrown, the straw comes in handy for bedding. Oats vary greatly in composition and should be purchased on grade and weight. Oats should be crimped, crushed, or ground to a medium fineness for dairy cows. This is not necessary for calves up to 8 months of age, as they seem to prefer the whole grain.

## Rye

Ground rye may be used in place of corn in a dairy ration. It is nearly equal to corn in its food value, but it is not very palatable. There may be danger in feeding it when it is infected with ergot. In large amounts this may cause abortion. When fed in large quantities it tends to produce butterfat with a hard body. It can be fed safely up to one-fifth of the grain ration when free from ergot.

## Wheat

When the price will permit, wheat may be substituted for corn in a dairy ration. It is slightly higher than corn in feeding value and is quite palatable. Because of the small size and hardness of the kernels, the wheat should be ground before being fed. Wheat, when fed alone, forms a paste in the cow's mouth. In a great many sections the cost prohibits its use in the dairy ration. Wheat should not make up more than one-third of the grain ration for heavy-producing cows.

## Milo or Sorghum Grain

Milo or sorghum grain is similar in composition to corn, being slightly higher in protein and slightly lower in energy. It can be used interchangeably with corn in dairy grain mixtures. The grains should be ground for dairy cows. Its use is greatest in the Plain States area.

## LEGUMINOUS SEEDS

### Soybeans

The soybean is one of the important crops grown in this country. It is extremely high in good-quality protein and fat and rather low in fiber. It is also higher in total digestible nutrients than any of the cereals. It is not high in minerals or vitamins, and is not too palatable. It should be ground for dairy cows, an operation that is often difficult on account of its sticky nature. However, when mixed with other grains it will grind easily. It is likely to turn rancid after grinding if stored too long, especially during warm weather.

### Field Peas

Field peas sometimes are grown in the cooler parts of the United States as feed, but because of their low yields are not extensively used. They are medium high in protein of only fair quality. They are low in calcium and vitamins, but are quite palatable and can be used satisfactorily.

## Cowpeas

Cowpeas are not extensively used because the seed ripens unevenly. They can, however, be fed satisfactorily to dairy cattle. In composition and characteristics they are very similar to field peas.

## BY-PRODUCTS

Many industries have by-products that can be used as cattle feed. Some of these by-products are high in quality and food value; others have but limited value.

### Milling Industry

**Buckwheat Middlings.** Buckwheat middlings is a by-product of the buckwheat-flour industry. It is fairly high in protein and makes a very good dairy feed. It should not be mixed with the hulls, which have little value as a dairy feed.

**Wheat Bran.** Wheat bran is one of the best of the milling products for feeding dairy cattle. It is palatable, bulky, and mildly laxative. It is the highest of the common feeds in phosphorus content. It is desirable in any ration for dairy cows when the price is competitive. It is of special value to cows just before and after calving. It is also a good feed for calves and heifers. Standard bran contains some of the finely ground screenings.

**Wheat Middlings.** In the manufacture of flour the finer particles of bran and some of the flour and germ are collected and sold as standard middlings, red dog, or shorts. They are higher in both protein and energy than wheat bran. They are not so bulky or so palatable as bran. They are rich in phosphorus but low in calcium and the vitamins. They are quite heavy and have a tendency to form into a sticky mass when wet. Middlings can constitute a higher percentage of the grain mixture if the feed is pelleted, also when it is used in a complete ration.

### Oil Industry

**Cottonseed Meal.** Cottonseed meal is the residue from the extraction of oil from the cottonseed, and is sometimes sold in cake form, although the cake is usually ground and the product sold as meal. Several grades of cottonseed meal are on the market. Where transportation is an item of cost, the highest grade is usually the cheapest and best to purchase. Formerly this feed was the most commonly used high-protein feed. In the South it usually furnishes protein in the cheapest form of any of the concentrates. It is not too palatable and is low in calcium and vitamin A. It contains a poison known as gossypol, which makes it injurious when fed in large amounts to young calves and other animals. With dairy cattle, however, cottonseed injury seems to be caused by a lack of calcium and vitamin A. When good alfalfa hay or plenty of good quality hay of other kinds is fed, cottonseed injury will not occur.

Sometimes the hulls are ground and mixed with the meal and sold as cottonseed feed. This is not nearly so nutritious as the meal.

**Soybean-Oil Meal.** The production of soybeans has been increasing greatly, so much so that soybean-oil meal is now the most important high-protein supplement. Soybean-oil meal is the residue after the oil has been extracted from the bean. It is rich in protein of high quality, palatable, and slightly laxative. Soybean-oil meal made by the expeller or hydraulic process is preferred to that made by the solvent process, for it is considerably higher in fat. This is especially true if other feeds in the ration are low in fat. Soybean-oil meal is being fed in large amounts to dairy cattle and is popular with the feeders.

**Soybran Flakes.\*** Owing to a demand for a soybean-oil meal containing up to 50% protein for high-energy rations, some of the soybean hulls that normally went back with the meal have been diverted into mill by-product feeds. The soybean hulls, commercially called soybran flakes, are laxative and cannot be fed in too large a quantity. They have been found in feeding trials to equal oats or citrus pulp for milk production, and to equal ear corn for growth.

**Peanut-Oil Meal.** Peanut-oil meal, a by-product of the manufacture of peanut oil, is well liked as a dairy feed in sections of the country where it is available. It varies considerably in composition, but first-grade peanut-oil meal is very high in protein, palatable, and slightly laxative. It tends to become rancid in warm weather if stored very long.

**Coconut-Oil Meal.** Coconut-oil meal (copra-oil meal) is the residue from the manufacture of oil from the coconut. Coconut-oil meal is higher in feeding value than wheat bran, but not so high as many of the other oil meals. It also has a tendency to turn rancid in warm weather.

* *J. of Dairy Sci.*, **42:** 185(1959); *Proc. of the Cornell Nutrition Conf.*, 1960; and *Feedstuffs*, **34:** 28:16(1962).

**Linseed-Oil Meal.** Linseed-oil meal is the residue left after extracting linseed oil from flaxseed. It is more palatable than cottonseed meal and is laxative. When fed in large amounts it tends to produce soft butter. Linseed-oil meal has added value in fitting cattle for shows or sales. It contains a substance, called mucin, which tends to put more bloom or condition in the hide and hair of the animals, as is desired in fitting them.

**Old- and New-Process Meals.** The oil meals are processed by two methods. The old process consists of crushing the seed, heating it to a high temperature, and removing the oil by high pressure. This includes the expeller and the hydraulic processes. Some fat remains in the meal; consequently, the energy or TDN value is higher than in the solvent-process meal.

The new-process meal is produced by treating the bean in a way similar to the old process, but extracting the oil with a fat solvent. The solvent-process meal is lower in fat and TDN. It is usually higher in protein.

Because of its greater efficiency and economy of operation, more of the meals are now produced by the solvent process.

### Brewing and Distilling Industry

**Brewer's Grains and Malt Sprouts.** In the manufacture of beer, barley is allowed to germinate, during which process some of the starch is converted to sugar. The roots and sprouts are separated from the grain and dried. These are sold as malt sprouts. The grain is further fermented to change the starch more completely to sugar. The residue is wet brewer's grains. These are dried and sold as dried brewer's grains. When the brewery is nearby, dairymen often feed the wet brewer's grains.

**Distillers' Grains.** In the manufacture of distilled liquors and alcohol from grain, the coarser particles are strained out of the watery residue (stillage), forming wet distillers' grains, which, when dried, are called distillers' dried grains without solubles, or light grains. The water-soluble material that passes through the strainer, known as thin stillage or thin slop, is sometimes condensed and dried, forming dried distillers' solubles. The condensed solubles are sometimes added to the wet distillers' grains, and this combination is then dried and called dried distillers' grains with solubles, or dark grains. Both the dark grains and light grains are medium high in protein and high in fat. They are highly digestible, although fairly bulky, and are equal to corn in energy. The chief difference between the light and dark grains is that the latter is much richer in the B-complex vitamins, which, however, is of little importance in dairy feeds. The rye distillers' dried grains are much lower

SIMPLIFIED FLOW DIAGRAM

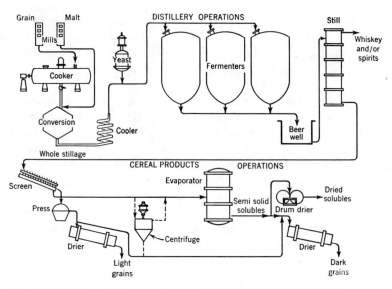

**Fig. 7-1.** Diagram of the manufacture of distiller's feeds (Feedstuffs).

in protein and fat, higher in fiber, and less digestible than the corn distillers' dried grains, and hence not nearly so good a feed.

Distilleries are now using other sources of starch and sugar, such as molasses, grain sorghum, wheat, and potatoes. Distillers' dried grains from these sources are less valuable than from corn and should be purchased on the basis of their guaranteed analysis.

**Starch Industry**

**Corn-Gluten Meal and Corn-Gluten Feed.** Corn-gluten meal and corn-gluten feed are by-products in the manufacture of starch and glucose. The gluten of corn, separated from the starch, is dried and sold as corn-gluten meal; but when mixed with the corn bran it is known as corn-gluten feed. The meal is therefore richer in protein and energy than the feed, but not so rich in calcium or vitamin A. Both are satisfactory when they can be purchased at a price that makes them economical enough to include in the ration. The meal is heavy and should be mixed with some more bulky feed. Neither is so palatable as some of the other feeds. As their protein content varies considerably, they should always be purchased on guaranteed analysis.

## Sugar Industry

**Dried Beet Pulp.** Beet pulp is the residue left after the sugar has been extracted from the sugar beet. It is high in carbohydrates but low in protein and fat. When silage or other succulent feeds are not available, dried beet pulp soaked with about three times its weight in water for about 12 hours before feeding makes an excellent substitute. When the cows are given free access to water, however, equally good results are obtained if the beet pulp is fed dry. It is highly palatable and has a very good physiological effect upon the cow. Dried beet pulp is especially valuable in feeding cows for heavy production.

Beet pulp is used extensively for fitting animals for sales and shows and to feed while on the show circuit. It provides a succulent feed to use under these conditions.

**Molasses.** Molasses is a by-product of the manufacture of sugar from both beets and cane. A crude syrup is produced that is often used in feeding dairy cows. The molasses from cane is known as blackstrap. Since these products are rather low in protein, they are essentially energy feeds. Molasses is very laxative and should not be fed in a quantity greater than 5 or 6 lb. daily for each animal. The customary way of feeding it on the farm is to pour it over silage or hay. It is sometimes fed in liquid form in vats. It can be added to dusty feeds to improve their physical property. Molasses should be limited to 10 to 15% of the feed mixture. Table 7-1 lists the absorption power of certain feeds. These may be used to absorb the molasses before being included in a mixed feed. Molasses is often mixed with feeds in order to give them palatability, which is one of its greatest values. It is commonly found in many of the commercial mixed feeds. It is used as a preservative in the ensiling of grasses and legumes.

Corn molasses is the residue from corn-sugar manufacture. Wood waste has been hydrolyzed with acids to produce wood molasses. It has been fed to dairy cows and other animals experimentally, and within certain limitations may have a feed-replacement value.

**Liquid Supplements.** Liquid supplements are usually commercial mixtures that are delivered to the farm in tank trucks and pumped into feeder tanks from which the cattle eat. A wooden paddle wheel or slatted frame in the tank allows the cattle to eat by licking the material from the frame or wheel. The material may also be added to silage, chopped hay, or to complete rations.

The usual ingredients in liquid supplements are molasses, urea or other

**Table 7-1** Molasses Absorption Power of Certain Feeds[a]

| Ingredient | % of Absorption |
|---|---|
| Dried distillers' grain | 5 |
| Extraction-process soybean meal | 5 |
| Corn gluten feed | 7 |
| New-process linseed meal | 7 |
| Old-process linseed meal | 9 |
| Dried brewer's grains | 9 |
| Expeller-type soybean meal | 10 |
| Wheat middlings | 13–19 |
| Bran | 15 |
| Cottonseed meal | 15 |
| Cottonseed hulls | 15–19 |
| Corn meal | 15 |
| Hominy | 22 |
| Malt sprouts | 25–27 |
| Coconut-oil meal | 33 |
| Dried alfalfa meal | 30–40 |
| Oat-mill feed | 40 |
| Screenings | 30–40 |
| Ground corn cobs (depending on grind) | 40–65 |
| Flax-plant by-product | 45 |

[a] *Feedstuffs* **34**:13:34 (1962).

nonprotein nitrogen-containing ingredients, phosphoric acid, minerals, and vitamins. Other ingredients that will go into solution or remain in suspension may be used. The amount of urea, phosphoric acid, and other unpalatable ingredients, as well as the distance from the liquid feed supplement to the water supply, limit the amount that a cow will consume. There are large variations in the amount that individual cows will consume; therefore cows may not necessarily balance their nutritional needs with liquid supplement fed free choice in addition to other feeds.

The protein equivalent varies with the composition of ingredients, it is usually about 32%. The cost per unit of protein equivalent generally is higher than when urea is purchased in dry form, but the convenience may justify it. It provides an alternative to adding urea to silage.

This feed can be fed to cattle above six months old, the age at which they ruminate and have rumen organism activity sufficient to utilize nonprotein nitrogen.

## Hominy Grits Industry

**Hominy Feed.** In the manufacture of hominy grits from corn a by-product known as hominy feed is obtained. It has practically the same protein and carbohydrate content as corn, and may be used as a substitute for it. Practically the same things can be said of it as of corn for the feeding of dairy cows, although it is a little more bulky because it contains considerable amount of corn bran. When it can be purchased as cheaply as corn, it may well be used in the ration.

## Oatmeal Industry

**Oat-Mill Feed.** Oat-mill feed is a by-product in the manufacture of oatmeal. The oat grain, when removed from the hull, leaves a tip attached to the hull. This, together with the hull and imperfect and broken grains, is ground into a feed that is high in fiber but low in protein and energy. It is often used as a substitute for roughage, but its analysis is not much higher than that of oat straw. When fed for a concentrate it must be supplemented with some high-protein feed. Sometimes it is mixed with molasses to make it more palatable.

## Citrus-Canning Industry

**Citrus Pulp.** Citrus pulp is a by-product of citrus-canning factories, where fruit juices, fruits, and other products are canned. It consists of the peels, inside portion, and seeds, and sometimes the entire cull fruit. This is dried, ground, and sold as dried citrus pulp. It is sometimes used as a substitute for beet pulp, which it resembles in composition, although it is slightly higher in total digestible nutrients. Cows may require a short time to learn to eat it. It is high in calcium but low in phosphorus. It can be fed dry or soaked with water, as can the beet pulp. Dairymen living near the canneries sometimes get it fresh and either feed it within a short time or ensile it; citrus pulp is a standard dairy feed ingredient in Florida and other citrus canning areas.

The liquid portion of the citrus residue is often separated from the other portion and concentrated. It is called citrus molasses and has about the same composition and feeding value as blackstrap or beet molasses.

## Packing Plant and Fishing Industries

**Tankage.** Tankage is a by-product of the packing plant. It is a high-protein feed of animal origin and has been used as a protein supplement for dairy cattle. It is high in protein of an excellent quality and in calcium and phosphorus. It is not particularly palatable to dairy cattle, but they soon learn to eat it if it is kept fresh. When fed in limited amounts it does not seem to impart undesirable flavors to the milk.

**Blood Meal and Flour.**   Blood collected at the packing plant is dried and sold as blood meal. It is a very high-protein feed, containing as much as 80% protein, but it is low in calcium and phosphorus. It is not very palatable. A special process in which the blood is dried at a somewhat lower temperature has been used in the manufacture of a product known as a soluble blood flour, which makes an excellent feed for calves. It is more palatable and more digestible than the ordinary blood meal. It seems to have value in reducing scours in young calves.

**Fish Meal.**   Fish meal is made either from the entire fish, as is the menhaden fish meal produced in the eastern United States, or from damaged fish and fish heads, as is done with cod, sardine, salmon, and other fish of various regions. Fish meal is a high-protein feed and is rich in calcium and phosphorus. The protein is of high quality. When vacuum dried it seems to be more digestible and contains more vitamins A and D. It can be used for dairy cattle satisfactorily, although it should not be fed in excessive amounts as it may give the milk a fish flavor. Many cows do not like it but usually will eat it mixed with other feeds. It makes a good protein supplement for calves because of the high quality of its protein. It is usually too costly to be used in dairy feeds.

**Animal Fats.**   With the surplus of fats of all kinds on the market, some animal fats have been used in feeds. They help in building the energy in high-energy feeds. Some animal fat is used in calf feeds, little in dairy feeds. To be economical, its cost should not be greater per pound than 2½ times that of corn.

## PROTEIN SOURCES

The natural high-protein concentrates are largely the oil meals, distillery by-products, and gluten feed and meal.

There are some differences in their productive value for milk production. They are all good feeds, and under most feeding practices little or no differences may be noticed.

### Synthetic Feeds

**Urea.** Synthetic urea has been manufactured and sold as a nitrogenous fertilizer for some time. More recently, as a result of research at many experiment stations, large quantities of it are used in ruminant feeds. It does not contain protein but has a high nitrogen content.

The urea on the market as a feed ingredient is a white, crystaline product that is mixed with other materials to prevent its caking. It is standardized to contain 42 to 45% nitrogen, which calculates to 262.5 to 281.25% protein equivalent (N% × 6.25).

It is currently recommended that urea feeding be limited to no more than 200 g. (0.2 kg.) per day, or that nonprotein nitrogen replace no more than 20 to 25% of the daily protein requirement.

Urea should not be included in the ration of young calves while they are still functioning as simple-stomach animals.

The ruminant secures benefit from urea through the action of rumen bacteria. The bacteria combine the urea with carbohydrate feeds and convert these into protein in their cells. As the bacteria cells are digested further on in the digestive tract, the protein becomes available for the animal.

Urea does not supply energy for the animal as do protein supplements. To determine the relative economy, consider that 1 lb. of urea plus 6 lb. of corn or other grains should not cost more than 7 lb. of soybean-oil meal.

It is recommended that urea not be mixed on the farm, but by efficient mixing equipment with small amounts of ingredients well controlled. Too large an amount in a mixture is unpalatable, and large amounts are dangerous to the animals.

## COMMERCIAL FEED INDUSTRY

The commercial feed manufacturing industry supplies more than two-fifths of the concentrate feeds fed to dairy cattle as ready mixed feeds. In addition, some of the ingredients that dairymen purchase and mix on the farm or that are used in custom mixing are purchased from this source.

As in the case of other industries, mechanization and automation are being used in handling and feeding operations. Bulk delivery to the farm has largely replaced the use of the feed bag.

This industry is composed of both cooperatives and private corporations. Each have well-trained nutritionists to supervise their laboratories and to formulate the feeds.

Most commercial feeds are made of high-quality ingredients milled and mixed intelligently. Some competitively priced feeds include low-quality ingredients and are low in energy value. Most states have feed laws requiring the labeling of each bag of feed with a tag, or, if bulk feed, that the tag accompany it; this tag guarantees a minimum percentage of protein and fat, the maximum percentage of fiber, and a list of the ingredients the feed contains. The amount of each ingredient need not be listed, but some manufacturers, especially farmer cooperatives, do list the amount of each ingredient. Such feeds are called open-formula feeds. This information is helpful in calculating digestible protein and energy in the feed. The open-formula system does not lend itself so readily to quick changes in feed formulas to take advantage of price changes.

The decision as to whether to mix the feed on the farm or to buy ready-mixed feeds must depend upon several factors. The quantity of farm grains grown on the farm, the relative cost of buying ingredients, the labor available for grinding and mixing, the cost of equipment, and the volume of feed used are all to be considered. (See Table 7-2.)

## High-Protein Concentrates

When ready-mixed feed is not used, some high-protein ingredient is needed to mix with farm grains. This may be any of the oil meals, other high-protein ingredients, or a high-protein mixture. High-protein mixtures are made as supplements containing 32, 40, 50, and even 65% protein.

**Table 7-2** Feeds and Their Percentages Used in Dairy Feeds, 1975[a]

| Ingredient | Percentage |
|---|---|
| Corn | 32.3 |
| Oats | 9.5 |
| Barley | 3.4 |
| Sorghum | 2.2 |
| Soybean meal | 3.1 |
| Cottonseed meal | 0.7 |
| Wheat bran and middlings | 0.7 |
| Molasses | 1.1 |
| Alfalfa meal or pellets | 0.5 |
| Minerals | 0.8 |
| Miscellaneous | 3.4 |
| Commercial mixed feeds | 42.3 |

[a] USDA Statistical Reporting Service, Crop Reporting Board, *Dairy Production,* November 1975.

## Custom Mixers

Feed dealers and cooperatives often offer farmers the service of grinding the farmer's grains and mixing these with ingredients purchased from the dealer. To carry this service to the farm, there have been developed mobile feed mills mounted on heavy trucks that go to the farm to grind and mix feed.

The method used in securing or processing feeds must be based on the productive value of the feeds and on the economy of their use.

## REFERENCES FOR FURTHER STUDY

Church, D. C., and Pond, W. G., *Basic Animal Nutrition and Feeding,* D. C. Church, Corvallis, Ore., 1975.

Crampton, E. W., and Harris, L. E., *Applied Animal Nutrition,* 2nd ed., W. H. Freeman & Co. San Francisco, 1969.

Cullison, *Feeds and Feeding,* 1975.

Helmar, L. G., and Bartley, E. E., "Progress in the Utilization of Urea as a Protein Replacer for Ruminants." A Review. J. Dairy Sci., **52**:25 (1971).

Larson, H., "High Moisture Corn—And What You Should Know About It," *Silo News,* National Silo Assoc., Fall 1972.

Moe, P. W., et al., "Physical Form and Energy Value of Corn Grain," J. Dairy Sci., **56**:1298 (1973).

Morrison, F. B., *Feeds and Feeding,* Morrison Publishing Co. Ithaca, N. Y., 1959.

National Research Council, *Composition of Concentrate By-Products Feeding Stuffs,* Pub. 449 (1956) and *Composition of Cereal Grains and Forages,* Pub. 585 (1958).

National Research Council, *Nutrient Requirements of Dairy Cattle,* revised ed., Washington, D. C., 1971.

# Formulating Rations for Dairy Cattle

**8**

A ration is the feed allowance for one animal for one day; a balanced ration is the feed allowance for one animal for one day that meets that animal's nutrient requirements. The objective of the feeding program for the dairy herd is to provide a ration that will encourage optimum economical production of milk of acceptable composition while being conducive to the health of the cows. Achieving this objective is one of the greatest challenges in dairy herd management. There is no one essential feed or best feeding program for all dairymen, but rather there are many combinations of forages, concentrates, and feeding systems that can result in high milk production, reasonable feed costs, and good health. The challenge to the dairyman is to design a feeding program that makes optimum use of his resources to get this job done for his herd.

Previous chapters have discussed the anatomy and physiology of digestion, nutrients and their use in the animal body, nutrient requirements, and forages and concentrates for dairy cattle. This chapter will be devoted to using this information to design practical feeding programs for the milking herd and for dry cows.

## DESIGNING THE FEEDING PROGRAM FOR THE MILKING HERD

The initial procedure in designing the feeding program for any individual herd should be to determine the most economical sources of nutrients for that herd. In most cases this involves evaluating the land resources to determine which crop or combination of crops to grow and harvest to produce an adequate supply of high-quality forage at minimum cost per unit of nutrient (see Chapter 6). In some situations, when dairying is concentrated on very limited land, this step may involve evaluating available supplies of forages or other fibrous feeds that are available for purchase.

The second procedure involves supplementing the available forage with a combination of other feeds that provide the cows with a balanced ration at reasonable cost. In addition to meeting the cow's nutrient requirements at reasonable cost, the ration should:

1.  Contain ad-lib amounts of clean, fresh water.

2.  Contain a total-ration crude-fiber level on a dry matter basis of 15 to 24%, 15 to 19% for early-lactation or high-producing cows, and 19 to 24% for late-lactation or lower-producing cows.

3. Contain a total-ration crude-protein level on a dry matter basis of 12 to 16%, 15 to 16% for early-lactation or high-producing cows, and 12 to 14% for late-lactation or low-producing cows.

4. Be within the dry matter intake capacity of the cow. This varies widely with individual cows but in general ranges from 2.5 to 2.8% of body weight for lower-producing cows and from 3.3 to 3.6% of body weight for higher-producing cows.

A logical step-by-step procedure can be used to formulate rations for milking dairy cattle. The essentials are:

1. Estimate the daily nutrient needs or requirements.

2. Determine the nutrient content of the available forage.

3. Determine the forage intake.

4. Calculate the nutrient intake from the forage.

5. Calculate the nutrient need from concentrates.

6. Balance the ration for energy with concentrates.

7. Balance the ration for protein, calcium, phosphorus, and the other micronutrients by including proper amounts of these nutrients in the concentrate mixture.

8. Blend or purchase a mixture of supplemental feeds that provide the needed amounts of the various remaining nutrients at least cost per unit.

### Estimating Dairy Nutrient Requirements

In order to feed a nutritionally balanced ration, one must first estimate the amounts of each nutrient that are needed for the various body functions (maintenance, growth, production, and reproduction). The amounts of each nutrient needed vary widely depending primarily on the size of the animal and the amount and composition of the milk produced (see Chapter 5). In the herd situation nutrient requirements should be calculated for cows at varying levels of production or, in the case of grouping of cows, for the average or slightly above level of production of each of the various groups.

### Determining the Nutrient Content of the Available Forage

The available forage provides the base feed for most dairy rations. In order to determine which nutrients to provide, and the quantity of each that will be needed to supplement the forage to provide a balanced ration, one must have an accurate estimate of the nutrient composition of the forage. Forages vary widely in nutrient content depending on the moisture content, stage of maturity when harvested, curing and storage conditions. Because of this variation in nutrient content between forages and the inaccuracy of the "eyeball" test of forage quality, most state extension services offer or promote the use of a forage testing service (see Figure 8-1). Many feed companies also offer this service. The purpose of forage testing is to determine the nutrient content of forages more accurately than can be achieved by estimating it from physical appearance, cutting data, or tabular values such as those presented in Appendix Table E. This, in turn, makes possible more precision in balancing rations.

Chemical analyses most widely offered by forage testing services include dry matter (DM), total or crude protein (TP or CP) and crude fiber (CF). In addition, estimates of total digestible nutrients (TDN), estimated net energy (ENE), and digestible protein (DP) are often included.* Mineral analyses are also available in some programs. Accurate determination of the dry matter, energy, and protein content are of major importance in forage analysis. Forages, especially silages and haylages, vary widely in dry matter content. Forages are fed primarily as sources of energy and protein. Thus accurate determination of these values is important. Accurate determination of mineral content is also becoming more important as level of production increases and soils become depleted in some mineral elements.

In summary, laboratory forage testing is a tool that can be and should be used to increase the precision of ration formulation. Accurate determination of the nutrient composition of forages can enable the dairy herd manager more precisely to supplement the forage with the nutrients needed to provide a balanced ration for the dairy herd.

Accurate sampling of the material to be tested also is essential if the results are to be most useful. The sample to be analyzed should be representative of the material being tested. For sampling hay, core or grab samples should be taken from several bales from each cutting from each field. For sampling silage, grab samples from various locations should be taken and placed immediately in a plastic bag to keep them from drying out. If drying occurs before analysis, the dry matter figure will be inaccurate. Silage should be resampled whenever significant changes in the silage occurs. This may result from varying cutting dates or silage from different fields or varieties. Many dairymen are currently sampling silage as it is being ensiled. Research at the Virginia Research Station** indicates that very little change occurs in

* E. C. Coppock, Forage Testing and Feeding Programs. *J. Dairy Sci.*, **59:** 175 (1976).
** W. R. Murley et al., *V.P.I. & S. U. Dairy Guideline 156, Pre-Silage Analysis: An Aid in Planning your Feeding Program*, 1972.

dry matter, crude protein, or crude fiber content from the time of ensiling to the time of feeding, under proper storage conditions. The primary advantage of this presilage analysis is that it enables dairymen to anticipate forage composition changes and adjust rations before production drops occur.

### Determining the Forage Intake

Accurate determination of forage intake coupled with determination of nutrient composition of the forage provides precision in estimating nutrient intake from forage. The ideal situation would be to weigh the forage intake for each cow in the herd regularly or periodically. This is practically impossible to do. In spite of this, the average forage intake rate for the herd or groups of cows within the herd is still a valuable tool in developing a balanced ration for the herd. There are several methods that can be used in various types of herd feeding systems to accurately estimate forage intake. Some of these are described below.

**Baled Hay.** Weigh at least 5 and preferably 10 bales of hay to obtain an average weight per bale. Multiply this by the number of bales fed to the herd or to a group of cows within the herd. Make corrections for refused hay and divide this figure by the number of cows in the herd or group.

**Silage Fed with Mechanical Unloaders and Feed Bunk Distributors.** Mark several locations of feed bunk into sections of 3 ft. each. After filling the bunk, remove and weigh the silage in each section. Convert this to pounds per foot of feed bunk and multiply by the length of the feed bunk. Divide this figure by the number of cows in the herd or group. If silage is fed more than once daily, multiply the figure obtained by the times per day silage is fed. One may also estimate silage dry matter intake from the values presented in Table 8-1.

**Silage Self-Fed from a Horizontal Silo.** Determine the weight per cubic foot of silage by weighing the silage from an area 3 ft. wide by 3 ft. high by 1 ft. deep (9 cu. ft.) and dividing by 9. Determine the cubic feet of silage consumed by the herd or group in one week (average width of the silo times the average height of settled silage times the depth eaten in one week), subtract the wastage, and multiply this by the weight per cubic foot.

**Forages Fed in Complete or Blended Rations.** Mixer wagons should be equipped with scales or an electronic weighing device so that precise measurement of forage intake may be obtained on a daily basis. This adds greatly to feeding precision and is one reason for the increasing popularity of this type of feeding system.

Name _____  Farm Sample No. _____
Address _____  Date Submitted _____
                              Laboratory Sample No. _____
County _____  _____ (leave blank)
                              Date received by lab _____ (leave blank)

## SAMPLE (check one):

☐ Hay (includes all forms of air dried forages)

☐ Silage ( ) direct cut or ( ) wilted (any forage sampled when removed from silo including wilted hay-crop silage or "haylage")

☐ Green-Chop (any forage sampled when harvested)

☐ Miscellaneous (includes samples that have added grain or complete rations—describe grain additions below)

DATE FORAGE HARVESTED _____  CUTTING _____
                          (month and day)                    (1st, 2nd, etc.)

## TYPE OF LIVESTOCK FOR WHICH FORAGES ARE INTENDED:

☐ Dairy (1)                    ☐ Beef cow and calf herd (2)        ☐ Sheep (3)
☐ Growing animals (4)          ☐ Fattening animals (5)             ☐ Other _____ ( )
                                                                      (name)

## CROPS: (Give percentage of each)

%                          %                              %
____ Alfalfa               ____ Corn                      ____ Barley
____ Red clover            ____ Grain sorghums            ____ Oats
____ Orchardgrass          ____ Forage sorghums           ____ Rye
____ Timothy               ____ Sorghum-sudan hybrids     ____ Wheat
____ Other _____ ____ Other _____
          (name)                      (name)

178

VARIETY _____ (Give name if sample is single forage)

SILAGE: (List preservatives or additives contained in silage samples)

_____ at _____ lb/ton _____ at _____ lb/ton

TYPE OF SILO:

☐ Tower (1)            ☐ Tower, epoxy coated (2)       ☐ Bunker (3)
☐ Trench (4)           ☐ Stack (5)                     ☐ Temporary, vacuum sealed (6)
☐ Permanent, gas-tight; give name _____ (7)

ANALYSIS (for laboratory use);

AS FED:
DM____% TDN____% NE Index____ CP____% DP____%

DRY BASIS:
CP____% CF____% TDN____% ENE____(T/100 lb)

_____
Extension Agent's Signature

**Fig. 8-1.** Forage information and evaluation report (V.P.I. & S.U.).

179

**Table 8-1** Pounds of Dry Matter in 2-in. Layers of Tower Silos of Different Diameters[a]

| Silo diameter (feet) | Pounds dry matter[b] (in 2-in. layers) |
|---|---|
| 12 | 282 |
| 14 | 384 |
| 16 | 502 |
| 18 | 636 |
| 20 | 785 |
| 22 | 950 |
| 24 | 1131 |
| 26 | 1329 |
| 28 | 1539 |
| 30 | 1767 |

[a] Reprinted with permission from Dairy Housing and Equipment Handbook, Midwest Plan Service, Ames, Ia, 1976.
[b] Assuming 15 lb. DM/cu. ft.

Accurate determination of forage intake is relatively easily achieved in some feeding systems and difficult and time consuming in others. It should be done as it can add significantly to the precision of the feeding program.

## Determine the Nutrient Intake from Forage

One determines the nutrient intake from forage by multiplying the nutrient composition of the forage times the forage intake.

## Determine the Nutrient Need from Concentrates

To determine the nutrient need from concentrates, subtract the nutrients supplied by the forage from the nutrients required by the animal. The difference must be supplied with supplemental feeds, usually concentrates.

## Balance the Ration for Energy

Based on the available concentrate energy sources (Chapter 7), either homegrown or purchased, determine the amount of concentrates needed to balance the energy requirement. Divide the megacalories of net energy needed from concentrates by the net energy content of the concentrate source available or by an average energy value for dairy concentrates to determine the quantity of concentrate needed to balance the energy requirement. Most dairy concentrate mixtures contain about 0.7 to 0.75 Mcal. of net energy per pound (1.5 to 1.7 Mcal./kg.) on an as-fed basis of approximately 90% dry matter, and 0.75 to 0.85 Mcal. of net energy per pound (1.7 to 1.9 Mcal./kg.) on a dry matter basis. More precise estimates of the energy value of various concentrate feeds are given in Appendix Table E.

## Balance the Ration for Other Nutrients

Balancing the ration for other nutrients can be achieved by dividing the amount of each nutrient needed in the concentrate by the amount of concentrate to be fed. This calculation will yield the percentage of each of the nutrients that needs to be included in the concentrate mixture.

## Blending or Selecting the Mixture of Supplemental Feeds

The previous calculations will define the amount of supplemental concentrates needed and the percentage of each nutrient needed to balance the ration. If homegrown grains are available, these should be used for the major energy source. Protein, mineral, and vitamin sources will usually need to be purchased. These purchases should be based on the cost per unit of nutrient needed.

Economic evaluation of concentrate feeds is usually made on the basis of their ability to provide needed energy or protein. There is a wide variety of feeds that can be used to provide energy; another group of feeds are primarily protein supplements (see Appendix Table E); and there are commercially mixed dairy concentrate feeds. If the protein content and cost are known and the energy value can be estimated accurately, one can calculate the cost per unit of protein or energy and purchase those feeds or feed mixtures that provide the needed nutrients at least cost per unit of nutrient. Table 8-2 illustrates the cost per kilogram of protein and per Mcal. of $NE_{lactation}$ of some common feeds for dairy cattle at the listed prices.

Protein and energy values on an as-fed basis were calculated by multiplying the protein (%) and energy (megacalories/kilogram) values on a DM basis, as listed in Appendix Table E, times the DM content of that feed.

**Table 8-2** Cost of Nutrients in Feed

| | Protein as fed, (%) | Estimated $NE_{lact}$ as fed, (Mcal./kg.) | Price per ton | Price per 100 lb. | Price per 100 kg. | Cost per kg. Protein | Cost per Mcal. $NE_{lact}$ |
|---|---|---|---|---|---|---|---|
| **Energy feeds** | | | | | | | |
| Corn | 8.9 | 2.15 | $100.00 | $5.00 | $11.00 | $1.24 | $0.051 |
| Barley | 11.6 | 1.90 | 100.00 | 5.00 | 11.00 | 0.95 | 0.058 |
| Hominy feed | 10.7 | 2.32 | 100.00 | 5.00 | 11.00 | 1.03 | 0.047 |
| Oats | 11.7 | 1.71 | 100.00 | 5.00 | 11.00 | 0.94 | 0.064 |
| **Protein feeds** | | | | | | | |
| Linseed meal | 35.1 | 1.74 | 160.00 | 8.00 | 17.60 | 0.50 | 0.101 |
| Peanut meal | 45.8 | 1.78 | 160.00 | 8.00 | 17.60 | 0.38 | 0.099 |
| Soybean meal | 45.8 | 1.84 | 160.00 | 8.00 | 17.60 | 0.38 | 0.096 |
| **Commercial mixes** | | | | | | | |
| 20% dairy mix | 20.0 | 1.60 | 140.00 | 7.00 | 15.40 | 0.77 | 0.096 |
| 24% dairy mix | 24.0 | 1.60 | 144.00 | 7.20 | 15.84 | 0.66 | 0.099 |
| 32% dairy mix | 32.0 | 1.60 | 154.00 | 7.70 | 16.94 | 0.53 | 0.106 |

**Example:**

$$89\% \text{ DM, } 10\% \text{ protein DM basis} = \frac{.89 \times .10}{100} \times 100$$

$$= 8.9\% \text{ protein, as-fed basis}$$

$$89\% \text{ DM, } 2.42 \text{ Mcal./kg. DM basis} = \frac{.89 \times 2.42}{100} \times 100$$

$$= 2.15 \text{ Mcal./kg. NE, as-fed basis}$$

The price per 100 kg. was calculated by multiplying the price per 100 lb. times 2.2 (1 lb. = 2.2 kg.). Cost per kilogram of protein was calculated by dividing the cost per 100 kg. of protein by the number of kilograms of protein per 100 kg. of the feed.

**Example: Corn**

$$\$11/100 \text{ kg.} \div 8.9 \text{ kg. per } 100 \text{ kg.} = \$1.24 \text{ per kg. of protein}$$

Cost per megacalories of $NE_{lactation}$ was calculated in a similar manner.

**Example: Corn**

$$\$11/100 \text{ kg.} \div 215 \text{ Mcal./100 kg.} = \$.051/\text{Mcal. of NE}$$

This procedure can be used to evaluate the cost per unit of nutrient of any feed. Of the feeds listed in Table 8-2 it can be seen that hominy feed costs the least per unit of energy, with peanut and soybean meal being the most economical sources of protein.

Many dairymen purchase commercial dairy feed concentrate mixtures to supplement homegrown forages. These are very satisfactory provided they contain the nutrients needed to balance the ration at reasonable cost. Comparisons of the costs of commercial grain mixes with home-formulated mixes of similar nutrient composition should be made.

### Calculating Rations for Individual Cows

The following example illustrates the step-by-step procedure for calculating rations on an individual cow basis. This is followed by an explanation of how to apply this to a herd situation.

**Example:** Old Betsy
Body weight = 600 kg.
Production (daily) = 20 kg. of milk that contains 4.0% butterfat.
Age = milking in her second lactation.
Reproductive status = not pregnant.

1. Estimate her daily nutrient requirements by calculating these from Appendix Table B. (In our example we will calculate only energy, protein, calcium, and phosphorus and illustrate later how to balance her ration for the other nutrients.)

|  | $NE_{lact}$ (Mcal.) | Protein (g.) | Ca (g.) | P (g.) |
|---|---|---|---|---|
| Maintenance | 10.3 | 734 | 22 | 17 |
| Growth (+10% of maintenance) | 1.0 | 73 | 2 | 2 |
| Production | 14.8 | 1,560 | 54 | 40 |
| **Total nutrient requirement** | **26.1** | **2,367** or **2.37 kg.** | **78** | **59** |

2. Determine the nutrient content of the forage.
Betsy is being fed corn silage and alfalfa hay with the following nutrient analysis (Appendix Table E):

|  | DM (%) | Protein (%) | NE (Mcal./ Kg.) | Fiber (%) | Ca (%) | P (%) |
|---|---|---|---|---|---|---|
| Corn silage | 40 | 8.1 | 1.7 | 24.4 | .27 | .20 |
| Alfalfa hay | 90 | 18.4 | 1.25 | 29.8 | 1.25 | .23 |

3. Determine the forage intake.
Betsy is consuming 25 kg. of the silage and 3 kg. of the hay daily.

4. Calculate the nutrient intake from the forage:

**Corn silage**
25 kg. × 40% DM = 10 kg. DM
10 kg. DM × 8.1% Prot. = 0.81 kg. protein
10 kg. DM × 1.70 Mcal./kg. = 17 Mcal. NE
10 kg. DM × 0.27% Ca = 27 gm Ca
10 kg. DM × 0.20% P = 20 gm P

**Alfalfa hay**
3.0 kg. × 90% DM = 2.7 kg. DM
2.7 kg. DM × 18.4% Prot. = 0.50 kg. protein
2.7 kg. DM × 1.25 Mcal./kg. = 3.4 Mcal. NE
2.7 kg. × 1.25% Ca = 34 gm Ca
2.7 kg. DM × 0.23% P = 6 gm P

Total = 1.3 kg. protein; 20.4 Mcal. NE; 61 gm Ca; 26 gm P

5. Calculate the nutrient need from concentrates:

Nutrient requirements (1) = 2.4 kg. prot., 26.1 Mcal. NE, 78 gm Ca, 59 gm P
Forage nutrients (4) = 1.3 kg. prot., 20.4 Mcal. NE 61 gm Ca, 26 gm P

Needed in concentrates = 1.1 kg. prot., 5.7 Mcal. NE, 17 gm Ca, 33 gm P

6. Balance the ration for energy:
For our example, assume concentrate energy of 1.6 Mcal./kg. on an as-fed basis or 1.8 Mcal./kg. on a dry basis.

Energy need = 5.7 Mcal. ÷ 1.6 Mcal./kg. = 3.6 kg. of concentrate needed (as-fed basis) to balance the ration for energy, or 3.2 kg. concentrate needed on a DM basis.

7. Balance the ration for protein, Ca, P, etc.:
   a. Percentage of protein needed in the concentrate =
      (as-fed basis)

      $$\frac{1.1 \text{ kg. protein}}{3.6 \text{ kg. concentrate}} \times 100 = 30.5\% \text{ protein}$$

      (DM basis)

      $$\frac{1.1 \text{ kg. protein}}{3.2 \text{ kg. concentrate}} \times 100 = 34.4\% \text{ protein}$$

   b. Percentage of calcium needed in the concentrate =
      (as fed basis)

      $$\frac{.017 \text{ kg. Ca}}{3.6 \text{ kg. concentrate}} \times 100 = 0.47\% \text{ calcium}$$

      (DM basis)

      $$\frac{.017 \text{ kg. Ca}}{3.2 \text{ kg. concentrate}} \times 100 = 0.53\% \text{ calcium}$$

   c. Percentage of phosphorus needed in the concentrate =
      (as-fed basis)

      $$\frac{.033 \text{ kg. P}}{3.6 \text{ kg. concentrate}} \times 100 = 0.92\% \text{ phosphorus}$$

      (DM basis)

      $$\frac{.033 \text{ kg. P}}{3.2 \text{ kg. concentrate}} \times 100 = 1.03\% \text{ phosphorus}$$

## Example summary

|  | DM (kg.) | NE$_{lact-}$ (Mcal.) | Protein (kg.) | Ca (g.) | P (g.) | Fiber (kg.) |
|---|---|---|---|---|---|---|
| Nutrient requirements Ration | | 26.1 | 2.4 | 78 | 59 | |
| 25 kg. corn silage | 10.0 | 17.0 | 0.8 | 27 | 20 | 2.4 |
| 3 kg. alfalfa hay | 2.7 | 3.4 | 0.5 | 34 | 6 | 0.8 |
| 3.6 kg. concentrate | 3.2 | 5.8 | 1.1 | 17 | 33 | 0.2* |
| Total | 15.9 | 26.2 | 2.4 | 78 | 59 | 3.4 |

DM intake as % of body weight $= \dfrac{15.9}{600} \times 100 = 2.65\%$.

Percentage of ration protein on a DM basis $= \dfrac{2.4}{15.9} \times 100 = 15.1\%$.

Percentage of ration fiber on a DM basis $= \dfrac{3.4}{15.9} \times 100 = 21.4\%$.

---

[a] Concentrate fiber = 6% on an as-fed basis.

8. Select the mixture of supplemental feeds. Formulating a combination of concentrate feeds or selecting a commercial concentrate mixture to provide the nutrients not supplied by the forages is the final step in balancing the ration. In our example let us assume that the dairyman has an ample supply of homegrown shelled corn and can purchase soybean meal, salt, dicalcium phosphate, and limestone to mix with his corn. We previously determined that the concentrate mixture should contain 30.5% protein, 0.47% calcium, and 0.92% phosphorus on an as-fed basis, or 34.4% protein, 0.53% calcium, and 1.03% phosphorus on a DM basis. Concentrate mixtures for cows being fed corn silage as the primary forage should contain 1% salt, preferably trace mineral salt, and 2.0% dicalcium phosphate or other sources of calcium and phosphorus. With this information, and by using the analysis figures in Appendix Table E, we can formulate the concentrate mixture needed.

Analysis of the feeds (DM basis):

|  | DM (%) | NE$_{lact-}$ (Mcal./ kg.) | Protein (%) | Fiber (%) | Ca (%) | P (%) |
|---|---|---|---|---|---|---|
| Corn | 89.0 | 2.42 | 10.0 | 2.2 | .02 | .35 |
| Soybean meal | 89.0 | 2.07 | 51.5 | 6.7 | .36 | .75 |
| Dicalcium phosphate | | | | | 23.10 | 18.70 |

## FORMULATING THE MIXTURE

As stated earlier, 3% of the concentrate mixture should be composed of mineral supplements to provide the needed salt, calcium, and phosphorus. The remaining 97% must be a mixture of corn and soybean meal that contains enough protein so that the entire mixture contains 30.5% protein on an as-fed basis. As there is no protein in the salt and dicalcium phosphate, the remaining 97% of the mixture must contain 31.4% protein (30.5 ÷ 97%). As analysis of grains and protein supplements are usually stated on an as-fed basis, the analysis figures for corn and soybean meal must first be converted from a DM to an as-fed basis. This is done as follows:

Corn:　100 kg. corn contains 8.9 kg. protein (10% of the 89 kg. DM).

$$\frac{8.9}{100} \times 100 = 8.9\% \text{ protein on an as-fed basis}$$

Soybean meal:　100 kg. soybean meal contains 45.8 kg. protein (51.5% of the 89 kg. DM).

$$\frac{45.8}{100} \times 100 = 45.8\% \text{ protein on an as-fed basis}$$

　　The proportion of corn and soybean meal needed to contain 31.4% protein on an as-fed basis can be determined by the Pearson's square method. To use this method, construct a square and insert in the center the percent protein needed (in our example, 31.4). On the upper left corner, place the percentage of protein in the protein supplement, which in our case is 45.8, and on the lower left corner the protein percentage of the grain to be used, which here is 8.9. Then take the difference along the diagonal lines as illustrated:

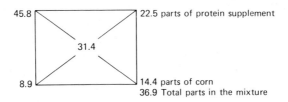

```
45.8                    22.5 parts of protein supplement

        31.4

8.9                     14.4 parts of corn
                        36.9 Total parts in the mixture
```

These figures can then be converted to percentages:

$$\frac{22.5}{36.9} \times 100 = 61\% \text{ soybean meal}$$

$$\frac{14.4}{36.9} \times 100 = 39\% \text{ corn}$$

The concentrate mixture needed to balance the ration for our example would contain:

|  | % | Kg./t. | Lb./t. |
|---|---|---|---|
| Salt | 1 | 9.1 | 20 |
| Dicalcium phosphate | 2 | 18.2 | 40 |
| Soybean meal (61% of the nonmineral part) | 61 | 537.7 | 1,183 |
| Corn (39% of the nonmineral part) | 39 | 344.0 | 757 |
| **Total** | | **909.0 kg.** | **2,000 lb.** |

Nutrient content of 1 ton or 909 kg. concentrate mixture:

|  | DM (kg.) | NE$_{lact-}$ (Mcal.) | Protein (kg.) | Ca (kg.) | P (kg.) | Fiber (kg.) |
|---|---|---|---|---|---|---|
| 9.1 kg. salt | 8.7 | — | — | — | — | — |
| 18.2 kg. dicalcium phosphate | 17.5 | — | — | 4.05 | 3.26 | — |
| 537.7 kg. soybean meal | 478.6 | 990.7 | 246.5 | 1.72 | 3.59 | 32.1 |
| 344.0 kg. corn | 306.2 | 741.0 | 30.6 | .06 | 1.10 | 8.0 |
| **909.0 kg.** | **811.0** | **1731.7** | **277.1** | **5.83** | **7.95** | **40.1** |
| Analysis (DM basis) | 89.2% | 2.14 Mcal./kg. | 34.2% | .72% | .98% | 4.9% |
| Analysis (as-fed basis) | 100.0% | 1.9 Mcal./kg. | 30.5% | .64% | .87% | 4.4% |

Feeding Betsy 3.6 kg. of this concentrate mixture along with 25 kg. of corn silage and 3 kg. of alfalfa hay meets or exceeds the nutrient requirements calculated in our example, except for 2 g. of phosphorus (Table 8-3). This deficiency can be corrected by feeding free-choice salt and dicalcium phosphate in a mineral feeder. This practice is highly recommended to ensure sufficient intake to allow for the part that is not digested.

Sources of vitamins A and D should also be included in the concentrate mixture in amounts sufficient to ensure adequate intake. The cow in our example requires about 40,000 I.U. of vitamin A, or 100 mg. of carotene, and 6,000 I.U. of vitamin D daily. To ensure adequate intake of these vitamins, a supplemental vitamin source(s) should be included in the concentrate mixture to provide a minimum concentration of 12,000 I.U. of vitamin A per kilogram, or 5,000 I.U. of vitamin A per pound and 2,000 I.U. per kilogram of vitamin D, or 1,000 I.U. of vitamin D per pound.

It is also important that dairy rations be economical. If one knows the cost of each of the feeds in the ration, the daily milk production, and the value of the milk, then economic parameters such as cost per unit of concentrate, feed cost per day, feed cost per 100 lb. of milk, and income over feed cost per hundredweight of milk can be calculated. Assuming the following values for feeds used in our example, we can calculate the following parameters.

|  | Value/cwt. | Value/t. | Value/kg. |
|---|---|---|---|
| Corn silage | $ 1.00 | $ 20.00 | $0.022 |
| Alfalfa hay | 3.00 | 60.00 | 0.066 |
| Corn | 5.00 | 100.00 | 0.110 |
| Soybean meal | 8.00 | 160.00 | 0.176 |
| Salt | 4.00 | 80.00 | 0.088 |
| Dicalcium phosphate | 10.00 | 200.00 | 0.220 |
| Milk | 10.00 |  | 0.22 |

| a. **Concentrate cost** | Per cwt. | Per t. of Feed | Per kg. |
|---|---|---|---|
| 757 lb. corn × | $ 5.00 | $ 37.85 |  |
| 1,183 lb. soybean meal × | 8.00 | 94.64 |  |
| 20 lb. salt × | 4.00 | 0.80 |  |
| 40 lb. dicalcium phosphate × | 10.00 | 4.00 |  |
| Grinding and mixing | 0.50 | 10.00 |  |
|  | **$ 7.36** | **$147.29** | **$0.162** |

| b. **Daily feed cost** | |
|---|---|
| 25 kg. corn silage | $0.55 |
| 3 kg. alfalfa hay | 0.20 |
| 3.6 kg. concentrate | 0.58 |
|  | **$1.33** |

c. **Feed cost/cwt. milk**
Feed cost/day ÷ lb. milk produced/day × 100

$$\$1.33 \div 44(20 \text{ kg.} \times 2.2 \text{ kg./lb.}) \times 100 = \$3.02$$

d. **Income over feed cost per cwt. milk**
Value of milk ($10.00/cwt.) minus feed cost/hundredweight milk ($3.02) = $6.98

    This example has been used to demonstrate the basic principles of balancing a ration to meet a lactating cow's nutrient requirements for the major qualitative nutrients. The ration should also be examined to determine if it contains adequate amounts of carotene, vitamin D, and other micronutrients. If it is deficient in any of these, appropriate supplements should be added to the concentrate mixture. If a commercial concentrate mixture is fed, it must also contain sufficient amounts of the various nutrients needed to meet the animal's nutrient requirements.

    Performing the illustrated calculations for each cow in the herd on a

**Table 8-3** Summary of Requirements and Ration Nutrients

|  | DM (kg.) | $NE_{lact}$ (Mcal.) | Protein (kg.) | CA (g.) | P (g.) | Fiber (kg.) |
|---|---|---|---|---|---|---|
| Nutrient requirements |  | 26.1 | 2.4 | 78 | 59 |  |
| Forage | 12.7 | 20.4 | 1.3 | 61 | 26 | 3.2 |
| Concentrate (3.6 kg.) | 3.2 | 6.8 | 1.1 | 23 | 31 | 0.2 |
| Total | 15.9 | 27.2 | 2.4 | 84 | 57 | 3.4 |

$$\text{DM intake} = \frac{15.9}{600} = 2.65\% \text{ of body weight}$$

$$\text{Protein intake} = \frac{2.4}{15.9} = 15.1\% \text{ of DM}$$

$$\text{Fiber intake} = \frac{3.4}{15.9} = 21.4\% \text{ of DM}$$

weekly or a monthly basis is impractical because it is laborious and time consuming. It is important, however, to apply these principles in formulating guidelines for a profitable herd feeding program, one that encourages high production at reasonable feed cost per unit of production.

## Application of the Basic Principles to Herd Situations

In herds where forages are primarily homegrown and forages and concentrates are fed separately, the following guidelines should be followed:

1. Feed ad-lib amounts of high quality forage. The type of forage fed will vary with resources such as land and labor.

2. Periodically (monthly) formulate a ration for a high-, medium-, and low-producing cow in the herd, as illustrated in our example. These calculations will yield an accurate estimate of the ratio of concentrate to milk that needs to be fed at varying levels of production (usually narrower ratios at higher levels of production and wider ratios at lower levels), and percentages of protein, calcium, phosphorus, etc., that need to be included in the concentrate mixture.

3. Formulate or purchase a concentrate mixture that balances the available forage based on the least-cost mixture available. Careful shopping for energy and protein sources is very important as costs are variable. Check the ration to ensure that all nutrient requirements and other essentials of a dairy ration (fiber level, protein level, etc.) are being met.

4. Feed recently fresh and high-producing cows liberally. It is very difficult

to maintain positive energy balance in high-producing cows, and most will utilize body stores of energy in early lactation. Feed enough protein during this period to meet protein requirements, as body stores of protein are very limited. Many dairymen individually feed additional high-protein concentrate (soybean meal, etc.) during this period.

5. Feed younger cows, first- and second-calf heifers, for growth as well as maintenance and production. It is a good practice to increase concentrate intake about 20% for first-calf heifers and 10% for second-calf heifers. This is especially true if heifers are calving at 24 to 26 months of age.

6. Utilize available tools to reduce the time and inaccuracies of hand calculated ration balancing and in inaccuracy of estimating forage quality. In many areas computerized least-cost ration programs, least-cost concentrate mixtures, and maximum-income-over-feed-cost programs are available.* Use of these programs and forage-testing programs can add greatly to dairy herd feeding precision.

In herds where cows are grouped according to level of production, and complete or blended rations are fed, formulate a ration for about one standard deviation above the average production of each of the groups and about 10% higher than average for the high-producing and first-calf heifer groups. Reformulate the ration whenever there are significant changes in forage or concentrate composition. Careful measurement and thorough mixing of each load of feed is essential.

In herds where part of the concentrate is fed with the forage and the remainder in the milking parlor or barn, similar formulation procedures can be followed. The major difference is that the concentrate fed with the forage should be included in the base forage calculation. An example of how to balance a ration for protein and energy in this situation follows.

|  | DM (kg) | Prot. (kg.) | NE$_{lact-}$ (Mcal./kg.) |
|---|---|---|---|
| Nutrient requirement |  | 3.00 | 32.0 |
| 25 kg. corn silage | 10.0 | .81 | 17.0 |
| 3 kg. alfalfa hay | 2.7 | .50 | 3.4 |
| 3 kg. corn | 2.7 | .27 | 6.5 |
| Forage nutrients |  | 1.58 | 26.9 |
| Needed in concentrate |  | 1.42 | 5.1 |

Amount of concentrate needed = 5.1 ÷ 1.6 = 3.2 kg. concentrate

% protein needed in concentrate $= \dfrac{1.42}{3.2} = 44.4\%$.

* D. L. Bath et al., *J. Dairy Sci.*, **55:** 1607 (1972); Chandler, P. T., and Walker, H. W., *J. Dairy Sci.*, **55:** 1741 (1972); Bath, D. L., *J. Dairy Sci.*, **58:** 226 (1975).

These examples are just three of the possible situations. They illustrate how the feeding program can be managed by applying the basic principles illustrated earlier to the individual set of circumstances. The basic principles of balancing a ration have not changed appreciably in many years. The feeds, methods of feeding, knowledge of nutrient requirements, and tools available to formulate a more precise ration have changed and will continue to do so in the future. The good manager will make maximum use of current knowledge and tools and constantly seek new knowledge, improved tools, and techniques to do even a more precise job of feeding the milking herd. He or she realizes the importance of the following key factors in developing a profitable feeding program.

1. High intake of high-quality forage; maximum use of land resources for the production of high-quality forage.

2. Feeding a balanced ration; meeting the cow's nutrient requirements for high levels of production.

3. Maximum nutrient intake in early lactation.

4. Purchasing feeds on a cost per unit of nutrient basis.

## FEEDING DRY COWS

The purpose of the dry-cow feeding program is to have dairy cows in good condition, healthy, and with good appetites at freshening. The recommendations for feeding dry cows have changed drastically in recent years due to an increased incidence of metabolic disorders such as milk fever, ketosis, and displaced abomasum at or shortly after calving. It is no longer recommended that dry cows be fed heavily to regain weight and body stores of minerals lost during the lactation. Rather, it is recommended that they be overfed a bit during late lactation so that they are in good condition, but not fat or overconditioned when they are ready to be turned dry. It is further recommended that they be fed only to maintain themselves during the dry period and that excessive intake of energy, protein, and minerals be avoided.

In herds fed complete rations formulated for the lactating cows, the dry cows must be separated from the milking cows. This is also true in herds where ad-libitum amounts of high-quality forages (especially high-quality corn silage, which is very palatable and high in energy) are being fed to the milking herd. If the dry cows are not separated, excessive weight gains will occur and the incidence of metabolic disorders is likely to increase.

### Formulating Dry Cow Rations

In comparison with milking cows, nutrient requirements (maintainance and pregnancy) for dry cows are quite low. Current N.R.C. requirements are listed in Appendix Table B. These were published in 1971; many nutritionists feel they are somewhat high and that dry cows should be fed for maintenance only rather than for maintenance and reproduction. At the time of preparing this text, evidence is not available to justify major deviation from the N.R.C. recommendations.

The essential steps in formulating a dry cow ration are the following.

1.  Estimate her nutrient requirements.
2.  Formulate a ration to meet those requirements, with the following restrictions or modifications.
    a.  Total DM intake restricted to 1.25 to 1.50% of body weight.
    b.  Energy intake restricted to 1.60 to 1.80 Mcal./100 kg. of body weight.
    c.  Ration crude fiber level of 25 to 35% on a DM basis.
    d.  Ration crude protein level of 8 to 12% on a DM basis.

The table that follows gives an example of rations for a 600-kg. pregnant dry cow.

| | DM (kg.) | Prot. (kg.) | NE$_{lact}$ (Mcal.) | Ca (g.) | P (g.) | Carotene (mg./kg.) | Vit. A (I.U./g.) | Crude fiber (% of DM) |
|---|---|---|---|---|---|---|---|---|
| Nutrient requirements | | | | | | | | |
| (Appendix Table B) | 10 | .91 | 13.5 | 34 | 26 | 114 | 46 | |
| (with restrictions) | (9) | | (10.8) | | | | | |
| Ration 1 (acceptable) | | | | | | | | |
| 10 kg. orchardgrass hay | 8.8 | .85 | 11.0 | 40 | 33 | 295 | N.A. | 34.0 |
| Ration 2 (excessive energy) | | | | | | | | |
| 30 kg. corn silage | | | | | | | | |
| (40% DM, well-eared) | 12.0 | .97 | 20.4 | 32 | 24 | N.A. | N.A. | 24.4 |

Feeding 10 kg. per day of orchardgrass hay with free-choice trace mineral salt would provide an acceptable ration for a 600-kg. dry cow (ration 1). It can be seen that if a 600-kg. dry cow were offered free-choice good-quality corn silage (ration 2 with estimated intake at 2.0% of body weight on a dry matter basis) she would consume about 50% more energy than needed. Ration 2 would not be an acceptable dry-cow ration. This ration would likely result in excessive weight gains and could lead to problems at and after calving.

## FEEDING AT CALVING TIME

The purpose of the feeding program just before calving, at calving, and just after calving is to get the cow on full feed within a relatively short period of time. This is necessary to prevent rapid loss of weight, which is conducive to a high incidence of ketosis. It is also important not to feed too high a proportion of concentrate feed with concurrent reduced ration fiber level during this period, as recent evidence indicates this can increase the incidence of displaced abomasum. Too-rapid increases in concentrate feeding can lead to acid rumen and temporary off-feed conditions.*

Feeding concentrates at a maximum level of 1% of body weight along with high-quality forage just before and at calving is recommended. One must remember, however, that high-quality corn silage is 50 to 60% grain, and if corn silage is fed as the forage, additional concentrate intake must be limited.

Below are examples of rations for cows at freshening (600 kg. cow).

| | DM (kg.) | Protein (kg.) | $NE_{lact}$ (Mcal.) | Crude fiber (kg.) |
|---|---|---|---|---|
| Ration 1 (acceptable) | | | | |
| 10 kg. orchardgrass hay | 8.8 | 0.85 | 11.0 | 2.99 |
| 6 kg. 20% concentrate (6% CF) | 5.4 | 1.20 | 9.6 | .36 |
| **Total** | **14.2** | **2.05** | **20.6** | **3.35** |

$$\text{Ration protein (DM basis)} = \frac{2.05}{14.2} \times 100 = 14.4\%$$

$$\text{Ration fiber (DM basis)} = \frac{3.35}{14.2} \times 100 = 23.6\%$$

| | DM (kg.) | Protein (kg.) | $NE_{lact}$ (Mcal.) | Crude fiber (kg.) |
|---|---|---|---|---|
| Ration 2 (high energy, low fiber) | | | | |
| 30 kg. corn silage | 12 | .97 | 20.4 | 2.91 |
| 6 kg. 20% concentrate (6% CF) | 5.4 | 1.20 | 9.6 | .36 |
| **Total** | **17.4** | **2.17** | **30.0** | **3.27** |

$$\text{Ration protein (DM basis)} = \frac{2.17}{17.4} \times 100 = 12.4\%$$

$$\text{Ration fiber (DM basis)} = \frac{3.27}{17.4} \times 100 = 18.8\%$$

Feeding ration 1 along with supplemental minerals prior to and at calving would maintain adequate ration fiber level to help avoid a high incidence of displaced abomasum and off-feed problems. It is also sufficiently high in energy and protein to encourage moderate production without rapid weight loss. Feeding a ration of this type for one to two weeks before calving and at calving should meet the purpose of the program. After calving the cow should be fed increasing amounts of high-quality forage and concentrates to gradually

* Mich. St. Bul. E-702, 1973

bring her to full feed with a ration that meets the specifications for a good ration for lactating cows.

Feeding ration 2 during this period may lead to problems, as nearly two-thirds of the dry matter intake is from grain and concentrate feed (50% of the corn silage DM plus 100% of the concentrate DM). This results in high energy and lowered fiber intake, which is conducive to the problems mentioned previously.

## FEEDING SYSTEMS

The feeding system includes the land resources for feed production, the feeding program, the feed storage area(s), and method of feeding. The basic purpose of management of the feeding system is to utilize the available resources to provide each cow with a nutritionally balanced ration as economically and labor-efficiently as possible.

Dairymen have many alternatives in choosing how to use land resources to raise and store forages (Chapter 6) and concentrates (Chapter 7). Silage is the most popular forage fed to dairy cows in most of the United States, with the exceptions being the Pacific and southern Plains states (Table 8-4). Most dairymen who feed silage also feed hay, but usually limited amounts. Some also utilize pasture during certain seasons. However, the ease of mechanization, the reduced risk of harvesting losses in most of the United States, the reduced wastage by trampling and manure contamination, and the high yield of nutrients per acre of silages, especially corn silage, have contributed to a trend of increased silage feeding with an accompanying decrease in hay feeding and pasturing in areas where suitable land resources are available.

The choice of a feeding program depends largely on the type and amount of feeds that are available (home produced or available for purchase), whereas the choice or method of feeding depends largely on the type of housing arrangement.

In conventional barns (stanchion or tie stalls) choice of a method of feeding is somewhat limited, as all feeds are usually hand fed in the same manger. Typically, forages are fed two to four times per day and concentrates individually fed twice daily. Hay storage areas are usually overhead or in an area adjacent to the stanchion barn. Silos are usually adjacent to the stanchion barn and connected to it. Concentrate storage bins or areas are usually located in or adjacent to the barn. Moving feeds from storage areas to the manger is carried out largely by hand. This is time consuming and can be costly, although it allows individual cow feeding according to level of production. Some newer stanchion barns are equipped for mechanized feeding of silage or complete rations.

In loose housing arrangements (loose, free stall, and corral), however,

**Table 8-4** Tons of Total Silage Crops Stored per Cow and Farm and Percentage Distribution by Type of Storage, by Major Areas, 1974 Estimate[a]

| Area | Tons stored/cow | Tons stored per farm reporting | | | Percentage of silage stored in: | | |
|---|---|---|---|---|---|---|---|
| | | Conventional tower | Sealed tower | Bunker or trench | Conventional tower | Sealed tower | Bunker or trench |
| Northeast | 8.6 | 369 | 411 | 583 | 64 | 13 | 23 |
| Lake States | 8.4 | 357 | 410 | 358 | 78 | 12 | 10 |
| Corn Belt | 7.5 | 362 | 442 | 360 | 62 | 15 | 23 |
| Appalachian | 10.8 | 554 | 455 | 600 | 46 | 10 | 44 |
| Delta | 7.7 | 720 | 620 | 720 | 32 | 12 | 56 |
| Southeast | 12.7 | 910 | 710 | 1,140 | 42 | 10 | 48 |
| Northern Plains | 8.3 | 366 | 455 | 478 | 40 | 7 | 53 |
| Southern Plains | 4.1 | 578 | 382 | 1,080 | 23 | 9 | 68 |
| Mountain | 7.3 | 266 | 459 | 1,120 | 3 | 4 | 93 |
| Pacific | 3.6 | 347 | 362 | 1,484 | 12 | 4 | 84 |

[a] C. R. Hoglund, *Res. Rept.* 275, Mich. St. Univ., 1975.

many alternative feeding programs and methods are available. Most of the arrangements are well adapted to mechanized feeding and, in some, to self-feeding of part of the ration.

There are two basic methods of feeding in loose systems: feeding complete or blended rations and feeding the forage(s) and concentrates separately. Typical feeding programs and methods for loose housing arrangements include: (1) group feeding of hay and silage in separate feeders, and individual feeding of concentrates in the milking area; (2) group feeding of one forage (hay or silage) in one area and individual feeding of concentrates in the milking area; (3) group feeding of forage(s) and part of the concentrate in one area, and individual feeding of the remainder of the concentrate in the milking area; (4) group feeding of forage(s) and part of the concentrates and individual feeding of the remainder of the concentrate with the use of magnetic feeders; and (5) feeding complete feeds in one area. The following is a brief discussion of some of the advantages and limitations of each of these five systems.

### Hay, Silage, and Concentrate Feeding in Separate Areas

The major advantage of this system is that it provides the opportunity of utilizing and feeding a variety of feeds to the cows. The feeding of some hay, especially with primarily corn silage rations, has been reported to be helpful in avoiding some metabolic disturbances. There is also less risk of micronutrient deficiencies if more than one forage is fed. In addition, many existing facilities are designed and equipped to handle all three feeding materials.

Some of the major limitations of this system are: (1) inadequate concentrate intake for high producers, (2) inefficient use of labor in the feeding operation and (3) high building and equipment costs.

Cows can consume dry concentrates at the rate of about 0.5 to 0.6 lb./min. and pelleted concentrates at the rate of 0.9 to 1.0 lb./min. High-producing cows may require 25 or more pounds of concentrate feed daily to meet nutrient requirements. Milking parlor efficiency often results in the cow spending 5 to 10 min. per milking in the parlor; this is inadequate time for some cows to consume enough concentrate feed and results in failure to meet one of the basic purposes of the feeding system. For this reason many existing systems have changed to group feeding part of the concentrate in another area or the use of magnetic feeders for high-producing cows. Feeding large amounts of very high-quality forage can effectively reduce the quantity of concentrates needed and alleviate this problem. Preferential eating by individual cows (some prefer silage and some hay) can result in nutritionally unbalanced rations for some cows. Most often the hay in such a system is hand fed because mechanized handling of hay is difficult and expensive, especially when limited amounts are fed. Three feeding areas (bunks for silage, mangers for hay, and concentrate feeders in the parlor), three feed-storage areas (silos, hay barns, and concentrate bin) and two mechanized feeding systems must be constructed and maintained. These factors increase building and equipment costs and usually decrease labor efficiency.

## Single Forage Plus Concentrate

Hay as the single forage plus concentrates in the milking facility is a popular method of feeding dairy cows in the Pacific and southwestern United States. In these areas hay is of high quality and weather is such that little or no hay-storage facilities are needed. Feeding can be largely mechanized by cubing or wafering the hay. Many dairymen in other parts of the United States feed corn silage or a combination of corn silage and wilted grass or legume silages or haylage, all of which can be fed in one area by the same feeding equipment. This system effectively reduces the number of feeding areas and forage-storage facilities and amount of feeding equipment needed, resulting in lower building and equipment costs. Opportunity for mechanization is high, especially in silage-feeding systems, increasing labor efficiency.

Parlor grain feeding is still required and this results in added equipment costs and, if the forage is not of high quality, inadequate concentrate intake for high producers can be a problem. Risk of metabolic disorders and micronutrient deficiencies is also higher if some single forages, such as corn silage or forages grown on nutrient-deficient land, are fed. Precise analysis of forages and nutrient balancing with concentrates is required for good results.

## Feeding Part of the Concentrate with the Forage

With this system a base amount of concentrate is mixed with or spread over the top of the silage and group fed. The major reason for group-feeding part of the concentrate in this manner is to alleviate the problem of inadequate parlor intake of concentrate feeds. The amount and type depend on the nutrient content of the forage and the type of concentrate fed in the parlor. Some dairymen feed the same concentrate mixture both places, while others feed a lower-protein mixture or single concentrate feed such as a cereal grain or by-product with the forage and supplement this with a high-protein mixture in the parlor. This allows more precise feeding of protein, which is more expensive per unit of nutrient. This system works more effectively when the group-fed part of the concentrate can be mixed with the silage. Devices are available that can meter a specified amount of concentrate into the silage as it is being augered into the feeder. Hand spreading of the concentrate on top of the silage or feeding it in a bunk by itself often results in dominant or boss cows consuming a disproportionate part and timid cows receiving little or no concentrate.

The major advantage of the system is that it can alleviate the problem of inadequate concentrate intake for high-producing cows. Its major limitations are that additional labor and/or equipment is needed, and in many cases it is difficult to feed as precisely as desired. Often more dominant, lower-producing cows consume unneeded concentrate and become overconditioned, while more timid, higher-producing cows consume less than they need. The system works best when the concentrate is mixed with silage or other succulent feed and in systems where the cows are grouped according to level of production.

## Use of Magnetic or Electronic Feeders

In this system a magnetically activated feeder(s) is located in the lounging or feeding area. The feeder delivers grain to a feed box when activated by a magnet attached to the neck chain of high-producing cows. This offers high-producing cows free access to concentrate feed in addition to that she receives in the parlor. Magnets are usually attached to cows at freshening, allowed to remain until the cow decreases in production to a level at which her nutrient requirements can be met by parlor feeding, and then removed. This system has the advantage of providing high-producing cows with additional concentrates with little extra labor and without the need to group cows or shift cows from group to group. The limitations of the system include possible overconsumption of grain, with accompanying lowered fat test and off-feed problems by some cows when magnets are first attached. Some cows may not use the feeder. There is the additional expense of the feeders and magnets. Boss cows also may be a problem, especially boss cows that recently have had magnets removed. This limitation can be controlled to some extent by providing an adequate number of feeders (one per 20 to 25 cows) and constructing a feeding stall that protects cows using the feeders. A recent report* indicates that dairymen using magnetic feeders were generally pleased with the results. In this report, which represented a survey of 20 dairy farms using magnetic feeders, the average concentrate consumption from the feeder was 11.5 lb. per cow per day. Sixteen of the 20 dairymen indicated they felt the system worked well or excellently, and only one had discontinued the use of the feeder. The author of the report recommended removing magnets from mature Holstein cows when production dropped to 60 lb./day and from young

* M. F. Hutjens, *Hoards' Dairyman*, April 25, 1976.

cows when production dropped to 50 lb./day. The system seems best suited for smaller-sized operations or in systems where grouping of cattle is difficult.

Electronic concentrate feeders that will deliver specified amounts of concentrate as indicated by the calibrated activating device worn by the cow are also available. The activating device is equipped with a dial that allows dairymen to adjust the amount of concentrate a cow can receive from the feeder in a 24-hour period. This has the potential of allowing dairymen to feed concentrates individually without feeding in the parlor.

## Complete or Blended Rations

Complete rations for dairy cattle are mixtures of feedstuffs formulated to meet the nutrient requirements of the cows being fed that mixture and fed ad libitum as a single feed. If rations are properly formulated and thoroughly mixed, this system offers several advantages. Some of these advantages are listed below.

1. Each bite of feed contains the same balance of nutrients; this prevents selective consumption of various feeds by individual cows, which could result in an unbalanced-ration problem such as milk fat depression if fiber intake were inadequate, depression of energy intake if too much fiber were consumed, etc.

2. Feeding in 2 to 3 separate areas is eliminated, thus reducing building and equipment costs and improving labor efficiency.

3. The flavor of less palatable, but highly nutritious, feeds can be masked by mixing them with more palatable feeds.

4. Major changes in feedstuffs, to take advantage of price fluctuations, can be made with less risk of inducing off-feed problems.

5. Cows can consume their daily feed intake in numerous small meals throughout the day rather than in a few large meals, and this may result in more efficient utilization of nutrients in some feeds (urea for example).

6. Competition for feed is reduced, so boss cows are less of a problem and less manger space is needed per cow (12 instead of 24 in./cow) if the mixture is available throughout the day.

7. Properly formulated complete rations have been shown to support high levels of production in research trials and in dairy herds.

The system also has limitations. Some of these are the following.

1. A mixing device capable of precise measuring of ration components and thorough mixing is necessary, creating additional expense.

2. Dry cows must be separated from milking cows to avoid overconditioning during the dry period.

3. Further separating cows into groups based on nutrient needs (level of production) is necessary for most efficient results.

4. When the program is initiated, cows previously fed concentrate in the milking parlor may be reluctant to enter the parlor.

5. The system may not be easily adaptable to some herds because of existing building layout or small herd size that makes grouping of cows and mixing several rations difficult or inefficient.

6. Long or baled hays must be chopped before they can be thoroughly blended with other ration ingredients.

This feeding system is most adaptable to large herds whose primary forages are silages, haylages, or chopped, cubed, or wafered hays, and with building arrangements convenient for grouping of cows into at least 3 to 4 production groups plus a dry-cow group so that several complete rations can be formulated to more precisely meet group nutrient needs. Drive-through or fenceline feeding arrangements are also desirable, as they facilitate ease of feeding. Satisfactory results can be obtained in smaller herds, but usually with some sacrifice in ration precision if only one or two rations are used, or in labor efficiency if numerous small batches of several rations must be mixed. Regardless of where it is used, there are several key factors involved in successful use of the system. Some of these are the following.

1. Accurate weighing or measuring of each of the feedstuffs in the ration, as inaccuracies can result in a ration that is not as formulated and can contain an imbalance of nutrients.

2. Accurate feed analysis, especially of the forages in the ration, to allow ration formulation within reasonable minimum–maximum nutrient requirement limits.

3. Thorough mixing of ingredients to prevent selective eating by the cows.

4. Separation of dry cows from the milking herd.

Those who have ignored one or more of these factors have obtained poor results both in level of production and in efficiency.

**Table 8-5** Nutrient Specifications of Complete Rations for Dairy Cows[a]

| Specifications | Levels of production | | | |
| --- | --- | --- | --- | --- |
| | High | Medium | Low | Dry cow |
| Production (lb.) | over 60 | 40–59 | below 40 | — |
| Dry matter intake (lb.) | 46 | 40 | 34 | 15 |
| Crude protein (% of DM) | 16 | 15 | 14 | 8.5 |
| Crude fiber (% of DM) | 16–18 | 19–21 | 22–24 | 30–34 |
| Calcium (% of DM) | 0.7 | 0.6 | 0.5 | 0.4 |
| Phosphorus (% of DM) | 0.5 | 0.45 | 0.4 | 0.3 |
| Magnesium (% of DM) | 0.3 | 0.3 | 0.2 | 0.2 |
| Sulfur (% of DM) | 0.3 | 0.3 | 0.2 | 0.2 |
| Salt (TM) (% of DM) | 0.5 | 0.4 | 0.3 | 0.25 |

[a] W. R. Murley, Feeding Quidelines for Dairy Cattle, *V.P.I. & S.U. Ext. Pub.* 630, 1974.

Some suggested nutrient specifications for complete rations and capacities of mixer wagons are presented in Tables 8-5 and 8-6.

## SUMMARY

A good feeding system for dairy cattle is one that utilizes available resources to provide each cow with a ration that meets her nutrient requirements for maximum or near-maximum production, and is palatable, economical, and conducive to a good state of health. A high degree of labor efficiency is also desirable. There is no one essential or best feeding program for all dairymen. There are many feeds, combinations of feeds, feeding programs, and methods of feeding that are quite acceptable.

Keys to developing a practical and profitable feeding program include the following.

1.  Maximum intake of high-quality forage, as forages are generally the most economical source of nutrients for dairy cattle. High quality ensures high intake and high utilization.

2.  Accurate forage analysis, which makes possible greater precision in ration balancing.

3.  A balanced ration—one that contains all the needed nutrients and in correct proportion.

4.  Avoidance of overfeeding and overconditioning of stale and dry cows.

Recent developments in computerized ration formulation programs can

**Table 8-6** Capacity of Mixing Wagons for Different-Sized Herds[a]

| Loads/day | Mixer wagon size, cu. ft.[b] | | | | |
| | 170(136) | 218(175) | 280(216) | 320(256) | 380(370) |
| --- | --- | --- | --- | --- | --- |
| | (Herd or cow group size) | | | | |
| 1 | 30– 40 | 40– 55 | 50– 70 | 60– 80 | 70– 95 |
| 2 | 60– 80 | 80–110 | 100–140 | 120–160 | 140–190 |
| 3 | 90–120 | 120–165 | 150–210 | 180–240 | 210–285 |
| 4 | 120–160 | 160–220 | 200–280 | 240–320 | 280–380 |

[a] W. R. Murley, Feeding Guidelines for Dairy Cattle, *V.P.I. & S.U. Ext. Pub.* 630. (1974). Capacity is calculated at 15 to 20 lb. per cu. ft. of 50% dry matter material, 40 lb. of dry matter per cow daily, and allowing space for mixing.
[b] The numbers in parentheses ( ) indicate bushel capacity.

and should be used to overcome the tedious task of hand calculations and increase precision in developing rations for dairy cattle.

## REFERENCES FOR FURTHER STUDY

Akinyle, I. O., and S. L. Spahr, Stage of Lactation as Criterion for Switching Cows from One Complete Feed to Another, *J. Dairy Sci.*, **58:**917 (1975).

Bath, D. L., et al., Evaluation of Computer Program for Maximizing Income above Feed Cost for Dairy Cattle, *J. Dairy Sci.*, **55:**1607 (1972).

Bath, D. L., Maximum Profit Rations: A Look at the Results of the California System, *J. Dairy Sci.*, **58:**226 (1975).

Brown, L. D., et al., Feeding High Energy Rations for Various Lengths of Lactation, *J. Dairy Sci.*, **57:**459 (1974).

Carr, S. B., and W. R. Murley, Forage Testing—A Tool for Improved Feeding of Dairy Cows. *V.P.I. & S.U. Dairy Guideline Series* 153, 1969.

Carr, S. B., and W. R. Murley, Measuring Forage Intake Rates, *V.P.I. & S.U. Dairy Guideline Series* 154, 1970.

Chandler, P. T., and H. W. Walker, Generation of Nutrient Specifications for Dairy Cattle for Computerized Least-Cost Ration Formulation, *J. Dairy Sci.*, **55:**1741 (1972).

Coppock, C. E., Forage Testing and Feeding Programs, *J. Dairy Sci.*, **59:**175 (1976).

Hillman, D., et al., *Basic Dairy Cattle Nutrition, Mich. St. Bul.* E-702, 1973.

Lodge, G. A., et al. Influence of Preparation Feed Intake on Performance of Cows Fed Ad Libitum During Lactation, *J. Dairy Sci.*, **58**:696 (1975).

Murley, W. R., Feeding Guidelines for Dairy Cattle, *V.P.I. & S.U. Ext. Publication* 630, 1974.

National Academy of Sciences, *Nutrient Requirements of Dairy Cattle*, 4th ed., 1971.

Pardue, F. E., et al. Effect of Complete Ensiled Ration on Milk Production. *J. Dairy Sci.*, **58**:901 (1975).

Stoddard, G. E., Group Feeding of Concentrates to Dairy Cattle, *J. Dairy Sci.*, **52**:844 (1969).

# Principles of Dairy
# Cattle Breeding

**9**

The goals of a breeding program are (1) to improve the herds' genetic ability for high performance in the economically important traits, as rapidly as possible; (2) to do this as economically as possible; and (3) to get the cow bred. All three goals are interrelated, but this chapter is devoted primarily to goals (1) and (2)—improvement of genetic ability for high performance at reasonable cost. To achieve these goals it is necessary to (1) understand the basic principles of selection (breeding); (2) know and understand the tools and information that are available to identify genetically superior animals; and (3) know how to apply these principles and to use available information to develop a profitable breeding program.

## BASIC PRINCIPLES OF
## BREEDING OR SELECTION

Selection can be defined as choosing the parents for the next generation. It is the primary tool for genetic improvement in dairy cattle. Genes are the basic unit of inheritance. A calf receives one-half of its genes from its sire and one-half from its dam. These genes, which together are called the animal's *genotype,* determine the upper limit of a cow's performance ability. Environmental factors, such as feeding, milking practices, and so on, will determine how much of this ability the cow achieves. Total performance, then, is a product of both genotype and environment: genotype gives the animal the ability to perform and environment provides the opportunity to perform. (The performance, or manner in which the genotype is expressed, is called the *phenotype.*) Both are important, and either can be the limiting factor in total performance.

### Type of Inheritance

Traits can be inherited in either a qualitative or a quantative manner. Qualitative traits such as haircoat color or presence or absence of horns are controlled by one or a few pairs of genes and are referred to as *discontinuous* traits. An animal is either horned or polled and has either a dominant or a recessive color. Environment does not have a major effect on the expression of these traits. Quantitative traits such as milk production and physical conformation are controlled by many pairs of genes, each with a small additive effect, and are continuous traits. All cows have conformation and the genetic ability to produce milk, but to varying degrees. The expression of these traits is heavily influenced by environment. For example, if a cow theoretically has the genetic ability to produce 20,000 lb. of milk per lactation, but she receives only enough nutrients to produce 8,000 lb., less than one-half of her genetic ability is expressed because of environmental limitations.

Most of the economically important traits of dairy cattle are inherited in a quantitative manner; hence dairy cattle breeding programs must utilize the basic principles of quantitative genetics. No new genes are made through selection; rather, herd genetic improvement is brought about by increasing the number of desirable genes and decreasing the number of undesirable genes in the herd.

All breeders of livestock, including dairymen, have been trying since animals were first domesticated to improve animals' genetic ability for high performance. Until recently progress was slow in dairy cattle because we lacked the tools to identify accurately genetically superior animals, because the generation interval in cattle is long, and because many performance traits such as milk production are expressed only in the female. In spite of these limitations, considerable progress has been made in increasing the frequency of desirable genes and decreasing the frequency of the undesirable ones in dairy cattle. However, undesirable genes are still present; they will continue to appear and to result in low-performance animals. Careful selection practices will increase the chances for offspring to acquire desirable genes and decrease the chances of acquiring undesirable ones. In the meiotic division of germ cells (producing sperm and ova), literally millions of combinations of genes are possible. Because of this, all sperm or ova from the same parent are not identical in genetic composition. Full sisters may be genetically very dissimilar for many performance traits. The exception to this is monozygotic twins (offspring arising from a single newly fertilized egg; "identical" twins).

Results of dairy cattle breeding, then, are not entirely predictible. Mating superior males to superior-performance females does not always result in high-performance offspring because the offspring may receive a poor sample of genes from one or both parents. The probability of this happening, however, can be greatly reduced by mating animals with a high proportion of desirable genes. This can be illustrated by analogy to dealing cards. If a deck of cards has the usual 13 cards in each of four suits, the probability of dealing a hand with three or four aces is very low. But if the deck contains 26 aces and 26 other cards, the probability of dealing a hand with three or four aces is much higher. If the deck contains 39 aces and only 13 other cards, the probability is even higher; however, the possibility of dealing less than three aces still exists. In this illustration, as in dairy cattle breeding, we can increase the probability of desirable genes (aces) by increasing the proportion of desirable genes (aces) in the parents (deck of cards). We still can expect some poor offspring, and we will need to cull these.

Even though 50% of the genes in the offspring come from the male and

50% from the female parent, most of the genetic improvement in dairy cattle is the result of male selection. This is true because much more selection pressure, fewer are needed thus more can be culled, is possible in the male, and, through progeny testing, the accuracy of estimating a bull's genetic value is much higher than for a cow because bulls can have many more offspring than cows. With the use of artificial insemination, only one in thousands of males born is needed for breeding. In contrast, 70% or more of the females born are needed as they are the producing, as well as the reproducing, units of the herd.

## DEVELOPING A PROFITABLE BREEDING PROGRAM

A logical procedure can be used to develop a profitable breeding program—one that will result in maximum or near-maximum improvement in genetic ability for economically important characteristics of dairy cattle. The essential steps in such a procedure are the following.

1. Identify and record the parentage of each animal in the herd.

2. Enroll the herd in a performance-testing program (DHI; see Chapter 2).

3. Define the goals or objectives for the individual herd.

4. Identify those traits that can or should be improved by selection.

5. Determine the current status of the herd in regard to these traits.

6. Minimize nongenetic female culling so that genetic culling can be maximized.

7. Evaluate the available sources of information to identify genetically superior animals.

8. Breed a high proportion (70 to 80%) of the herd females to sires with proved ability to transmit superior performance in the desired traits.

9. Breed a small proportion of the herd females to highly selected young sires.

10. Avoid breeding to bulls known to be carriers of genetic defects or abnormalities.

11. Breed dairy heifers to dairy sires using artificial insemination when practical.

## Identify and Record Parentage

Each animal born into the herd that is to be reared as a herd replacement or for sale should be permanently identified, whether grade or purebred, before it is removed from its dam. A variety of methods are available for doing this, such as a photograph, sketch, tattoo, freeze brand (see Chapter 24). Permanent identification and recording of parentage are essential if the animal is to be registered in the breed registry or if it is to be enrolled in the Verified Identification Program (VIP). Grade animals should also be sire identified so that their performance records can be used in sire summaries (see Chapter 2). This basic management procedure takes only a few minutes, but it is essential if performance records from that animal are to be useful in accurately identifying animals genetically. Yet nearly 50% of the dairy cows in the United States that are enrolled in DHI are not sire identified. The percentage of cows in a DHI herd that are sire identified is often an accurate index of the management of that herd. Often, managers who overlook this small detail of herd management also overlook many other details, with the overall result of poor herd performance.

## Performance-Testing Enrollment

The basic performance information needed for both female and male selection is the result of performance testing. Individual cow production provides the information needed for culling low producers and for estimating the transmitting ability of both females and males. As mentioned in Chapter 2, the development and wide-spread use of the DHI program probably has been the single most important factor in the improvement of dairy cattle in the United States. Dairymen simply cannot afford not to performance test their cows by enrolling their herds in the DHI or a similar program.

## Define the Goals—Identify the Traits Desired

It is highly unlikely that any two dairymen will have precisely the same goals. It is therefore difficult to develop a standard program to satisfy the needs of each dairy herd manager. It is, however, very important that each dairyman define the goals for the breeding program so that the principles of breeding can be applied and the available tools can be used to help achieve those goals. In spite of the improbability of any two dairymen having precisely the same goals, we can identify several desirable characteristics of dairy cattle that we believe every dairyman should aim for regardless of breed of cattle or whether purebred or grade. They form the basis for efficient production of milk, and milk is the basis for the existence of the dairy industry. Desirable traits of dairy cattle include the following.

1. High milk production.

2. Acceptable milk composition (fat and solids-not-fat).

3. Acceptable utility type (strongly attached, high-quality udders; sound feet and legs; and dairy character).

Many dairymen may have additional goals; for example, high classification score, stature, and others.

## Identify the Traits to Improve by Selection

There are several criteria that can be applied to a trait to help define those desirable characteristics of dairy cattle that we can and should improve by selection. These are the following.

1. Is the trait economically important? Does it contribute significantly to the profitability of the dairy herd operation? Certainly high milk production of acceptable composition meets this criterion. Acceptable utility-type characteristics also meet this criterion if dairy cows are going to be milked by machine and handled in confinement housing systems. For some breeders of registered cattle, high classification scores, show-ring type or color of markings, and many other traits may also have economic importance.

2. Is there variation among individuals for the trait? If no variation exists, the population is *homozygous* for that trait, and the opportunity for improvement through selection does not exist. Conversely, the greater the variation, the greater the potential for genetic improvement. Fortunately, there is wide variation in the economically important traits of dairy cattle. One needs only to observe cows or their performance data to ascertain this.

3. Is the trait heritable? Does it have a moderate to high heritability estimate? Total performance, or phenotype, is a combination of genotype and environment. Only the part that is due to genes can be improved by selection. A heritability estimate is an estimate of that part of the variation that results from genetic composition. Therefore the higher the heritability of the trait, the greater the opportunity for improvement of that trait by selection. Heritability estimates of several traits in dairy cattle are listed in Table 9-1. As can be seen from the

**Table 9-1**  Heritability Estimates of Certain Traits of
Dairy Cattle[a]

| Trait | Heritability estimate | Range |
|---|---|---|
| Production characteristics | | |
|   Milk yield | .25 | .20–.35 |
|   Fat yield | .25 | .20–.35 |
|   Solids yield | .25 | .20–.35 |
|   Fat percentage | .50 | .45–.60 |
|   Solids-not-fat percentage | .50 | .45–.60 |
|   Protein percentage | .50 | .45–.65 |
| Management characteristics | | |
|   Disposition | .25 | .05–.40 |
|   Services/conception | .03 | .00–.10 |
|   Calving interval | .04 | .00–.10 |
|   Longevity | .04 | .00–.10 |
|   Milking rate | .25 | .20–.30 |
| Type characteristics | | |
|   Overall type score | .30 | .15–.35 |
|   General appearance | .25 | .20–.35 |
|   Stature | .40 | .35–.50 |
|   Dairy character | .25 | .15–.35 |
|   Body capacity | .25 | .20–.30 |
|   Mammary System | .20 | .15–.25 |
|     Support | .20 | .13–.30 |
|     Quality | .05 | .04–.06 |
|     Depth | .20 | .15–.25 |
|     Teat placement | .20 | .15–.25 |
|     Rear legs | .15 | .07–.26 |
|     Rump | .20 | .17–.24 |

[a] Adapted from J. M. White, Genetic Parameters of Conformational and Management Traits, *J. Dairy Sci.,* **57**:1267 (1973), and Others.

table, milk production, fat production, and SNF production are moderately (.20 to .39) heritable, whereas milk composition is highly heritable (above .40). This means that there is opportunity for significant improvement in genetic ability for these traits through selection. Most type characteristics, with the exception of rear legs and udder quality, also fall in the moderately heritable category, as does disposition and milking rate. Reproductive efficiency and longevity are very lowly heritable, hence selection pressure to improve these characteristics is likely to be unrewarded.

**4.**  Can the trait be measured accurately and practically? Many traits of dairy cattle can be objectively weighed or measured easily (milk, fat, fat %, etc.). Others can be subjectively measured with a fair degree of ease and accuracy (most type traits), whereas others are difficult to measure. Disposition, udder quality, and perhaps rear legs are three traits that are difficult to measure objectively and accurately. A bad disposition is difficult to define, and it is difficult to determine whether it is inherent or the result of environment. Udder quality is extremely variable, depending on stage of lactation, incidence of mastitis, and time of evaluation in relation to last milking. Rear legs look different depending on the surface on which the cow is standing and how she is standing at the time of evaluation. One questions whether effort should be exerted in trying to improve traits that are not accurately and easily measured.

5.  What is the association of the trait with other economically important traits? Some traits of dairy cattle are positively associated, some have no association, and some are negatively associated. If two economically important, desirable traits have a high positive association, selection pressure need only be applied to one to improve both, thus reducing the number of traits selected for simultaneously. If the traits have no association, selection pressure for one will have no effect on the other, and if they are negatively associated, selecting for improved performance in one will result in decreased performance in the other. Some of the associations between various traits of dairy cattle are presented in Table 9-2. Work reported by Cassell et al.* indicated that overall type score had high and positive correlations with all scorecard traits and overall classification score in Holsteins ranging from +.93 for general appearance to +.64 for dairy character. This essentially means that selection for final type score will include selection for scorecard traits as well. Conversely, the association between milk production and type score and between milk production and fat percentage is negative. This means that if most or all of the selection pressure is based on milk production, then one can expect lower type score and decreased butterfat percentage in succeeding generations. This need not necessarily happen, however, as there is considerable variation in the population of dairy cattle. If one selects first for milk production, and then screens high-producing animals for type score and/or butterfat percentage, genetic progress for negatively associated traits can be achieved concurrently. This will be more difficult for traits with a high degree of negative association such as milk production and fat percentage (−.40) than for

* B. G. Cassell et al., Relationships among Type Traits in Holstein Cattle, *J. Dairy Sci.*, **56**:1171 (1973).

**Table 9-2** Estimated Association of Various Traits of Dairy Cattle[a]

| Traits | Phenotypic correlation | Genotypic correlation |
|---|---|---|
| Milk (kg.) and fat (kg.) | .80 | .70 |
| Milk (kg.) and fat (%) | −.25 | −.40 |
| Milk (kg.) and SNF (kg.) | .90 | .90 |
| Milk (kg.) and SNF (%) | −.25 | −.40 |
| Fat (%) and SNF (%) | .75 | .60 |
| Milk (kg.) and overall type score | −.10 | −.20 |
| Milk (kg.) and general appearance | −.10 | −.20 |
| Milk (kg.) and mammary system | −.15 | −.20 |
| Milk (kg.) and dairy character | .30 | .40 |
| Milk (kg.) and body capacity | −.10 | −.10 |
| Milk (kg.) and udder support | −.10 | −.10 |
| Milk (kg.) and fore udder | −.10 | −.20 |
| Milk (kg.) and rear legs | −.10 | −.15 |
| Final classification and general appearance | .80 | .90 |
| Final classification and mammary system | .75 | .80 |
| Final classification and dairy character | .45 | .60 |
| Final classification and body capacity | .50 | .80 |
| Milk (kg.) and herd life | .30 | .75 |
| Milk (kg.) and gross feed efficiency | .70 | — |
| Milk (kg.) and milking speed | .05 | .10 |

[a] Adapted from J. M. White, Genetic Parameters of Conformational and Management Traits, *J. Dairy Sci.,* **57**:1267 (1973), and numerous other sources.

traits with a smaller negative association such as milk production and type score (−.20).

6. What is the effect of number of traits selected for simultaneously? Assuming equal selection pressure for each trait, the relative genetic progress for any one trait decreases as the number of traits selected for increases (Table 9-3). This fact stresses the importance of applying economic and other criteria to the decision concerning which traits to select for. The potential for genetic progress in any one is drastically reduced as the number of traits selected for increases from one to 30 or more traits.

## Current Herd Status

Determining the current status of the herd in regard to those traits that one has identified as goals of the herd and that can be improved by selection is an important step in developing the breeding program. It identifies the problem

**Table 9-3** Effect of Number of Traits Simultaneously
Selected For on Relative Genetic Progress for Milk
Production

| Traits selected for[a] | Genetic progress for one trait (milk production) | |
| --- | --- | --- |
| | (%[b]) | (kg[c]) |
| 1 trait (milk) | 100% | 500 |
| 2 traits (milk + 1 other trait) | 71 | 355 |
| 3 traits (milk + 2 other traits) | 58 | 290 |
| 4 traits (milk + 3 other traits) | 50 | 250 |
| 5 traits (milk + 4 other traits) | 45 | 225 |
| 10 traits (milk + 9 other traits) | 32 | 160 |
| 20 traits (milk + 19 other traits) | 22 | 110 |
| 30 traits (milk + 29 other traits) | 18 | 90 |

[a] Assuming equal selection pressure for all traits.

[b] Percentage of relative progress for any one trait is equal to $\frac{1}{\sqrt{N}}$ of the progress possible by selection for one trait only, where $N$ = number of traits.

[c] Assuming that selection for milk production only will result in progress of 500 kg.

areas. Three of the most important objectives for most dairymen are milk yield, fat percentage, and type score. The basis of evaluating the current status of the herd as a whole, as well as of individual cows, in milk yield and fat percentage is performance testing in the DHI program. Performance for milk and fat yield may be expressed in several ways.

**D.H.I. Rolling Herd Average for Milk Yield, Fat Yield, and Percentage.** This is the total pounds of actual milk and fat produced by the herd during the last 12 months, divided by the average number of cows in the herd during that period. The fat percentage is calculated by dividing the total pounds of fat by the total pounds of milk. These herd-average figures are excellent evaluations of overall herd performance. They can be compared to breed averages or individual goals to help determine current status.

**Individual Cow Performance Records.** Cow performance is affected by many variables, such as age at freshening, times milked per day, and season of freshening. To evaluate a cow's performance accurately it is necessary to adjust for the many factors that can influence that performance.

FACTORS AFFECTING PRODUCTION PERFORMANCE (RECORDS).

1.  Times milked per day.

2.  Length of lactation or days milked.

3.  Breed of cow.

4.  Age of calving.

5.  Season (month) of calving.

6.  Region.

    In the United States, performance records have been standardized to a twice-a-day milking, 305-day lactation, mature equivalent (2× 305 ME) basis to adjust them for these factors so that milk and fat yield of cows of different ages milked under varying environmental conditions can be compared.

    Appendix Table M contains age-conversion factors for representative ages at calving for the months of January and July for region I for the six U.S. dairy breeds. A complete set of age conversion factors for all breeds, ages, months, and regions is contained in the USDA publication ARS-NE-40, 1974.

    Appendix Table M-1 contains factors for standardizing records of cows milked three or four times daily for varying parts or a complete lactation to a twice daily (2×) milking basis. A complete set of these conversion factors is contained in the USDA publication ARS-52-1, 1955.

    Appendix Table M-2 contains factors for projecting partial lactations (less than 305 days milked) to a 305-day lactation basis. These factors are used only on records terminated in less than 305 days because of environmental influences having no relation to the cow's genetic ability to complete a normal 305 day lactation. When a cow is milked more than 305 days, her yield for the first 305 days is used as her lactation yield. A complete set of factors for projecting incomplete records is contained in the USDA publication ARS-44-239, 1972.

7.  Year of freshening. Records of individual cows can be compared to records of all other cows of that breed that calved in that herd during the same year and season.

8.  Temporary conditions affecting one lactation.

9.  Permanent conditions affecting all lactations. Average all records and

weight in favor of multiple records. Single lactation records have a repeatability of 0.5; therefore one record indicates about one-half of what there is to know about a cow's producing ability, two records about two-thirds, three records three-fourths, and so on, as determined by the formula $\dfrac{N}{N+1}$, where $N$ = number of records.

| $N$ | 1 | 2 | 3 | 4 | 5 | 6 | 7 | 9 |
|---|---|---|---|---|---|---|---|---|
| Weight | 0.5 | 0.67 | 0.75 | 0.80 | 0.83 | 0.86 | 0.88 | 0.90 |

The estimated producing ability (EPA) or estimated relative producing ability (ERPA) of a cow is an estimate of her ability to produce either above or below her herdmates. These figures for milk and fat represent the cow's average difference from herdmates, weighted by the number of lactations in the cow's average ME production expressed in plus or minus deviation from herdmate average. These are excellent figures to use in evaluating cows regardless of length of lactation, age, season of freshening, and so on. The DHI program calculates EPA's for each cow in the herd and periodically (two or three times per year) distributes these to the herd owner or manager. They are valuable tools to determine which cows to cull for low production as well as to identify current status of all cows in the herd.

The EPA is not a very accurate index of a cow's ability to transmit her producing ability, because it does not consider the degree to which production is inherited. An index that includes the performance of a cow's relatives as well as the cow's own performance is a more precise way of evaluating a cow's transmitting ability. Various indices have been developed to evaluate a cow's transmitting ability. The two most commonly used are the USDA cow index (CI) and the estimated transmitting ability (ETA) or estimated average transmitting ability (EATA). The USDA CI includes performance information on the cow herself and her paternal sisters. ETAs or EATAs are calculated by several DHI processing centers. Some include, in addition to performance information on the cow and her paternal sisters, performance information on dams, daughters, and maternal sisters. These indices are valuable tools in evaluating the genetic transmitting ability of cows and can be used to identify which cows should produce the highest performance heifer calves or which cows might be considered as dams of young sires.

Current herd status for milk yield, fat yield, and fat percentage, and in some areas protein and SNF yield and percentage, can be evaluated accurately through DHI performance testing programs. From the raw data collected by DHI, various indices to evaluate cow performance (EPA) and cow genetic transmitting ability (USDA CI, EATA, etc.) can be calculated and be used to more accurately evaluate current status and develop better breeding programs.

Evaluation of herd type status is more difficult because (1) it is subjectively rather than objectively measured, and (2) there is no uniform nationwide program available to both grades and purebreds of all breeds such as the DHI program. Type in purebred herds can be evaluated through breed classification programs. In addition, some DHI organizations and AI organizations have type appraisal programs designed to identify herd and individual cow type problems. Many dairymen also evaluate their own cows to determine type problems. In view of the work published by Cassell et al.,* it appears that final classification score is an accurate method of evaluating type in Holsteins because of its high correlation with all components of type. It is envisioned that more uniform and widely acceptable methods of evaluating type (especially functional aspects of type) will be developed and used in the near future.

## Culling

Intensity of selection in cows is relatively low, as about 70 to 75% of the females are retained to maintain herd size. The 25 to 30% culled annually can contribute significantly to the genetic improvement of the herd if they are culled for genetic reasons (low production because of lack of genetic ability). If, however, most or all of them are culled for nongenetic or management reasons (nonbreeders, those with mastitis, those that have died, etc.), less genetic progress is possible. Rate of culling can also be increased if the generation interval is shortened (heifers freshening at 24 rather than 30 or 36 months of age), reproductive efficiency is high (less cows culled as nonbreeders, and more calves), and calf mortality is low. Genetic progress through female culling or selection then is affected by rate of culling, reason for culling, number of heifer calves born, heifer mortality, and age of heifers at freshening, as well as the genetic ability of the herd replacements. Various combinations of these factors can result in wide variation in genetic progress from female selection.

Mature females should be culled primarily on the basis of their EPA (ERPA), as this is the best index of their ability to produce milk. Young females should be culled on the basis of their pedigree index (PI) or estimated breeding value (EBV). This index is based on pedigree information as is calculated as follows:

$$EBV = \frac{\text{sire's PD} + \text{dam's cow index (or EATA)}}{2}$$

* B. G. Cassell, J. M. White, W. E. Vinson, and R. H. Kleiwer, *J. Dairy Sci.*, **56**:1171 (1973).

The sire's PD is the best estimate of his transmitting ability and the dam's cow index (or EATA) the best estimate of her transmitting ability.

The same indices can and should be used to evaluate contemplated purchases of females. If cows are to be purchased primarily for their milk-producing potential, prospective buyers should insist on knowing their EPA, as it is a much more accurate tool in determining their producing ability than are actual performance records. In the case of heifer purchases, prospective buyers should insist on having pedigree indices to evaluate or should calculate them from the dams' indices and PDs of the sires.

### Evaluate Information to Identify Genetically Superior Animals

Genetic improvement in herd performance for desired traits can be made through female selection, cows or heifers either entering or leaving the herd, or through sire selection. The three primary sources of information to evaluate regarding selection decisions are the pedigree, individual performance, and progeny test information.

**Pedigree.** A pedigree is a record of the ancestry of an animal and should contain, in addition to the ancestry, performance information on those ancestors. The value of the pedigree in evaluating animals is dependent on the amount and accuracy of the information it contains. If it contains PDs for milk, fat, type, and so on, on the male ancestors; EPAs, cow indices, and, for purebreds, type classification, on the female ancestors; and EBVs on immature animals, it can be a valuable tool in selection (Figure 9-1). If it does not contain this sort of information, it is of little or no value (Figure 9-2). It is the primary source of information in the selection of immature animals. Other than the physical appearance of the animal, the pedigree is the only information on which to base immature male or female selection. The pedigree indicates what the animal should be, based on ancestor performance.

**Performance.** Performance results from both genotype and environment and indicates the degree of expression of various traits (phenotype) of the animal, such as milk production, which can be measured objectively, and type conformation, which can be measured subjectively. Performance information is a valuable tool in evaluating milking age cows for performance ability. Use of indices such as EPA or ERPA is a much more accurate performance measure than are actual milk records, as these indices adjust for environmental differences. Performance data indicate what an animal is and are a valuable tool in milking-age female selection.

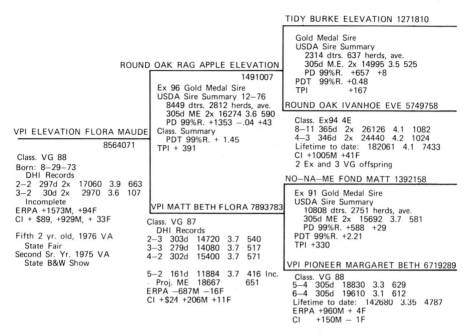

**Fig. 9-1.** An excellent pedigree, showing complete information.

TIDY BURKE ELEVATION 1271810

Gold Medal Sire
USDA Sire Summary
   2314 dtrs. 637 herds, ave.
   305d M.E. 2x 14995 3.5 525
   PD 99%R. +657 +8
PDT 99%R. +0.48
TPI     +167

ROUND OAK IVANHOE EVE 5749758

Class. Ex94 4E
8—11 365d 2x 26126 4.1 1082
4—3 346d 2x 24440 4.2 1024
Lifetime to date: 182061 4.1 7433
CI +1005M +41F
2 Ex and 3 VG offspring

ROUND OAK RAG APPLE ELEVATION
                    1491007

Ex 96 Gold Medal Sire
USDA Sire Summary 12—76
   8449 dtrs. 2812 herds, ave.
   305d ME 2x 16274 3.6 590
   PD 99%R. +1353 —.04 +43
Class. Summary
   PDT 99%R. + 1.45
TPI + 391

NO—NA—ME FOND MATT 1392158

Ex 91 Gold Medal Sire
USDA Sire Summary
   10808 dtrs. 2751 herds, ave.
   305d ME 2x 15692 3.7 581
   PD 99%R. +588 +29
PDT 99%R. +2.21
TPI +330

VPI PIONEER MARGARET BETH 6719289

Class. VG 88
5—4 305d 18830 3.3 629
6—4 305d 19610 3.1 612
Lifetime to date: 142680 3.35 4787
ERPA +960M + 4F
CI   +150M — 1F

VPI ELEVATION FLORA MAUDE
                 8564071

Class. VG 88
Born: 8—29—73
DHI Records
2—2 297d 2x 17060 3.9 663
3—2 30d 2x 2970 3.6 107
Incomplete
ERPA +1573M, +94F
CI + $89, +929M, + 33F

Fifth 2 yr. old, 1976 VA
  State Fair
Second Sr. Yr. 1975 VA
  State B&W Show

VPI MATT BETH FLORA 7893783

Class. VG 87
  DHI Records
2—3 303d 14720 3.7 540
3—3 279d 14080 3.7 517
4—2 302d 15400 3.7 571

5—2 161d 11884 3.7 416 Inc.
. Proj. ME 18667    651
ERPA —687M —16F
CI +$24 +206M +11F

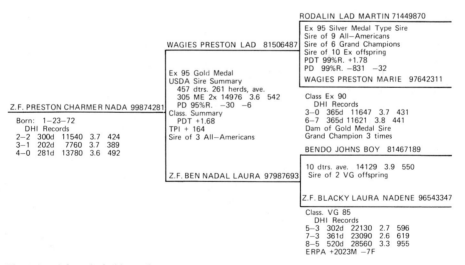

**Fig. 9-2.** A less desirable pedigree.

RODALIN LAD MARTIN 71449870

Ex 95 Silver Medal Type Sire
Sire of 9 All—Americans
Sire of 6 Grand Champions
Sire of 10 Ex offspring
PDT 99%R. +1.78
PD 99%R. —831 —32

WAGIES PRESTON MARIE 97642311

Class Ex 90
  DHI Records
3—0 365d 11647 3.7 431
6—7 365d 11621 3.8 441
Dam of Gold Medal Sire
Grand Champion 3 times

WAGIES PRESTON LAD 81506487

Ex 95 Gold Medal
USDA Sire Summary
  457 dtrs. 261 herds, ave.
  305 ME 2x 14976 3.6 542
  PD 95%R. —30 —6
Class. Summary
  PDT +1.68
TPI + 164
Sire of 3 All—Americans

BENDO JOHNS BOY 81467189

10 dtrs. ave. 14129 3.9 550
Sire of 2 VG offspring

Z.F. BLACKY LAURA NADENE 96543347

Class. VG 85
  DHI Records
5—3 302d 22130 2.7 596
7—3 361d 23090 2.6 619
8—5 520d 28560 3.3 955
ERPA +2023M —7F

Z.F. PRESTON CHARMER NADA 99874281

Born: 1—23—72
DHI Records
2—2 300d 11540 3.7 424
3—1 202d 7760 3.7 389
4—0 281d 13780 3.6 492

Z.F. BEN NADAL LAURA 97987693

219

**Progeny Test.** A progeny test is the measurement of the performance of an animal's offspring. It is the most accurate method of identifying the transmitting ability of an animal. If adequate numbers of unselected progeny are included in the test, it will accurately differentiate between differences resulting from environment and those arising from genotype. The purpose of a progeny test is to determine the true breeding value—the ability to transmit to the next generation—of an animal. The pedigree indicates what an animal should be; performance indicates what it is; and progeny test shows what its transmitting ability is.

Progeny tests are not so widely used in female selection as in male selection. This is primarily because any one cow has relatively few offspring. Progeny tests have been used since the 1930s as a method of evaluating dairy sires. Initially, progeny testing of bulls consisted of comparing the average production of the daughters of a bull with that of their dams (daughter-dam sire proofs). This progeny-test index lacked accuracy because of the possibility that improved management practices (environment) were adopted between the time the dams and their daughters were in the herd. A major part of the positive proof of many sires, in fact, was from improved management (feeding, milking, etc.) rather than improved genetic ability. The need for an index that would more accurately measure a sire's ability to transmit genetic improvement was apparent.

During the early to mid 1960s the USDA began comparing the production of daughters of a bull with their herdmates (other cows producing in the same herd at the same time). This herdmate comparison was used to calculate a bull's predicted difference (PD). PD is an estimate of the genetic superiority of a bull based on the comparison of the performance of his daughters with their herdmates. There is an adjustment for number and distribution of daughters. It indicates the amount of milk (PD milk), fat (PD fat), butterfat percentage (PD% test) and the dollar value per lactation (PD$ value) by which future daughters of a bull are expected to differ from their herdmates in breed-average herds. The basic formula for calculating a bull's PD is

$$PD = R[D-HM + .1(HM-BA)]$$

where

> R = Repeatability. Repeatability is a measure of the confidence one can have in the accuracy of the proof. It is based on the number and distribution of daughters and herds represented in the progeny test. The more daughters and herds represented in the progeny test, the more accurate the proof and the higher the repeatability. The higher the repeatability the more confidence one can have in the accuracy of the progeny test (Tables 9-4 and 9-5). Repeatability is not an indication of the fertility level of the bull or his daughters.

D = Daughter average production on a 2× 305 ME basis.

HM = Herdmates' average production on a 2× 305 ME basis.

D-HM = Average deviation of daughter production from herdmates.

BA = Breed average production on a 2× 305 ME basis.

.1(HM-BA) = An adjustment for genetic differences among herds to account for the level of competition provided by the herdmates. This adjustment is based on the assumption that 20% of the differences between a herd average and the breed average are genetic differences and that one-half of this difference (.1) is from the sire being evaluated and the other one-half is from the dams.

Actual PD values of a sires' daughters are most precise in breed-average herds. One can expect the deviation from herd average for daughters of a +PD bull to be lower in above-breed-average herds and higher in below-breed-average herds. Contrasting results can be expected of daughters of a −PD bull. The effect of herd average on expected deviation from herdmates using a PD +1000 lb. milk and a PD −1000 lb. milk bull are presented in Table 9-6. These figures are based on the assumption that approximately 10% of the superiority or inferiority of herds above or below breed averages results from genetic differences.

In the early 1970s it became apparent that the use of PDs as a tool in dairy sire selection had become widespread, but variable in use by dairymen. A sire's PD value declined over time from genetic trend. For these and other reasons, in the fall of 1974 the USDA initiated the use of the modified contemporary comparison (MCC) method of indexing the genetic merit of

**Table 9-4** Variation in Repeatability, Number and Distribution of Daughters

| No. daughters | No. herds | Daughters per herd | % repeatability |
|---|---|---|---|
| 20 | 1 | 20 | 21 |
| 20 | 4 | 5 | 37 |
| 20 | 20 | 1 | 48 |
| 40 | 1 | 40 | 23 |
| 40 | 8 | 5 | 55 |
| 40 | 40 | 1 | 66 |
| 80 | 1 | 80 | 25 |
| 80 | 16 | 5 | 71 |
| 80 | 80 | 1 | 80 |

**Table 9-5** A Comparison of 60% and 80% Confidence Intervals by Breed for Predicted Differences at Given Repeatability Levels[a]

| Repeatability (%) | 60% confidence interval (lb.) | 80% confidence interval (lb.) |
|---|---|---|
| Brown Swiss and Holstein | | |
| 20 | ±414 | ±630 |
| 30 | ±387 | ±589 |
| 40 | ±358 | ±546 |
| 50 | ±327 | ±498 |
| 60 | ±293 | ±446 |
| 70 | ±253 | ±386 |
| 80 | ±207 | ±315 |
| 90 | ±146 | ±223 |
| Ayrshire, Guernsey, Jersey, Milking Shorthorn, Red Dane, and mixed breeds | | |
| 20 | ±301 | ±458 |
| 30 | ±281 | ±429 |
| 40 | ±261 | ±397 |
| 50 | ±238 | ±362 |
| 60 | ±213 | ±324 |
| 70 | ±184 | ±281 |
| 80 | ±150 | ±229 |
| 90 | ±106 | ±162 |

**Source.** USDA, *DHIA Sire Summary List*, ARS-NE-24-2, Spring 1975.

dairy sires. Use of the MCC results in a PD for dairy sires that is labeled "PD 74"; 74 indicates the year used as a base. If base years change in the future, 74 will be changed to show the new base year. The formula for PD 74 is

$$PD\ 74 = R(D\text{-}MCA + SMC) + (1\text{-}R)\ GA$$

where:

R = Repeatability.

D = Daughter average production on a 2× 305 ME basis.

MCA = Modified contemporary average production (groups of herdmates that are of a similiar age; group 1 = first lactations and group 2 = second and later lactations) on a 2× 305 ME basis.

SMC = Average genetic merit of the sires of the contemporaries (an adjustment for the level of competition from the contemporaries).

GA = Genetic group average (an adjustment for the pedigree value of the sire based on the genetic transmitting ability of his sire and maternal grandsire). This adjustment is slight for high-repeatability bulls, but significant for low-repeatability bulls.

The major differences between the PD 74 program and earlier PD values are the following.

1. Use of new factors to adjust for age and month of calving.

2. Use of records in progress of 40 or more days in milk.

3. The former average correction for the genetic level of the herd is replaced by a correction that divides herdmates into two groups, first lactation and later lactations, and gives more weight to herdmates of similar age.

4. A slight difference in the calculation of R values, which gives less value to large numbers of daughters in one herd and more emphasis to new daughters and new records in herds other than those already containing daughters of the bull.

5. Each bull's modified contemporary deviation will be regressed to the mean of a genetic group rather than to breed average. This grouping will

**Table 9-6** Expected Average Deviation of Daughters of a Bull From Herdmates in Herds with Various Herd Averages

| | Expected deviation of herdmates from herd average | |
| Herd average | Bull A: PD = +1000 lb. milk | Bull B: PD = −1000 lb. milk |
| --- | --- | --- |
| 18000 | +600 | −1400 |
| 17000 | +700 | −1300 |
| 16000 | +800 | −1200 |
| 15000 | +900 | −1100 |
| 14000 (breed average) | +1000 | −1000 |
| 13000 | +1100 | −900 |
| 12000 | +1200 | −800 |
| 11000 | +1300 | −700 |
| 10000 | +1400 | −600 |

be based on the PD of the sire and maternal grandsire plus information as to whether the bull is sampled in AI service.

These changes in the calculation of the PD 74 should improve the accuracy of estimate of the genetic transmitting ability of dairy sires, particularly younger sires whose repeatability is low. It should not affect the use of the PD index as a selection tool. Undoubtedly, there will be future changes in the calculation of estimates of a sire's genetic transmitting ability that will be even more accurate than PD 74; however, it is the most accurate tool we have at the time of preparation of this text and should be used as the major tool in sire selection until a more precise tool is developed.

## Breed Most of the Herd Females to Sires with Proved Ability to Transmit Superior Genetic Ability

Approximately 70 to 75% of the breeding herd, including heifers of breeding age, should be bred to bulls with high PDs for milk of acceptable composition that are transmitting other traits that meet the objectives or goals of the individual dairyman. There are various methods of identifying such bulls. The following system of selecting for production and screening for other traits has proved effective for many dairymen.

1. From the USDA sire summaries that are published three times a year, make a list of all the bulls available that have high PDs for milk or dollar value, with at least 50% repeatability. The definition of high can vary among individual dairymen depending on present herd average, goals, etc., but should be a minimum of +800 to 900 for Holsteins, +600 to 700 for Brown Swiss, and +500 to 600 for the other U.S. dairy breeds.
2. From this list of high +PD milk bulls, eliminate those bulls that are much below stud average in conception rate or are overpriced in relation to production transmitting ability. This recommendation may be modified for individual dairymen who wish to breed certain cows to high-priced or low-conception bulls because of the salability of resulting offspring.
   Determining how much a dairyman can afford to pay for semen is primarily controlled by three factors.
   a. The degree of performance superiority of the bull's daughters (PD for milk and fat or dollar value).
   b. The bull's fertility rate.
   c. The merchandising value of the bull's daughters (pedigree popularity, type, etc.).
   The first two factors are important to all dairymen; the importance of the third varies greatly depending on the individual dairyman's income from the sale of cattle. If one considers only the first two factors, degree of performance

superiority and fertility rate, one can readily arrive at a figure that answers the question, "How much can I afford to pay for the really good bull?" by using the formula:

$$\text{Amount you can afford} \atop \text{to pay for semen} = \frac{\$ \text{ value} \times .86 \times .50 \times 3 \times 2}{\# \text{ ampules needed per fresh heifer}}$$

$$= \frac{\$ \text{ value} \times 2.58}{\# \text{ ampules needed per fresh heifer}}$$

where

$ value = the USDA PD dollar value

.86 = correction to convert mature equivalent (ME) to actual milk

.50 = correction for extra costs or producing the additional milk (feed, etc.)

3 = number of lactations per animal (research indicates that the average cow has about three lactations in her lifetime)

2 = impact on future generations (½ of the cow's superiority passed on to her daughter, ¼ to her granddaughter, and ¹/₈ to her great granddaughter, etc.)

# ampules needed/fresh heifer = number of ampules of semen used to get one heifer ready to freshen and be placed in the milking string. Values to be used are given in Table 9-7.

For example, if your choice were between:

*Bull A.* PD $ value = +$10, 60–90 day NR = stud average
*Bull B.* PD $ value = +$100, 60–90 day NR = +5% above stud average

Amount you can afford to pay per ampule for Bull A:

$$\frac{\$10 \times .86 \times .50 \times 3 \times 2}{6.0} = \$4.30$$

Amount you can afford to pay per ampule for Bull B:

$$\frac{\$100 \times .86 \times .50 \times 3 \times 2}{5.5} = \$46.90$$

**Table 9-7**  Semen Usage at Varying Conception Rates

| ± Stud average 60–90 day nonreturn | Services per fresh heifer |
| --- | --- |
| +10% | 5.1 |
| + 5% | 5.5 |
| ± 0% | 6.0 |
| − 5% | 6.5 |
| −10% | 7.0 |
| −15% | 7.5 |

You can afford to pay $42.60 ($46.90 − $4.30) *more* per ampule for Bull B than for Bull A. Even though you will have $232.15 ($46.90 × 5.5 − $4.30 × 6.0), more investment in semen per replacement heifer, this extra cost will be recovered through increased net milk sales from the cow and her future offspring.

As mentioned earlier, using this formula to determine the value of semen considers only performance superiority of the daughters and fertility rate of the sire. For those who realize a significant part of their income from cattle sales, other factors such as type and pedigree popularity are also important and can add to the value of the semen.

Price lists on available bulls are readily available; however, information on bull reproductive efficiency is less available. The authors suggest that bull studs have an obligation to provide dairymen with such information on a within-stud between-bulls comparison basis. Dairymen should insist on receiving this information so that they can better evaluate which bulls to use in their herds and the economics of purchasing AI service.

3. Screen the remaining bulls for their ability to transmit other traits that are economically important to the goals of individual dairymen. Some of these are:

   a. Percent Incomplete. If a bull has a high percent incomplete record, this indicates a significant number of his daughters were removed prior to the completion of their first lactation. If this figure is relatively high (over 12 to 15%), the reasons for that culling should be determined before the bull is used heavily, especially if the bull also has a low repeatability. The high % incomplete may result from environmental reasons, or daughters of the bull may have been sold to dairymen whose herds were not on test. It may, however, result from a high incidence of some undesirable characteristic such as very poor functional type, bad disposition, slow milking, etc. If the unusually high % incomplete records cannot be attributed to nongenetic reasons, dairymen should not use the bull heavily.

   b. Type or Conformation. Type is moderately heritable, and although it is not positively correlated with production, certain characteristics of type appear to be positively correlated to yield and longevity (Table 9-8).* Dairy character also has a high genetic correlation (+.40) with milk yield. In view of the high correlation between type components and overall score mentioned earlier in this chapter, it seems that selection for type score or using PD for type (PDT) in Holsteins would be an acceptable method of screening for type. Recent analysis of Holstein data by Grantham et al.** indicate that if type is ignored in the selection process, it is likely to decrease at a rate of about .6 (final type score) per generation. Whether this decline would eventually result in serious type problems after a few years is not known for certain, but potential problems can be avoided by applying some selection pressure for type. Based on data analyzed in this same study, the following observations were made. From a group of 1,000 bulls, the 50 having the

* H. D. Norman, and L. D. Van Vleck, *J. Dairy Sci.,* **55**:1726 (1972).
** J. A. Grantham, et al., *J. Dairy Sci.,* **57**:1483 (1974).

**Table 9-8** Genetic Correlations of Yield with Some Udder Traits[a]

| Trait | First lactation milk yield | Lifetime milk yield |
|---|---|---|
| Udder halving | .26 | .64 |
| Rear udder length | .21 | .48 |
| Fore udder length | −.54 | .31 |
| Depth of udder | .36 | −1.08 |
| Height of rear udder | −.15 | .25 |
| Strength rear udder attachment | −.27 | −.02 |
| Strength fore udder attachment | −.71 | 2.23 |

[a] H. D. Norman and L. D. Van Vleck, reprinted from *J. of Dairy Sci.*, **55:**1726 (1972) (published by the American Dairy Science Association).

highest PD milk were selected. These averaged +1,130 PD milk and their daughters averaged 79.7 in final type score. From these 50 highest PD milk bulls, the 10 with daughters having the highest average type score were summarized. They averaged +1,312 PD milk and 81.7 final type score. This illustrates a very important point: even though type and milk production apparently have a slight negative association, genetic progress can be made in both concurrently if careful breeding management is applied.

c. Milk Composition. As mentioned earlier in this chapter, there is a negative association between milk yield and milk composition. The economic importance of acceptable milk composition is also well known; therefore, as was the case with type, some selection pressure (especially in the low fat breeds) must be exerted to maintain acceptable composition.

d. Body Size and Efficiency. There is recent evidence that although larger cows within a breed give more milk, they also do it less efficiently and less profitably. Body size also probably is associated with feeding speed during the first lactation, which has a high (nearly 1.0) genetic correlation with lifetime milk yield. The interrelationships of body size, efficiency of feed utilization, and yield are apparently complex and not well defined. It appears that little, if any, selection pressure should be exerted on body size until these interrelationships are more clearly defined. If yield and body size are highly correlated, selection for yield should result in increased size and, if not, why have it?

e. Other factors such as milking speed, mastitis resistance, productive efficiency, and reproductive efficiency are economically important to dairymen, but their interrelationships with other economically important traits are not well defined at this time. Selection pressure for these traits should not be great until they are more clearly defined. Pedigree popularity and show-ring type of daughters are also economically important to some dairymen who receive a significant part of their income from the sales of purebred stock. These individuals should also screen for these traits, but not at the sacrifice of yield traits.

Following this recommendation of breeding 70 to 75% of the herd to four to eight bulls selected and screened in this manner should ensure genetic progress for the traits defined in the goals or objectives of the breeding program at reasonable cost.

## Breed a Small Portion of the Herd to Highly Selected Young Sires

Breeding 25 to 30% of the herd to highly selected young sires (those with high EBVs or those with high PDs and low repeatability) should be a part of the breeding program to (1) ensure adequate sampling of young sires, and (2) to obtain the services of future "hot bulls" while their semen is still reasonably priced. The first reason is essential if rapid genetic progress in dairy cattle is to continue. It is so important that most or all bull studs arrange cooperator herd programs with a broad cross-section of dairymen whereby these dairymen sample young sires regularly. The second reason is less essential, but does offer the opportunity to take early advantage of recently summarized sires with low repeatability. Often semen from these bulls is more economically priced and, if their use is limited to 25 to 30% of the herd, the risk of significantly decreasing the herd's genetic ability is relatively low.

## Avoid Using Carriers of Genetic Defects and/or Abnormalities

There is a wide variety of recessive genetic abnormalities present in the dairy cattle population varying from lethals such as achondroplasia to abnormalities such as wry tail and wry face. Detection of carrier animals is difficult because the physical appearance of carriers is the same as normal animals. A carrier will provide one-half of his offspring with the undesirable gene. If an animal has an abnormal offspring, the animal can be identified as a carrier because one gene must have been acquired from each parent. Depending on the frequency of the gene in a population, carriers may be detected very quickly or may be undetected for a long period of time. Mating a bull to known carriers is a method of quick detection of his carrier status. Bull studs monitor carrier status of bulls very closely during progeny testing and generally make this information known to dairymen. Dairymen should report any occurrence of abnormalities to AI representatives.

## Breed Heifers to Dairy Sires

In a progressive breeding program (as discussed in this chapter) the first calf heifers should have higher genetic ability than the current cow herd. If these heifers are bred to beef bulls, dairymen are sacrificing 25 to 30% of potential genetic progress. Many dairymen feel that female calves from first-calf heifers are genetically inferior because first-calf heifers do not milk as heavily during the first lactation. This lowered yield is not from lack of genetic ability, but rather from the immaturity of the animal. If dairymen are to maximize genetic progress, all breeding females in the herd should be utilized, including first-calf heifers. The two other most commonly stated reasons for not breeding heifers to AI +PD bulls are calving difficulties and the inconvenience of restraining the heifer for insemination. The first problem can be greatly alleviated by feeding heifers well so they reach adequate size at first calving and by breeding them to AI bulls known to sire smaller calves. This information on calf size is known on most proved AI sires and is available from bull stud personnel. In the case of heifers, calf size is another economically important trait for which bulls should be screened. The second problem, that of inconvenience in catching and restraining heifers for insemination, can be alleviated in several ways. One of the most effective for many dairymen has been to build or fence off a small area near the main dairy operation. A holding pen or small corral with a breeding chute is constructed within this area and breeding-age or -size heifers are moved to this area. The heifers are kept in this area for seven or more weeks and bred artificially during this time. After this, they are returned to pasture in the company of a clean-up bull. If a clean-up bull is used, he should be a young dairy bull selected on the basis of his estimated breeding value or pedigree index. This practice results in 60 to 80% of the heifers being bred artificially. Some dairymen prefer to leave the heifers in the holding area until they are diagnosed pregnant. Other dairymen typically mix their breeding-age heifers with the milking herd until they are diagnosed pregnant. Other future possibilities include heat synchronization of heifers so that all or most of them can be bred within a few days. The use of progestational compounds and prostaglandins for this purpose is in the experimental stage, but should be closely observed for future use.

## SYSTEMS OF BREEDING DAIRY CATTLE

The pros and cons of inbreeding, linebreeding, outcrossing, and crossbreeding have been discussed at length by many authors.

## Inbreeding and Linebreeding

Inbreeding may be defined as the breeding of very close relatives such as son to dam, sire to daughter, or brother to sister. Linebreeding may be defined as the breeding of animals more closely related to each other than the average of the breed but less closely related than in inbreeding. Often linebreeding is referred to as mild inbreeding. The effect of these breeding systems is similar and varies with the closeness of the breeding. Inbreeding increases the likelihood of similar genes becoming paired, thereby increasing gene homozygosity and decreasing gene heterozygosity. This has the following effects in dairy cattle.

1. Exposing recessive genes in a population. This can be used to identify carrier animals.

2. Increasing similarity between animals within the same inbred group or line of animals.

3. Decreasing general vigor. Calf mortality is increased, growth rate is decreased, and mature body size is smaller.

4. Decreasing performance. Both milk production and reproductive performance are lower in inbred animals. It is estimated that milk production decreases about 0.4 to 0.5% for each 1% increase in inbreeding.

In view of these effects of inbreeding in dairy cattle, it is doubtful if it should be used by most dairymen. The authors do not recommend inbreeding as a practical system of breeding dairy cattle.

## Outcrossing

Outcrossing may be defined as the breeding of animals of the same breed but less closely related than the average of the breed. Evidence indicates this type of breeding system, when compared to inbreeding, results in the likelihood of increased heterozygosity, increased vigor, decreased calf mortality, higher productive performance, and higher reproductive performance. Most dairy geneticists as well as the authors currently recommend this system of breeding.

## Crossbreeding

Crossbreeding may be defined as the breeding of purebred animals belonging to two different breeds. Various combinations of crossbreeding in dairy cattle have been researched for many years and the expected results are fairly well defined. Crossbreeding in dairy cattle is likely to result in increased calf survival rate, growth rate, and fertility, as is the case with other species of livestock. Milk production of crossbreeds, however, has been between the averages of the breeds of the two parents (higher than the low one, but lower than the high one) rather than higher than either parent. Primarily for this reason crossbreeding of dairy cattle has not been a popular or highly recommended practice. With the advent of provisional registration by some breed associations, more breeders of non-Holsteins may crossbreed to Holsteins in an effort to more rapidly increase genetic ability for high milk production.

## SUMMARY

The goals of a dairy breeding program are to maximize genetic progress for desirable traits of dairy cows. These include the production of a large amount of milk of acceptable composition and acceptable utility type, as well as other individual variable goals. The development and use of the DHI program has provided the basic information needed to develop accurate tools (indices) needed to identify genetically superior dairy animals, both male and female. These indices should be used as the basis for selection of both dairy males and females. The development and widespread use of artificial insemination has provided a method of making widespread use of genetically superior males, yet at the time of preparation of this text only about 50% of the U.S. dairy herd was being bred artificially. Currently the dairy industry has the tools to make rapid genetic progress for most economically important traits. Other traits are being studied. More accurate tools to evaluate these traits will surely be developed. It is imperative that dairymen use the current tools and future ones as they are developed if genetic progress is to be maximized. Genotype does give the animal the ability to perform and puts a genetic ceiling on performance. Dairymen must constantly strive to raise the genetic ability of their cattle, as well as to manage them properly so they are given the opportunity to perform at or near their genetic ability. Based on current information, outcrossing within the breed will probably result in the most genetic progress for production for most dairymen.

## REFERENCES FOR FURTHER STUDY

Barr, H. L., and Ludwick, T. M. *Developing a Modern Breeding Program, Ohio State University Coop. Ext. Service Bul.* No. 485, 1967.

Buffington, R. *Breeding for All It's Worth, V.P.I. & S.U. Coop. Ext. Service Bul.* No. 659, 1975.

Cassell, B. G., et al. Genotypic and Phenotypic Relationships among Type Traits in Holstein-Friesian Cattle, *J. Dairy Sci.*, **56**:1171 (1973).

Dickenson, F. N. Using U.S.D.A.-D.H.I. Sire Summary Information in Herd Management, *J. Dairy Sci.*, **58**:233 (1975).

Dickenson, F. N., et al. Sire Summaries and Cow Indexes, *J. Dairy Sci.* **57**:977, 1974. Also see *J. Dairy Sci.*, **57**:951; **57**:955.

Grantham, J. A., White, J. M., Vinson, W. E., and Kliewer, R. H. Genetic Relationship Between Milk Production and Type, *J. Dairy Sci.* **57**:1483 (1974).

*Hoard's Dairyman.* September issues 1967–1977.

McDaniel, B. T. Why New Sire Summaries are Needed, *J. Dairy Sci.*, **57**:951 (1974).

McDowell, R. E., and McDaniel, B. T. Interbreed Mating in Dairy Cattle, Yield Traits, *J. Dairy Sci.*, **51**:767 (1968); Herd Health, **51**:1275; Economic Aspects, **51**:1649. Also see, **52**:1624; **53**:757; **57**:220.

Norman, H. D., et al. Type Traits and Production, *J. Dairy Sci.*, **57**:647 (1974).

Norman, H. D., et al. USDA-DHIA Factors for Standardizing 305-day Lactation Records for Age and Month of Calving, USDA-ARS-NE-40, 1974.

*Proceedings of the National Workshop on Genetic Improvement of Dairy Cattle,* April 6–7, 1976.

*USDA Cow Index List.* published three times/year.

*USDA Sire Summary List.* published three times/year.

White, J. M. Genetic Parameters of Conformational and Managemental Traits, *J. Dairy Sci.*, **57**:1267 (1974).

White, J. M., and Nichols, J. R. Relationship between First Lactation, Later Performance and Length of Herd Life, *J. Dairy Sci.*, **48**:468 (1965).

# 10

# Maintaining Breeding Efficiency

**10** Profitable milk production and genetic improvement of dairy cattle are dependent on a high degree of reproductive efficiency. The production of milk is a secondary sex characteristic; hence milk production is dependent on reproduction. Genetic improvement is dependent on an adequate supply of high-genetic-potential heifers to replace cows culled for low production or other reasons. This supply of heifers is again dependent on a high rate of reproductive efficiency, as well as a good calf and heifer rearing program.

Most dairymen and reproductive specialists agree that a 12-month calving interval is ideal to maximize production and profit although there is some recent evidence that a slightly shorter (11½ month) or slightly longer (12½ month) calving interval may be ideal for some individual cows. Failure to maintain this high degree of reproductive efficiency is a major economic loss to the dairy industry, estimated in 1974 in the United States at $540 million annually.* These losses occur from decreased milk production, decreased feed efficiency, decreased number of calves, decreased value of valuable animals, and increased treatment costs.

Sterility, which is the complete absence of reproductive ability, describes animals that cannot reproduce. Such animals are usually easy to identify and should be culled from the herd. These include freemartin heifers, bulls who do not produce live sperm cells, and other animals that, through inherent abnormalities, injury, or disease are sterile. Infertility or lowered fertility, which is subnormal breeding efficiency, describes animals that are not sterile but are not normally fertile. These animals are not so easily detected and are a more serious and costly problem for most dairymen than are sterile animals. A normal cow can be described as one that becomes pregnant on the first or second service and produces a live healthy calf every 12 to 13 months.

Sterility and infertility are not a single problem, but rather a very complex one and may be the result of any one or more of a broad spectrum of factors that are known to influence fertility as well as many other factors that are yet to be understood. These include management, genetic, physiological, and disease factors.

## GOALS OF REPRODUCTIVE MANAGEMENT
## AND MEASURES OF
## BREEDING EFFICIENCY

We have said that the goal of the reproductive management program should be to have each cow calve every 12 months. Perhaps a more complete goal

* J. A. Lineweaver and G. W. Spessard. *J. Dairy Sci.*, **58**:256 (1975).

would be "to have each heifer freshen by 24 to 25 months of age and every 12 months thereafter." There are several ways to monitor the rate of reproductive efficiency of dairy herds. It is important to have adequate records so that one or more of these measures can be calculated. Deficiencies or problems can then be identified and corrective measures can be implemented. Some of these measures of reproductive efficiency are the following.

1. Age of heifers at first freshening. Most researchers agree that an average age of 24 months at first freshening is ideal in terms of maximizing production per day of life of the animal. If the age exceeds 27 months it is a costly problem and the cause(s) should be identified and corrected.
2. Calving interval (CI). A 12-month average calving interval is thought to be ideal. This is seldom achieved in practice, but a good practical goal is 12½ months. If the average herd calving interval exceeds 13 months, it indicates a serious problem and the cause(s) should be identified and corrected. DHI monthly herd summaries list projected minimum CI to the nearest tenth of a month.
3. Days open (DO). An average days open (number of days between freshening and conception) of 85 is thought to be ideal. A good practical goal is an average of 100 days open. If the average days open exceeds 110 to 115, it indicates a serious problem and the cause(s) should be identified and corrected. DHI monthly herd summaries list average days open on a per-day basis.
4. Services per conception (S/C). An ideal, but impossible, goal for average S/C would be 1.0. A more realistic goal is 1.5 S/C. If this figure exceeds 1.75 on a herd-average basis, it indicates a serious problem and the cause(s) should be identified and corrected.
5. Percentage of nonreturns (NR). The percentage of NR is the measure used in most artificial breeding associations to measure the breeding efficiency in bulls. An AI nonreturn is an animal that has been bred and for which there is no request for another breeding. This measure obviously is not a very exact one as there are many reasons why another service is not requested other than the cow becoming pregnant. Some cows may die, others may be sold, still others may be bred naturally or be bred to a bull from another stud, or the breeding receipt may not be completed properly. The length of time after breeding that the nonreturns are counted will also affect the percentage of NR. Most AI studs calculate a 30- to 60-day and a 60- to 90-day NR. The latter allows more time for repeat services and is usually 5 to 9% lower than the former. It also includes a greater proportion of early embryonic mortality than a 30- to 60-day NR. In spite of these limitations, the 60- to 90-day NR rate is a useful measure of fertility and should be used by dairymen to compare fertility rate of AI bulls within the following limitations.*
   a. Compare only bulls within the same stud and whose semen is similarly priced.
   b. Use a plus or minus from bull stud average for the within stud comparison.
   c. Compare only bulls with sufficient number of services and preferably compare NR rates over a 3–6 month period of time.

* R. G. Saacke, *Hoards Dairyman,* Dec. 25, 1972.

6. Percent conception to first service, first two services, first three services. Practical goals can be stated as 60% for first service, 80% for the first two services and 90% for the first three services. Herd average figures of less than 55%, 75%, and 85% indicate a serious problem and the cause(s) should be identified and corrected.

7 Percentage of cows that calve within a year (% calf crop). A fertile herd should produce a 90% calf crop annually if cows only are considered or a 110% calf crop annually when heifers are included and turnover rate is 25 to 30%.

8. The number of days per year that a cow carries a calf has been used as a measure of reproductive efficiency. If a cow carries a calf 9 months of a year, she is rated as having a reproductive efficiency (RE) of 100%.

Gilmore and others** have developed the following for estimating the reproductive efficiency of dairy animals.

$$RE = 12 \times \frac{\text{No. calves born}}{\text{Age of cow (mo.)} - \text{age at first breeding (mo.)} + 3} \times 100$$

For example, a 5 year-old cow (60 months) that was successfully bred at 15 months of age and had dropped 4 calves would have a 100% reproductive efficiency.

$$RE = 12 \times \frac{4}{60 - 15 + 3} \times 100 = 100\%$$

If, however, she was 6 years old (72 months) the results would be only 80%.

$$RE = 12 \times \frac{4}{72 - 15 + 3} \times 100 = 80\%$$

With such a system the reproductive efficiency of each animal in the herd can be determined. It is practically impossible to get a herd with a reproduction efficiency of 100%, and so it should be used only as a goal.

There are also many other measures of reproductive efficiency that can be used, such as percent of cows in heat within 50 days of calving, percent of cows open over 100 days, percent of cows with abnormal estrus cycles, etc., and many of these as well as those discussed in this chapter can be useful in identifying reproductive efficiency problems. A summary of some goals and measures of reproductive efficiency is presented in Table 10-1.

## PHYSIOLOGY OF REPRODUCTION

Sterility or infertility can have a broad spectrum of reasons or problems. In order to better understand and identify the problem areas, it is essential to

** L. O. Gilmore, *Dairy Cattle Breeding*, J. B. Lippincott Co., Chicago, 1952.

**Table 10-1**  Goals and Measures of Reproductive Efficiency

| Goals | Ideal | Practical goal | Serious problem |
|---|---|---|---|
| Age at first freshening (mo.) | 24 | 25 | Over 27 |
| Calving interval (mo.) | 12 | 12.5 | Over 13 |
| Days open | 85 | 100 | Over 115 |
| Services per conception (no.) | 1.0 | 1.5 | Over 1.75 |
| First service conceptions (%) | 100 | 60 | Under 55 |
| First 2 service conceptions (%) | 100 | 80 | Under 75 |
| First 3 service conceptions (%) | 100 | 90 | Under 85 |
| Calfcrop—cows only (%) | 100 | 90 | Under 85 |
| Calf crop—heifers included (%) | 120 | 110 | Under 100 |
| 60–90 day nonreturn, bulls (%) | Above stud Average | Stud Average | More than 5% Below stud average |

understand the factors involved in successful reproduction. It is not the purpose of this text to give a thorough treatise on the anatomy and physiology of reproduction, as excellent texts are available for this purpose (see selected references); however, we will review briefly the basic principles of successful reproduction and the characteristics of reproductive patterns in dairy cattle.

Successful reproduction involves a complex series of physiological functions by both male and female. The male's functions are too:

1. Produce large numbers of viable male germ cells (spermatazoa).

2. Ejaculate these spermatozoa into the vagina of the cow or, in the case of artificial insemination (AI), into an artificial vagina.

The reproductive organs of the bull are presented in Figure 10-1. Spermatazoa are produced in the seminiferous tubules of the testes, stored in the epididymus, and at the time of ejaculation pass through the vas deferens, ampulla, urethra, and penis. During ejaculation, fluids are added to the spermatazoa from the accessory sex glands, the ampullae (a small amount), seminal vesicles, prostate, and Cowper's glands. This fluid, seminal fluid or plasma, serve as a medium for sperm transport, sperm activation (sperm are relatively immotile until ejaculation), and as a supply of nutrients for the sperm. Volume of semen per ejaculate varies from 2 to 15 ml., with an average of 5 to 6 ml. Sperm concentration ranges from 1.0 to 3.0 billion sperm per ml., with an average of about 2.0 billion per ml. Percent motile cells ranges from 0 to 85%, with an average of 70%. Based on these figures, an average ejaculate contains about 7.0 billion motile cells (5 ml. semen $\times$ 2.0 billion/ml. $\times$ 70% motile cells). In natural service, maximum fertility rates

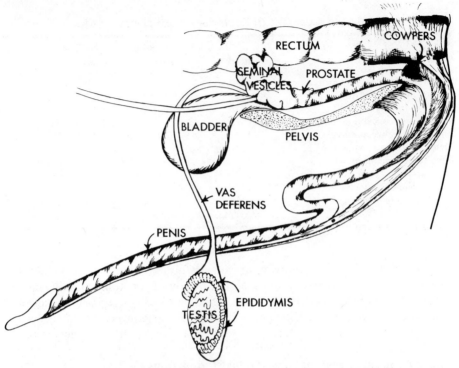

**Fig. 10-1.** Diagram of a bull's reproductive organs. (Reproduced by permission from "Dairy Cattle Fertility and Sterility." Copyright 1973 by W. D. Hoard and Sons Co., Ft. Atkinson, Wisconsin 53538).

should be obtained by bulls whose semen characteristics meet or exceed these average values and contain not more than 15% abnormal sperm (head or tail abnormalities). Semen from bulls used in AI is monitored carefully at each ejaculation for volume, concentration, motility, and abnormal sperm. Bulls used naturally should be collected and their semen checked before they are widely used in the herd to avoid prolonged breeding delays due to low-quality semen. As mentioned earlier, nonreturn rates from bulls in AI should also be checked, and the use of bulls that are 5% or more below stud average should be limited.

The development and function of the male reproductive system is under the control of the pituitary gland. Follicle stimulating hormone (FSH) and leuteinizing hormone (LH) from the anterior pituitary gland stimulate the testis to produce sperm and the male sex hormone. This process usually begins at 6 to 9 months of age and dairy bulls usually reach puberty (the time at which they are capable of producing viable sperm and mating) at 9 to 12 months of age. Sperm production is directly related to testis size so young bulls, because of smaller testes, produce a lower volume of semen. Male libido (the desire to copulate), as well as the development of the secondary sex characteristics, is largely controlled by the male sex hormone.

The function of the female in successful reproduction is more complex than that of the male because her role continues after fertilization. The female's functions are the following.

1. Produce viable female germ cells (ova).

2. Deliver the ova to the site of fertilization (the oviduct).

3. Provide optimum environment for fertilization, embryo development, implantation, and development of the fetus to term.

4. Deliver a live, healthy calf at the end of the gestation period.

5. Deliver the afterbirth and involute the reproductive tract back to normal size and condition so the process may be repeated.

The reproductive organs of the cow are presented in Figures 10-2 and 10-3. Ova are produced in the follicle of the ovary, released from the follicle (ovulation), picked up by the infundibulum of the oviduct, moved into the oviduct, fertilized in the oviduct, moved into the uterine horn 4 to 6 days after fertilization, and implanted in the uterus 30 to 33 days after fertilization. The embryo is carried in the uterus while developing into a full-term calf, and the calf is discharged through the cervix, vagina, and vulva. The process of ova production and release begins at puberty (usually 6 to 10 months of age) and continues on a 21-day cyclic basis until pregnancy. The cycle is normally reestablished within 40 to 50 days after calving and continues until pregnancy occurs again. The development and function of the female reproductive process in the female is also under the control of the pituitary gland. The anterior pituitary hormones FSH and LH stimulate the development and maturation of the follicle that contains the ova. The maturing follicle secretes estrogen, which causes the animal to exhibit the symptoms of estrus (heat). LH causes the rupture of the follicle. It also causes the luteinization of the follicular cells and the formation and function of the corpus luteum. The corpus luteum secretes progesterone, which prepares the uterus to receive the fertilized ova and is essential in maintaining pregnancy.

Thus successful reproduction involves the presence of viable male and female germ cells in the right place at the right time, in combination with a normal, healthy female to provide suitable environment for the growth, development, and delivery of a healthy calf. Synchronization—proper timing—of this series of physiological functions is as critical to successful reproduction as are viable germ cells and normal healthy reproductive tracts.

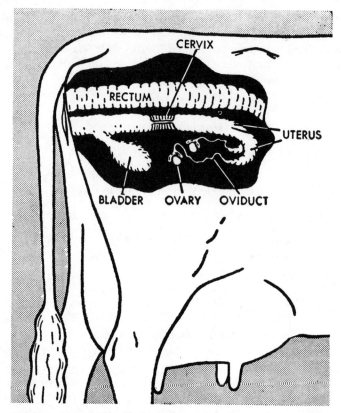

**Fig. 10-2.** Diagram of a cow's reproductive organs. (Reproduced by permission from "Dairy Cattle Fertility and Sterility." Copyright 1973 by W. D. Hoard and Sons Co., Ft. Atkinson, Wisconsin 53538).

Problems can and do arise at any point. This is why there is no one way to solve all reproductive problems; as the specific problem differs, so does the proper corrective measure. This is also why some basic understanding of the components of successful reproduction is essential to identify problems and correcting them.

## REPRODUCTIVE PATTERNS OF DAIRY CATTLE

The dairy cow is a polyestrous animal, that is, she cycles regularly regardless of season of the year. Characteristics of the reproductive patterns of dairy cattle are given in Table 10-2.

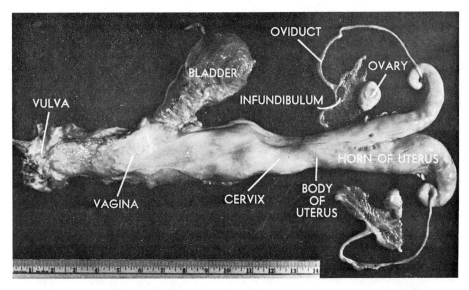

**Fig. 10-3.** Reproductive tract of a cow. (Reproduced by permission from "Dairy Cattle Fertility and Sterility." Copyright 1973 by W. D. Hoard and Sons Co., Ft. Atkinson, Wisconsin 53538).

## CAUSES OF STERILITY AND INFERTILITY

As mentioned previously, the causes of inefficient reproductive performance are many and varied. A few of these can be attributed to the male, many to the female, and a large number can be attributed directly to the person in management. As the heritability estimate of female reproductive efficiency is quite low (0.05–0.10), one might conclude that reproductive efficiency or the

**Table 10-2** Characteristics of the Reproductive Patterns of Dairy Cattle

| Characteristic | Average | Normal range |
|---|---|---|
| Age at puberty (mo.) | 10 | 6–14 |
| Length of estrous cycle (days) | 21 | 18–24 |
| Length of estrus (heat) period (hr.) | 18 | 10–24 |
| Time of ovulation (hr. after heat) | 11 | 5–16 |
| Gestation (days) | | |
|    Ayrshires | 278 | |
|    Jerseys and Holsteins | 278 | |
|    Milking Shorthorn | 282 | |
|    Guernsey | 283 | |
|    Brown Swiss | 288 | |

lack of it is almost entirely under the control of management. This implies that to improve reproductive efficiency, management of the reproductive program must improve. The following discussion will concentrate on the identification of some of the more common reproductive problems and some alternatives that might help solve some of these problems in both males and females.

### Inheritance

Many dairymen have reported differences in reproductive efficiency of certain cow families. These differences undoubtedly occur, but they have not been verified by research. Selection for high reproductive efficiency is not likely to be successful, as the heritability estimate is very low, especially for such measures as days open, services per conception, and nonreturn rates. There are, however, some inherited conditions that do affect reproductive performance.

**Inbreeding.** Inbreeding causes lowered reproductive efficiency in most species of animals including dairy cattle. The effect is more noticable as inbreeding percentage increases. High levels of inbreeding reduce fertility rate, increases the incidence of expression of recessive lethal genes and other abnormalities, and increase calf mortality. All of these factors directly or indirectly affect total herd reproductive performance.

**Multiple Births.** Twins occur in about 2 to 4% and triplets in about 0.2% of dairy cattle births. Multiple births are inherited, although some multiple births are caused by hormone treatments. Multiple births reduce fertility in several ways. Following multiple births, the incidence of retained placentas is higher, the interval to first heat is increased, and conception rates are lower. In addition, females born twin with a male in multiple births are freemartins over 90% of the time. Freemartins are intersexes that may be mostly female, mostly male, or in between male and female to varying degrees. They do not have completely developed female reproductive systems and may have rather well-developed male sex organs. The development of freemartins in heifers born twin with a bull may relate to the effect of male hormones in the blood system, but it is not well understood. Suspected freemartins can usually be detected at a relatively early age by palpation of the udder and/or reproductive tract or by blood typing. Multiple heifer births, either mono- or dizygotic, are normal as far as reproductive function is concerned.

**Male Abnormalities.** Recently, four male reproductive deficiencies have been reported to be inherited.* These are (1) underdevelopment of the testes

* R. G. Saacke, *Hoard's Dairyman,* Sept. 10, 1974.

and other reproductive organs; (2) cryptorchidism (failure of one or both testes to descend into the scrotum); (3) defective sperm formation (a high incidence of abnormal sperm); and (4) lack of libido. Genes that control these factors appear to be sex limited; that is, they are present in both the male and female, but are expressed only in the male. Thus they only affect male reproductive performance but they should be considered when selecting dams of sires.

**Miscellaneous Inherited Defects.** A wide variety of reproductive-tract abnormalities are thought to be inherited to some extent. These include white heifer disease (a membrane over the cervix), ovarian hypoplasia, cystic ovaries, single-horned uteri, etc. The mechanism of inheritance of these conditions is not well understood, and fortunately the incidence of these abnormalities is quite low. Many of these conditions can be detected by a combination of good records and rectal palpation. If a heifer does not begin to cycle regularly by 12 to 14 months of age or does not conceive by the second or third service, palpation by a competent veterinarian should identify most of the problems.

### Infectious Diseases

Several infectious diseases can cause infertility or sterility. These are caused by various microorganisms and some are contagious as well as infectious. Most of these will be discussed in depth in the chapter on health of dairy animals and are only mentioned in passing here.

**Brucellosis, or Bang's Disease.** This disease is contagious as well as infectious. It is caused by the organism *Brucella abortus*. It affects reproductive efficiency by causing a high incidence of abortions, usually in the last third of pregnancy, retained placentas, uterine infections, and lowered conception rate. It can effectively be prevented by calfhood vaccination.

**Trichomoniasis.** This disease is a contagious venereal disease that is caused by the protozoan *Trichomonas fetus*. This organism lives in the sheath of the bull and the vagina of the cow. It causes a high incidence of abortions, usually in the first third of pregnancy, uterine infections, irregular heat periods, and lowered conception rate. As it is a true venereal disease (transmitted at sexual contact) it can effectively be prevented by using AI.

**Vibrio Fetus, or Vibriosis.** Vibriosis is a contagious disease caused by the organism *Vibrio fetus*. It typically causes a high incidence of abortion, usually in the middle third of pregnancy, irregular heat periods, and lowered conception rates. It can be prevented by vaccination, and its spread effectively prevented by the use of AI.

**Leptospirosis.** Leptospirosis is a contagious disease that can be caused by any one of three spirochete organisms—*Leptospira pomona, L. hardjo,* and *L. grippotyphosa.* The most common causative organism in cattle is *L. pomona,* although the others have been known to infect cattle. It impairs reproductive function directly by causing abortions at any time during pregnancy, and indirectly by its systemic effect on the animal. It can effectively be prevented by vaccination. It can be transmitted between different species of animals (cattle, dogs, cats, goats, sheep, humans, deer, and other wild animals). Contact with other species of animals should be eliminated insofar as possible.

**The Triple Viruses, Infectious Bovine Rhinotracheitis (IBR), Bovine Viral Diarrhea (BVD), and Para-Influenza 3 (PI-3).** Any one or combination of these three viruses can affect reproductive efficiency by causing infections in the reproductive tract or, more often, by infecting the developing fetus. If the infection is severe, it may cause the death of the fetus and consequent abortion. In other cases it causes defects such as blindness, brain defects, or other abnormalities in the calves. These virus diseases can effectively be prevented by vaccination.

**Metritis.** Metritis is defined as any inflammation of the uterus. It can be caused by any one or combination of several organisms. Its highest incidence usually occurs shortly after calving. Major causes of metritis include poor sanitary practices at calving, such as allowing the cow to calve in a filthy environment, inserting dirty hands or equipment into the uterus to assist in the delivery or to insert medication after the delivery, and retained placenta. A high proportion of the cases of metritis, then, can be prevented by following good sanitary practices at calving. These include providing a clean environment (outside grassy lot, freshly bedded clean box stall) for calving, thorough washing and disinfecting of equipment used to assist the delivery (it is impossible to thoroughly clean and disinfect a piece of rope, so stainless steel pull chains should be used), and thorough washing and disinfecting hands and arms before inserting them in the reproductive tract (or, even better, the use of disinfected rubber or plastic gloves and sleeves). Patience at calving is also helpful in avoiding tearing of the reproductive tract by too much assistance too quickly.

The cervix is the barrier between internal and external environment and is normally closed. It is, however, open during and just after calving and during the heat period. With this natural disease barrier open during these periods, utmost care and caution must be exerted to avoid metritis.

Metritis is characterized by a cloudy or pussy discharge from the vulva as the result of the infection of the lining of the uterus. The symptoms may vary from a few flakes of pus to large quantities of yellowish pus with a foul odor, depending on the degree and severity of the infection. Prompt detection and treatment is essential if repeat breeders and delayed conception are to be avoided. Cows with retained placentas have a high incidence of metritis. Prophylactic treatment of these cows to prevent the problem is often recommended by veterinarians.

**Other Infectious Diseases.** Vaginitis, cervicitis, and salpingitis may also affect breeding efficiency. These diseases require treatment under the direction of a competent veterinarian.

### Physiological or Functional Disturbances

When reproduction is not normal but the animal has no structural defect and is not infected with a recognized infectious disease, the problem is classified as a physiological or functional one. Recalling the series of physiological events that results in successful reproduction from formation and release of viable ovum to delivery of a healthy calf at term, and the complexity of the control of these events, one can more easily understand where and why the problems can arise.

**Anestrus.** Anestrus is the absence of estrus, a common reproductive problem in which there is a temporary or permanent absence of ovarian function. It is often caused by a retained corpus luteum. In a normal, nonpregnant cow, the corpus luteum that is formed after ovulation begins to diminish in size and function about 16 to 18 days after heat. The progesterone secreted by the corpus luteum, among other things, keeps the cow out of heat. However, a cystic corpus luteum does not recede in size and function. It remains on the ovary and is capable of preventing estrus if it can produce sufficient amounts of progesterone. Normally, a cystic corpus luteum occurs singly on one ovary only. Often it contains a fluid-filled cavity. The cause of cystic corpora lutea is not well understood, but certain conditions appear to be associated with higher incidence of this problem. Some of these are the following.

1. It is more prevalent in dairy than in beef cows. Some reports indicate the incidence to be as high as 25%.

2. The incidence is higher in high-producing cows, especially during periods of peak production.

3. It occurs more often during the winter months.

4. It can occur in dairy cattle any time after puberty, but is observed more in second-lactation and older cows.

5. It occurs more often when there is an infection of the uterus.

Methods of treatment of this problem have been debated for many years. Some recommend manual removal of the cyst, others recommend hormone therapy, and still others recommend that the cow be allowed to recover spontaneously. There is no general agreement concerning the best therapy. If the incidence of anestrus is above 10 to 12% in a herd, the authors recommend a more careful heat-checking program as we sincerely believe that a major share of the cows thought to have cystic corpus luteum are cycling but have not been detected in heat because of so-called "silent heats" or lack of careful observation.

**Nymphomania.** Nymphomania is characterized by frequent and/or prolonged heat periods. It is caused by a cystic follicle or follicles. In a normal cow, the follicle ruptures (ovulation) about 5 to 16 hours after the end of heat. When a cow has a cystic follicle, ovulation does not occur and the follicle continues to grow, often reaching a diameter of 1 to 2 in. A normal follicle reaches a diameter of ⅝–¾ in. before it ruptures. Often one or more additional cystic follicles develop. The condition occurs most often one to two months after calving. If the condition is not corrected, or if the cow does not recover spontaneously, often she develops a high tailhead, her voice deepens, and she may exhibit other masculine symptoms. The cause of cystic follicles is thought to be insufficient LH to rupture the follicle. The injection of LH is a common treatment. The incidence of spontaneous recovery is relatively low, and hormone therapy is less effective if the condition has persisted for a considerable length of time. In view of these factors, early diagnosis and treatment as soon as the cystic condition is confirmed is critical to a high recovery rate. This problem is often associated with higher-producing cows and is more prevalent in dairy than in beef cattle. Estimates of 5 to 10% incidence in dairy cattle have been reported.

**Fertilization Failure.** It has been estimated by various authors that 5 to 15% of the matings in dairy cattle do not result in fertilization of the ova, where no obvious problem is present. This failure may be caused by poor timing of insemination or breeding, early or late ovulation, poor-quality semen, an abnormal ovum, poor uterine environment, or some other unknown factor. Precise reasons for this problem are not well defined, but perhaps future research into sperm and ova morphology will shed new light in the not-too-distant future.

**Early Embryonic Mortality.** Losses due to the death of the embryo less than 30 days after conception have been estimated at 10 to 40%. Abortions after 30 days, other than those caused by infectious diseases, probably do not exceed 5% of the pregnancies in dairy cattle. The lower part of this range is more common among first-service cows and the higher part among repeat-breeder cows. The average early embryonic mortality probably lies in a range of 15 to 20%. Probable causes of this problem are similar to those associated with fertilization failure and are not well understood. The combined effect of fertilization failure and early embryonic mortality averages about 25 to 35% of all matings. This means that these two factors alone impose a maximum conception rate of about 65 to 75%. This point illustrates why we need additional research in this area and why we have not been able to significantly improve conception rates in dairy cattle during the past 25 to 50 years.

**Retained Placenta.** A retained placenta can be defined as one that is not expelled within a few hours (1 to 12) after calving. The incidence of this problem is extremely variable within and between herds. Individual herds have been known to vary from less than 5% during one year to over 35% the next year. The precise cause of the problem is not well defined but the reported incidence does appear to vary with several factors. Some of these are the following.

1. The incidence is higher following multiple births.

2. The incidence is higher following difficult calvings.

3. The incidence is higher following premature births.

4. The incidence is higher in higher-producing cows.

5. The incidence is higher during the colder months in temperate climates.

6. The incidence is lower in beef cattle and pastured dairy cattle (1.96% incidence in New Zealand).

Most researchers in dairy cattle physiology agree that, regardless of the cause, the factors that result in retained placenta are loss of continued muscle contraction after delivery of the calf, failure of the placental attachment to loosen from the wall of the uterus, or a combination of these two factors.

Retained placentas impair reproductive efficiency because there is a high degree of association between retained placentas and metritis and delayed conception. Recommended preventative measures include calving in a clean and dry area; allowing the calf to nurse the cow after calving, or at least to remain in the presence of the cow for a period of time; giving the cow ample time to deliver without premature excessive pulling; and giving the cow a reasonable length of time to clean herself. Recommended treatments vary among veterinarians; however, currently most follow a practice of antibiotic and/or hormone therapy rather than manual cleaning.

**Dystocia.** Dystocia, on abnormal calving, occurs in about 5 to 6% of dairy calvings. This problem can be caused by a large calf, a small pelvic opening in the cow, or an abnormal birth position. For a large calf or small pelvic opening, some assistance may be in order after the cow has had ample time (at least 1 to 2 hours after the feet are visible) to deliver herself. If the calf has not been born within several hours after the cow has been in labor, or within 1 to 2 hours after the legs are visible, an examination is recommended. The exterior of the cow (the vulva and surrounding area) and the person doing the examining should be kept strictly sanitary. If the examination reveals that the calf is in a normal position (lying on stomach with forelegs extended and head resting on its forelegs when the cow is standing—see (Figure 10-4) and appears to be small enough to pass through the pelvic girdle, some assistance may be in order. Clean, sanitized obstetrical chains should be attached to the legs of the calf and pressure exerted during contractions at a 45° downward angle from the cow's topline. If the calf is in an abnormal position (Figure 10-5) or appears to be too large to pass through the pelvic girdle, a veterinarian's assistance is indicated.

**Other Malfunctions.** As the reproductive process is primarily under the control of the endocrine system, there are undoubtedly other abnormal conditions. Should there be a deficiency or overabundance of any hormones, a hormone imbalance can occur that can impair reproductive efficiency. In addition, the precise mechanisms of endocrine control of many aspects of reproduction are not well understood. Perhaps there are some hormones involved that have not as yet been discovered.

### Nutritional Causes of Lowered Fertility

Under normal conditions, animals that are well fed for growth and/or milk production have few breeding problems caused by faulty nutrition. There are some nutritional conditions, however, that can seriously impair reproductive function.

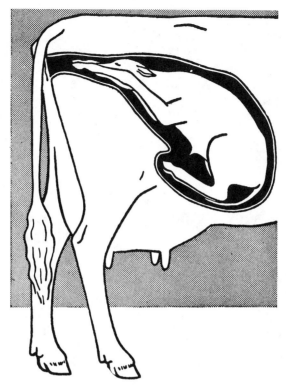

**Fig. 10-4.** Normal birth position. (Reproduced by permission from "Dairy Cattle Fertility and Sterility." Copyright 1973 by W. D. Hoard and Sons Co., Ft. Atkinson, Wisconsin 53538).

**Level of Nutrition.** A low level of feeding has been shown to delay the onset of puberty in dairy heifers. Studies at the New York Station* showed that heifers fed normal rations come in heat for the first time at an average age of 11.3 months; heifers fed a subnormal ration (65% of Morrison's Standard) averaged 17.3 months of age; and those on an above-normal (140% of Morrison's Standard) averaged 9.4 months. Average services per conception for the three groups were 1.22 for the normal group, 1.56 for the subnormal group, and 1.33 for the above normal group. These results have been substantiated by other studies and field reports, and it is generally accepted that gross underfeeding delays puberty in dairy heifers.

It has been suggested that overfeeding heifers is a cause of infertility because of fat deposits around the ovaries. This was not substantiated in the

* J. T. Reid et al., Effect of Feeding During Early Life on Reproductive Performance of Dairy Cattle, *J. Dairy Sci.*, **34:**510 (1951).

**CALF UPSIDE DOWN.** A veterinarian will usually turn the calf. Occasionally, such a calf must be delivered backward.

**HEAD BENT BACK.** If the veterinarian can push the calf back into the cow he often can bring the head into position.

**FORELEG RETAINED.** In this presentation, leg must be reached, and brought forward, joint by joint, so calf can be delivered.

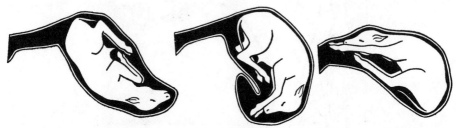

**BACKWARD—UPSIDE DOWN.** The calf must be turned in the uterus until it assumes normal position for delivery at birth.

**CALF BACKWARD.** Veterinarians try to straighten hindlegs of calf found in this position, then deliver calf backward.

**HINDLEGS FORWARD** is serious abnormal position if delivery is far advanced. Fetus frequently must be dismembered.

**Fig. 10-5.** Abnormal birth positions. (Reproduced by permission from "Dairy Cattle Fertility and Sterility." Copyright 1973 by W. D. Hoard and Sons Co., Ft. Atkinson, Wisconsin 53538).

New York work or by other research; however, overconditioning of heifers has been shown to decrease milk production, so it should be avoided for that reason.

Infertility is rarely caused in cows by a low level of nutrition. The level of nutrition must be very low to cause infertility because of the cow's nutrient priority system. Milk production and growth will be greatly depressed before fertility is impaired.

**Specific Nutrient Deficiencies.** As the cow is able to synthesize protein and most vitamins (except A, D, and E), specific nutrient deficiencies that can affect fertility are few.

VITAMIN A DEFICIENCY. Animals deficient in vitamin A have a higher-than-normal incidence of dead or weak calves and some abortions in late pregnancy. These animals also have a higher incidence of retained placenta. This deficiency can be easily avoided or remedied by feeding good-quality forage or supplemental vitamin A.

VITAMIN D DEFICIENCY. Deficiency of this vitamin results in partial or total anestrus. Calves are often born with rickets. Sunlight, feeding sun-cured hay, or vitamin D supplementation can easily avoid or remedy this problem.

VITAMIN E. Deficiency of vitamin E has been shown to impair reproductive function in rats, and wheat germ oil (an excellent source of vitamin E) has been promoted as an antisterility vitamin. Based on current evidence, the feeding of vitamin E has not been shown to improve dairy cattle fertility.

PHOSPHORUS DEFICIENCY. Phosphorus deficiency has been shown to delay puberty in heifers, impair the reproductive cycle, and, in extreme cases, to cause the ovaries to become nonfunctional. This deficiency is easily corrected by feeding phosphorus-rich feeds or phosphorus supplementation.

OTHER. Many other mineral, and combinations of mineral and vitamin deficiencies or excesses, have been postulated as causing fertility problems in dairy cattle. These claims are largely field reports and have not been substantiated by research at the present time. As production levels become higher and soils become more depleted in some minerals, the authors can foresee impending problems in mineral deficiencies. Potential effects need to be identified by continued research in this area.

## Management Factors in Breeding Efficiency

It is difficult to separate management from many of the previously mentioned factors affecting breeding efficiency, as management certainly is responsible for disease prevention and nutrition and can control to some extent some of the functional disturbances. In this section, however, we will discuss those factors directly under the control of management on a day-to-day basis. Perhaps "cowmanship factors" would categorize this area more correctly.

Proper management of the reproductive program involves several basic concepts.

**Keeping and Using Accurate Reproductive Records.** Heat dates, breeding dates, calving dates, abnormal conditions (when and what), pregnancy and other diagnosis, and treatments must be recorded and used to help anticipate and solve reproductive problems. Listings of cows ready to be bred, due in heat, problem cows, etc. can be compiled from accurate records. Special management lists including this information are available from DHI and other computerized record services.*

**An Effective Heat-Detection Program.** With the advent and widespread use of AI, the problem of heat detection has greatly increased because man

---

* J. A. Lineweaver and G. W. Spessard, Development and Use of a Computerized Reproductive and Management Program in Dairy Herds, *J. Dairy Sci.*, **58**:256 (1975).

has replaced the bull as the heat detector. Increased herd size has further complicated the problem. Heat detection is probably the number one reproductive management problem in herds using AI. Numerous studies report that 30 to 40% of the breedable heats in dairy cattle may go undetected. A similar or higher percentage of cows suspected of having a retained corpus luteum are probably cycling but are not detected in heat. There have been volumes written concerning this problem and potential solutions to the problem, but many dairymen still have the problem. Why? The authors propose that the major reason is that many dairymen simply do not give the problem high priority because the effect is not immediately apparent. If heats are missed, the amount of milk in the bulk tank is not decreased the next day, nor does the cow become ill or die. The economic effects often are not apparent for months or even a year or more, so the priority assigned to the task of heat detection is low. The following program will help solve this problem.

1. Assign one man, preferably the herdsman, the primary responsibility for heat detection and insist that he give the job top priority. In small herds this may be the owner; in large herds more than one person may have responsibility. Incentives based on cows pregnant before 100 or 110 days after calving are often helpful. One of the authors is familiar with an incentive program devised by a large herd manager that proved very effective. Initially, incentives based on heat detection were paid to all workers. This was not successful, as too many cows *not* in heat were reported in heat. The incentives were changed and based on cows pregnant within 100 days; this program effectively reduced days open by 30+ days in less than one year.

2. Know the symptoms of heat, not just the apparent ones such as standing when being ridden, but the less obvious ones as well. Some of these are restlessness, bawling, walking or running the fences, bothering other cows, riding other cows, swollen vulva, clear mucous discharge from the vulva, decreased appetite, and lowered production. Cows vary widely in the display of symptoms of heat. Persons also vary widely in their ability to detect these symptoms. Some persons can detect slight abnormalities in cattle, others cannot. Cowmanship is extremely important in heat detection.

3. Know the reproductive status of the cows. Based on accurate records, keep a calendar of expected heats (over 90% cycle between 18 and 24 days), of cows that have been fresh for 20 to 30 days and have not cycled, etc., and pay particular attention to these cows.

4. Check the cows frequently (average length of time in standing heat is

about 18 hours but ranges from 4 to 40) and at times other than at feeding and milking time. Recent reports indicate that mounting behavior is much more frequent between 6 p.m. and 6 a.m. (nearly 70% of the mounts) than between 6 a.m. and 6 p.m. Nightime activity is also more prevalent in hot weather, so late evening and early morning heat detection is highly recommended.

5. Consider the use of heat detectors, especially in large herds and for problem breeders. These heat detectors are not a substitute for good management, but properly used, can be a valuable adjunct to it. Numerous reports have indicated high degrees of effectiveness of various devices. Some of the more commonly used ones are The KaMaR Heatmount Detector, the chin ball mating device, chalking the tailheads of cows daily during milking, and the use of altered males (penis amputated, penis deflected, or preputial obstruction).*

**Time Breeding to Obtain Optimum Conception Rates.** The time of optimum fertility of the ova is very short, 2 to 4 hours. Ovulation time varies from 5 to 16 hours after the end of standing heat, with an average of 10 to 11 hours. Fertile life of sperm is limited to about 28 hours in the female reproductive tract. This indicates that timing of breeding is critical to optimum fertilization rates. Optimum conception rates have been reported by the Nebraska workers and others by breeding during the last one-half of the standing heat period. (Table 10-3) Breeding earlier or later than this reduces conception because of reduced viability of either sperm or ova. Based on these facts and conception-rate data, the following recommendations have been followed for some years: if standing heat is first observed in the morning, breed in the afternoon or evening; if standing heat is first observed in the afternoon or evening, breed the next morning. These recommendations are based on an average length of heat period of 18 hours, average ovulation time of 10 to 11 hours after the end of standing heat, and the assumption that the first observed standing heat was near the beginning of the standing heat period. These are sound recommendations if the conditions and assumptions are true. However, field reports and current research in progress in Virginia indicate that the latter assumption is not true in many herds. Many of the cows detected in standing heat in the morning apparently were in the last half of their standing heat period but had not been detected because there had been no night-time observation. The same situation is apparently true for evening-observed cows because of lack of observation or mounting activity because of warm weather between morning and evening milking times. These

---

* R. H. Foote, Estrus Detection and Estrus Detection Aids, *J. Dairy Sci.*, **58:**248 (1975).

**Table 10-3** Conception Rate at Various Stages of Heat[a]

| Time of breeding | No. of cows | Cows conceiving from one service | |
|---|---|---|---|
| | | No. | % |
| Start of heat | 25 | 11 | 44.0 |
| Middle of heat | 40 | 33 | 82.5 |
| End of heat | 40 | 30 | 75.0 |
| 6 hours after heat | 40 | 25 | 62.5 |
| 12 hours after heat | 25 | 8 | 32.0 |
| 18 hours after heat | 25 | 7 | 28.0 |
| 24 hours after heat | 25 | 3 | 12.0 |
| 36 hours after heat | 25 | 2 | 8.0 |
| 48 hours after heat | 25 | 0 | 0.0 |

[a] G. W. Trimberger, and H. P. Davis, *Nebraska Agr. Exp. Sta., Res. Bull.* 129, 1943.

factors have led to reports that many dairymen are breeding too late in the heat period, and again stress the importance of precise heat detection. If cows are observed for heat detection only twice daily at approximately 12-hour intervals, perhaps immediate breeding should be practiced. A better solution is to follow a more thorough and precise heat-detection plan.

With artificial insemination, the period of breeding can be extended to 6 hours after the end of heat with satisfactory results. Beyond 6 hours after heat, fertility declines rapidly (Table 10-3).

Using standing heat as the indication, a general timetable can be used to aid in breeding during the latter half of the heat period (Table 10-4).

**Breeding after Calving.** The traditional recommendation is to rebreed cows at the first heat following the sixtieth day after calving. Recent reports[*] indicate earlier breeding (40 to 50 days after calving) may be effective in shortening average herd calving interval. Based on current evidence, it appears that rebreeding at the first heat 50 days after calving is reasonable for cows who have not experienced calving problems or retained placenta and whose reproductive tracts have been palpated and found to be normally involuted. This is especially true for average- or below-average-producing cows. It also seems advisable to recommend waiting until 60 to 70 days for higher-producing cows and those that have experienced problems at or since last calving.

* J. H. Britt, Early Postpartum Breeding of Dairy Cows, A Review, *J. Dairy Sci.*, **58:**266 (1975); H. L. Barr, Breed at 40 Days To Reduce Days Open, *Hoard's Dairyman*, Oct. 10, 1975.

**Table 10-4** Timetable for Best Results in Breeding Dairy Cows[a]

| Standing heat first observed | When to breed | Usually too late to breed |
|---|---|---|
| In the morning (a.m.) | The same day, preferably late in day | The next day |
| In the afternoon or evening (p.m.)[b] | Forenoon next day and up to midafternoon | After middle of afternoon the next day |

[a] If symptoms of heat other than standing to be mounted are used, this table does not apply. This table is based on the average standing heat period of 18 hours.
[b] Cows definitely not in heat in the morning.

**A Competent Veterinarian.** The service and cooperation of a competent veterinarian, skilled in palpation and knowledgeable in diagnostic, treatment, and preventive medicine is a valuable asset in maintaining breeding efficiency. The authors recommend contractual services that include periodic visits (preferably monthly or less in large herds) as well as emergency service. Periodic visits should include the following reproductive services.

1.  Pregnancy diagnosis for all animals bred six weeks or more.

2.  Reproductive tract palpation for all animals fresh 30 days or more and treatment of all diagnosed pathological conditions.

3.  Problem diagnosis for all cows fresh 40 days or more that have not been in heat. Treatment as needed.

4.  Problem diagnosis for all animals bred three or more times. Treatment as needed.

5.  Problem diagnosis for all animals that display any reproductive abnormality (abnormal heat cycles or discharges).

6.  Routine vaccination for infectious diseases that affect reproductive efficiency. The veterinarian should also be on call for emergency reproductive problems such as dystocia and retained placenta.

The four keys to successful management of the dairy herd reproduction program, then, are (1) giving priority to the program by keeping and using accurate records, (2) detecting heats with accuracy and precision, (3) timing breeding properly, and (4) including the services of a competent veterinarian in the program. Such a program shows a low calving interval, low number of services per conception, minimal losses from sterility and infertility, and maximal profit.

## Other Factors

There are many other factors that are thought to affect reproductive efficiency and merit additional research. There are also new techniques under experimentation that bear close observation for future use. Some other factors that may be associated with reproductive efficiency are level of production (higher-producing cows have been reported to have longer intervals between calving and first heat, higher incidence of repeat breeding, and longer calving intervals), and higher temperatures and solar radiation, which appear to reduce fertility.* New techniques include estrus cycle control through the use of hormones or prostaglandins, induced lactation (lactation without calving), embryo transfer, and induced parturition.

## REFERENCES FOR FURTHER STUDY

Barr, H. L., *J. Dairy Sci.*, **58**:88 (1975).

Barr, H. L. Influence of Extrus Detection on Days Open in Dairy Herds, *J. Dairy Sci.*, **58**:246 (1975).

Bozworth, R. W., et al., Factors Affecting Calving Interval in Dairy Cattle, *J. Dairy Sci.*, **55**:334 (1972).

Hafez, E. S. E., Reproduction in Farm Animals, Lea and Febiger, 2nd ed., Philadelphia, 1973.

Hafs, H. D., and Boyd, L. J., Dairy Cattle Fertility and Sterility, *Hoard's Dairyman*, 1973.

*Hoard's Dairyman*, regular columns and numerous articles by Dr. R. G. Saacke or Dr. E. A. Woelfer, 1974 and later.

Lineweaver, J. A., Potential Income from Increased Reproductive Efficiency, *J. Dairy Sci.*, **58**:780 (1975).

Louca, A., and J. E. Legates, Production Losses in Dairy Cattle Due to Days Open, *J. Dairy Sci.*, **51**:573 (1968). Also see **50**:975, 1967; and **57**:629, 1974.

Speicher, J., and L. H. Brown, The Cost of Delayed Conception in Your Herd, *Hoard's Dairyman*, **116**:140 (1971). Also see **120**:1239 (1975).

---

* F. C. Gwazdauskas et al., Environmental and Management Factors Affecting Conception Rate in a Subtropical Climate, *J. Dairy Sci.*, **58**:88 (1975).

# 11

# Artificial Insemination of Dairy Cattle

The breeding of dairy cattle by artificial insemination (AI) has given the dairy industry the opportunity to make widespread use of superior genes for improving the performance of dairy cattle. The development of effective techniques of collecting, extending, freezing, storing, and transporting semen and methods of inseminating dairy cows and heifers has paralleled the development of accurate methods of identifying genetically superior dairy bulls. These two factors have combined to result in rapid genetic improvement of U.S. dairy cattle during the last 30 years. They have been largely responsible for the doubling of production per cow during that period and should continue to be even more effective in the future as techniques improve in both areas. Currently, approximately 50% of the U.S. dairy herd is bred artificially. In some countries (Finland, for example) 95% of the dairy cattle are bred artificially. This is a reasonable goal for the U.S. dairy industry over the next 10 years.

## HISTORICAL OVERVIEW

The earliest use* of artificial insemination of animals that has been reported was in 1322, when an Arab chief artificially inseminated a mare. In 1782 an Italian used the technique successfully with dogs. Workers in Denmark and Russia** began investigating the possibilities of the practice shortly before 1900. The Russians were the pioneers in its use with cattle; from 1930 to 1940 they made rapid advances in using the technique with horses, cattle, and sheep.

The first cooperative artificial-breeding association was organized in Denmark in 1936. Perry* studied the methods and organizational setup used in Denmark, and in 1938 was instrumental in organizing the Cooperative Artificial Breeding Association in New Jersey. Although this was the first large organization for artificial breeding in this country, it was preceded by a small experimental unit in Minnesota. The method had already been practiced in several herds, and research work was getting under way at several agricultural experiment stations.

## ADVANTAGES OF AI

The advantages of AI are greatest when combined with careful selection of bulls to be used, strict adherence to recommended methods of processing and

* *USDA Circ.* 567 (1940).
** E. J. Perry, *Artificial Insemination of Farm Animals,* 4th ed., 1968.
* Perry, *ibid.*

storing semen, insemination and a sound reproductive management program. Some of the advantages of AI are the following.

1. It provides greater opportunity for genetic improvement through the use of proved genetically superior bulls at reasonable cost.

2. It makes more widespread use of genetically superior bulls by making more efficient use of their semen. In natural service a bull may service 50 to 100 cows per year. In AI it is not unusual for a bull to service 10,000 to 20,000 cows per year.

3. It eliminates the danger involved in keeping a herd sire.

4. It offers dairymen a wider selection of bulls, thus avoiding the danger of "putting all his eggs in one basket." It also makes it easier to avoid inbreeding.

5. It reduces the risk of acquiring and spreading infectious reproductive diseases.

6. There is less chance of using poor semen. Semen from bulls in AI is checked before and after processing and, if properly stored and handled, there is little risk of poor semen.

7. It eliminates problems of mating large bulls to small heifers and small bulls to large cows.

### Limitations of AI

Some of the limitations of AI are the following.

1. Heat detection. This is probably the biggest limitation. In AI a person, instead of the bull, must detect heats. A person must also catch and restrain the animal for breeding. The difficulty of the former and the inconvenience of the latter, if not carried out conscientiously, can result in decreased reproductive efficiency and profit.

2. A skillful, conscientious technician is required. Poor semen-handling techniques, lack of sanitation, poor insemination practices, and poor timing of insemination can result in decreased reproductive efficiency, whether the inseminator be an AI organization technician or the owner or other farm personnel.

3. Fewer bulls are needed; consequently, the sale of bulls from purebred herds is reduced.

U.S. dairymen have the available alternatives of breeding their herds AI, naturally, or some combination of the two. If one carefully evaluates the advantages and limitations, it is easily seen that AI offers a great opportunity for herd improvement at reasonable cost. It also demands more precise management of the reproductive program. The authors strongly recommend AI. We believe dairymen have or can develop the ability to manage an AI reproductive program successfully. If 95% of Finnish dairymen can do so, why can't 95% of U.S. dairymen?

Many dairymen do retain a cleanup bull, a bull to use naturally on problem breeders. This practice has been of benefit in maintaining reproductive efficiency for many dairymen, though it does limit genetic progress. If a herd cleanup bull is used, he should be a high pedigree index (PI) young sire to maximize his chances of transmitting high performance. Even if his PI is +1000 lb., however, the odds of such a bull having a PD of +500 lb. of milk or more is rarely above 50% and the odds of his being a − PD bull are as high as 20%.

An alternative to the use of a natural cleanup bull is to use an AI cleanup bull. An AI cleanup bull is one whose conception rate in AI is among the top 20 to 25% of the bulls in that stud. His nonreturn rate is typically 5 to 10% above stud average. Most AI organizations have several of these bulls and are now listing or making conception data available to dairymen. Because several are available, dairymen can select from these high-conception bulls for +PD as well as other economically important traits. This gives dairymen the opportunity to continue genetic improvement while maximizing the odds of getting the cows bred. As mentioned earlier, the primary goal of the breeding program is genetic improvement. However, reproductive efficiency is also very important, and after the second or third service it becomes the primary goal. The use of an AI cleanup bull offers a viable alternative to the use of a natural cleanup bull. If the use of AI results in a high incidence of problem breeders in a herd and these cows conceive readily to natural service, the cows are not the problem. Rather, poor AI techniques are the problem. There may be an inadequate heat detection program, poor insemination techniques, poor semen handling, or poor timing of breeding. These problems should be corrected rather than substituting natural service, as the latter decreases opportunity for genetic improvement and increases risk of disease transmission and the dangers of keeping a bull.

## SEMEN COLLECTION, PROCESSING, AND HANDLING

The AI industry in the U.S. has changed dramatically during the past 35 to 40 years. Currently, most bulls used in AI are handled in bull studs or breeding centers. Most of these are cooperative. Some are privately owned. Semen collection and handling techniques have also changed. Currently, most semen is frozen in ampules or straws and can be stored indefinitely in liquid nitrogen tanks. It can be transported worldwide in containers that will maintain temperatures low enough to preserve its viability. It can be stored on farms in relatively inexpensive liquid nitrogen tanks indefinitely if proper cautions are followed. These factors have combined to offer dairymen the opportunity to select herd sires from essentially a worldwide supply.

## Semen Collection and Processing

As nearly all semen collection and processing is handled in breeding centers by highly trained personnel using sophisticated equipment, only the basic procedures will be discussed here. Semen is normally collected by the use of an artificial vagina. Volume is recorded and the ejaculate is evaluated for concentration (spermatozoa per ml.), motility, and morphology. Inferior-quality semen is discarded as its use is likely to result in lowered conception rates. The semen is diluted with an egg yolk-citrate extender or milk-base extender; the extenders contain antibiotics and, if the semen is to be frozen, glycerol. Dilution rates are calculated to yield a final motile spermatozoa per insemination of 10 to 12 million. An average ejaculate of 5 ml. of semen that contains 1.25 billion live cells per ml. after freezing contains enough spermatozoa for 500 to 600 inseminations. The semen is packaged in ampules or straws, frozen and stored (usually in liquid nitrogen storage tanks) at a temperature of $-300$ to $-320°F$ ($-184$ to $-196°C$). As long as the semen is maintained at this temperature, it remains viable for 10 to 15 years or more.

## Inseminating the Cow

Currently most U.S. dairymen have the alternatives of using AI technician insemination service or of purchasing semen and inseminating the animals themselves. The technique of artificially inseminating a cow is essentially the same in either case. The technique normally used is the rectovaginal method (Figure 11-1). A plastic or rubber glove and sleeve is used on one hand, which is inserted into the rectum. The cervix is grasped by this hand through the rectal wall and the inseminating tube is inserted into the vagina and through the cervix with the other hand. On first service the semen is deposited in the uterus at the anterior part of the cervix (Figure 11-2). On second and subsequent services the semen is deposited in mid-cervix. Regardless of whether insemination is done by a technician or by herd personnel, careful training, good sanitary practices, and continued practice are essentials to developing and maintaining good inseminating techniques and high conception

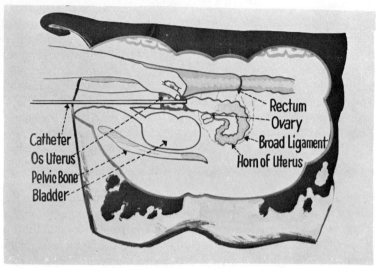

**Fig. 11-1.** Artificial insemination of a cow.

rates. Just as important is proper handling of the semen. It is essential to follow the procedures recommended by the semen supplier in regard to transferring semen from shipping container to farm tank, thawing rates and methods, and nitrogen levels in the storage tank. Periodic microscopic evaluation of stored semen is advisable, especially if it is suspected that the semen has warmed because of low tank nitrogen level or poor handling. Storage temperatures are very critical as semen quality deteriorates rapidly at temperatures of −110°F (−79°C) or warmer. Even though the semen is still frozen at these temperatures, it has deteriorated in quality and recooling will not restore its quality.

There are several factors that need to be evaluated before deciding whether to use technician service or to purchase semen and use herd personnel to inseminate the animals (direct herd service). Some of the advantages of direct herd service are the following.

1. It provides an opportunity for more precise timing of insemination.

2. It provides the opportunity for a wider selection of bulls in most areas.

3. It ensures a semen supply from a particular bull or bulls desired in the breeding program.

Some of the limitations of direct herd service follow.

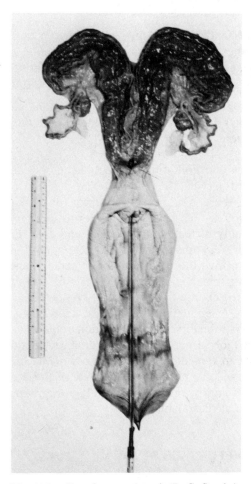

**Fig. 11-2.** Site of semen deposit (R. G. Saacke).

1.  The inconvenience of having to be on the farm when cows need to be inseminated. This is a particular problem in smaller herds where only one person is trained to inseminate cows. This person needs to be there essentially 365 days a year. Some dairymen have overcome this problem by combining technician service with herd service.

2.  The responsibility for the time-consuming details normally provided by the technician. These include: purchasing semen, liquid nitrogen, inseminating supplies, and equipment; monitoring semen quality; completing breeding receipts and semen transfers; and keeping up to date on semen handling and insemination techniques.

3. The hazard of handling liquid nitrogen. Liquid nitrogen can cause severe injury if it comes in contact with the skin. Extreme care must be used to avoid spillage and, if there are children on the farm, care must be taken in selecting a storage place for the liquid nitrogen tank.

4. Maintaining good technique. It is doubtful if many people can maintain a high degree of professional inseminating skill if they perform less than 100 inseminations per year.

The advantages and limitations of technician service are the inverse of those of direct herd service. In regard to the decision of technician versus direct herd service, there is no one best answer for all dairymen. Advantages and limitations need to be evaluated in terms of herd goals, herd size, availability and quality of technician service, and cost. Either system can be very satisfactory or very unsatisfactory. If direct herd service is used, it is very important to receive thorough training in semen handling and inseminating techniques (most semen suppliers hold periodic training sessions); inseminate enough cows to retain a high degree of competency; keep up to date on semen handling and inseminating techniques; keep up to date on regulations pertaining to semen transfers and registration requirements; and do a thorough, conscientious job of maintaining breeding receipts and records.

It is important to remember the objectives of the breeding program and then evaluate the available alternatives in view of these objectives. They are to maximize genetic improvement, maintain high reproductive efficiency, and do these two things at reasonable cost.


## REFERENCES FOR FURTHER STUDY

Hafez, E. S. E., *Reproduction in Farm Animals,* 2nd ed., Lea and Febiger, Philadelphia, 1973.

National Association of Animal Breeders, *The A.I. Digest,* published monthly. See especially **XXIV: 6** (1976).

Perry, E. J., The Artificial Insemination of Farm Animals, 4th ed., Rutgers University Press, New Brunswick, N. J., 1968.

Salisbury, G. W. and Van Demark, N. L. Physiology of Reproduction and Artificial Insemination of Cattle. Freeman and Company, 1961.

USDA, Agri. Res. Serv., *Dairy Herd Improvement Letter,* published twice yearly. See especially **51:2**:ARS-NE-60-1; **52:2**:ARS-NE-60-4, 1976.

# 12

# Milk Secretion and Milking

Milk is the basic product of the dairy industry; the secretion of large amounts of high-quality milk is the basic purpose of the modern dairy cow. Milk is secreted by, stored in, and removed from the udder, or mammary glands, of dairy cows and other mammals. The removal of large amounts of high-quality milk from the udders of dairy cows with minimum damage to the udders is the single most important job of dairymen. Improper or careless milking practices can result in decreased letdown, increased incidence of udder diseases, decreased milk quality and ultimately decreased productivity and profitability.

To achieve the goal of the milking program requires a basic understanding of the anatomy and physiology of the bovine udder, the process and control of milk synthesis and milk let-down, and the function of milking machines. It also requires thorough knowledge of managed milking practices that result in maximum let-down of a high-quality product with minimum udder damage.

## ANATOMY OF THE BOVINE UDDER

The bovine udder is complex. Each mammary gland is composed of secretory tissue (alveoli), a duct system, two cisterns, and a teat. It is serviced by the circulatory and nervous systems, suspended in the inguinal region by ligaments, and covered with skin. Each part has a vital function in the secretion or ejection of milk. The udder starts to develop in the fetus about 30 days after conception; continues through birth, puberty and pregnancy; and reaches its functional state at calving. Improper milking, handling, or disease can damage various parts of the udder and thus impair its degree of function. This results in lowered production, decreased product quality, and decreased profitability.

The udder of a cow is normally composed of four mammary glands, two (the fore and rear quarter) in each half of the udder. There may be one or more supernumerary, or extra, teats, with small glands, but these glands usually do not develop. The right and left halves of the udder are separated by a distinct septum (membranous wall), called the median suspensory ligament. The front and rear quarters are divided by a very thin connective septum, irregular in outline and not easily identified by the naked eye. As viewed from the side, the udder should be level, of moderate length, and firmly attached to the abdominal body wall. As viewed from the rear, it should be of uniform width, have a definite cleavage between the halves, with high and wide attachments. The rear quarters usually are larger than the fore quarters and secrete about 60% of the milk. Milk from each gland is removed through the teat. Teats should be of moderate size, evenly spaced under the udder, and hang perpendicular to the ground when the udder is full.

## Udder Support

The udder is supported primarily by the median suspensory and lateral suspensory ligaments (Figure 12-1). The median suspensory ligament is composed of strong, elastic tissue that extends between the halves of the udder, and has numerous lamellae that extend into the glands. Weakness of this central udder support will allow the udder to become pendulous and the teats to stick out, or strut, causing difficult milking and making the udder more susceptible to contamination and injury. The lateral suspensory ligaments, one on each side of the udder, are composed primarily of fibrous, inelastic tissue. They extend between the skin and gland tissue on both sides of the udder and have numerous lamellae that extend into the glands. Weakness of these ligaments allows the udder to become pendulous.

## Teats

Milk is removed from each gland through the streak canal (teat canal), which is ¼- to ³/₈-inch in length. The canal is kept closed by a circular sphincter

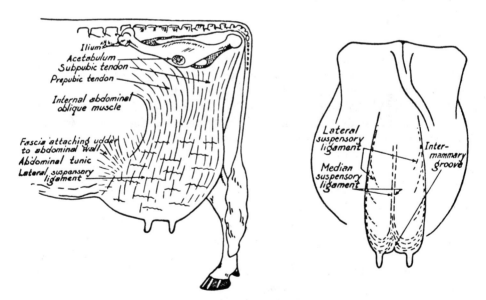

**Fig. 12-1.** Suspensory apparatus of the udder. (Reprinted by permission from PHYSIOLOGY OF LACTATION by Vearl R. Smith, Fifth Edition © by Iowa State University Press, Ames, Iowa 50010).

muscle near its outer end. This muscle serves to keep the milk in the udder between milkings and to keep bacteria and other foreign materials out of the udder. If the canal is small or the sphincter muscle is unusually strong, the cow is a hard and slow milker. If the canal is large or the sphincter muscle is unusually weak, the cow will leak milk between milkings and is more susceptible to invasion of mastitis-causing organisms.

The teat cistern, located just above the streak canal, is the cavity of the teat into which milk drains from the gland and from which milk is removed during the milking process. It normally holds ½ to 1½ ounces of milk, depending on the size of the teat.

### Gland Cistern

The gland cistern, located just above the teat cistern and partially separated from it by a circular fold, acts as a reservoir for the milk. The circular fold does not ordinarily interfere with the draining of the milk into the teat cistern, but sometimes it becomes so thick that milk drains very slowly. The gland cistern is a cavity of considerable size, varying somewhat in size and shape with different individuals and between quarters within the same individual. An average capacity would probably be about 1 pt., although it may vary from ¼ pint to as much as 1½ to 2 pt. (Figure 12-2).

### Ducts

A number of large ducts branch off from the gland cistern. The number of these ducts may vary from 12 to as many as 50 in each quarter. These ducts branch and rebranch into smaller and smaller ducts and finally into the terminal ductules that drain each alveolus (see below). The function of the duct system is to collect the milk from the secretory tissue, store part of the milk between milkings, and transport it to the gland cistern.

### Secretory Tissue

The alveolus (Figure 12-3) is a microscopic structure almost spherical in shape and lined with a single layer of epitheleal cells (Figure 12-4). The epitheleal cells are the milk-secreting cells. They remove nutrients from the blood and transform them into milk. The milk passes from these cells into the lumen of the alveolus and out into the duct system. Each alveolus is surrounded by a capillary network, which provides nutrients, and myoepitheleal cells, which contract during milking, causing milk let-down. The alveoli are grouped together into lobules. A group of lobules forms a lobe (Figure 12-2).

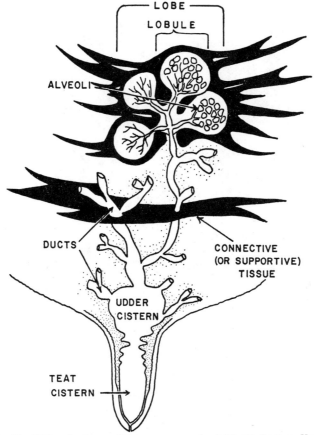

LOBE

LOBULE

ALVEOLI

DUCTS

CONNECTIVE
(OR SUPPORTIVE)
TISSUE

UDDER
CISTERN

TEAT
CISTERN

**Fig. 12-2.** Anatomy of the mammary gland (C. W. Turner, *Harvesting your Milk Crop*, 3rd edition, Babson Brothers Co. 1973).

## Circulatory System

The circulatory system of the udder is very extensive as it takes 400 to 500 volumes of blood passing through the udder for each volume of milk produced. The major arteries of the udder are the external pudic arteries, each one supplying arterial blood to one-half of the udder. They are branches of the external iliac arteries, and enter the udder through the inguinal canal. Just inside the udder, these arteries divide into the caudal (rear) and cranial (front) branches and redivide many times, finally to form capillaries that deliver blood nutrients to the cells. The perineal arteries also supply some arterial blood to the udder. The major veins draining the udder are the two

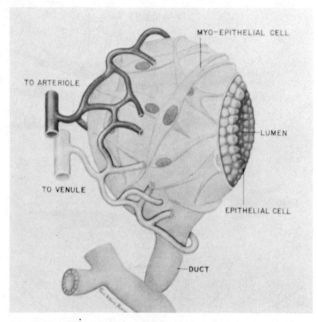

**Fig. 12-3.** Diagrammatic sketch of an alveolus, its blood supply and surrounding myoepitheleal cells. (C.I.B.A. Veterinary Monograph Series/One. C.I.B.A. Pharmaceutical Co. 1964).

external pudic and the two subcutaneous mammary veins. The external pudic veins lie adjacent to the external pudic arteries. The subcutaneous mammary, or milk, veins emerge at the front edge of the udder, run along the abdominal cavity and reenter the body cavity at the milk wells. Some venous blood is also drained through the perineal veins.

The udder also contains lymph ducts and nodes. Lymph is a colorless tissue fluid that arises from arterial blood and surrounds the udder tissue. Part of this fluid is reabsorbed by the venous system and part leaves the udder via the lymph ducts. Lymph ducts are difficult to identify because they are very thin walled and carry a colorless fluid, but they are present in the glandular tissue and teats and between the udder skin and tissue. Lymph flows from the udder, through the supramammary lymph nodes, and eventually back into the blood system. The lymph nodes serve to remove bacteria and foreign material from the lymph and as a source of lymphocytes. These functions help localize and combat infections.

### Nervous System

The udder is innervated with afferent sensory and efferent sympathetic nerve fibers. The sensory nerves carry the sensory stimuli from the udder to the central nervous system, an important step in milk let-down. The sympathetic nerve fibers regulate mammary blood flow and innervate mammary smooth muscles.

## MILK SYNTHESIS

The site of milk synthesis is the epitheleal cells lining the alveoli. Some milk components, vitamins, minerals, and some proteins, are not synthesized in the epitheleal cells but rather are filtered from the blood through the cells into the milk. Others, lactose, fat, and most of the protein, are synthesized in the epitheleal cells from blood nutrients. The blood precursor of lactose is blood glucose. The major blood precursors of milk fat are acetate, B-hydroxybutyrate, glucose, and fatty acids; the major synthesized protein precursors are free amino acids.

The filtered components and the blood precursors of the synthesized components of milk pass into the cell through the plasma membrane. This membrane, when functioning normally, exerts considerable selectivity concerning which blood substances pass into the cell. Some blood substances pass with ease and others are excluded; thus the plasma membrane is often called a selective semipermeable membrane. It is believed that during the first

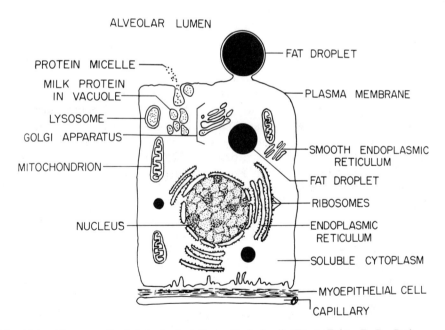

**Fig. 12-4.** Diagrammatic sketch of a lactating epitheleal cell. (From: Foley, R. D., Bath, D. L., Dickenson, F. N. and Tucker, H. A., Dairy Cattle: Principles, Practices, Problems, Profits, Lea and Febiger, 1972).

few days of lactation and during udder infections the secretory cells and the membrane are less functional, so that the composition of colostrum and mastitic milk more closely resembles that of blood serum.

Milk secretion is a continuous process in lactating cows. It reaches its maximum rate just after milking and its lowest rate just prior to the next milking. As milk accumulates in the lumen of the alveoli and the ducts between milkings, the epitheleal cells must secrete milk against an increasing pressure gradient. This results in a decreasing rate of milk secretion until the cow is again milked. If the pressure is not reduced by milking, milk secretion will cease and the milk constituents will be resorbed by the blood. The secretion rate is high and relatively steady for the first 9 to 11 hours after milking. It then starts decreasing at a rapid rate and if the cow is not milked will cease at about 35 hours after the last milking. This is why cows, especially high-producing cows, should be milked twice a day at regular intervals of 12 hours or 11 and 13 hours. It is well established that a major reason for the higher production of cows milked three times a day is that they secrete milk against a lower pressure gradient more of the time than cows milked twice a day.

## MILK LET-DOWN

As the secretion of milk is a continuous process, most of the milk secured at any milking is present in the udder when the milking starts. A small amount is in the teat and gland cisterns, but most of it is in the alveoli and duct system. The process of milk let-down, or ejection, is that of moving the milk from the lumina of the alveoli and ducts to the cisterns, where it may be removed. Milk ejection is an involuntary act on the part of the cow. It is a neural hormonal reflex that is very important if maximum milk production is to be obtained. Milk ejection is initiated by a stimulus such as washing the udder, manipulating the teats, the nursing of a calf, or other factors that the cows relate to milking. These stimuli cause a nervous impulse to travel via the afferent sensory nerves to the spinal cord and brain. The brain causes the release of the hormone oxytocin from the posterior pituitary gland into the blood system, which carries it to the mammary glands. There it causes contraction of the myoepitheleal cells that surround the alveoli and ducts and results in milk ejection. This process occurs in about 45 to 60 sec. after stimulation. The maximum effect lasts only 7 to 8 min., so prompt initiation of milking (1 min. after stimulation) and rapid milking are important in obtaining maximum milk yields.

Milk ejection can be inhibited even though the cow has been properly stimulated. It is often said that a cow "holds up her milk." It is impossible for a cow to "hold up her milk." The inhibition is involuntary and caused by the release of another hormone, adrenaline, often called the "fear" or "fright" hormone. Release of this hormone is also the result of nervous stimulation such as rough treatment of the cow, loud noises, pain, and irritation. Continual overmilking can be irritating and painful to the cow and can result in the cow's anticipation of pain, which may cause the release of adrenaline. The action of adrenaline can neutralize the effect of oxytocin by constricting blood vessels, which reduces blood flow and oxytocin presence in the udder, and by interfering with oxytocin release from the posterior pituitary. If adrenaline release is stimulated before the milk-ejection stimulus, it almost completely blocks milk ejection, whereas if it is released after milk ejection is started, it only partially inhibits milk ejection. Poor ejection results in larger amounts of residual milk retained in the udder, which in turn results in more rapid buildup of intramammary pressure between milkings. This decreases rate of milk secretion and eventually reduces production. Cows handled gently and milked carefully at regular intervals rarely experience this problem. Milking should be a pleasant experience for the cow if maximum ejection of milk is to be obtained.

## THE DEVELOPMENT AND
## FUNCTION OF THE MAMMARY GLAND

Initial development of the mammary glands is discernible in the 30- to 40-day bovine fetus. At birth the teats are well defined and teat and gland cisterns are present. Growth and development of the teats and duct system continues relatively slowly until the animal becomes pregnant. Estrogen, from ovarian follicles, is thought to be a primary stimulus in the growth and development of the duct system during this period. After conception a rapid growth and extension of the duct system to all parts of the udder occurs. Alveolar development begins at this time and continues rapidly. Significant amounts of fluid begin to accumulate in the glands at about the seventh to eighth month of pregnancy. Enlargement of the udder after this time is largely due to the accumulation of colostrum in the duct system and alveoli. Hormonal control of growth and development of the udder during this period is not well defined, but estrogen, progesterone, growth hormone, and prolactin play important roles. The composition of the secretion called colostrum is considerably different from normal milk for the first three to four days. This colostrum is essential for the newborn calf (see Chapter 14).

Mammary growth continues after parturition occurs and lactation begins. The number of alveoli continues to increase during early lactation and probably reaches a maximum at about peak lactation, then declines for the remainder of the lactation. Additional growth occurs during the early parts of subsequent lactations.

Initiation of lactation at parturition is believed to be stimulated by a marked rise in blood prolactin and adrenocorticotropic hormone (ACTH) levels, concurrent with a marked decrease in blood progesterone levels. After parturition the level of milk production rises rapidly, until peak lactation is reached in two to six weeks. Level of milk production then declines slowly until drying off, usually after 10 to 12 months. The degree to which lactation is maintained is called persistency of lactation. After peak lactation is reached, succeeding monthly milk yields should equal or exceed 90% of the previous month's milk yield. Persistency percentages over 90% are desirable and result in high lactation yield when compared to cows with equal peak yield but lower persistency. (Table 12-1). Monitoring degree of persistency is often a method of identifying problems. If a large proportion of the cows do not maintain high persistency of lactation, especially if milk yield decreases rapidly during the third or fourth month in a large number of cows, herd milking and feeding programs should be evaluated carefully. This problem could be caused by incomplete milking, poor milk ejection, or inadequate nutrient intake. Maintainance of lactation is also affected by the hormones prolactin, adrenal corticoids, and growth hormone. In dairy cows, milking stimulates release of prolactin and ACTH from the anterior pituitary. ACTH in turn stimulates release of adrenal corticoids. Thus, as well as reducing intramammary pressure, milking provides an important hormonal mechanism for maintaining lactation.

**Table 12-1**  Effect of Persistency on Lactation Production

| Month | Cow A: persistency = 90% production | | Cow B: persistency = 85% production | | Cow C: persistency = 80% production | |
|---|---|---|---|---|---|---|
| | Daily | Monthly | Daily | Monthly | Daily | Monthly |
| 1 | 60.0 | 1,800 | 60.0 | 1,800 | 60.0 | 1,800 |
| 2 | 70.0 | 2,100 | 70.0 | 2,100 | 70.0 | 2,100 |
| 3 | 63.0 | 1,890 | 59.5 | 1,785 | 56.0 | 1,680 |
| 4 | 56.7 | 1,701 | 50.6 | 1,518 | 44.8 | 1,344 |
| 5 | 51.0 | 1,530 | 43.0 | 1,290 | 35.8 | 1,074 |
| 6 | 45.9 | 1,377 | 36.5 | 1,095 | 28.6 | 858 |
| 7 | 41.3 | 1,239 | 31.0 | 930 | 22.9 | 687 |
| 8 | 37.2 | 1,116 | 26.4 | 792 | 18.3 | 549 |
| 9 | 33.5 | 1,005 | 22.4 | 672 | 14.6 | 438 |
| 10 | 30.1 | 903 | 19.0 | 570 | 11.7 | 351 |
| 10-month total | | 14,661 | | 12,552 | | 10,881 |

Development and function of the mammary gland is affected by genetics, the endocrine system, and the environment. Proper nutrition and management play a large role in the normal development of the udder as well as maintaining a high degree of function.

## MILKING MACHINES

Modern milking machines are capable of milking cows quickly and efficiently, without injuring the udder, if they are properly installed, maintained in excellent operating condition, and used properly. The milking machine performs two basic functions.

1. It opens the streak canal through the use of a partial vacuum, allowing the milk to flow out of the teat cistern through a line to a receiving container.

2. It massages the teat, which prevents congestion of blood and lymph in the teat.

### Essential Parts

The essential parts of a milking machine are: a source of vacuum, a vacuum line, a receptable for the milk, a pulsator, and milking units (a teat cup assembly including a shell and liner for each teat). The system should also contain a vacuum reserve tank, one or more vacuum controllers, a vacuum gauge, and a milk line in parlor or barn pipeline systems.

The milking machine is the most used equipment on the dairy farm. Proper installation, maintainance, and use are essential to good milking. An understanding of the basic components of the milking machine and their function is necessary for their proper use.

### Function

**Vacuum.** Milk is removed from the teat by vacuum when vacuum is in both the teat cup liner and the space between the liner and the shell as shown in Figures 12-5B and 12-6, milking phase. The milk flows from the teat, through the milk line and into the receiver. When vacuum is present in the teat cup liner, but not in the space between the liner and the shell, as shown in Figures 12-5A and Figure 12-6, rest phase, the liner collapses and massages the teat. In normal operation, vacuum is constantly present in the liner, while vacuum and atmospheric pressure alternate in the space between the liner and the shell because of the action of the pulsator.

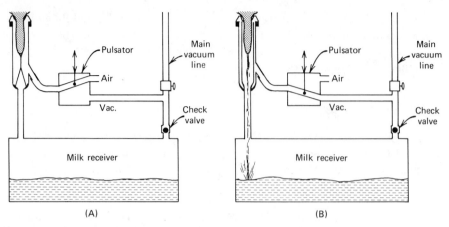

**Fig. 12-5.** The basic components of the milking unit *(A)* and their operation *(B)*. [*Penn. State Sp. Circ.* 74 (1965)].

A vacuum pump, lines, regulator, reserve tank, pulsator, vacuum gauge, and milk lines that are functioning to provide a steady source of partial vacuum at the teat end and intermittent vacuum in the space between the liner and the shell are essential for proper milking. All milking machines operate by vacuum.

The source of vacuum in a milking machine is a vacuum pump. Its function is to create a partial vacuum in the system by removing air from a confined space (the lines, teat cup liner, and reserve tank). Vacuum pumps are rated by their capacity to remove air in terms of cubic feet per minute (CFM) at a given vacuum level. When determining the size of vacuum pump needed in any system, the following points must be considered: (1) the vacuum level at which the machine operates; (2) the CFM requirements of the entire system, including milker units, vacuum regulator, measuring devices, etc.; and (3) the CFM capacity of the vacuum pump at the milking vacuum level. The pump CFM capacity can be determined from the pump rating or by measuring with an air flow meter.

The vacuum line transfers the vacuum from the pump to the milking claw assembly. The size of line needed is largely determined by the number of milker units in the system and the length of the line. Vacuum lines should not be less than 1¼ in. in diameter and should be sloped 1 in. per 10 ft.

The vacuum regulator(s) or controller(s) regulates the vacuum level in the system. When the vacuum in the system reaches recommended operating level, air is admitted through the controller and vacuum level is stabilized. It should have sufficient capacity to admit air equal to the vacuum pump capacity at operating vacuum level. It should be sufficiently sensitive to allow no more than 1 in. of mercury fluctuation in vacuum level under any operating condition. The controller should be located near the pump in a place easily accessible for servicing.

A vacuum gauge measures the vacuum level in the system. It should be used to monitor the vacuum level to ensure that recommended vacuum levels are maintained and vacuum fluctuations avoided. It is usually mounted in the vacuum line near the end farthest from the vacuum pump. Often two gauges are mounted in the vacuum line, with the second mounted near the pump. Many dairymen keep an extra gauge on hand, usually equipped with a rubber hose (a discarded piece of milk hose is fine), so that they may periodically check the vacuum in various parts of the system. This practice is highly recommended.

The function of the vacuum reserve tank is to cushion any inrush of air, such as happens when a milker unit falls off, that would result in a large drop

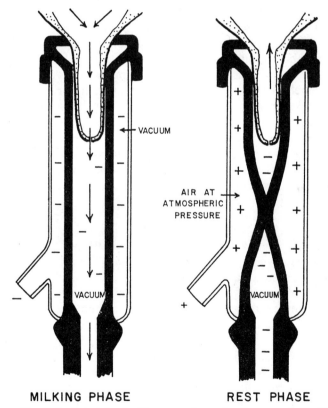

MILKING PHASE    REST PHASE

Fig. 12-6. Principle of milker operation and milk removal. [G. M. Jones, *V.P.I. & S.U. Pub.* 633 (1975)].

in vacuum level. It may also function as a trap for liquid that might get into the system accidentally or during cleaning. This protects the vacuum pump from damage. The tank should self-drain and be located near the vacuum pump.

**Pulsator.** The pulsator alternately directs atmospheric air (resting phase) and vacuum (milking phase) into the space between the teat cup liner and the shell (see Figure 12-6). During the milking phase, partial vacuum is present between the shell and the liner. The liner assumes its normal shape and the vacuum inside the liner opens the streak canal and withdraws the milk. During the rest phase, atmospheric air is admitted into the space between the liner and the shell. The liner collapses because of the vacuum in the milk line and massages the teat, thus preventing congestion in the teat.

Pulsators may be of the unit type (one at each milking unit) or the master type (one for more than one unit). They also may be operated by vacuum or electrically. One pulsation is defined as one opening and one closing of the liner, or the combination of one milking and one rest phase. Pulsation rate refers to the number of pulsations per minute. Pulsation ratio is the time ratio between the milking and resting phases of the pulsation cycle.

Vacuum level, pulsation rate, and pulsation ratio vary among brands of milking machines. Each manufacturer recommends the combination of vacuum level, pulsation rate, and pulsation ratio that works best for that brand of milking machine. These recommendations should be followed if one expects the machine to do its job efficiently without damaging the teats or causing discomfort to the cow. Many cows have been seriously injured by dairymen who felt they could milk faster if they increased the vacuum level or the pulsation rate or both.

**Milk Line.** In many pipeline (barn or parlor) milking systems, the milk line is also part of the vacuum system. It moves the milk from the claw assembly to the milk receptacle and supplies vacuum to the teat end. These lines must be large enough to allow the vacuum pump to move air fast enough to avoid vacuum fluctuation at the teat end. The size of the line needed is largely dependent on the number of milker units in the system.

Vertical lines (risers) should be avoided; rather, a continuous slope of at least 1½ in. per 10 ft. toward the receiver should be maintained. This allows gravity flow of milk without flooding and facilitates complete draining of cleaning and sanitizing solutions. Risers contribute to rancidity problems and line flooding. The latter may contribute to vacuum fluctuation problems. Milk inlet valves should enter the top one-half of the milk line, above the level of the milk moving in the line, to prevent milk flowing back down the milk hose.

## Characteristics of Efficient Milking
## Machine Systems

Regardless of milking machine manufacturer, certain characteristics are desirable to all pipeline milking systems. Some of these are the following.*

1. Air flow capacities for vacuum pumps of 9 to 12 cu. ft./min. (CFM) for each milking unit ASME standard or 18 to 24 CFM per unit New Zealand (N.Z.) standard.

    The revised (1977) "3-A Accepted Practices"* recommends the following points for calculating vacuum pump requirements (15 in. vacuum).

| | Requirement (CFM) | |
|---|---|---|
| Component | ASME | N.Z. |
| Milker unit, each | 6.0 | 12.0 |
| Weighing device, each | 1.0 | 2.0 |
| Sanitary couplings, per 20 | 1.0 | 2.0 |
| Pulsated vacuum line, per 10 ft. | 1.0 | 2.0 |
| Inlets, per 10 | 1.0 | 2.0 |
| Vacuum-operated releaser | 5.0 | 10.0 |
| Regulator reserve, each | 3.0 | 6.0 |

* 3-A Accepted Practices For the Design, Fabrication and Installation of Milking and Milk Handling Equipment, Serial No. T 606-01, amended 1977 (unpublished). (American Society of Mechanical Engineers)

The following is a sample calculation of vacuum pump requirements (CFM) for a double-4 milking parlor with 8 milking units equipped with milk meters (ASME standard).

| Component | Number | CFM requirement |
|---|---|---|
| Milking units | 8 | 48 |
| Milk meters | 8 | 8 |
| Sanitary couplings | 60 | 3 |
| Inlets | 16 | 2 |
| Subtotal | | 61 |
| Reserve for regulators | 3 | 9 |
| **Total** | | **70** |

* Adapted from Jones, G. M. and Troutt, H. F. Guidelines for Milking Systems Analysis. V.P.I. and S.U. Ext. Pub. 742, 1977.

2. Air flow reserve of at least 5 CFM (ASME standard) or 10 CFM (N.Z. standard) per milking unit. Reserve air flow is the additional air-moving capacity of the vacuum pump after satisfying the requirements of the milker units, bleeder vents, operating accessories, and other sources of air loss. Sufficient reserve air flow helps maintain a stable milking vacuum, especially when air is admitted into the system. Air flow reserve should be measured with all units operating but not attached and at the receiver jar in looped systems or at the end of the milk line in dead-end milking line systems.

3. Milking and vacuum lines inside diameter (ID) of the following minimums.

   a. Milking pipline: For conveying milk and supplying milk vacuum.

   | Number milking units per slope (maximum) | Milking pipline size (ID) | |
   |---|---|---|
   | | In. | Cm. |
   | 2 | 1½ | 3.8 |
   | 4 | 2 | 5.1 |
   | 6 | 2½ | 6.4 |
   | 9 | 3 | 7.6 |

   b. Main vacuum supply line: Lines from the vacuum supplier through the sanitary trap to the receiver.

   | Number of units | Pipe size (minimum ID) | |
   |---|---|---|
   | | In. | Cm. |
   | 2–3 | 1¼ | 3.2 |
   | 4–5 | 1½ | 3.8 |
   | 6–10 | 2 | 5.1 |
   | 11–13 | 2½ | 6.4 |
   | 14 or more | 3 | 7.6 |

   c. Vacuum pulsator lines: Supply vacuum to pulsators. These lines should be looped to a vacuum balance tank or a vacuum pulsator header line.

   | Number of units | Pipe size (minimum ID) | |
   |---|---|---|
   | | In. | Cm. |
   | 2–4 | 1¼ | 3.2 |
   | 5–7 | 1½ | 3.8 |
   | 8 or more | 2 | 5.1 |

4. Weigh jars or low-level milking pipelines.

5. Large volume claws with air-bleeder vents in the claws for most systems.

6. Vacuum level at the teat end should not exceed 13 in. (33 cm.) during milking.

7. Vacuum fluctuation at the teat end should not exceed 2 in.

8. Milk-rest ratio should not exceed 65:35.

**Milking Unit.** The purposes of the milking unit or claw assembly are to remove the milk from the teat to the milk receptacle or pipeline and to massage the teat. Milking unit assemblies usually contain the same basic components: a teat-cup claw, teat-cup shells and teat-cup liners (inflations), a milk hose, and air hoses. Some also contain a pulsator.

The claw functions as a point of attachment for the four teat cup assemblies and a milk receiver for the milk as it is removed from the teats by the liner. The milk-receiving portion of the claw should be large enough to handle the milk flow from high-producing cows without flooding the liner with milk from other liners or the milk hose. Proper claw size can also help prevent vacuum fluctuation at the teat end due to overfilling with milk and blocking air flow. The claw should be equipped with an air vent about $1/32$ in. in diameter to increase the rate of milk flow away from the teat end. Some claws are made partially of glass so that visual determination of milk flow can be determined.

The teat-cup shells are of rigid construction and provide a base for the teat-cup liner and the short air hose (from the shell to the claw). Lightweight shells should not be used with lightweight claws or the teat cup may crawl up on the teat and impede the passage of milk from the gland cisterm to the teat cistern.

The function of the teat cup liners is to remove the milk from the teat during the milking phase of the cycle and to massage the teat during the rest phase of the cycle. They are usually constructed of rubber. They should be flexible enough to close completely and with sufficient force to massage the teat during the rest phase, yet be rigid enough not to balloon during the milking phase. The use of narrow-bore liners is recommended, as they open and close the teat with the lease amount of distortion, expose the smallest amount of the teat end to the vacuum, and prevent the teat cup from crawling up on the teat. Liner size, however, must be matched to the size of the teat-cup shell. Standard narrow-bore liners are usually effective for 1,500 to 2,000 individual cow milkings before they start to bulge and enlarge. Their effective life may be increased by alternating two sets of liners weekly.

The functions of the milk hose are to move the milk from the claw to the milk receptacle or pipeline and to allow air passage from the vacuum line or bucket to the claw and teat-cup liner. Suspended-bucket milkers do not have a milk hose, as the claw assembly is attached to the lid of the bucket. Milk hose length and "lift" in the case of pipeline milkers should be kept to a minimum (a maximum of 9 ft. hose length and 7 ft. "lift"). A pipeline installed below the level of the cow's udder is preferable. A vacuum drop of 3 in. of mercury is normal in barn pipeline installations where the pipeline is mounted above the stanchions. A drop of 6 to 8 in. or more is not unusual in this type of installation if the cow is a rapid milker, the CFM capacity of the system is low, vacuum reserve is small, or the diameter of the milk pipeline is small. If a fluctuating vacuum problem is suspected, it can be checked by removing one teat cup from a cow during peak flow and inserting a vacuum gauge equipped with a rubber hose into the teat cup to measure the vacuum level. The authors have performed many such checks during which vacuum fluctuations of 3 to 11 in. of mercury were measured. Vacuum fluctuations can be minimized by adequate air movement (CFM capacity) and vacuum reserve, short milk hoses, and pipelines installed below the level of the cow's udder or no more than 2 to 3 ft. above the claw assembly.

The long and short air hoses function to supply air or vacuum from the claw to the space between the teat-cup liner and shell. If they crack or leak, their function is impaired and they should be replaced.

**Filters.**  Milk filters are a normal part of pipeline milking systems and can be a common source of trouble. Filters that are placed so that vacuum pulls or sucks the milk through them (in the claw assembly, milk hose, or, in systems that use vacuum bulk tanks, in the milk line) must be checked regularly to detect clogging, which can reduce the milking vacuum level. In-line pressure filters, located between the milk pump and tank, do not affect vacuum level of the system. If, however, they become clogged, the pressure of the milk being pumped through them may cause them to rupture. The less troublesome type of filter is a gravity-flow filter mounted in the line between a receiver and the tank. Clogging of these types of filters do not result in reduced vacuum to the system, and they do not tend to rupture. However, when clogging occurs the filter must be changed or milk will back up in the system. All filters should be changed periodically, depending on the number of cows milked.

**Automatic Milkers.**  Milking machines are available that are equipped with devices that sense changes in milk flow and remove the teat cups from the cow. Some of these remove all four teat cups at once and another removes them individually. Another available system reduces vacuum level and pulsation rate and ratio when milk flow decreases below a designated rate but does not remóve the teat cups. Because more mechanization is involved, proper maintenance is even more important than usual.

Decisions regarding brand of milking machine and degree of automation should be made on an individual basis after considering the alternatives in view of the purposes of the milking program, availability of service, and personal preference.

## MILKING

The process of milking is similiar to that of harvesting a crop. The calf has been grown to a cow. The feed has been grown or purchased. Buildings and equipment have been built and purchased. The cow has been fed and cared for. The cow is milked and the returns for all these operations, the money to pay for them and for family living expenses, depend on the amount and quality of milk she yields. The goal of the milking program is to harvest the maximum amount of high-quality milk with minimum udder irritation and with reasonable labor efficiency. We have previously discussed the structure and function of the udder, the location of milk at the time of milking, the milk-ejection reflex, the basic components of the milking machine, and how a milking machine removes milk from the udder. The remaining ingredient needed to achieve the goal of the milking program is a person or persons to apply these principles to "getting the job done."

Milking is the most important single job on a dairy farm. Careless milking techniques can result in udder damage or disease, decreased production, and decreased profitability. Milking also requires more time than any other one job in the production of milk. Most studies report over 50% of the total labor time involved in the dairy herd operation is involved in milking. Milking requires a large quantity of a high-quality effort.

### The Milker

A good milker is one who likes cows and is gentle with them. He or she is a highly skilled worker—rapid, efficient, gentle, and clean. A good milker can recognize abnormal conditions of the teats and udder by sight and feel, and can move about the cow gently and quietly to avoid making her nervous and afraid. A good milker also knows the operation details of the machinery and is able to maintain it in peak operating condition. Such a person also knows the essentials of a good managed milking program and follows them to get the job done properly and efficiently.

## Managed Milking Procedures

A managed milking program should be designed to remove all of the milk obtainable after milk ejection, prevent that milk from being contaminated with dirt or other foreign material, avoid irritation and injury to the teats and udder, prevent the spread of pathogens from infected to noninfected glands, and do these things efficiently. Each step of the following program is designed to promote these objectives.

1. Use a milking machine that is properly designed and installed—one that has adequate air flow capacity, large enough vacuum and milk lines with proper slope, etc., as discussed in the Milking Machine section of this chapter and as specified in 3-A Accepted Practices (see References). Many older installations have numerous inadequacies in design and installation that contribute to inefficient, poor-quality milking. These should be identified and corrected (see Appendix Table G).

2. Maintain the milking machine properly. Routine preventive maintainance of milking equipment and systems helps ensure top machine operating efficiency, minimizes the risk of breakdown during milking, and reduces emergency service or repair calls. Machine breakdowns during milking are costly in milk, time, damaged udders, reduced milk quality, and inconvenience. Most of these can be avoided with regular preventive maintainance. Preventive maintainance guidelines have been developed by the Milking Machine Manufacturers' Council of the Farm and Industrial Equipment Institute. Important points for consideration are presented in Appendix Table G.

3. Establish a regular milking routine. Milkers should have their work organized so they do not have other jobs to look after during milking. Good milking requires devoting 100% of the milkers' time and concentration to that job. Cows are creatures of habit and they quickly adjust to a regular routine. Some become nervous and irritable when the routine is not followed regularly. This and any other unpleasant experience that might interfere with milk ejection should be avoided.

4. Milkers' hands and the milking equipment should be cleaned and sanitized. Organisms multiply rapidly on milkers' hands, especially if the hands are chapped or cracked, and can be spread from cow to cow. Milk from healthy cows is a high-quality product. To keep it in this condition, clean, sanitized milking equipment is necessary.

5. Wash and massage the cow's teats and udder, preferably with warm water containing a sanitizing solution. (Figures 12-7 and 12-8) If a bucket

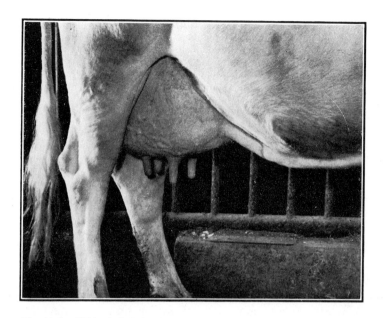

**Fig. 12-7.** Udder before let down of milk (W. Va. Univ.).

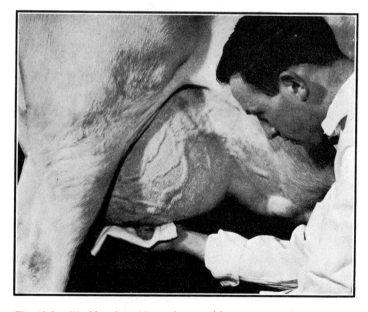

**Fig. 12-8.** Washing the udder and teats with warm water (W. Va. Univ.).

285

of warm water is used for this, individual disposable paper towels are preferred to a common sponge or cloth. Many modern systems are equipped with spray nozzles that spray warm, sanitized water on the teats and udder. This operation usually takes 15 to 30 sec. and serves as a major stimulus for milk ejection. It also protects the quality of the milk.

6. Thoroughly dry the teats and udder, preferably with an individual paper towel. This provides additional stimulus for milk ejection and removes contaminated water, which can get into the milk supply if not removed from the surface of the teats and udders.

7. Strip a few streams of milk from each quarter on to a strip cup (care must be taken not to splatter the milk to the other teats) and examine it for abnormal appearance. (Figure 12-9). This step serves several functions. It can help detect clinical cases of mastitis and thereby avoid contaminating the saleable milk with milk from infected cows. It eliminates the small amount of the first milk from uninfected cows that contains the highest concentration of bacteria. It also helps stimulate milk ejection. When clinical cases of mastitis are detected, the cow

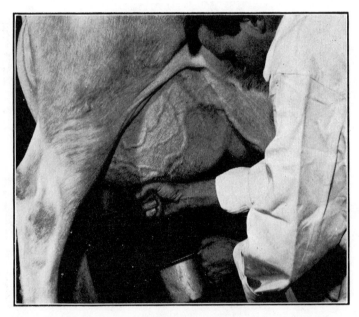

**Fig. 12-9.** Foremilking into a strip cup (W. Va. Univ.).

should be milked by a separate unit and the milk discarded. If a separate unit is not available, thorough cleaning and sanitizing of the unit after milking the cow is essential.

8. Attach the machine gently after milk ejection occurs. (Figure 12-10) If procedures 5, 6, and 7 are performed thoroughly, about 45 to 60 sec. will have elapsed since cow preparation started. This is the normal time it takes for milk ejection to occur. Attaching the machine too soon can result in teat-cup crawling, which inhibits milk flow from the gland cistern to the teat cistern and irritates delicate tissues. More than a few minutes delay in attaching the machine can result in a decreased amount of milk removed from the udder because of decreased effect of the milk-ejection reflex. A normal cow usually requires 3 to 5 min. to milk out, and the maximum effect of oxytocin lasts only 7 to 8 min.; therefore proper timing of attachment is important to maximum removal of milk from the udder.

9. Adjust the teat cups as necessary during milking to ensure complete milking of all quarters. This may not be necessary for many of the cows, especially if the milker knows the cows well enough to position properly the milking unit when it is attached.

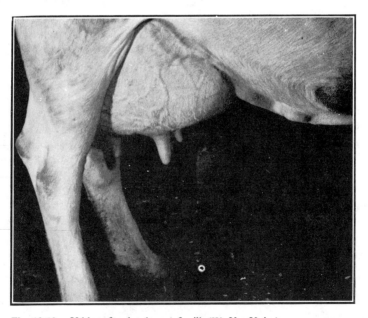

**Fig. 12-10.** Udder after let down of milk (W. Va. Univ.).

10. Remove the machine gently as soon as milk flow stops. Machine stripping just before removal of the machine is also recommended if it is done quickly. Research has not indicated that this practice significantly increases milk yield. It is advisable, however, to check the udder quickly to make sure all the available milk has been removed before detaching the machine. Teat cups should be removed gently by breaking the vacuum at the mouth of one teat-cup liner and shutting off the vacuum supply to the milking unit.

11. Disinfect the teat ends by dipping them in an antiseptic solution specifically approved for this purpose immediately after removal of the machine. This procedure removes milk that may remain on the teat end after milking and sanitizes the teat end. It has been reported by several researchers to be a very effective method of preventing new cases of mastitis.

Total man time per cow involved in procedures 5 through 10 is about 1¼ to 1½ min. per milking unless the cows are very dirty or other problems occur. The milker's routine must be well organized and needed equipment and supplies readily available. The man time can be reduced by the use of more automated equipment such as prep stalls and automatic machine-removal devices enabling one person to milk more cows per hour. Increased labor efficiency as the result of increased automation must be evaluated carefully in economic terms and quality of the milking program.

Following these 11 managed milking procedures should achieve the goals of the milking program. They basically involve two major factors—human and machine. There is little excuse for the machine to operate at much less than 100% efficiency. Human beings usually are more difficult to motivate at a high degree of efficiency. Many milkers know what they should do, but do not do it.

Cows like to be milked, especially when things they associate with milking are pleasant. Under normal and pleasant conditions their reactions are shown by their general willingness to be milked. When there are conditions associated with milking that they dislike or are painful such as rough treatment, overmilking, etc., they often respond by exhibiting signs of nervousness such as kicking or not standing quietly. The milk-ejection reflex may be inhibited and less than all of the available milk may be available for removal. This rarely occurs when milking procedures are performed by a conscientious milker who likes cows and is gentle with them.

## Drying Off the Cow

The purpose of allowing a cow a dry period is to give her a rest before the next lactation. Repair and regeneration of the secretory cells of the udder is believed to be involved. Allowing dairy cows a dry period has been shown to result in significantly higher production during the succeeding lactation. Conversely, continuous lactation has been shown to impair succeeding lactation production. The recommended length of dry period is 45 to 60 days. The drying-off period is a critical period, as cows appear to be more susceptible to mastitis during drying off and during the early part of the dry period.

Two methods of drying off the cow are commonly used: complete cessation of milking and intermittent milking. Complete cessation of milking is the recommended method for mastitis-free cows. It can also be used for mastitis-infected cows if they are treated with an effective antibiotic. Pressure build-up in the udder is rapid and milk synthesis stops because of the increased pressure. If intermittent milking is practiced, the periodic removal of milk relieves the pressure and stimulates continued milk synthesis. This prolongs the drying-off period, but this practice may be necessary in some cows. A few dairymen use another method, incomplete milking. This is not a recommended practice.

Cows whose level of production has dropped to 20 lb. or less per day by drying-off time are usually not a problem. In cessation of milking, milk secretion stops and the milk is resorbed. This is usually all that is needed. Many high-producing cows, however, are still producing 30 to 50 lb. of milk daily at drying-off time. Such cows require much more care and management. Milking such cows once daily, while restricting feed intake (usually of grain), until production drops to less than 20 lb./day and then ceasing milking is used effectively by many dairymen. Some cows, however, continue to milk at high levels and the drying-off process becomes very long. Reducing roughage and water intake during the once-daily milking period is an effective method for further decreasing milk production. An increasingly popular method of drying off these high-producing cows is to severely restrict feed intake (eliminate concentrates and reduce forage) and water intake, but to continue milking the animal twice a day until production drops to about 20 lb./day, and then cease milking. This method has been a safe and effective one for many dairymen.

Regardless of the method used, it is recommended that the cow be milked for the last time about 5 to 7 days after the cessation of milking to check on the condition of the udder. Treating cows after this milking with an antibiotic specifically formulated for this purpose is a relatively new, but increasingly popular, method of controlling new and existing mastitis infections. Some of the advantages of dry-cow treatment are: (1) new infection rates are high during the dry period, especially during the first three weeks, and this treatment can be an effective preventive measure; (2) it is effective in reducing the number of pathogenic organisms (except coliforms) in the udder; (3) as the cow is not being milked, preparations of stronger concentrations and slower release can be used to maintain high levels of therapeutic agent in the glands for long periods of time; and (4) it reduces the level of infection at the beginning of the next lactation when daily milk production is highest. Some authorities recommend treating all quarters of all cows at drying off and others only those quarters that have exhibited clinical or subclinical symptoms of mastitis during the previous lactation.

## UDDER EDEMA

Udder edema is a disorder of cows characterized by an accumulation of fluid (lymph) around and adjacent to the mammary gland. The udder, especially the lower part, appears swollen, and sometimes the swelling may extend in front of the udder along the belly to the navel or beyond and nearly to the vulva in the rear. The disorder appears just before or at the time of calving and is more prevalent in first-calf heifers and cows with deep udders. Udder edema is not caused by a pathogenic organism, but rather by more fluid being filtered out of the capillaries than is being returned to the circulatory system. This fluid then accumulates in the intercellular tissue spaces and the result is swelling of the udder and surrounding area. Edema causes the cow some discomfort, inhibits her movements, places excessive strain on udder attachments, and in severe cases can also make milking more difficult. Chafing and irritation of the tissue of the upper part of the udder and inside of the thigh often occurs and should be treated promptly.

The precise cause of the fluid imbalance is not known. Genetic, hormonal and management factors such as feeding and exercise are probably involved. Moderate exercise during the prepartum period and feeding moderate amounts of salt during the dry period may be helpful. High grain feeding of heifers prior to calving seems to contribute to a higher incidence of the problem. Prepartum milking does not appear to alleviate the condition. The swelling normally recedes within a few days after calving and completely disappears within two to three weeks. Massaging the udder to stimulate lymph circulation is helpful, but often not practical in many herds. In severe cases diuretics, alone or in combination with glucocorticoid hormones, are effective in reducing the swelling. They should, however, be used with caution and not in excess of the manufacturer's or veternarian's recommendations.

## REFERENCES FOR FURTHER STUDY

3-A Accepted Practices for the Design, Fabrication and Installation of Milking and Milk Handling Equipment, *J. of Milk and Food Tech.* **31**:355 (1968).

Dairy Guideline No. 275, V.P.I. & S.U. Ext. Service, 1970.

Guest, R. W., and Spencer, S. B., The Milking Machine System, *Penn State University Spec. Cir.* 74, 1965.

Jones, G. M. and Troutt, H. F. Guidelines for Milking Systems Analysis. V.P.I. and S.U. Ext. Pub. No. 742, 1977.

Schmidt, G. W., *Biology of Lactation,* W. H. Freeman, San Francisco, 1971.

*The Modern Way to Efficient Milking,* published by the Milking Machine Manufacturers' Council of the Farm and Industrial Equipment Institute.

# 13

## Herd Health

One of the greatest ravages of the profits of dairy farms is disease. Each cow culled and sent to slaughter because of a disease or injury represents a loss equal to the difference between beef value and replacement cost. Each animal that dies represents a loss equal to her value as a dairy animal. An even greater loss may occur when cows are not milking up to inherited capability and their environment because of some level of disease or impaired health. A summary of some herd health factors and number of cows removed from DHI dairy herds during the 1975 testing year whose records were processed at the Southern Regional Dairy Records Processing Center is presented in Table 13-1. Over 435,000 cows from 4,586 herds in 13 states are included. Nineteen percent of all dairy cows, or 67.9% of all cows culled, left the herd for reasons other than low production. In the 13-state region this amounted to 84,708 cows of the 121,808 culled. Removals for reproductive failure alone amounted to over 39,000 cows. These data indicate losses of considerable magnitude for these 13 states. These figures are probably similar for most of the United States.

The goals or objectives of a herd health program are to minimize nongenetic culling (culling resulting from disease, injury, etc.) and mortality while maintaining a healthy herd with a high degree of reproductive efficiency. To do this requires a herd health program centered around the prevention of disease and other health problems rather than the sporadic treatment of various conditions. This can best be achieved through the cooperative efforts—the teamwork—of a conscientious, knowledgeable, interested herd manager and a competent large-animal veterinarian. Keeping adequate health records, day-to-day observation of the cattle, feeding adequate amounts of a nutritionally balanced ration, providing a healthy environment, and other aspects of good cowmanship are responsibilities of the herd manager. Appropriate vaccination programs, accurate diagnosis of problems as well as pregnancy tests, and prompt effective treatment as needed are the major responsibilities of the veterinarian. The program should be evaluated in terms of costs and returns, just as are other aspects of herd management. Effectiveness of the program should be evaluated in measureable terms and if deficiencies exist, changes should be made. Some measures of the effectiveness of the program might be the following.

1. Calf mortality less than 5%.

2. Cow mortality less than 2%.

3. Nongenetic culling less than 10%.

4. Average days open less than 110.

**Table 13-1** Health Factors and Cow Removals in Dairy Herds

|  | Average (%) | Highest state (%) | Lowest state[a] (%) |
|---|---|---|---|
| Total left herd | 28 | 30 | 23 |
| Culled for low production | 9 | 12 | 5 |
| Culled for reproduction | 7 | 10 | 5 |
| Culled for other reasons | 12 | 14 | 10 |

[a] All but one state fell within a range of 27 to 30.

**5.** Incidence of clinical mastitis less than 5%.

**6.** Milk production 1,500 lb. above breed average.

Health-care programs will vary from farm to farm; however, the basic principles of an effective health-care program apply to all farms. Some of these are noted below.

1. Prevention of disease and problems is more effective and more profitable than treatment. Prevention can be achieved by the following practices.
   a. Preventing exposure to disease-producing organisms. This can be achieved by good sanitation and cleanliness, isolation of incoming animals, and eradication of certain diseases.
   b. Maintaining a high level of resistance. This can be achieved by vaccination for those diseases for which there is an effective vaccine, maintaining all animals at a good level of nutrition, and providing a comfortable environment.
2. Reducing the spread of diseases that do occur. This can be achieved by the following practices.
   a. Isolation of animals that contract or are suspected of contracting a contagious disease. Quarters inhabited by that animal should be thoroughly cleaned and disinfected before being reused.
   b. Rapid, accurate diagnosis and prompt treatment of disease problems. This may involve blood testing and other laboratory diagnostic tests as well as post mortem examination.
   c. Keen observation by herd personnel to detect minor abnormalities before they become serious problems. Good cowmen can detect minor problems by the characteristics and behavior of the animals. A cow that is slightly off feed, stands uneasily, has a roughened haircoat, passes feces that are too soft or too hard, or stands by herself in the corner of the lot or barn one day may be seriously ill the next. Her condition may be easily remedied if treated immediately; if she has a contagious disease, immediate isolation may prevent spreading the disease through the herd.

3. Maintaining and using an accurate health record system. Good health records can aid in the diagnosis of problems, help alert dairymen to potential problems (milk fever, for example has a high repeat incidence), and ensure that important details are not overlooked such as missing vaccinations on some animals. The record system should be one that contains the needed information, yet is not so exhaustive that it is not kept accurately.

4. Cooperation and mutual respect among the owner, manager, herdsman, veterinarian, and state and federal disease control personnel. A good working relationship among these persons is essential to the success of the program. Each has vital areas of responsibility, and the continued success of the program depends on each fulfilling his or her responsibility in the total effort.

Changes in dairying that have occurred over the last several decades have increased the need for applying these basic principles. Two of these changes and their effects on herd health are increased herd size and increased production per cow.

Increased herd size has been accompanied by increased concentration of cattle in smaller areas and increased labor efficiency. Partial or total confinement systems for the milking herd are common; fewer cattle are pastured; number of cows per person has increased; and more cattle are handled by hired labor rather than the owner or family. Increased concentration of cattle increases the opportunity for incidence and spread of disease because of the closer proximity of the animals, the larger number of susceptible animals, and increased organism build-up. The latter occurs because the number of organisms in a given area is greater and organisms that pass through animals often increase in virulence. The higher the concentration of animals, the greater the opportunity for this to occur. Increased labor efficiency results in less man time per cow and less frequent observation of individual animals. This, coupled with less personal interest in the cattle by some hired labor, means lack of early detection of many diseases and problems. Numerous studies have reported increased calf mortality, cow mortality, days open, incidence of mastitis, etc., in larger herds.

The modern dairy cow is a very different animal from her ancestors of a few decades ago. She possesses the genetic ability for producing huge amounts of milk. Dairymen have learned how to feed and manage dairy cattle to produce at or near genetic potential. The net results are that per cow production has nearly doubled in the last 20 years. This increased performance has increased susceptibility to and incidence of a wide variety of metabolic and functional disturbances and stress diseases. This trend is likely to continue as performance continues to climb. Control will be difficult because precise physiological mechanisms involved are not well defined. The development of more precise ration formulation methods and improved diagnostic procedures, coupled with keen observation, will be necessary if these problems are to be controlled.

The combination of these and other factors have led to the development and use of comprehensive preventive health-care programs by many veterinarians and dairymen, rather than emergency veterinary care only. The specific details of a comprehensive health-care program will vary among farms and veterinarians depending on personal preference, economics, and local variation in incidence of certain diseases and problems. The value of such programs has been reported by numerous authors, but perhaps one of the most thorough studies reported was that by Barfoot,* who compared the effect of five levels of manager response to a preventive medicine program (control—emergency case only—and minimum, average, above average, and maximum) on five variables (milk production, days open, calf mortality, cow mortality, and culling rate). Costs of each program were recorded and average estimated return on the investment in the health-care program at varying levels of production and milk prices were determined. A brief summary of the findings is presented in Table 13-2. The program included monthly herd visits, at which time medical checks such as diagnosis of reproductive status and udder health, vaccinations, dehorning, and extra teat removal were performed and recorded. Preventive and corrective treatments as needed were also administered and recorded. Emergency service was provided on a call basis.

The benefits of such a program are readily identified in every measurement parameter. Predictably, as the degree of positive response to the program increased, so did the costs. More important, however, was the response in increased net value of the additional milk. Undoubtedly, additional savings were realized through decreased losses of valuable calves and cows. These and other similar studies emphasize the value of such preventive health-care programs and contractual agreements with competent veterinarians. Details may vary from farm to farm and from veterinarian to veterinarian, but the basic importance and value of such a program is well justified. A contractual agreement should include the following elements.

1. Periodic (monthly) herd visits, when the following services are performed.
   a. Reproductive check (as detailed in Chapter 10).
   b. Vaccinations (recommendations concerning specific diseases and vaccines will vary with location and availability of vaccines).
   c. Preventive and corrective treatment as needed.
   d. Recording of all diagnoses, treatments, and vaccinations.
   e. Consultation and recommendations concerning other health problems, preventive measures, and routine management practices that can affect the health of the herd. This might include training personnel in the proper methods of using medical equipment and supplies for routine or emergency treatments.
2. Furnishing equipment and supplies at additional charge.

* L. W. Barfoot, Master's thesis, University of Guelph, Ontario, Canada.

**Table 13-2** Response to Five Levels of Health Care[a]

| Health plan and description | Annual cost per cow | Average days open | Calf mortality (%) | Cow mortality (%) | Culling rate (%) | Milk prod. per cow (lb.) | Addit. milk (lb.) | Value of addit. milk at $10.00/cwt. | Net value of addit. milk |
|---|---|---|---|---|---|---|---|---|---|
| A Control—emergency care only | $5.00 | 138.5 | 11.8 | 2.5 | 13.50 | 11,830 | | | |
| B Minimum response | $20.00 | 135.0 | 11.0 | 2.15 | 11.50 | 12,110 | 280 | $28.00 | $13.00 |
| C Average response | $25.00 | 130.6 | 9.0 | 2.00 | 9.96 | 12,530 | 700 | $70.00 | $50.00 |
| D Above average response | $30.00 | 127.1 | 7.9 | 1.95 | 9.00 | 12,810 | 980 | $98.00 | $73.00 |
| E Maximum response | $35.00 | 123.6 | 7.5 | 1.90 | 8.10 | 13,020 | 1,190 | $119.00 | $89.00 |

[a] Adapted from L. W. Barfoot, Master's thesis, University of Guelph, Ontario, Canada.

3. Postmortem examinations.
4. Emergency on-call service at extra charge.

## SPECIFIC PROGRAMS AND PRACTICES*

### Calf and Heifer Program

1. Dip navel in 7% tincture of iodine as soon as possible after birth.

2. Make sure calf receives 4 to 6 lb. of colostrum within 1 to 2 hours of birth. In case of weak calves, the use of an esophageal tube is recommended.

3. Administer Large-Animal Serogen, mixed bacterin and vitamins A, D, and E or other prescribed proplylactic treatments when calf diseases and mortality are a problem.

4. Dehorn with caustic paste at 3 to 14 days of age.

5. Maintain strict cleanliness of quarters and equipment used in feeding calves milk or milk replacer. Dry and draft-free calf quarters are essential.

6. Observe calves closely twice a day for appetite, attitude, and condition of bowel movements. If scours occur, replace milk with an electrolyte solution.

7. Check all heifers for extra or supernumerary teats at 3 to 5 months of age. Remove any if necessary at this time.

8. Vaccinate heifers for *Brucella abortus* (bangs), leptospirosis, vibriosis, IBR, BVD, PI-3, blackleg and malignant edema as indicated by local conditions and the recommendation of the veterinarian.

9. Maintain calves in groups, by age if possible, for internal parasite control. Do fecal exams and worm when it is necessary.

10. Practice good external parasite control. Treat all heifers from 6 months through breeding age for warbles in fall.

11. Observe older heifers daily and record all heats and abnormal conditions.

12. Examine all heifers not in heat by 13 to 14 months of age.

13. Cull heifers not pregnant by 22 months of age.

* Adapted from recommendations of T. Bibb, Vet. Sci. Dept., VPI & S.U.

## Adult Cows

### Calving Area and Procedures.

1.  Provide a clean, dry, well-bedded stall or pasture.

2.  Don't be impatient when parturition begins, but be ready to assist or summon professional assistance when needed. If labor persists 4 to 6 hours without progress, the position of the calf should be checked and assistance summoned as needed. Heifers normally need a longer time then older cows.

3.  Check udder for open teats, acute mastitis.

4.  Give cow warm water after parturition as this may reduce the incidence of retained placenta.

5.  Observe for passage of placenta and remove from the calving area. The cow should not be allowed to eat the placenta. It will not be digested and may cause digestive disturbances.

### Reproductive Program.

1.  Maintain accurate records, including calving dates, calving difficulties, retained placenta, abnormal vaginal discharges, heat dates, irregular estrus cycles, breeding dates, sires used, and medical and hormonal treatments.

2.  Check for heats at least 2 or 3 times daily.

3.  Examine all cows 30 to 40 days postpartum (preliminary exam) for condition of the reproductive tract.

4.  Examine all cows with abnormal vaginal discharge or abnormal estrus cycle.

5.  Reexamine at 50 to 60 days postpartum all cows that have not been in heat.

6.  Breed all cows on first heat following 50 to 60 days postpartum if reproductive status is deemed good.

7.  Examine all cows and heifers for pregnancy 40 to 60 days after last breeding.

8.  Examine all cows and heifers for abnormal reproductive tract conditions after the second or third service if they return to heat.

9.  Examine all cows and heifers that abort.

## Mastitis Control Program.

1. Ensure proper installation and function of the milking machine (Chapter 12).

2. Follow a proper milking routine that encourages maximum let-down with minimum teat and udder damage (Chapter 12).

3. Routinely use mastitis screening tests—strip cup, CMT, etc.

4. Dip teats after milking.

5. Treat dry cows.

6. When mastitis occurs, identify causative organisms and treat as prescribed by a veterinarian.

7. Reduce udder edema by exercise and/or medication as prescribed by a veterinarian.

## New Additions to the Herd.

1. Bring in animals from herds with a good health history.

2. Have the animals tested and cleared for tuberculosis, brucellosis, leptospirosis, IBR, BVD, and PI3 before arrival in the herd.

3. Isolate new animals for 30 days whenever possible. Observe closely and vaccinate as needed during this period.

## Miscellaneous.

1. Practice good external parasite control programs, especially for flies and lice. This helps to reduce eye problems and prevent anemia.

2. Have fecal exams performed periodically to determine if a worming program for internal parasites is needed.

3. Keep lots and pastures free of broken equipment, wet spots, and trash. This helps to reduce foot problems, injuries, and hardware disease.

4. Trim feet when needed.

5. Monitor carefully the condition of dry cows (not too thin or too fat) to help reduce calving troubles, ketosis, milk fever, and the fat cow syndrome.

6. Make sure animals are receiving a balanced ration, including adequate minerals and vitamins.

7. Be on alert for occurrence of displaced abomasum in recently fresh cows. This is one condition for which early diagnosis is critical to successful treatment.

8. Have all animals that die autopsied.


## SPECIFIC DISEASES OF DAIRY CATTLE

Dairy cattle are susceptible to a wide variety of diseases, disorders, and parasites. A brief discussion of the causes, symptoms, preventive measures, and recommended treatment of some of these follows. For a more complete and thorough discussion of diseases and disorders of dairy cattle, the authors recommend reading *The Merck Veterinary Manual* (see reference list).


### Milk Fever, Also Called Parturient
### Hypocalcemia or Parturient Paresis

Milk fever is a metabolic disorder that usually occurs within 72 hours after calving (approximately 75% of the cases within 24 hours of calving), but occasionally it occurs before, during, or some months after calving. Its incidence is estimated at 6 to 8% in the United States. It is rare in first-calf heifers and occurs more frequently in older high-producing cows. Jerseys have a higher incidence than the other U.S. dairy breeds. Over 50% of the cases are repeats, and there is approximately a 30% relapse rate.

**Cause.** Milk fever is caused by low blood serum calcium level, but the exact cause of the decreased serum calcium is not known. Symptoms usually appear when the blood calcium drops from a normal 10 mg./100 ml. to 3 to 7 mg. A high calcium-phosphorus ratio (2:1 or higher) and/or high calcium intake (over 100 to 125 g./day) during the dry period, change in vitamin D metabolism, and hormone imbalance (parathormone) are predisposing factors.

**Symptoms.** Initial symptoms of unsteadiness and falling temperature appear when serum calcium falls below 6.5 to 7 mg. As serum calcium falls to 5.5 to 6 mg., the cow usually lies down, but in a normal position. She becomes partially paralyzed and is unable to rise. Often her head is turned back toward her side. As the condition progresses, the cow will usually stretch out on her side. When this occurs she should be propped in an upright position to help her avoid inhaling food material flowing back from the stomach, which may lead to pneumonia, and to avoid bloating. Milk fever usually results in death of the animal in a short time unless treatment is given promptly.

**Prevention.** Because of the relatively high incidence of milk fever and its severe nature, numerous methods of prevention have been studied.* Some of these are (1) feeding low-calcium or low-calcium, high-phosphorus diets during the dry period; (2) feeding massive doses of vitamin D (20 to 30 million units/day) three to seven days prepartum; (3) prepartum administration of 25-hydroxycholecalciferol, a vitamin D metabolite; (4) feeding acidic diets, mineral acids, or ammonium chloride prepartum; and (5) short-term administration of 90 to 100 g. of calcium chloride daily. The feeding of massive doses of vitamin $D_2$ has proved effective in preventing milk fever if fed at least three days before parturition. The feeding must be terminated after seven days, however, or toxicity occurs. The preventive effect continues only three days after the treatment is discontinued. Administering 25-hydroxycholecalciferol is also effective, but the substance has not been cleared for use at this time. Based on these factors, current recommendations are to avoid calcium intake over 100 to 125 g./day during the dry period, feed adequate phosphorus and vitamin D during the dry period, avoid overconditioning, and keep on feed at freshening. In herds with a very high incidence of milk fever, consider feeding massive doses of vitamin D as recommended,** or the vitamin D metabolite when it becomes available.

**Treatment.** Milk fever is treated by injecting intravenously 250 to 500 ml. of a 20% calcium gluconate solution. Treatment usually results in rapid recovery; however, there is approximately a 30% relapse rate, so careful observation for return of the symptoms is essential and prompt retreatment is often needed.

An older treatment, in which the udder was inflated with air, proved effective. Treatments were sometimes followed by udder infections from unfiltered air and the use of equipment that had not been cleaned and sterilized. The air in the udder builds up back-pressure that stops milk secretion and the withdrawal of calcium from the blood. This process allows the blood to rebuild its calcium supply. Response after this treatment is not so rapid or regular as it is after the injection of calcium solutions.

Medicine should never be given a cow by mouth when she has milk fever, as her throat is partially paralyzed and the medicine may go down the windpipe and cause pneumonia.

It is a recommended practice not to milk a suspected cow completely dry for several milkings after calving. This will reduce the amount of milk secreted and consequently the amount of calcium taken from the blood. It has

* N. A. Jorgensen, Combatting Milk Fever, *J. Dairy Sci.,* **57**:933 (1974).
** J. W. Hibbs and H. R. Conrad, Studies of Milk Fever in Dairy Cows, *J. Dairy Sci.,* **43**:1124 (1960).

been observed that cows with caked udders at time of calving seldom have milk fever. This may be because not all the milk can be removed from such udders.

### Acetonemia, Also Called Ketosis

Acetonemia is a metabolic disturbance caused by decreased blood glucose and an abnormal concentration of the ketone bodies in the blood, milk, and urine of affected animals. These ketone bodies are the products of incomplete utilization of body fat, and the main ones are acetone, acetoacetic acid, and beta-hydroxybutyric acid. The disease usually occurs within six weeks after calving, with three weeks after calving the most critical period. The incidence is higher when cows have been overconditioned and then lose weight rapidly. Reduced feed intake just before and after calving seems to be a contributing cause. There seems to be a high incidence also during the late winter and in higher-producing older cows. The incidence is approximately 10% in cows over 36 months of age.

**Symptoms.** When present in its typical and uncomplicated form, acetonemia may cause the following symptoms: abrupt drop in the milk flow, loss of appetite, rapid loss in weight, and incoordination. Other symptoms may include excitement, noted by circling in the stall or yard; pressing forward against the manger; and depraved appetite.

**Diagnosis.** A test can be made of the urine for ketone bodies. This test may not always indicate the presence of the disease, as any fasting animal will have ketone bodies in the urine.

**Prevention.** Maintaining animals on a reasonably high energy intake just before calving and rapidly increasing intake after calving to prevent unusual demands on body fat stores is helpful in preventing ketosis. Addition of sodium propionate or propylene glycol (.25 to 1.0 lb./day) to the diet for six weeks after freshening has also been helpful. The disease is often associated with cases of mastitis, metritis, and milk fever, so prevention of these problems is also important.

**Treatment.** The intravenous injection of glucose is temporarily effective in alleviating the depressed blood glucose. It is often combined with injection of glucocorticoids for a longer-lasting effect. The oral administration of sodium propionate or propylene glycol for 10 days is also effective, though the response is slower. Often these oral remedies are used following the injection of glucose and/or glucocorticoids.

## Displaced Abomasum

Abomasal displacement appears to be occurring with increasing frequency. It usually occurs near calving, and its incidence is higher in large, high-producing, heavily fed, older dairy cows. It is more prevalent in winter and spring months and in cows fed high grain and corn silage rations, especially when corn silage is the only forage. It is often associated with one or more of the following problems, milk fever, ketosis, or metritis.

**Cause.** The primary cause is believed to be decreased tone of the abomasum, often accompanied by abomasal enlargement and accumulation of gas and/or fluids. This leads to migration of the abomasum either to the right (RDA) (approximately 15% of the U.S. cases) or the left (LDA) (approximately 85% of the U.S. cases) and upward within the abdominal cavity. The rumen usually descends to trap the abomasum in the abnormal position. In cases of RDA, passage of digesta is completely or nearly completely blocked and the condition requires immediate treatment. In LDA, passage of digesta is reduced and a chronic condition often ensues.

**Symptoms.** Symptoms include depressed appetite, decreased defecation, rapid loss of weight, and diminished milk production. Often a fullness in the left flank is observable, and the animal will show a positive urine-ketone reaction.

**Prevention.** Abomasal displacement is believed to occur because of the coincidence of a variety of predisposing factors including high production. Dairymen obviously do not wish to decrease production to help prevent the disease; however, prevention of the other predisposing factors is advisable. These include feeding prepartum high-forage (other than corn silage) low-grain diets, avoiding overconditioning, providing clean calving quarters, and using sanitary calving techniques.

**Treatment.** Some cows respond temporarily and a few permanently to mechanical manipulation. This consists of turning the cow on her back, leaving her in this position for a few minutes and then allowing her to rise. Corrective surgery is a more common and effective treatment.

## Foreign Bodies, or Hardware Disease

A large number of dairy cattle are lost every year from foreign bodies. Metal objects, especially those with a sharp point such as nails, pieces of wire, and staples, are dangerous when swallowed by a cow. Because the cow does not thoroughly masticate her feed before swallowing, these materials are some-times taken in with the feed. When swallowed they tend to move forward into the reticulum, which may be only an inch or less from the heart. If the object passes through the wall of the reticulum it may result in peritonitis, which causes the cow to drop suddenly in milk and to go off feed. Unless the foreign body is removed by surgery, the peritonitis may become more severe and be followed by abscesses in various parts of the body and eventually death. The metal may penetrate the pericardium, or heart sac, a condition for which there is no treatment.

**Prevention.** Prevention consists of eternal vigilance in picking up all pieces of wire, nails, or other metal that might get into the feed. Most feed companies pass feed over a magnet in order to remove any pieces of metal. Some farm mills have a magnet build into the feed chute.

In some situations when there are an unusual number of cases, small magnets 2 in. × ½ in. in size are placed in the cows rumen to hold the metal, which will decrease the amount of damage. Some of the metal will gradually be disintegrated.

## Bloat

Bloat is most common in cattle pastured on damp, rapidly growing legume pastures although it can occur when cattle are being barn fed.

**Cause.** Bloat is caused by excessive accumulation of gas in the rumen and reticulum. In severe cases this causes great distention of these organs, extreme discomfort to the animal, labored breathing, and collapse. Death usually occurs within a few minutes after collapse.

The gases formed in the rumen of bloated animals are also formed in the rumen of normal animals under normal feeding conditions, but with normal cows the gases, as they are formed, are either belched or given off through the excretory organs. Bloat is more likely to occur when the animals are on legume pastures, such as the clovers and alfalfa. Heavy grain feeding sometimes causes bloat. There does not seem to be complete agreement as to the conditions under which bloat will occur. Some years it gives little trouble, whereas in other years, under apparently similiar conditions, it will be troublesome. Some report more bloat in the morning, others more in the late evening, and still others more during the hottest part of the day. Some report more bloat when the dew is on the pasture; others feel that these conditions have no effect on bloat.

Excess gas production in the rumen is not the cause of bloat, per se, since the ruminant can normally belch more gas than is likely to be produced.

The cause of the accumulation of gas is the animal's inability, under certain conditions, to belch the gas from the rumen. Evidence indicates that belching may result from coarse roughage in the diet, and that bloat on green legumes results from a lack of scabrous material to stimulate nerve fibers in the wall of the rumen. For this reason it is advisable to give animals a feed of hay before turning them on legume pasture. Often gas accumulation is from the trapping of gas into a froth rather than the simple accumulation of gas in the rumen above the rumen ingesta.

**Prevention.**    Certain precautionary measures will aid in reducing the number of cases of bloat caused by legume pastures. Some of these are the following.

1. Give the cattle a good feed of dry hay before turning them on a legume pasture. This is very important, especially the first time they are on the pasture. Do not turn them on the pasture when they are hungry.

2. Do not allow them on the pasture for a long period of time during the first few days. Continue to barn feed until they become accustomed to the change.

3. Do not seed legumes by themselves when they are to be used for pasture. Have some grasses such as brome grass, orchard grass, or timothy in the mixture.

4. Have some dry, coarse hay available for the cows, in either a rack or stack in the pasture, or feed them before turning them out.

5. Provide water in the pasture or nearby.

6. A wide variety of antifoaming agents have been used successfully, providing an effective concentration is maintained throughout the danger period.* Some of these are animal, vegetable, and mineral oils; fats; turpentine; and synthetic detergents.

**Treatment.**    In mild form, the gas may be released by placing a wooden stick or a bridle bit in the cow's mouth and placing the animal with its' front feet higher than the rear feet. With the cow in this position, the gas will have more opportunity to be expelled and the act of chewing on the stick or trying to get it out of its mouth may be effective in expelling the gas. Another way to remove the gas is to insert a long rubber tube, through which the gas may escape, down the esophagus into the rumen. The animal may be drenched with various remedies in order to reduce the amount of fermentation. Two to

---

* R. T. J. Clarke and C. S. W. Reid, Foamy Bloat of Cattle: A Review, *J. Dairy Sci.,* **57:**753 (1974).

4 oz. of turpentine or kerosene in 1 qt. of water, given as a drench, may inhibit further fermentation and allow the gas to escape normally. Antiferment, or debloating, medicines have been developed and are effective treatments.

Relieving severe cases of bloat consists of puncturing the rumen on the left side of the animal, using a trocar and cannula; in emergencies, a knife may be used. The place to be tapped is the center of a triangle formed by the hipbone, the last rib, and the backbone. This method should be used only as a last resort. The cannula should be allowed to remain in the wound until all the gas has passed off. As long as the animal is standing, the use of a stomach tube is just as effective and much safer.

## Tuberculosis

Tuberculosis is caused by *Mycobacterium tuberculosis*. At one time, tuberculosis was the most important disease of dairy cattle. An effective control program, carried on by the federal government in cooperation with the various states, has reduced infection in dairy cattle to less than 0.5%. The control program consists of testing the animals with tuberculin, and any reactors are slaughtered.

An accredited tuberculosis-free herd is one that has had two clean annual tests. To retain this status, the herd must continue to be tested yearly, with all animals passing the test. A modified accredited area may be a county or a state in which all cattle are tested and where not over 0.5% reactors are found. Also, the herds in which reactors have been found must be retested at intervals until all animals in it are found to be negative. All health ordinances dealing with the sanitary production of milk require that the herd be tested periodically.

## REPRODUCTIVE DISEASES

### Brucellosis (Bang's Disease)

Brucellosis, sometimes referred to as Bang's disease or infectious abortion, has caused heavy losses to the dairy-cattle industry. It is caused by an organism known as *Brucella abortus*. Many abortions in cattle are caused by this organism, although abortions may be caused by other diseases, injury or conditions. This organism may also cause brucellosis, or undulant fever, in man. Cases of the disease contracted from raw milk, from meat products, or from contact with infected animals have been reported.

In dairy herds the loss due to brucellosis is not from the loss of the calf alone, but also from such factors as the loss of many cows on account of their weakened condition; retained afterbirths and metritis, which often leads to sterility; shy breeding; loss of flesh, loss in milk flow, lowering the sales value of the animal; and extra labor required for handling an infected herd. It is particularly costly to purebred breeders, since they lose not only from the interference with their milking and breeding program but also from the sale of breeding stock.

The disease is very infectious. The chief mode of infection is through the digestive tract. The organisms usually enter by means of the feed and water. From the digestive tract the organisms are taken into the blood and may localize in the genital organs or in the udder. Infected animals discharge the bacteria at time of abortions, at normal calvings, and for a period thereafter as long as there is any uterine discharge. The organisms may also be present in the milk.

The blood-agglutination test is used to diagnose the disease. The federal government, in cooperation with the various states, has a program of brucellosis eradication. Formerly it was based on blood testing and slaughtering the reacting animals. More recently a screening test, the milk ring test, is conducted on herd milk samples at three- four-month intervals and, when positive tests occur, is followed by blood agglutination tests of each cow in the herd. This technique has proven both effective and efficient. Requirements for accredited or certified herds are similar to tuberculosis requirements.

Vaccination of animals against brucellosis is being used as a control program. In this method, heifer calves between the ages of two and six months of age are vaccinated. A low-virulence strain (Strain 19) of the *Brucella* organism is used almost exclusively. When a calf is vaccinated it will react for a time to the agglutination test. The older the calf when vaccinated, the longer the time that is required for it to return to a negative blood test.

When older animals are vaccinated they may carry a positive reaction indefinitely or throughout their lifetime. The reaction cannot be differentiated by routine testing from that due to natural infection. Adult vaccination with Strain-19 vaccine is not generally recommended, although it may have merit in some specific cases.

As a result of the effectiveness of the test-and-slaughter program and the widespread use of calfhood vaccination, many areas of the United States became essentially free of the disease. Because of this many dairymen discontinued vaccination during the early to mid-1970s. This proved to be a serious error, as the incidence of the disease began rising soon thereafter and in some states has reached alarming proportions (over 10%). There is a safe, effective vaccine for brucellosis. It should be used as the preventive measure.

## Other Reproductive Diseases

Other diseases also cause abortions in cattle. In some the causative organism locates in the uterus and causes the death and expulsion of the fetus. Others cause abortion indirectly by affecting the general condition of the animal. Most can be effectively controlled by AI, vaccination, or both.

**Trichomoniasis.** Trichomoniasis is a venereal disease of dairy cattle, causing abortion and other breeding difficulties such as failure to conceive and pyometra. It is caused by a protozoan, *Trichomonas fetus,* first reported in this country in 1932. The disease is spread during service by infected bulls and occasionally by close contact with infected cows. A bull may become infected by serving an infected cow.

A definite diagnosis of the disease can be made only by finding the causative organism by direct microscopic examination of material from the sex organs of the cow or bull. This disease may be extremely difficult to diagnose positively. A study of the breeding history of the herd and of individual animals in the herd may give an indication of the presence of the disease.

When the herd is known to be infected, artificial insemination from clean bulls should be practiced exclusively. Cows that abort may recover if given a period of sexual rest of about three months, as the organisms will not live longer than three months in the nonpregnant cow. There is no effective vaccination for this disease.

**Vibriosis.** Vibriosis is a venereal disease characterized by infertility and early embryonic abortion. It interferes with normal function and attachment of the placenta. It is caused by *Vibrio fetus,* a comma-shaped bacterium. It can be effectively controlled by artificial insemination with *Vibrio-fetus*-free semen and by vaccination before breeding. Diagnosis is based on examination of the aborted fetus, isolation of the organism from the genital tract, or a vaginal mucus agglutination test.

**Leptospirosis.** Leptospirosis is a contagious disease of man and many species of animals, wild and domestic. Infections in cattle are usually caused by *Leptospira pomona,* but are sometimes caused by *L. grippotyphosa* and *L. hardjo.* Infections can be transmitted by contact or by contaminated feed, water, etc., between cows and between species. Causative strains of leptospirosis are not species specific. The combination of these factors yields a difficult problem.

SYMPTOMS. In calves the disease causes a high fever and depressed appetite, often followed by anemia. In older cattle the disease and the symptom varies from mild to severe. Infections caused by *L. hardjo* are particularily obscure and the disease is difficult to diagnose. Unusual symptoms in older cattle include a sharp drop in milk production; milk often becoming thick and yellowish in color; depressed appetite; discolored and often blood-tinged urine; and a high incidence of abortion two to five weeks after infection. There is no inflammation of the mammary gland. The disease can be diagnosed with blood test or isolation of the organism from infected animals or aborted fetuses.

PREVENTION. Annual vaccination with *L. pomona* bacterins have proved effective in preventing abortions and death caused by infections by this organism. Vaccination against *L. hardjo* is required every six months. Recently a vaccine effective against all three strains has been developed. Good management practices such as controlling exposure to rodents, infected wildlife, and other livestock, contaminated streams, and ponds is also helpful.

TREATMENT. If diagnosed early, antibiotic therapy (streptomycin, chlortetracycline, and oxytetracycline) can be used to treat infected animals.

**Infectious Bovine Rhinotracheitis (IBR).** IBR is an acute, contagious infection of the upper respiratory tract. It is caused by a virus. The disease can cause abortion and stillbirths two to three months after the respiratory infection. The organism often invades the placenta and fetus. The virus also causes infectious pustular vulvovaginitis. Usual observable symptoms include increased temperature, nasal and eye discharge, increased respiration, and mildly depressed appetite. Mortality rate in adult cattle is low, and most recover without treatment in 10 to 14 days. Calf mortality is high. The disease can be effectively prevented by vaccination.

**Bovine Viral Diarrhea (BVD).** BVD is a virus-caused infectious disease of cattle. It is characterized by lesions and ulcerations of the digestive tract. Usual symptoms are profuse diarrhea, dehydration, elevated temperature during the first day or two of the infection, and depressed appetite. It is more commonly observed in yearling heifers but can be observed in calves as well as adult cattle. It can cause abortion in pregnant cattle. BVD can be effectively prevented by vaccination. Antibiotic therapy is ineffective in treating the disease but may be helpful in controlling secondary infections.

**Parainfluenza 3 (PI3).** PI3 is an influenza type of infection with respiratory trouble and fever. It is a virus infection and is often complicated with bacterial infections. Good sanitation and vaccination are the preventive measures. The vaccine should be administered in two treatments at least three weeks apart. This should be done two or three weeks before calves are put under stress such as moving, changing locations, changing groups, etc., so that there is time to develop resistance.

## OTHER DISEASES

### Mastitis

Mastitis is the reaction of milk secreting tissue to injury—resulting from any condition or combination of factors leading to infection of the internal tissue of one or more quarters of the cow's udder. In more concise terms, mastitis is any inflammatory condition of the udder.

Mastitis reduces milk yield and shortens productive life of affected cows. It is prevalent wherever milk is produced and in the United States alone it causes an estimated loss of $900 million per year. This monetary loss is from decreased milk production, loss of valuable cows, cost of treatment, and production of poor-quality milk. Mastitis may occur in several forms, acute, gangrenous, chronic, and subclinical.

Acute mastitis is characterized by hot, painful, and swollen quarter or quarters. The cow is usually off feed and has a high temperature. The milk contains flakes, shreds, clots, or blood. Severe depression of milk secretion accompanies an acute attack.

In gangrenous mastitis the quarter or quarters show bluish discolorations. They are cold to the touch. The discoloration proceeds from the teat upward as the case advances. If the part becomes necrotic, it sloughs off from the body.

Chronic mastitis is characterized by mild, repeating attacks. During the attack the quarter or quarters may be swollen. The milk contains flakes, shreds, or clots. Repeated attacks cause induration of the gland, which feels hard on palpation.

Subclinical mastitis normally precedes the other three types; no clinical symptoms are evident. Diagnosis of mastitis at this stage can be achieved by various leucocyte screening tests (CMT, WMT, Whiteside, etc.) or by bacteriological culture.

**Cause.** Mastitis may be caused by any of literally dozens of organisms; however, most mastitis is caused by one or more of the following organisms: *Staphylococcus aureus; Streptococcus agalactiae; Streptococcus uberis;* coliform organisms such as *Escherichia coli; Klebsiella* and *Aerobacter aerogenes; Pseudomonas aeruginosa; Corynebacterium pyogenes;* and mycoplasma organisms. These and other organisms gain entry to the mammary gland through the teat opening. Predisposing factors include any factor that enhances entry of the organism or lowers the resistance of the animal to growth of the organism. Some of the predisposing factors are the following.

HEREDITY. Selected cow families appear to have a relatively high or low incidence of mastitis. The volume of milk produced does not appear to be related to susceptibility to mastitis. It has been found, however, that fast milkers tend to become infected more readily than slow milkers, which possess a tighter teat sphincter muscle. Inheritance may determine the shape and structure of the teats, the potency of the bacteriostatic substance in the milk, and the conformation of the udder. Poor udder attachment can be a major cause of udder injuries and therefore increase the chances of mastitis. Cows with teats of a larger diameter have a higher incidence of mastitis.

ENVIRONMENT. Housing conditions can lead to an increased incidence of mastitis. Contributing factors may be short or narrow stalls, wet or dirty bedding, improper ventilation, high door sills on which cows may bruise their udders, calves sucking, slippery floors, etc.

Weather may also be a predisposing factor for mastitis. The lower the mean temperature, the higher the rate of mastitis in both high- and low-incidence herds. Rain aggravates the effect of cold weather. The peak incidence of mastitis often coincides with storm periods.

Feed is thought to be a predisposing factor. There is no evidence that the kind or amount of concentrate fed has any effect on the incidence or severeity of mastitis. However, it is wise to follow regular feeding schedules and above all avoid sudden changes in feed. Udders of high-producing cows are subject to more stress than those of low producers, and this may have some influence on the incidence of mastitis.

MILKING MACHINES. Milking machines can be a major predisposing factor of mastitis if they are not properly installed and kept in good operating condition.

The following four malfunctions of milking machines tend to irritate the udder and may lead to mastitis.

1.  Fluctuating vacuum—leaking lines, clogged lines, worn out pump, loose belt, etc., cause the vacuum level to fluctuate. Such fluctuation may irritate the teats.

2.  High vacuum level—may cause teat cups to crawl up on the teats toward the end of milking. This irritates the delicate membrane lining inside the teat, and the pressure of the teat cups may hinder the milk flow. Bacteria inside the udder may infect the irritated glandular substance, causing mastitis. Extremely high vacuum level causes sore teats and may cause inversion of the ends of the teats.

3.  Pulsation—if the pulsator is not working properly, massaging of the teat is not carried out properly. This may lead to sore teats.

4.  Teat cup liners—flabby and rough liners may irritate teats. Cracked liners can harbor bacteria that may infect the udder.

MILKING PRACTICES. Improper milking practices also can predispose to mastitis. Some of the more common faults are the following.

1. Attaching the machine too quickly, before let-down, improper stimulation, or being in too much of a hurry. Milk acts as a lubricant in the milking process. If no milk is present, the machine can cause membranes to rub together and produce irritation.

2. Overmilking produces a similar situation; that is machine operation in the absence of milk flow.

3. Not dipping teats in a suitable disinfectant after milking. After milking, the teat sphincter is relaxed and readily accessible to the entry of mastitis-causing organisms.

4. Unclean or contaminated equipment and operators' hands can spread mastitis.

OTHER FACTORS

1. Lice and mange.

2. Exposure to cold and dampness.

3. Dirty cows.

4. Dirty premises, especially those conductive to fly breeding.

5. Unclipped udders and flanks.

6. Muddy areas, swamps, and stagnant ponds accessible to cows.

7. Stage of lactation. The rate of new infection is usually the highest during the first month of lactation and during the dry period.

8. Improper use of teat dilators or tubes. This may spread infection and injure the very thin protective skin of the teat canal. Temporary inflammation may reduce the resistance of the entire gland, thus permitting an infection to become established.

9. Age. The rate of udder infection generally increases with age. Cows that have been treated for mastitis may become reinfected. With few exceptions first-calf heifers are free from mastitis at the time of parturition. The exceptions are usually because of infection acquired during calfhood when calves have been fed infected milk or were allowed to suckle each other.

**Prevention.** A significant proportion of the cases of mastitis (perhaps as high as 75 to 80%) could be prevented. Preventive measures are those that reduce the predisposing factors that facilitate the entry and multiplication of the organisms capable of causing mastitis. Some highly recommended preventive measures are:

1. Proper installation, maintenance, and sanitation of the milking equipment to ensure steady vacuum and pulsation and rapid removal of milk from the teat end.

2. Proper sanitation of milking equipment and milkers' hands, especially after milking infected cows.

3. Use of proper milking procedures such as:
   a. Regularity of milking times and procedures.
   b. Proper cleaning and stimulation of cows with a warm disinfectant solution.
   c. Attaching the machine as soon as let-down occurs.
   d. Removal of the machine when milk flow ceases.
   e. Dipping the teats in an approved teat dip after milking.
   f. Insofar as possible, milking noninfected cows first and infected cows last.

4. Treating dry cows after the last milking when turning dry.

5. Reducing environmental bacterial build-up by providing clean sanitary quarters for cattle and keeping the cows clean and flanks and udders clipped during winter months.

6. Discarding mastitic milk directly into a drain.

7. Using screening and bacteriological tests to identify subclinical cases and identify causative organisms in clinical cases.

8. Culling cows with chronic mastitis that do not respond to treatment.

**Treatment.** Clinical infections, especially acute cases, should be treated with appropriate systemic and/or mammary infusion treatments. Treatment is more effective during the dry period for subclinical and chronic cases. Laboratory cultures and antibiotic sensitivity tests are recommended to help identify causative organisms and susceptibility of these organisms to specific antibiotics to increase effectiveness of treatment.

**Summary.** Effective control of mastitis is probably the number one disease-control problem for most dairymen. Reasons for this are the wide variety of organisms that can and do cause the disease, the many predisposing factors associated with the disease, and the contagious nature of the disease. Effective control requires the continued conscientious efforts of all herd personnel. Effective control is, however, prerequisite to profitable dairying.

## Foot Rot, Also Called Foul Foot and Foot Evil

Foot rot is an infectious disease generally appearing between the claws of cattle. Affected cows move around with difficulty, decrease severely in milk production, and may develop an ill-shaped hoof as a result of this trouble. Most dairy herds are troubled to a certain extent by this disease. Certain conditions, such as wet, muddy barnyards or pasture, and foot injuries such as wire cuts and bruises, are predisposing factors to infection, as the causative organisms usually enter through a break in the skin or horny tissue of the hoof. The disease spreads rapidly in muddy yards and when established is difficult to eradicate. The infection may also enter the body through the digestive tract or by other routes. The first symptom is lameness, which, together with the characteristic odor, is sufficient to show its presence.

**Treatment.** The first thing to do is to clean the affected part thoroughly, washing it with a strong solution of disinfectant, and to keep the animal in a clean, dry place. Powdered copper sulfate, dusted over the affected part or dissolved to form a paste or a 5% solution and applied has proved satisfactory. If several cows are affected, a solution of formaldehyde or of copper sulfate can be made in which the animal can stand for several minutes every day.

Intravenous injection of sulfapyridine or sulfamethazine or a combination of the two has been effective in its treatment when given in the early stages of the infection. Antibiotics are also used in its treatment.

## Scours, Also Referred to as Calf Scours

Scours is a major disease problem in dairy calves, especially those less than two weeks of age. It is often complicated by or predisposing to pneumonia. Depending on the causative organism and severity of infection, the mortality rate from scours can vary from low to very high.

**Cause.** Scours can be caused by overfeeding of milk or milk replacers, and a variety of microorganisms including *E. coli, salmonellae,* and viruses.

**Symptoms.** Scours is characterized by diarrhea, dehydration, loss of appetite, and lethargy. The severity of the symptoms is dependent on the causative organism and the condition of the calf.

**Prevention.** The single most important factor in preventing scours is to feed 4 to 6 lb. of first colostrum within one to four hours of birth and continue feeding colostrum for three days. This practice enables the calf to acquire passive immunity to many infectious organisms. Colostrum is also very high in vitamins, protein, and minerals that increase the general resistance of the calf.

Prevention of nutritional or noninfectious scours can be achieved by avoiding overfeeding of milk or milk replacers and, if milk replacers are used, using milk-based replacers. Vegetable-based milk replacers have proved not to be suitable for newborn dairy calves.

Prevention of infectious scours and its spread is more difficult. In addition to those measured recommended for the prevention of nutritional scours, proplylactic measures such as the administration of LA Serogen, mixed bacterin, and vitamin A, D, and E have been used. Clean, dry, and draft-free quarters and the use of clean feeding utensils are also essential. In cases of severe outbreaks, cleaning, sanitizing, and vacating the calf quarters is often recommended.

**Treatment.** Preventing dehydration is an important aspect of therapy. This may be achieved by feeding an electrolyte solution or, in cases when the calf refuses to drink, administration of the solution via an esophogeal tube. Intravenous antibiotics and sulfonamides are also used.

### Pneumonia

Pneumonia is a respiratory disease, usually secondary to some other disease. It has a relatively high incidence in dairy calves, especially those exposed to damp, drafty conditions and those whose resistance is lowered because of scours or nutrient deficiencies.

**Cause.** Pneumonia can be caused by a wide variety of bacteria (chiefly *Pasteurella*), mycoplasmas, and viruses (IBR, PI-3 and other respiratory viruses).

**Symptoms.** Early symptoms include coughing and nasal discharge, accompanied by dullness and lack of appetite. This is followed by elevated temperature, rapid and shallow breathing, rough haircoats, and dry skin.

**Prevention.** Preventive measures include those that maintain a high level of resistance. These include dry and draft-free quarters, avoiding scours, and good nutrition.

**Treatment.** Antibiotic therapy is effective in treating bacterial pneumonia and also usually is used in viral pneumonia. It does not control the virus, but does help prevent ensuing bacterial infections. Administration of electrolytes to help prevent dehydration is also recommended. If the pneumonia persists for very long a considerable portion of the lung tissue may be destroyed. After recovery from the primary symptoms the calf often has a poor appetite and impaired breathing. It is often recommended that these animals be disposed of as they are likely to remain stunted and unhealthy because of the permanent lung damage.

**Naval ill also called joint ill.** Navel ill is an infection of the navel and sometimes of the joints. It is caused by a number of organisms. Prenatal infection may occur in herds with uterine infections. Most infections are postnatal entering through the umbilicus or by mouth. Infection is highest in premature calves.

Preventive measures are complete sanitation, clean calving stalls or lots, and treating the navel with tincture of iodine soon after birth.

Injectable penicillin—streptomycin mixtures may be effective as a treatment.

## Other Diseases and Conditions

The preceeding discussion covered some of the more common and economically significant diseases and problems of dairy cattle. There are many other less prevalent diseases and conditions affecting dairy cattle. Some of these are presented in summary form in Table 13-3; others are described below.

**Parasites.** Cattle are susceptible to a variety of external and internal parasites. They are relatively easily controlled and do not present a major problem in most herds.

LICE. Two species of lice commonly affect cattle, especially young cattle. Their presence is indicated by rubbing, loss of hair, and eventually loss in weight. One of these species is a blood-sucking louse, the blue louse, and the other is a biting louse, the red louse. Lice spend their life cycle on the host. The female lays 30 to 35 eggs when she is about 12 days old, and these eggs (nits) cling to the hair and hatch in 11 to 18 days. The red lice are usually found on the side of the neck, withers, and tail head, but they may be found

**Table 13-3** Diseases and Problems Affecting Dairy Cattle

| Disease/cause | Symptoms | Prevention | Treatment |
|---|---|---|---|
| Grass tetany<br>Low serum magnesium often low serum calcium; higher occurrence when on lush pasture | Nervousness, stiff gait, staggering, hypersensitive to touch and sound, frequent urination, convulsions, and death | Increase magnesium intake when on lush pasture; feed magnesium oxide, 0.5 to 1.5 oz./day | IV magnesium sulfate, 200 ml. of 25% solution followed by supplemental feeding of 1 to 2 oz./day |
| Cowpox<br>Virus | Blisters, sores, ulcers, and scales on teats and udder skin, spreads rapidly | Prevent exposure by avoiding entry into the herd with replacements | Antibiotic ointments to prevent secondary infections |
| Pinkeye<br>*Moraxella bovis* | Eye discharge, conjunctivitis, herditis, corneal lesions, blindness in severe cases | Control flies, feed adequate vitamin A. | Early application of antibiotics |
| Actinomycosis, or lumpy jaw<br>*Actinomycosis bovis* | Lumps on the jaws or other bony tissues of the head, lumps may abcess | Remove objects that may cause punctures of the head such as nails, etc. | Streptomycin, sulfas |
| Paratuberculosis, or Johne's disease<br>*Mycobacterum paratuberculosis* | Recurrent diarrhea, weight loss | Rigid sanitation, separate calves to prevent spread from cows to calves, test and slaughter cows | None |
| Blackleg<br>*Clostridium chauvoei* | Swollen muscles, lameness, prostration, and tremors. High mortality rates | Vaccination after 3 months of age | Penicillin in early diagnosis |
| Malignant edema<br>*Clostridium septiceum* | Infection and swelling around wounds, high fever | Vaccination | Penicillin in early diagnosis |
| Ringworm<br>Fungus | Round, scaly, crusted spots on the skin, especially the head, neck, and shoulders | Dry quarters, sunshine, adequate intake of vitamin A. | Remove scales and crusts and apply fungicidal drug. |

elsewhere. The blue lice attach themselves in patches on the side of the neck, brisket, back, inner surface of the thigh, and around the nose, eyes, and ears.

Good herd management goes a long way in controlling lice. Well-fed animals seem to resist infection, and when infected they do not suffer as severely as poorly fed animals. The trouble may be controlled by spraying the affected animals with rotenone. Rotenone powder is also effective, and it can be applied in cold weather when spraying would not be advisable.

GASTROINTESTINAL PARASITES. Various types of internal parasites are found in cattle, depending on the age of an animal, the climate, the season, and the management system. If present in large numbers, they cause decreased feed efficiency, unthriftiness, and anemia. The effects can be very serious in growing animals. Pasture rotation helps break the cycle of many internal parasites. Degree of infestation can be diagnosed by fecal counts; if problems are suspected, this should be done. Broad-spectrum anthelmintics such as thiabendazole and levamisole are effective treating agents.

CATTLE GRUBS (HEEL FLIES OR OX WARBLES). Cattle grubs damage meat and hides. In the heel fly stage in the spring they lay eggs on the hair of the legs of cattle. Cattle get excited and run from the flies, sometimes to the extent of stampeding and running through fences. When the eggs hatch, the grubs penetrate the skin and move through the body to the gullet and up to the back, where they are present through the winter. The grubs cut holes in the skin. In early spring they emerge, fall to the ground, and pupate to complete the life cycle.

Rotenone powder can be rubbed into the openings on the backs of the cattle or it can be applied as a spray with a power sprayer. Several other insecticides have been developed to use by applying on the back. These have been changing fast and new ones will probably be developed. Currently Coumaphos (Co-Ral), Famphur (Warbex), and Crufomate (Ruelene) are among those being recommended. These must not be used on lactating cows or dry cows within 21 days of freshening or on calves under three months of age.

**Poisons.** There are many poisonous chemicals and plants that may kill animals. Probably the worst offender is lead paint on fences or buildings to which animals have access and on discarded paint buckets and brushes. Fertilizers, especially nitrates, will kill cattle if they are where the animals can lick them from the bag or are spilled in a pile in the pasture. Sprays, rat poisons, and other poisonous materials should be stored so that animals cannot get to them.

A large number of plants are poisonous to animals. They are more dangerous when pastures are short and animals are not as selective in their grazing. Some of the common poisonous plants are larkspur, corn cockle, rubberweed, water hemlock, the poisonous milkweeds, jimson weed, rayless goldenrod, and locoweed. Immature sorghum and Sudan grass and wilted wild cherry leaves contain hydrocyanic or prussic acid; if very much is consumed, it is fatal. Acorns will cause milking cows to dry up. Cattle of any age are affected by severe digestive disturbances from eating acorns.

**Teat Troubles.** The teats of milking cows often become scratched or cut. Various ointments and skin antiseptics will hasten the healing of these wounds. Care should be taken to remove from the lots and pasture anything that might cause injury to the udder and teats of the cow.

Deep wounds that enter the teat canal, leaky teats, closed teats, spider teats, and various teat injuries should receive the attention of a veterinarian. Many of these may require surgical treatment if satisfactory results are to be secured. Surgical procedures and the use of teat plugs, dilators, and tubes should be used with extreme caution to prevent the entrance of infection into the udder, causing mastitis. Teat troubles are among the most difficult problems a veterinarian has to treat unless extreme care and cooperation on the part of the herdsman or owner are exercised at all times.

## HERD MEDICAL SUPPLIES

Many routine herd health practices and the regular and/or emergency treatment of many ailments can be handled by herd personnel. Those that the herdsman should perform regularly and in emergencies should be discussed by the herdsman and the veterinarian and clear understanding reached. Veterinarians should instruct the herdsman in the use of proper techniques and advise on purchase and proper use of equipment and medicines. As mentioned earlier, a good herd health program involves cooperation and mutual respect among herd owner, herdsman, and veterinarian. Lack of clear understanding and communication among these individuals concerning areas of responsibility in regard to health practices has led to many problems and strained relationships.

The following is a list of equipment that should be stocked on most dairy farms. Individual farm lists should be compiled and maintained with the advice of the veterinarian.

1. Thermometer, veterinary type, equipped with ring and cord for retrieval.

2. Balling gun.

3. Stainless steel bucket—10 to 12 qt. capacity.

4. Stomach pump, tube, and stainless steel speculum (Frick tube) to use with the stomach tube.

5. Drenching bottle—1-qt. capacity, rubber or plastic.

6. Stainless steel obstetrical chains (5-ft. length) and handles.

7. Intravenous outfit complete with needles (2 in., 14 gauge) and tubing.

8. Esophogeal tube and apparatus.

9. Surgical scissors.

10. Disposable plastic gloves and sleeves.

11. Disposable syringes—10- and 20-ml.

12. Hoof knife, chisel, and other hoof-trimming equipment.

13. CMT paddle and supplies.

14. Alcohol.

15. Calcium gluconate—500-ml. bottles.

16. Cotton.

17. Dehorning paste or substance.

18. Disinfectant.

19. Gauze bandages.

20. Iodine.

21. Paper towels.

22. Soap.

23. Other supplies as recommended by the veterinarian.

## FEDERAL AND STATE REGULATIONS AND ASSISTANCE

The USDA's Animal and Plant Health Inspection Service has many safeguards to prevent disease infections from coming into this country. In the event of outbreaks of certain diseases, this agency is charged with control. There are also regulations affecting interstate shipment of livestock. The various state departments of agriculture are working on the prevention and control of livestock diseases. There are regulations governing the importation of livestock into the state, as well as intrastate shipments. Many of these departments have one or several diagnostic laboratories for helping veterinarians and farmers in diagnosing diseases. Livestock farmers should acquaint themselves with federal and state regulations and facilities and make use of them in maintaining healthy herds.

## REFERENCES FOR FURTHER STUDY

Coppock, C. E., Displaced Abomasum in Dairy Cattle: Etiological Factors. *J. Dairy Sci.*, **57**:926 (1974).

*Current Concept of Bovine Mastitis.* National Mastitis Council. 965, 1972.

Galloway, I. H., *Farm Animal Health and Disease Control,* Lea and Febiger, 1972.

Janzen, J. J., Economic Losses Resulting from Mastitis—A Review, *J. Dairy Sci.*, **53**:1151 (1970).

Jones, J. M., *Managed Milking Guidelines, Publication No.* 633, *V.P.I. & S.U. Ext. Div.,* 1974.

Jorgenson, N. A., Combating Milk Fever, *J. Dairy Sci.*, **57**:933 (1974).

*The Merck Veterinary Manual,* 4th ed., Merck and Co., Inc. Rahway, N.J. 1973.

*Modern Mastitis Management,* TUCO Division of Upjohn Co., 1970.

Philpott, W. N., et al., various articles in *Hoard's Dairyman.*

Schalm, O. W., et al., *Bovine Mastitis,* Lea and Febiger, 1971.

# 14

## Herd Replacements

There are more than 11 million dairy cows in the United States, of which approximately 20 to 25%, or 2½ million, are culled annually. This creates a need for 2½ million replacements annually. The average age of freshening is about 30 months, so this amounts to a national replacement herd of nearly 6 million dairy calves and heifers. The need for a steady supply of replacement heifers is an essential fact of life in the dairy business.

The goal of the replacement program is to have an adequate supply of herd replacements that meet the following characteristics.

1.  Have high genetic potential for economically important traits. Genetic improvement of the herd is primarily dependent on achieving this goal. The PD of the sires of heifers entering the herd should average at least +600 to 800 lb. of milk.

2.  Are well grown but not overconditioned, so they are large enough to breed at 14 to 17 months of age, calve, and take place in the milking herd at 23 to 26 months of age.

3.  Are healthy, free of diseases and parasites, and immune against the diseases to which they are likely to be exposed and for which there is an effective vaccine.

4.  Are reasonable in cost.

Meeting these goals is a vital part of successful dairy operations. The manager is faced with many decisions in regard to achieving these goals. There are many alternatives to evaluate. There is no one best answer; rather, each dairy herd manager must evaluate the alternatives in view of his or her resources and goals and then design and implement a program to meet these goals.


## PURCHASING OR
## RAISING REPLACEMENTS

The decision as to whether to depend primarily or completely on either purchased or home-reared replacements depends on several factors that should be evaluated before the decision is made. Some of these are the following.

1.  Are the resources, feed, labor, housing, etc., available to raise replace-

ments, or could these resources be used more profitably to maintain more producing-age animals? In many large dairies in or near urban areas, the needed resources (especially land and labor) either are not available or are prohibitively expensive. In many cases milking-herd size can be increased by 20 to 30% if the feed, labor, and capital resources used for replacement rearing are applied to the milking herd. This may make the total operation more profitable. In other cases resources not readily usable to increase milking-herd size can be profitably applied to replacement rearing and increase total farm profitability. This is particularly true of small and medium-sized dairy farms with ample land and family labor resources, and where increasing milking-herd size is difficult because it would require significant capital resources for building, renovation, and equipment purchases.

2. Is there a reliable supply of the type of replacements needed at reasonable cost? Can high-genetic-potential, healthy, well-grown heifers or high-producing cows be purchased when needed at a cost comparable to rearing costs plus the market value of the heifer when born? Continuing high performance and genetic improvement of the herd is dependent to a large extent on having healthy cows with genetic ability for high production. If such replacements are not available for purchase, when needed, at reasonable cost, herd performance can be impaired.

3. Can the sale of surplus heifers contribute significantly to the profitability of the dairy operation? If a herd has the resources to raise all heifer calves and the level of management is such that all these heifers are not needed as replacements, surplus heifers can be an excellent source of income and add to the profitability of the operation. Healthy, well-grown daughters of high-PD bulls are usually in strong demand, especially as springing heifers.

The decision whether to buy or purchase depends primarily on these factors. Most U.S. dairymen raise all or most of their replacements because of the following advantages.

1. Raising replacements provides a better opportunity for steady and continuous genetic progress at less cost. It is usually more economical to buy semen from good bulls and raise these heifers than to buy the daughters of good bulls. It also allows dairymen to choose and follow a particular breeding program.

2. It minimizes the risk of bringing disease into the herd. Many dairymen have unknowingly purchased an outbreak of an infectious disease when they purchased replacements.

3.  It affords the opportunity to increase total profitability through the sale of heifers.

4.  It creates the opportunity to gain personal satisfaction. What breeder doesn't want to breed and develop an outstanding individual: a class leader for production, a local show winner, or an All-American? As we have often heard, "Anyone with enough money can buy an All-American, but I bred one."

Some of the disadvantages of rearing replacements are the following.

1.  There is no return on invested capital in the heifer and her feed and care for 2 to 2½ years.

2.  Using available resources to increase herd size may be more profitable.

Some of the following are major advantages of purchasing replacements.

1.  Purchasing allows dairymen to specialize in the milking herd. They can devote all their time and resources to handling milking cows, which are the major income producers on most dairy farms.

2.  It allows the discriminate buyer more opportunity for rapid genetic improvement. Some dairymen have the time and ability to find the really good cows and purchase them at a reasonable price.

3.  It allows an immediate return on investment when cows and springing heifers are purchased. There is no delay of 2 to 2½ years until the animal begins to generate income.

The major disadvantages of purchasing replacements are the following.

1.  There is a risk of bringing disease into the herd.

2.  It is often difficult to find good replacements and purchase them at a reasonable price at the time they are needed.

### Contract Heifer Rearing

Another alternative that merits consideration by some dairymen is contract heifer raising. In some cases this type of program can maintain most of the advantages and overcome most of the disadvantages of either home rearing or purchasing replacements. It can also afford other dairymen the opportunity to develop profitable specialized replacement-rearing operations. There are two general types of contractual agreements: (1) a grower contract and (2) a sell-purchase contract. In a grower contract the dairyman retains ownership of the heifer and pays the grower an agreed-on price for rearing the heifer. The time period varies. Some contracts cover only from birth to weaning or from breeding to two to four weeks prior to freshening, while others cover various parts or the entire period. The more usual period is from weaning to two to four weeks before freshening. The fee may be per pound of gain, per month, per heifer, or a cost-plus basis. The owner is usually responsible for costs of insurance, vaccinations, and other routine health costs, breeding, etc., and specifies the time the heifer is to be bred and the service sire. These conditions vary, but should be clearly stated in the contract. This type of contract has the advantages of allowing the dairyman to plan a genetic improvement program, to specialize in the milking herd, to reduce the amount of capital tied up in nonproductive units, and reduce risk of disease.

In a sell-purchase contract, the dairyman sells the heifer calf to the grower for an agreed on price and retains the first option of repurchasing the heifer as a replacement heifer, usually as a springing heifer. This type of contract has the additional advantages of minimizing risk of losses resulting from heifer mortality and further reducing the amount of capital tied up in nonproductive units. The dairyman, however, usually surrenders control of the breeding and preventive health program, and these can be serious disadvantages.

Contract heifer rearing programs have been successful for some dairymen, especially in large specialized commercial operations and in other situations when resources are limited or expensive. The success of these types of contract is primarily dependent on a thorough understanding of the contract conditions (a written contract is highly recommended) and the availability of a grower who has the resources and ability to get the job done correctly. He must prevent high mortality rates, get the heifers well grown without overconditioning, and get them bred to freshen when needed.

The choice of most dairymen is to raise all or most of their replacements. This fact, however, should not preclude thorough evaluation of the previously discussed alternatives. This is especially true for those herds with either limited resources to rear heifers or owners who do a very poor job of rearing heifers; that is, calf mortality of 25 to 50% or higher and heifers not large enough to breed until 24 to 30 months of age or freshen until three years of age or older.

## REARING REPLACEMENT HEIFERS

Assuming that the goal for establishing the genetic merit of the replacements was properly met at the time of breeding, the goals of the replacement rearing program might be stated as keeping an adequate number of heifers alive, growing, and healthy, and getting them bred to calve as needed for herd replacements.

## Number to Keep

The number of replacements needed is primarily dependent on the herd culling rate and the calf and heifer loss rate. Herd culling rates vary widely, but average about 25% nationally. The majority of these are culled for reasons other than low production or genetic culling. This lack of culling for low production results in a low level of female selection pressure and slows genetic progress. Many high-producing herds have culling rates of 35% or more. If the majority of these are culled for low production, this does increase female selection pressure. When these cows are replaced by high-genetic-potential heifers, the result is increased genetic progress in the herd. The authors recommend that all heifer calves be retained in cases where ample resources are available to rear these heifers and where a breeding program is being followed that results in the heifers having superior genetic performance potential to the cows. This practice affords the opportunity to increase genetic culling of the cow herd, replacement with superior heifers, and more rapid genetic progress.

Lack of resources to properly rear all heifers is a limiting factors in many herds. The question becomes, what is the minimum number that must be kept to maintain herd size? The answer depends on expected culling rate and the rate of calf and heifer losses. Assuming a herd size of 100 cows, with 50 heifer calves born per year, a culling rate of 25%, and combined calf and heifer losses of 30%, the answer can be calculated as follows.

| | |
|---|---|
| Heifer calves born per year = | 50%, or 50 heifers |
| Dead or abnormal at birth = | 8%, or −4 heifers |
| Remaining | 46 heifers |
| Die between birth and 6 mo. = | 10%, or −5 heifers |
| Remaining | 41 heifers |
| Die between 6 mo. and freshening = | 2%, or −1 heifer |
| Remaining | 40 heifers |
| Nonbreeders = | 6%, or −3 heifers |
| Remaining | 37 heifers |
| Loss at calving = | 4%, or −2 heifers |
| Remaining | 35 heifers |

Twenty-five, or approximately 70%, of the remaining heifers are needed to maintain herd size under these conditions. Retaining all heifer calves would result in 10 surplus heifers, or the opportunity to increase culling rate to 35%. Retaining 35, or 75%, of the heifers born alive would maintain herd size at a culling rate of 25%.

The loss figures may appear high. They are based on surveys conducted by researchers and extension personnel in several states. Based on the available reports, the authors are of the opinion that losses of one-third of the heifer calves born or one-fourth of the heifer calves born alive are realistic estimates. Many dairymen do not keep accurate records of calf and heifer losses and are unaware of the magnitude of these losses.

The previous example illustrates the importance of a good replacement program. If, for example, calf and heifer losses dropped from 25% of those born alive to 10% of those born alive, only 28, or 61% instead of 35, or 75% of the heifer calves born alive would need to be retained to maintain herd size. This, in turn would affect replacement herd size. Assuming an average age at freshening of 30 months or 2½ years, and 25% calf and heifer losses, the replacement herd would number 70 to 80 heifers of various ages per 100 cows. If losses were 10%, the replacement herd size needed would decrease to 55 to 65 calves and heifers per 100 cows and result in savings in costs of the heifer rearing program. Freshening the heifers at 24 instead of 30 months would further decrease size of the replacement herd and the costs associated with it.

This example also illustrates the importance of breeding replacement heifers to +PD proved dairy sires. If the replacement rate is 25% of the herd, it is logical to assume 25% of the heifer calves will be born to replacement heifers. If these heifers are bred to "cow fresheners," then nearly all the heifers from cows in the milking herd must be retained just to maintain herd size.

## Cost of Rearing Replacements

Costs of rearing replacement heifers are considerable. High loss rates increase the cost per heifer in the herd, as the cost of feeding and caring for the heifers that are lost is a part of the cost of rearing surviving replacements. They are further increased by delaying freshening age beyond 23 to 26 months. Depending on the program, this may vary from $10 to $30 per additional month. In addition, costs are just as high for poor-genetic-potential heifers as for good ones and their value at freshening is less. These statements summarize the reasons why high +PD sires should be used to breed dams of all possible herd replacements including heifers. Developing and implementing a replacement rearing program that minimizes calf and heifer losses and results in growth rates sufficient to allow freshening at 23 to 26 months of age is essential. A comparison of the costs of rearing heifers at varying management levels is presented in Table 14-1. One can substitute actual herd costs and values for the example figures and determine the effect of these factors based on Holstein and Brown Swiss calves and heifers at 1976 average costs and values. Costs should be decreased by 10 to 15% for Ayrshire and Milking Shorthorns and by 20 to 25% for Guernseys and Jerseys. Value of the calves and heifers should also be decreased. The important points of this comparison, however, are applicable to all breeds; these are: (1) calf and heifer losses increase costs of rearing replacement heifers, (2) delayed freshening increases costs of rearing replacement heifers, and (3) these factors, along with the genetic merit of the heifer, significantly affect profitability of the replacement-rearing enterprise.

## Calf and Heifer Mortality

Reports concerning the degree of calf and heifer losses and the factors associated with these losses are numerous. Two of the factors consistently reported to be associated with higher calf losses are: (1) herd size (Table 14-2) and (2) calves fed and cared for by hired help rather than operators, managers, or their family (Tables 14-3 and 14-4). The two factors are probably interrelated. Other factors reported to be associated with increased calf mortality include: (3) degree of calving difficulty (calves born to dams who experienced no calving difficulty had a higher survival rate),* (4) herd production level (higher-producing herds had a higher survival rate),* (5) type of liquid feeding program (survival of calves was higher when fed whole milk rather than milk replacer),** (6) breed of cattle (Ayrshires had the highest survival rate, Brown Swiss and Holsteins intermediate, and Jerseys and Guernseys the lowest),* and (7) season of birth (higher survival rates for calves born in the summer rather than the winter months).* Over 90% of calf and heifer mortality occurs during the first six months. Nearly one-half of the total occurs during the first week and most of the remainder during the first month. A survey of Virginia dairymen completed in 1974 indicated that over 46% of the calf losses (of those born alive) occurred during the first week and another 37% between one week and one month of age—or over 83% during the first month. The various reports illustrate the magnitude of the mortality problem in replacement programs and emphasize the importance of a high level of care and management needed to obtain high survival rates. The authors are familiar with many dairy operations, large and small, that have calf mortality rates of less than 5%, and we recommend a goal of 5% or less calf mortality (from birth to yearlings) for all dairymen. If mortality exceeds 5% the program needs to be evaluated and problems identified and corrected.

* *J. Dairy Sci.*, **58**:448 (1975).
** *J. Dairy Sci.*, **57**:578 (1974).

**Table 14-1** Effect of Three Levels of Management on Costs of Rearing Replacements

| Level of Management | High[a] | Average[b] | Low[c] |
|---|---|---|---|
| No. live calves to start per 25 replacements needed | 29 | 33 | 38 |
| Cost of rearing to 2 mo. of age, @ $60 each | $ 1,450.00 | $ 1,650.00 | $ 2,280.00 |
| No. calves surviving at 2 mo. | 28 | 30 | 32 |
| Cost of rearing from 2 mo. to 1 yr., @ $180 each | $ 4,620.00 | $ 4,950.00 | $ 5,760.00 |
| No. of calves surviving at 12 mo. | 27 | 28 | 29 |
| Cost of rearing from 12 mo. to freshening (32 mo.), @ $400 each | | | 11,600.00 |
| freshening (28 mo.), @ $320 each | | $ 8,960.00 | |
| freshening (24 mo.), @ $280 each | $ 7,560.00 | | |
| **Total cost of rearing 25 replacements** | **$13,630.00** | **$15,560.00** | **$19,640.00** |
| Nonbreeder salvage value, @ $300 each | $ 300.00 | $ 300.00 | $ 600.00 |
| Net costs of rearing 25 replacements | $13,330.00 | $15,260.00 | $19,040.00 |
| Net rearing cost per heifer | $ 533.20 | $ 610.40 | $ 761.60 |
| Original value of the calves | | | |
| Sire = unproved dairy bull (38 × $100 ea. ÷ 25 survivors) | | | $ 152.00 |
| Sire = PD +400 to 600 lb. milk (33 × $150 ea. ÷ 25 survivors) | | $ 198.00 | |
| Sire = PD +1,000 to 1,500 lb. milk (29 × $200 ea. ÷ 25 survivors) | $ 232.00 | | |
| Total cost per replacement heifer | $ 765.20 | $ 808.40 | $ 913.60 |
| Estimated value of replacement heifer[d] | $ 1,000.00 | $ 850.00 | $ 700.00 |
| Net gain or (loss) | $ 234.80 | $ 41.60 | ($ –213.60) |

[a] High level of management: 85% survival of heifers born alive, heifers freshening at an average of 24 mo. of age and sired by bulls with PDs of +1,000 to 1,500 lb. of milk.

[b] Average level of management: 75% survival of heifers born alive, heifers freshening at an average of 28 mo. of age and sired by bulls with PDs of +400 to 600 lb. of milk.

[c] Low level of management: 65% survival of heifers born alive, heifers freshening at 32 mo. of age and sired by unproved bulls.

[d] Based on estimated 1976 average prices for fresh, healthy grade Holstein heifers in the Southeast.

**Table 14-2** Herd Size and Calf Mortality[a]

| Herd size, number | Herds number | Died before 3 mo. (%) | Born dead or abnormal (%) | Total mortality (%) |
|---|---|---|---|---|
| 40–50 | 76 | 7.4 | 6.8 | 13.3 |
| 51–75 | 94 | 8.7 | 7.4 | 16.1 |
| 76–100 | 37 | 9.6 | 8.5 | 18.1 |
| 101–125 | 23 | 9.7 | 8.0 | 17.7 |
| 126–200 | 14 | 14.2 | 6.8 | 21.0 |
| 200–350 | 3 | 21.3 | 12.1 | 33.4 |

[a] Adapted from D. A. Hartman, from *J. of Dairy Sci.*, **57**:577 (1974). Published by the American Dairy Science Association.

## Growth Rate

Achieving growth rates that will result in heifers large enough to breed at 14 to 17 months of age without overconditioning is another serious problem on many dairy farms. Table 14-5 lists desirable ages and weights for breeding and freshening dairy heifers. Table 14-6 indicates recommended growth rates to achieve this objective. To achieve these growth goals from birth to breeding age, average daily gains of 1.05 to 1.15 lb. are necessary for Jerseys, 1.20 to 1.30 lb. for Guernseys and Ayrshires and 1.40 to 1.50 lb. for Holsteins and Brown Swiss. To do this without overconditioning requires steady growth throughout the period. Growth can be monitored by periodically measuring heart girth. Measuring withers height can also help differentiate skeletal growth from conditioning.

**Table 14-3** Person Feeding the Calves and Calf Mortality[a]

| Person feeding calves | No. of Herds | Died before 3 mo. (%) | Born dead or abnormal (%) | Total mortality (%) |
|---|---|---|---|---|
| Wife | 95 | 6.3 | 7.4 | 14.7 |
| Children | 98 | 8.4 | 7.4 | 15.8 |
| Owner or manager | 150 | 8.8 | 7.3 | 16.1 |
| Hired man | 72 | 11.7 | 6.3 | 18.0 |

[a] Adapted from D. A. Hartman, from *J. of Dairy Sci.*, **57**:577 (1974). Published by the American Dairy Science Association.

**Table 14-4**  Person Caring for Calves and Calf Mortality[a]

| Person providing care | No. of Herds | Calf mortality | | |
|---|---|---|---|---|
| | | Winter (%) | Summer (%) | Annual (%) |
| Mother or wife | 25 | 15.0 | 9.4 | 12.3 |
| Operator | 171 | 16.2 | 10.0 | 12.8 |
| Operator and others | 158 | 16.5 | 10.6 | 13.3 |
| Hired labor | 25 | 28.1 | 12.4 | 20.1 |

[a] Reprinted from *J. of Dairy Sci.*, **58**:448 (1975). Published by the American Dairy Science Association.

## Care of Pregnant Cows

Vigorous, healthy calves come from healthy, well-fed, and well-managed cows. Diseases that can affect the health of the calf must be prevented. An adequate amount of the various nutrients must be available in the cow's diet or body reserves to properly nourish the developing fetus. This is especially important during the last two months of gestation, when most fetal growth occurs.

Various diseases can affect the health of the fetus. In viral infections such as BVD, IBR, and PI-3 (See Chapter 13), the organism often infects the placenta and fetus. This can cause a high incidence of stillbirths. In addition, many calves are born weak and the mortality rate of these calves is high. Other infections can affect the uterus of the pregnant cow and may cause prenatal infections in the calf. Such calves are born weak and are more susceptible to calfhood diseases and mortality.

The nutrient priority system of the cow is such that the developing fetus has high priority on available nutrients from both the diet and from body

**Table 14-5**  Recommended Ages and Weights for Breeding and Calving Dairy Heifers[a]

| | Breeding | | Calving | |
|---|---|---|---|---|
| | Age, mo. | Weight, lb. | Age, mo. | Weight, lb. |
| Holstein and Brown Swiss | 14–17 | 780–880 | 23–26 | 1070–1150 |
| Ayshire and Guernsey | 14–17 | 640–740 | 23–26 | 930–1000 |
| Jersey | 13–16 | 550–620 | 22–25 | 770–850 |

[a] Adapted from V.P.I. & S.U. Ext. Newsletter, 1976.

**Table 14-6**  Recommended Growth Rates for Replacement Heifers[a]

| Age, mo. | Ayshire and Guernsey | | | Brown Swiss and Holstein | | | Jersey | | |
|---|---|---|---|---|---|---|---|---|---|
| | Heart girth (in.) | Weight (lb.) | Height at withers (in.) | Heart girth (in.) | Weight (lb.) | Height at withers (in.) | Heart girth (in.) | Weight (lb.) | Height at withers (in.) |
| Birth | — | 65 | 27 | 29 | 93 | 29 | — | 56 | 26 |
| 1 | 26 | 80 | 30 | 32 | 115 | 31 | — | 70 | 27 |
| 2 | 32 | 120 | 32 | 36 | 160 | 34 | 31 | 110 | 30 |
| 4 | 40 | 200 | 36 | 44 | 270 | 39 | 38 | 180 | 34 |
| 6 | 45 | 300 | 39 | 50 | 390 | 42 | 44 | 280 | 38 |
| 8 | 50 | 400 | 41 | 55 | 510 | 44 | 48 | 360 | 40 |
| 10 | 55 | 490 | 43 | 59 | 610 | 46 | 52 | 440 | 42 |
| 12 | 58 | 570 | 45 | 62 | 700 | 48 | 55.5 | 510 | 43 |
| 14 | 60 | 640 | 46 | 64.5 | 780 | 46 | 58 | 570 | 44 |
| 16 | 62 | 700 | 47 | 66.5 | 850 | 50 | 59.5 | 620 | 45 |
| 18 | 64 | 760 | 48 | 68 | 910 | 51 | 61 | 670 | 46 |
| 20 | 66 | 820 | 49 | 70 | 980 | 52 | 63 | 720 | 47 |
| 22 | 67 | 880 | 50 | 71.5 | 1050 | 53 | 64 | 770 | 48 |
| 24 | 69 | 950 | 51 | 73.5 | 1130 | 54 | 66 | 830 | 49 |

[a] *V.P.I. & S.U. Dairy Guideline Series 285, 1973.*

reserves. Energy levels are quite flexible for dry cows. Pregnant cow dietary energy levels should be adequate but not excessive. Overconditioning of dry cows and bred heifers can lead to serious postpartum problems for the cow. If cows and heifers are in reasonably good condition two months prior to calving, the fetus will develop normally, even if energy is somewhat low in the diet, by using the mother's body reserves. Body protein reserves are more limited than energy, so adequate protein must be fed during the dry period. Protein-deficient diets can result in lowered disease resistance and higher mortality in calves. Again, because of potential postpartum problems for the cow, excessively high protein intake should be avoided.

Mineral deficiencies can have a serious adverse effect on the newborn calf. Many deficiencies can cause deformed bones, enlarged joints, stiffness, and general weakness. Iodine deficiency can cause a high incidence of goiterous calves that are frequently hairless or dead at birth. Copper and selenium have also been implicated in a high incidence of stillbirths and weak calves (see Chapter 5). Feeding trace-mineralized salt (1 to 2 oz./head daily) plus recommended amounts of calcium and phosphorus is usually sufficient to avoid these problems.

Adequate dry cow intake of vitamins A, D, and E are essential if newborn calves are to be thrifty and healthy. Vitamin A deficiency can result in weak calves that are highly susceptible to infections and in high calf mortality. Lack of vitamin D impairs calcium and phosphorus utilization and makes the calf more susceptible to rickets. These vitamins are normally present in good-quality forage. If forages are of poor quality, especially if severely leached by weather damage or if overheating has occurred in storage, supplementation is advised.

The key to pregnant cow nutrition is to feed a balanced ration. Overfeeding or underfeeding of various nutrients can lead to serious problems for the cow or the calf or both (see Chapters 5 and 8).

## Care at Calving

Providing the cow with a clean, disease-free, easily observed area to calve is an essential part of the replacement program. In warm climates or during warm or moderate months in temperate climates, an outside grassy area close by for easy observation is probably best. Grassy sod affords a soft bed and good traction for the cow when she arises. The sun effectively controls many disease-causing organisms. In addition, the cow rarely lies in a corner where assisting her when needed is difficult.

During adverse weather, a clean, well-bedded, draft-free box stall should be provided. The use of sawdust and shavings for bedding should be avoided as these serve as a reservoir for bacteria because they usually contain more moisture than straw. If disease has been a problem, box stalls used for calving should be thoroughly scrubbed and disinfected between use. In cases of severe problems, leaving the box stall vacant for three to four weeks after disinfecting is recommended. As a cow begins to exhibit the symptoms of calving (swelling of the vulva, sinking on both sides of the tailhead, and uneasiness) she should be observed closely but not disturbed unless help is needed (see Chapter 10).

Many so-called stillbirths or deaths at birth can be prevented if someone is present at the time of birth. The initial requirement of the newborn calf is air. If the calf does not breathe within four to five minutes of birth, it usually dies. Brain damage is likely to occur if breathing doesn't start in two to three minutes. Immediately after birth, mucus or membranes around the nose and mouth should be cleared. If the calf does not begin to breathe immediately, it should be stimulated by alternately compressing and relaxing the chest walls (artificial respiration), tickling its nose with a straw or piece of hay, or picking it up by its rear legs and letting it hang in this position for a few minutes. Many dead calves labeled as stillbirths from cows unattended at calving were undoubtedly born alive, but died shortly thereafter because mucus or membranes prevented them from breathing.

Drying the calf with a clean cloth or sack is recommended in cold weather, especially if the dam doesn't arise immediately. The umbilical cord should be cleaned if necessary, the material squeezed from it, and dipped in 7% tincture of iodine. If the navel is bleeding, it should be tied off with a clean, disinfected (dipped in alcohol) thread or cord.

Probably the most critical factor in getting the newborn calf off to a good start is ensuring that it receives a feeding of colostrum as soon as possible after birth—within one hour is best, and certainly within the first two hours (Table 14-7). Calves of the larger breeds should receive 4 to 6 lb., the smaller breeds 3 to 4 lb. This may be fed by allowing the calf to nurse the cow after the teats and udder have been cleaned, or the colostrum may be milked from the cow and fed to the calf. Either is satisfactory. If the cow is milked, the amount of colostrum fed can be more precisely determined; however, if this system is used the calf should be fed in the presence of the dam. The calf nursing the cow or the presence of the calf while being fed is reported to have a beneficial effect on uterine contractions, which helps expel the placenta.

It is important that the calf be fed first-drawn colostrum, from the first milking after the dry period, at this initial feeding. It contains higher levels of antibodies and as the calf is born with a very low level of circulating antibodies, it is more beneficial to the calf. The calf's ability to absorb antibodies is high at birth, but decreases rapidly after birth, thus the importance of early feeding of colostrum. Colostrum is very nutritious, being high in vitamins and minerals as well as protein. It is also slightly laxative. In

**Table 14-7** Composition of Colostrum and Normal Milk

| Nutrients | Colostrum | Milk |
|---|---|---|
| Fat | 3.6% | 3.5% |
| Protein | 14.3 | 3.25 |
| Casein | 5.2 | 2.6 |
| Gamma-globulin | 6.2 | 0.1 |
| Lactose | 3.1 | 4.6 |
| Calcium | 0.26 | 0.13 |
| Phosphorus | 0.24 | 0.11 |
| Magnesium | 0.04 | 0.01 |
| Iodine | 0.20 | 0.04 |
| Vitamins (mg./100 lb.) | | |
| A | 73.6 | 12.7 |
| E | 198.9 | 31.8 |

case of weak calves that are unable to nurse or be fed from a bucket or nipple, an esophogeal tube and bag should be used to administer the colostrum. This initial feeding of first colostrum is so critical to the well-being of the calf that the authors recommend freezing and storing several half-gallon containers of first colostrum so that some is always available for emergency use. It can be thawed, warmed and used satisfactorily to feed the calf in cases of loss of the cow at calving or cows who calve with severe cases of mastitis or other diseases that prevent secretion or use of this first colostrum. It is also useful to feed to calves of cows who are premilked before calving.

## CARE OF THE CALF
## FROM BIRTH TO WEANING

This is the most critical period in the life of the calf. Resistance to digestive upsets and to diseases such as scours and pneumonia is lowest at this time. Mortality rates are highest. There is no one best way to feed, house, and prevent disease during this period, but there are many good ways. The common denominators of the good ways are: (1) they provide the calf with a nutritious diet without overfeeding, (2) they provide the calf with a dry, draft-free environment, and (3) they increase the resistance of the calf to diseases.

The calf should be permanently identified before it is removed from the dam. This is especially important in large herds where many calves are reared. Identification by means of an ear tag, tattoo, sketch, or photograph are all satisfactory. Ear tags are not so permanent as the other methods, as they can become loosened and lost. If they are used, the calf should also be more permanently identified at a later date by use of one of the other methods.

### Removing the Calf from Its Dam

There is a great variation in practice regarding the best time to remove the calf from the dam. Some successful calf raisers take the calf away from its dam at once, without allowing it to nurse at all. Others allow it to nurse once, and still others allow it to remain with its dam for two or three days, or until the congestion is out of the udder and the milk is suitable to put in the regular supply. It probably makes little difference when the calf is removed from its dam. The arguments for removing the calf from the cow early are that a calf may be taught to drink from the pail more easily; the amount of colostrum that the calf receives can be better controlled; and less fretting on the part of both the cow and the calf occurs. However, when the calf is allowed to remain for two or three days there is usually not much trouble in teaching it to drink, especially if a nipple pail is used. While the calf is with the cow it will usually consume the colostrum in small quantities at frequent intervals, in a manner most adapted to its digestive system. The usual practice is to remove the calf from its dam during or after the first day.

### Teaching the Calf to Drink Milk

The longer a calf nurses its dam, the most difficult it is to teach it to drink milk from an ordinary pail. By instinct, a calf stretches upward to receive its nourishment. In learning to drink from an ordinary pail, however, it must be taught to bend downward. There is no better method of teaching the calf to drink from an open pail than the simple one of putting one's finger in its mouth, bringing head and fingers into a pail containing a small amount of whole milk, and then carefully withdrawing the fingers. It will probably be necessary to crowd the calf into a corner, and to stand astride of its neck, in order to teach it in this way. Some calves learn to drink with the first attempt; with others it is a long process. Some dairymen omit the first feeding after taking the calf from the cow, thinking that the calf will be more eager for its feed and will be more easily taught to drink from the pail. Such a method may not be detrimental to a strong calf, but it will further weaken weak calves.

### Feeding the Calf

Liquids are the primary source of nutrients for the first few weeks. Because of its nutrient content as well as antibody content, colostrum should be fed at the rate of 8 to 10% of the body weight of the calf for the first two to three days. As long as the utensil used for feeding the liquid is clean and sanitary, either nipple pails, nipple bottles, or buckets have given satisfactory results.

After the two to three day colostrum feeding period any of several acceptable liquid feeds can be used. The three most commonly used are colostrum (either frozen or soured), whole milk, or properly formulated milk replacers. All have been shown to give satisfactory performance. The choice depends primarily on the comparative cost and convenience of feeding.

**Frozen Colostrum.** A cow normally produces 100 to 200 lb. of colostrum during the first three to four days of lactation. This colostrum can be frozen, stored indefinitely without appreciable change in nutrient value, thawed, and fed at the rate of about 8% of the body weight of the calf daily (Table 14-8). Because of its high solids content, it may be advisable to mix the colostrum at the rate of three parts colostrum to one part water and feed this mixture at a rate of 10% of body weight per day. The disadvantage of this system is the expense and inconvenience of freezing, storing, and thawing the colostrum.

**Soured (Fermented, Pickled) Colostrum.** Colostrum can be preserved for one month or longer by allowing it to ferment or acidify at room temperature. The fermenting process is similar to that which takes place in silage making. Large amounts of lactic acid are produced, which lowers the pH to 4.5 or lower, which preserves the material. Fermentation is completed in 10 to 14 days. Only normal colostrum should be used. Mastitic or bloody colostrum

**Table 14-8** Suggested Daily Schedule for Feeding Frozen or Soured Colostrum[a]

| Birth weight, lb. | Week of age | | | | | |
|---|---|---|---|---|---|---|
| | First,[b] (lb./day) | Second, (lb./day) | Third, (lb./day) | Fourth, (lb./day) | Fifth, (lb./day) | Total (lb.) |
| 50–64 | 5.0 | 6.0 | 6.0 | 5.0 | 3 | 175 |
| 65–74 | 6.0 | 7.0 | 7.5 | 5.0 | 3 | 200 |
| 75–84 | 7.0 | 8.0 | 8.5 | 5.0 | 3 | 220 |
| 85–94 | 8.0 | 9.0 | 9.5 | 5.0 | — | 220 |
| 95–104 | 9.0 | 10.0 | 11.0 | 5.5 | — | 248 |
| 105–114 | 10.0 | 11.0 | 12.0 | 6.0 | — | 273 |
| Above 114 | 11.0 | 12.0 | 13.0 | 6.5 | — | 297 |

[a] Adapted from G. M. Jones, Guidelines for Proper Calf and Heifer Nutrition, *V.P.I. & S.U. Ext. Leaflet,* 1976.
[b] Fresh colostrum should be fed for the first two to three days. Total mixture of three parts colostrum and one part water.

and colostrum from cows treated with antibiotics within two weeks of calving or after calving should be discarded.

The colostrum should be stored in plastic containers equipped with lids to exclude flies, rats, etc. Three containers or sets of containers should be used: one for feeding, one full and ready to feed, and one being filled. When one container is emptied, it should be thoroughly cleaned before refilling. The colostrum may be stored in the barn or calf barn. Temperatures of 50 to 60°F (10 to 15°C) are ideal. In very hot weather, fermentation proceeds very rapidly; the material may get too acid and putrefaction may occur. If this happens, the material should be discarded. During storage, the material should be stirred daily to prevent separation. Batches that are fermented can safely be mixed together. Fresh colostrum added to a fermented batch may raise pH temporarily and change fermentation. This practice should be avoided if possible.

The soured colostrum should be stirred before feeding, as the solids tend to separate. For best results, it should be mixed with water at the rate of three parts colostrum to one part water and the mixture fed at the rate of 10% of body weight daily, although it has been successfully fed undiluted at a rate of 8% of body weight or in a mixture of two parts colostrum to one part water at rates of 10 to 12% of body weight. Some dairymen have reported excellent response to slightly lower levels of feeding, 6 to 7% of body weight with a minimum of six lb. per day for Holstein calves. The soured colostrum should not be stored longer than 30 days because of problems with excessive acidity and putrefaction. It appears that the colostrum can be fed before fermentation is completed. Many dairymen start feeding it if needed, even if the fermenting has just started, with no adverse reactions. If the supply of colostrum is exhausted, calves can be shifted to whole milk without adverse effects. Additives are available to hasten the fermenting process and help preserve the material. The value of such additives is not well established at this time.

The major advantage of using soured colostrum is the cost advantage, although some reports also indicate more rapid growth of calves fed frozen colostrum. Two cows normally produce adequate colostrum to provide the liquid diet for one calf for four to five weeks (assuming disposal of bull calves), (125 to 150 lb./Holstein cow × 2 cows = 250 to 300 lb. ÷ 8 to 10 lb./calf/day = 25 to 38 days). An equivalent amount of whole milk would have a value of $30 to $35. Equivalent milk replacer is currently valued at $15 to $25.

**Whole Milk.** Whole milk, fed at the rate of 8% of body weight as is or partially diluted with water (two to three parts milk to one part water) is an excellent liquid diet for calves. The major disadvantage is the cost (Table 14-9).

**Table 14-9** Suggested Schedule for Whole Milk Feeding[a]

| Birth weight, lb. | Week of age | | | | | |
|---|---|---|---|---|---|---|
| | First,[b] (lb./day) | Second (lb./day) | Third (lb./day) | Fourth (lb./day) | Fifth (lb./day) | Total (lb.) |
| 50–64 | 5.0 | 5.5 | 6.0 | 4.5 | 3 | 168 |
| 65–74 | 6.0 | 6.5 | 7.5 | 5.0 | 4 | 203 |
| 75–84 | 6.5 | 7.5 | 8.5 | 6 | 4 | 228 |
| 85–94 | 7.5 | 8.5 | 10.0 | 7 | — | 230 |
| 95–104 | 8.0 | 9.5 | 11.0 | 8.0 | — | 256 |
| 105–114 | 9 | 10.5 | 12 | 8.5 | — | 280 |
| Above 114 | 10 | 11.0 | 13 | 9.0 | — | 301 |

[a] Adapted from G. M. Jones, Guidelines for Proper Calf and Heifer Nutrition, *V.P.I. & S.U. Ext. Leaflet,* 1976.
[b] Fresh colostrum should be fed for the first two to three days.

**Milk Replacers.** Properly formulated milk replacers, fed at recommended rates, have been used for many years as a satisfactory liquid diet for calves after one week of age. Costs are usually less than feeding whole saleable milk, but more than colostrum feeding. Calves fed milk replacers normally do not show as much bloom as calves fed whole milk or colostrum, but usually make comparable weight gains (Table 14-10).

To give good results, a milk replacer must be nutritionally adequate, palatable, easy to use, and economical.

To meet the calf's nutrient requirements, (appendix Table A) a milk replacer must supply sufficient energy, high-quality protein, minerals, and vitamins. The ingredients should be very low in fiber and highly digestable. They should contain a minimum of 22% total or 20% digestable protein. At least two-thirds of the protein should be from milk sources and not more than one-third from plant or fish sources. The highest-quality milk replacers contain all milk protein. They should contain a minimum of 10% fat, and there is considerable evidence that calves fed milk replacers with 15 to 20% fat are more resistant to diarrhea and make faster gains. Animal fats are preferable to plant fats in milk replacers. They should contain little or no fiber. They should also contain vitamins and minerals as specified by the National Research Council in Appendix Table C.

Milk replacers must be composed of feeds that are palatable to the calf; otherwise it may not consume enough to supply its requirements even though all essential dietary factors are present.

Milk replacers are normally mixed with water and fed as a liquid.

**Table 14-10**  Suggested Schedule for Feeding a Milk Replacer (10% or 20% fat)[a]

| Birth weight, lb. | Fat (%) | Milk (lb./day) | Week of Age | | | | | Total milk replacer (lb.) |
|---|---|---|---|---|---|---|---|---|
| | | | First, replacer (lb./day) | Second, replacer (lb./day) | Third, replacer (lb./day) | Fourth, replacer (lb./day) | Fifth, replacer (lb./day) | |
| 50–64 | 10 | 2.5 | 0.33 | 0.90 | 1.00 | 0.75 | 0.50 | 24 |
| | 20 | 2.5 | 0.33 | 0.90 | 1.00 | 0.70 | 0.50 | 24 |
| 65–74 | 10 | 3.0 | 0.40 | 1.10 | 1.25 | 0.90 | 0.60 | 29 |
| | 20 | 3.0 | 0.40 | 1.10 | 1.25 | 0.80 | 0.50 | 28 |
| 75–84 | 10 | 3.5 | 0.40 | 1.25 | 1.40 | 1.00 | 0.70 | 33 |
| | 20 | 3.5 | 0.40 | 1.25 | 1.40 | 0.90 | 0.60 | 32 |
| 85–94 | 10 | 4.0 | 0.50 | 1.40 | 1.60 | 1.20 | — | 33 |
| | 20 | 4.0 | 0.50 | 1.40 | 1.60 | 1.00 | — | 32 |
| 95–104 | 10 | 4.0 | 0.50 | 1.50 | 1.75 | 1.30 | — | 35 |
| | 20 | 4.0 | 0.50 | 1.50 | 1.75 | 1.2 | — | 34 |
| 105–114 | 10 | 4.5 | 0.60 | 1.70 | 1.90 | 1.4 | — | 39 |
| | 20 | 4.5 | 0.60 | 1.70 | 1.90 | 1.3 | — | 38 |
| Above 114 | 10 | 5.0 | 0.70 | 1.90 | 2.10 | 1.6 | — | 44 |
| | 20 | 5.0 | 0.70 | 1.90 | 2.10 | 1.4 | — | 42 |

[a] Adapted from G. M. Jones, Guidelines for Proper Calf and Heifer Nutrition, *V.P.I. & S.U. Ext. Leaflet*, 1976.
[b] Milk replacer should be diluted with 5–6 parts warm water/1 part replacer.

Because of this, any replacer should contain such ingredients and be of such a consistency that a part of it will go into solution and the remainder remain in suspension for a period of time sufficient for the calf to consume it. Ingredients that are not easily mixed with water but settle to the bottom are not satisfactory.

Unless the cost of milk replacement is less than the value of the milk that it replaces, it has not filled a need. The value of feeding it will depend upon the price of the milk that is saved and the cost of the milk replacement. Most milk replacers will replace milk at the rate of about 1 lb. of milk replacer for each 6 lb. of milk. If milk is valued at $10/cwt., or $.10/lb., then milk replacer must cost less than $.60/lb. to be economical.

**Method of Feeding.** The important factors are that: (1) feeding utensils, regardless of the type used, should be as clean and sanitary as the milking equipment, (2) the feeding program should be carefully planned and followed, and (3) drastic changes in the type, amount, or composition of the liquid diet should be avoided. Type of utensil used to feed, temperature of the liquid diet, and once- versus twice-daily feeding have been extensively studied and do not appear to affect performance significantly. Several types of automatic feeders are commercially available. These also appear to do a satisfactory job, providing they are kept clean and sanitary.

## Weaning

Early weaning systems of replacement heifer rearing have been practiced for many years. They offer the advantages of reduced feed costs, reduced labor needs, and decreased incidence of digestive upsets and scours. The dairy calf functions essentially as a monogastric animal at birth but is capable of becoming a functional ruminant by three to four weeks of age, and many are weaned from the liquid diet by this age. Few need to be fed liquid diets after six to eight weeks of age. Rather than recommend a specific age of weaning, the authors offer the opinion that the condition and dry feed consumption of the calf are better criteria. Normal, healthy calves should be weaned within the four- to eight-week range whenever they consume 1 to 1.5 lb. of calf starter daily for three consecutive days. Many calves of the larger dairy breeds meet these criteria by four or even three weeks of age. Calves of the smaller breeds usually require seven to ten days longer. Reducing liquid diet intake during the third or fourth week is a strong stimulus for the calf to increase consumption of the calf starter. Good cowmanship is important in determining when a calf should be weaned. Calves that have experienced problems and have not grown well should be continued on liquid diets for longer periods than normal calves.

**Dry Feeds.** Calves should be offered a calf starter concentrate mixture by the time they are five to seven days of age. Calf starters should be nutritious, highly digestible, and palatable. The mixture should contain 16 to 20% protein and the protein should be of high quality such as soybean protein. It should be high in energy and contain less than 15% fiber. Coarse grinding or pelleting increases palatability, as does the inclusion of molasses. Molasses is highly palatable and also controls dust. Urea should not be used in calf starter mixtures. Several suggested starter mixtures are listed in Table 14-11. High-quality mixtures are available from most commercial feed companies. Calves can be encouraged to begin consuming starter by placing a small amount in their mouth or by placing some on one's fingers and allowing them to suck it from the fingers. Starter should be fed to appetite until consumption reaches 4 to 5 lb./day.

Small calves do not need and will usually not consume much forage before weaning. They should become accustomed to it, however, by weaning, so high-quality hay (leafy, fine stemmed) should be offered to the calves within one to two weeks of age. They also may be offered wilted grass or legume silage, or haylage. High-moisture silages should not be fed. Leftover silage should be removed after each feeding.

**Water** Clean, fresh water should be available at all times. Consumption of water stimulates calf starter intake and growth rate.

**Table 14-11** Suggested Calf Starter Mixtures[a]

|  | Mixture[b] | | | |
|---|---|---|---|---|
|  | 1 lb./ton | 2 lb./ton | 3 lb./ton | 4 lb./ton |
| Corn: cracked or coarse grind | 800 | 760 | 560 | 900 |
| Oats: whole, rolled, or crimped | 500 | 600 | 500 | — |
| Barley: coarse grind | — | — | — | 600 |
| Wheat bran | — | — | 300 | — |
| Soybean meal | 480 | 400 | 600 | 460 |
| Molasses | 180 | 200 | — | — |
| Dicalcium phosphate | 20 | 20 | 20 | 20 |
| Trace-mineralized salt | 20 | 20 | 20 | 20 |

[a] G. M. Jones, Guidelines for Proper Calf and Heifer Nutrition, *V.P.I. & S.U. Ext. Leaflet*, 1976.
[b] Vitamins A and D added to provide 8 million and 2 million I.U./ton, respectively, or 4,000 and 1,000 I.U./lb.

## Condition and Growth

Just how fast calves should grow and how much bloom they should show is not a settled question. The rate of gain and condition of the calf may be satisfactory by one person while another would consider the same below standard. This difference is noted in research reports as well as in dairymen's calf barns.

A calf on liberal whole milk may gain up to 2 lb./day. This should make a top-grade veal; however, heifer calves being raised for herd replacements should not be grown like veal calves. They should be growthy, but not fat. They should be free of scouring and not pot-bellied. They should look thrifty, with their hair showing quality but not the sleekness of the veal.

A gain of about 1 lb./day up to six to seven weeks should be satisfactory for large breeds.*

## Housing

Housing for preweaning calves has three requirements. It must be dry, draft-free, and constructed in such a way as to prevent calves from nursing each other. It also must be kept clean and sanitary. It may be of either cold or warm design or have individual or group pens; it will be quite satisfactory if the above requirements are met. Failure to meet these requirements is a major contributing cause to calf mortality.

Cold housing can be described as a facility in which the temperature and humidity inside the structure are essentially the same as outside. Cold housing for calves may be in individual or group pens inside a three-sided barn, or an enclosed barn with windows for ventilation, or individual calf hutches. The varying and extreme temperatures in these facilities do not cause health problems, except possibly in the very coldest areas of the United States, if the calves are protected from drafts and have access to a dry place to lie down. Heat lamps over the calf beds may be necessary during the first two to three days in very cold weather. Reports indicate that calves reared in open-sided (open side away from the prevailing winds) rather than window-ventilated enclosed cold buildings have fewer health problems, primarily because of higher humidity in the enclosed buildings.* Individual pens with solid partitions on three sides are recommended as they eliminate drafts and prevent spread of disease and keep calves from sucking each other. If group pens are used, head restraints must be provided to prevent the sucking problem.

* *Proceedings of Cornell Nutrition Conference*, 1958.
* W. H. Collins *V.P.I. & S.U. Ext. Pub. 572*, 1973.

Groups pens have the advantage of being more labor efficient. Open-sided structures with outside exercise areas have become very popular in Virginia (Figure 14-1), primarily because of the economy of construction and good results obtained from their use.

The major advantages of cold housing for calves are that these types of facilities are economical to construct and maintain and they can provide a healthy environment for calves. Disadvantages include low labor efficiency, especially when individual hutches are used, uncomfortable working conditions for labor, and difficulty in cleaning and sanitizing.

Warm housing can be described as a facility in which the temperature inside the structure is controlled within a specified range, usually 40° to 70°F (4 to 21°C). An attempt usually is made to keep relative humidity under 75%. Calves are usually restrained in individual pens or elevated stalls with solid, slatted or mesh floors. Maintaining proper temperature and humidity conditions requires a closed, insulated structure equipped with ventilating fans, supplemental heat, and thermostatic controls, as heat produced by the calves is not sufficient to maintain building temperature and permit the ventilation system to exchange sufficient air for control of humidity.*

Warm housing facilities can provide an excellent environment for calves. They also provide more comfortable working conditions for labor, can be more labor efficient, and are easier to clean and disinfect, especially when metal tie stalls are used. They are also more expensive.

Housing can have a major effect on calf health and mortality. If a severe outbreak of disease strikes in a herd, moving to other facilities for one to three months may be necessary to break the cycle of the organism. Various types of housing have proved satisfactory if they have been properly constructed and maintained so as to provide the calf with a dry, draft-free environment. In choosing between the various available systems, costs of construction and maintainance and labor efficiency are probably the two major factors. Cold barn designs are probably best suited for herds located in moderate climates, and for small herds. Because of labor efficiency and ability to spread the construction costs over more calves, warm facilities may be more suited to large-herd operations.

Individual calf pens with a floor area of 20 to 25 sq. ft./calf and three solid sides 3½ to 4 ft. high or elevated stalls have disease- and sucking-control advantages and are recommended for calves prior to weaning. The number of these stalls should equal the maximum number of heifer calves born in an eight-week period, or 12 to 15 per 100-cow herd.

* W. H. Collins, *V.P.I. & S.U. Ext. Pub.* 572, 1973.

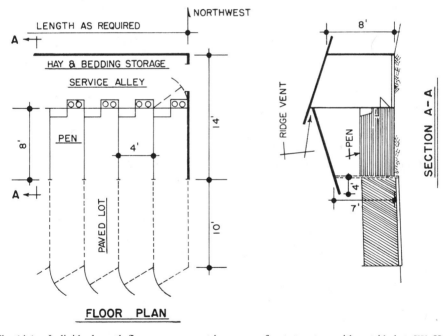

**Fig. 14-1.** Individual, earth-floor arrangement in an open-front structure with outside lots [W. H. Collins, *V.P.I. & S.U. Ext. Pub.* 572 (1973)].

349

## Disease Control

Lack of disease control for calves causes heavy death losses in many herds. The leading causes of calf mortality are probably scours and pneumonia; pneumonia often follows scours. Other common problems include navel ill and ringworm. The causes, prevention, and control of these problems are discussed in Chapter 13. If one has a good program of pregnant-cow care and care at calving, so that healthy vigorous calves are born, and if the calf receives colostrum early in life and is properly nourished and housed through weaning, most of these diseases can be prevented. High mortality and/or high incidence of sick or stunted calves can be avoided in most cases. Occasionally, however, severe outbreaks of scours or other infectious diseases do occur even in well-managed herds. In these cases a prophylactic program, designed in cooperation with a veterinarian, that increases the resistance of the calf at birth and throughout the period is advisable (see Chapter 13). It should be based on identification and antibiotic susceptibility of the causative organism(s). Thorough cleaning, sanitizing, and abandonment of calf quarters for one to three months or more may be necessary to break the cycle.

## CARE OF HEIFERS
## FROM WEANING TO BREEDING

The goal of the management program during this period are to maintain acceptable growth rates (Table 14-6) without fattening as fatness may inhibit future milk production and good health status. This is a period of rapid growth and increased disease resistance.

## Feeding

Heifers should be fed calf starter on a free-choice basis, with minimum consumption of 4 to 5 lb./day until they are three to four months of age. At this age the calf starter can be replaced gradually with a grower ration, or the herd ration if it is of suitable nutrient content. Grain feeding should be continued at a rate of 3 to 5 lb./day until the heifers reach 9 to 11 months of age, as ruminant function is inadequate for the heifer to maintain good growth on forage alone. The nutrient content of the grain mixture should be calculated to balance the forage being fed. As a general guideline, the grain mixture should contain 12 to 14% crude protein if the forage is over 50% legume, 18 to 20% if the forage is mostly grass hay or silage, and 20 to 24% if corn silage is the primary forage. Nutrient requirements and nutrient content of rations as specified by the National Research Council are listed in Appendix Tables B and C.

Most forages are acceptable for heifers during this period if properly balanced with concentrates. Two exceptions are all pasture and all corn silage. Calves under six months of age should be fed some hay or silage when they are pastured. Supplementing corn silage with some hay is advised until the heifer is three to six months old.

After the heifers reach 9 to 11 months of age, an acceptable rate of growth can be attained without concentrate feeding if good-quality forage is fed. This high-quality forage may be good pasture, hay, or grass and legume silage. If corn silage is fed as the only forage, it must be supplemented with additional protein, calcium, phosphorus, and trace mineral salt. If poor-quality hays or silages are fed, they must be properly supplemented with concentrate feeds to balance their deficiencies.

An adequate supply of clean, fresh water and trace mineral salt should also be provided during this period.

Steady growth during this period at an average of 1.2 to 1.3 lb./day for Jerseys and up to 1.8 to 1.9 lb./day for Holsteins and Brown Swiss is desirable. Overconditioning should be avoided.

## Housing

During this time heifers need access to an exercise area. They should also have access to a dry, draft-free area in which to bed until they reach 9 to 11 months of age. Heifers should be grouped according to size, and ample manger space should be provided to avoid exclusion of smaller or less aggressive heifers. More recently free-stall and confinement systems of rearing dairy heifers have come into use, especially in large herds with limited land resources. Some of these confinement systems (Figure 14-2) use as little as 20 to 30 sq. ft. of floor space per animal. Their suitability for rearing dairy heifers has yet to be determined, but they merit consideration in certain situations.

More traditionally, heifers are housed and fed in groups of four to six in box stalls, or pens with outside runs after weaning. Group size, pen size, and outside exercise area are usually increased as the calves progress in age and size.

Heifer housing needs after 9 to 11 months of age are minimal. Protection from extremely inclement weather (snowstorms, freezing rains, etc.) and from very hot weather during the summer is all that is needed. Three-sided pole barns are often used for winter protection, and shade trees suffice during the summer.

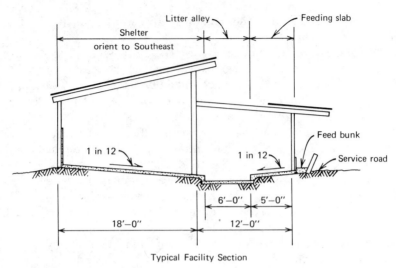

Typical Facility Section

**Fig. 14-2.** Young stock housing facility [*V.P.I. & S.U. Ext. Pub. 572* (1973)].

## Disease Control

Heifers are normally vaccinated for several diseases during this period (see Chapter 13). They should also be observed daily for symptoms of other problems such as ringworm, foot rot, overgrown hooves, digestive upsets, injury, or poor growth due to overcrowding. If some of these problems occur, they should be corrected promptly before they become serious. The heifers should be observed for symptoms of estrus during the latter part of this period, and heat dates recorded. Those not showing heats by 13 to 14 months of age should be checked to determine why.

This period, or in many cases the period from five to six months of age to breeding age, is a period of neglect on many dairy farms. Many dairymen do a conscientious job of feeding and caring for replacements while they are in or near the barn. They are often turned out to the back pasture at five to six months of age and, except for a periodic head count or feeding them a few bales of hay occasionally during the winter, they are literally forgotten until they reach breeding size. Often they do not reach this size until they are 24 months of age or older. This is a primary reason why many heifers are three years or older before they freshen. This practice is an inefficient use of resources and slows genetic progress of the herd. It should not be followed. The authors believe a primary reason for such practices is that the heifers are not income producing and the results are not immediately apparent; therefore care of these heifers is a low priority job. If they are neglected as very young calves, the effect is immediate and apparent: many die. If they are neglected after five to six months of age, few die and they are often far enough from the barns so that symptoms of nutritional deficiencies such as delayed puberty or stunted growth, parasitism, or injuries are not clearly observed. The effects of poor management of these heifers often isn't realized for a year or more after it occurs.

## CARE OF HEIFERS
## FROM BREEDING TO FRESHENING

This period should begin with breeding healthy, good-sized heifers at 14 to 17 months of age and culminate with the normal calving by strong, healthy heifers at 23 to 26 months of age.

### Feeding

Feeding practices as recommended for heifers from 10 to 11 months of age to breeding age should be continued until one to two months prior to calving. Because of increased size of the heifers, feed intake will increase, but growth rates will decrease somewhat during this period. Daily gains of .9 to 1.0 lb. for Jerseys, 1.0 to 1.1 lb. for Ayrshires and Guernseys, and 1.1 to 1.2 lb. for Holsteins and Brown Swiss are quite satisfactory (Table 14-6).

During the one to two months just prior to freshening, most heifers should receive some additional concentrate feed, especially if they are a bit on the thin side. Good judgment must be used, however. The objective is to have the heifers well grown and in good condition, but not fat, at freshening, and the diet can be altered during this period to meet this objective. Concentrate feeding during this period should be limited to no more than 1 percent of body weight daily because of potential calving and postcalving problems. Many dairymen routinely feed and house their springing heifers with the dry cows. This is a good practice, as requirements and recommendations are similar. They also become accustomed to these cows and their surroundings. Additionally there is advantage to getting them accustomed to going through the milking parlor.

### Housing and Disease Control

Requirements and recommended practices for housing and disease controls are the same as for the latter part of the previous period.

## Breeding

The main factors that should determine the times to breed heifers are: their size, their age, and the time at which there is greatest need for milk, as a basemaking period. If they have been fed and managed properly, most heifers will be large enough to breed at 14 to 15 months of age (Table 14-5). In most herds there will be a small number that, because of genetic composition or problems during rearing, will not be large enough to breed at that age and a few that will reach proper size earlier. Good judgment should be used in determining the time to breed these heifers. Early breeding of some heifers at 13 months or delaying breeding of others until they are 16 to 17 months of age to supply more milk when most needed is also an acceptable practice. Breeding age, however, should not be extended beyond 17 to 18 months for normal growing heifers as numerous reports have indicated that the optimum time for breeding is 14 to 17 months of age. Heifers calving at older ages normally produce slightly more milk (10 to 15% more if they are 30 to 36 months old at first freshening) during the first lactation, but this is more than offset by increased rearing costs (Table 14-1). There is also a delay in recovering the investment in the heifer and decreased total production per day of life to mature age. Heifers freshening at 23 to 26 months of age normally will have produced enough milk to pay for their rearing costs by the time they are 34 to 36 months old. Heifers freshening at 30 to 36 months of age normally will not have produced enough to repay these costs until they are about four years of age. Early freshening (23 to 26 months of age) can result in smaller mature cow size if they are not fed properly during the first lactation. They continue to grow during the first lactation and they must be fed extra nutrients to meet growth as well as production and maintainance needs (see Chapter 5). If these nutrient requirements are met, growth will continue and they will reach normal mature size.

Heifers should be bred artificially to sires known to sire smaller calves, to minimize calving problems (Chapter 9). Methods of handling and restraining heifers for artificial insemination are discussed in Chapter 10.

## OTHER CONSIDERATIONS
## IN CALF AND HEIFER REARING

### Dehorning

Dairy heifers should be dehorned at an early age. It is easier on the heifers and lessens the chance of the animal injuring other heifers and people working with the heifers. Three acceptable methods of dehorning young calves are: (1) by the use of commercially prepared dehorning solutions or paste, or the use of caustic sticks, (2) the use of an electric dehorner, and (3) the use of a Barnes dehorner. Which to use is largely a matter of personal preference, as any of these can do an acceptable job if properly used. Regardless of which is used, the calves should be carefully observed for a few days following dehorning, as problems with infections (particularily during the fly season) and complications can occur.

**Dehorning Solutions, Paste, or Caustic.** This method works best on calves from 3 to 14 days old (Figure 14-3). It can be used as soon as the horn button can be detected. The hair around the horn button should be clipped as short as possible. The substances will erode skin, so care must be exercised to avoid contact of the substance with the skin of the calf (or the person doing the dehorning). It is a good practice to isolate the calf after dehorning to avoid spreading the material to the skin of other calves or having other calves lick the substance. If the substance is improperly applied or is removed shortly after application, scurs will develop.

**Electric Dehorning.** This method works well on calves from one to five or six weeks of age, from the time the horn button is easily discernible until it

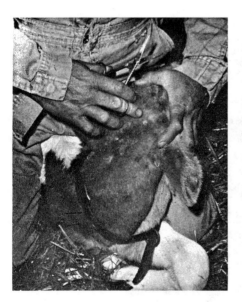 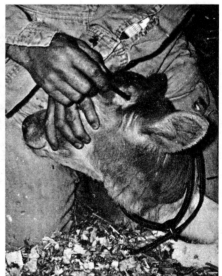

**Fig. 14-3.** Use of caustic compound to remove horns.

* V.P.I. & S.U. Dairy Guideline 287, 1972.

becomes attached to the skull (see Figure 14-4). The dehorner should be plugged in long enough to reach maximum heat before use and allowed to reheat between each calf dehorned if good results are to be obtained. The dehorner should be applied until a copper-colored ring completely surrounds the horn button or small horn. This usually takes 10 to 20 sec. Failure to do this can result in scurs. Calves must be rigidly restrained; the iron is very hot and a struggling calf can result in burns to the calf or operator. It is also hot enough to set fire to bedding or other flammable material.

**Barnes Dehorning.** This method works well on calves over a wide age range, from two or three to five or six months of age (see Figure 14-5). It is a very sure method of dehorning, but it does produce an open wound that must be properly cared for and closely observed, especially during the fly season. In this method the horn is physically removed after restraining the calf. The dehorner is placed over the horn and far enough from the horn to remove about ¼ in. of hair around the base of the horn. The horn is removed by a rapid, forceful spreading of the handles of the dehorner. Following horn removal, the arteries supplying the horn must be clamped or pulled by the use of a sterilized (dip in alcohol) surgical tool or needle-nosed pliers. After bleeding is controlled, antiseptic blood-clotting powder should be applied to

**Fig. 14-4.** Use of an electric dehorner.

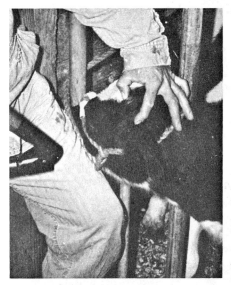
Locate the horn.

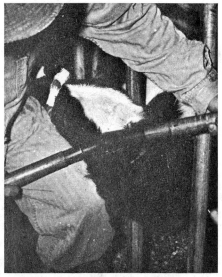
Remove with rapid, forceful movements.

Stop the bleeding arteries.

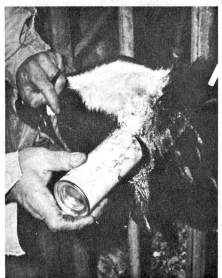
Use blood clotting powder

**Fig. 14-5.** Use of the Barnes dehorner.

357

the area. Animals dehorned in this manner should be closely observed for several days for excessive bleeding or infection.

If dehorning is delayed until older ages, this necessitates the use of a dehorning saw or a keystone dehorner. The hazards to the health of the animal are much greater, from both the horn removal and the possibility of other injuries while restraining the animal. This method should be used only by veterinarians or very experienced persons.

## Removing Extra Teats

Many calves are born with extra, or supernumerary, teats. Removal of these extra teats is usually a relatively safe and easy procedure if performed at an early age. It is more difficult and complicated if they are attached to or very close to other teats, or if the animal is older. In these cases they should be removed by a veterinarian, or not removed. Removal of extra teats can eliminate potential problems after freshening, as pathogenic organisms can enter the udder through these teats, especially at or near freshening. Removal will improve the appearance of the udder.

Extra teats are usually, but not always, located behind the four normal teats. They may also be located between or attached to the normal teats. It is a good practice to remove extra teats at a young age (one to six months), as the calf is easier to restrain, the teats are smaller, and the wound will be smaller. Many dairymen routinely check for and remove extra teats concurrently with some other operation such as dehorning or vaccinating. This is a good practice. After restraining the calf and identifying the extra teat(s), the udder should be thoroughly cleaned and disinfected. The extra teat should be grasped (a surgical clamp is excellent for this) and gently pulled away from the udder. It should be cut off with sterile surgical scissors and the wound treated with iodine or other suitable antiseptic. Usually, little bleeding occurs if the operation is performed at an early age. Emasculating clamps placed near the base of the teat before and during cutting prevents excessive bleeding (Figure 14-6).

## Vaccinations

A vaccination program should be developed in consultation with a veterinarian to meet the needs of individual herds, as vaccines needed will vary depending on the herd and location (Chapter 13). These vaccines should be administered for maximum effectiveness against potential disease problems and should be scheduled as a regular part of the replacement program.

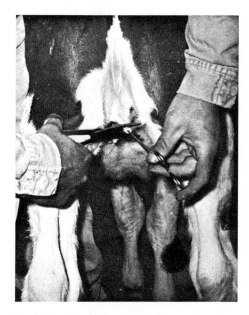

**Fig. 14-6.** Removing extra teats.

## Hoof Trimming

Calves and heifers with correct feet and legs that receive adequate exercise do not require regular hoof trimming. Many, however, do require regular or occasional hoof trimming to avoid foot and leg problems as heifers or problems that contribute to impaired ability to walk as cows. Careful observation of the condition of hooves and proper trimming is a sound practice.

## Records

Accurate records of identification, vaccinations, other health or abnormal conditions, and heat and breeding dates should be accurately recorded. Many systems are available (Chapter 2) and quite satisfactory.

## COMPLETE FEEDS FOR HEIFERS

Recent research* indicates that heifers from 2 to 24 months of age can be satisfactorily grown on complete rations if the bulk, and consequently the

* A. H. Rakes, and D. G. Davenport, *Hoard's Dairyman*, **119**, 13, 41 (1974).

palatability, are adjusted so that the heifers consume adequate but not excessive amounts of nutrients for normal growth. The bulk is adjusted by varying the ratio of fibrous feeds such as ground corn cobs, cottonseed hulls, coarsely ground hay, etc. to more nutritious feeds. Bulk should be lower at younger ages and higher at older ages for good results. The system can be used by itself or as a supplement to poor pasture. The major advantage of complete feeds for heifers is the saving in labor. Self-feeders can be used and filled periodically, rather than feeding heifers once or twice daily. The system also provides a method of utilizing low-quality by-products and coarse hay. It merits consideration where dry-lot feeding of heifers is practiced.

## ECONOMIC ASPECTS OF HEIFER REARING PROGRAMS

As can be seen from the previous discussions, there are many alternatives available in regard to the various aspects of the heifer rearing program. The most important aspects of the total program are to rear replacements with high genetic potential while avoiding losses from death and stunting, and growing them rapidly enough to reach sufficient size to freshen at 23 to 26 months of age. Also important is achieving these goals as economically as possible.

The potential effect on costs and returns of different levels of management in regard to genetic potential of the heifers, mortality rate, and age of freshening were discussed earlier in this chapter (see pp. 331–333 and Table 14-1). Other factors also can have a major effect on the economic aspects of the replacement program. Feed cost is usually the largest cost of the replacement program. Labor and the cost or value of the calf are usually the next largest. A typical budget for rearing a replacement heifer is presented in Table 14-12. Changing two practices between budget A and budget B resulted in cost reductions of over $90 per replacement heifer. This is just one example of how reducing costs can affect profitability of the replacement program. The important point is that the replacement program should be analyzed as an enterprise within the total dairy operation. It should contribute significantly to the profitability of the total operation. If, after analysis, it is determined that it does not, reasons for this should be identified (feeding, mortality, age at freshening, etc.) alternative methods evaluated, and either new practices implemented or alternative methods of acquiring replacements investigated.

## SUMMARY

The replacement program is a vital yet often neglected part of the dairy operation. Future genetic improvement depends primarily on an adequate

**Table 14-12** Estimated Costs of Raising Large-Breed Replacement Heifers to 24 Months of Age[a]

|  | Budget A[b] (per heifer) | | Budget B[c] (per heifer) |
|---|---|---|---|
| Feed: | | | |
| 230 lb. whole milk @ $10.50/cwt. | $24.15 | | — |
| 250 lb. calf starter @ $7.50/cwt. | 18.75 | | $18.75 |
| 2.5 tons mixed hay @ $70.00/cwt. | 175.00 | 1.5 T. | 105.00 |
| 6.0 tons corn silage @ $18.00/cwt. | 108.00 | 2.5 T. | 45.00 |
| Pasture @ $.25/day | — | 420 days | 105.00 |
| 1,000 lb. 12 to 14% grain @ $6.50 | 65.00 | | 65.00 |
| Salt, minerals & Vitamins | 5.00 | | 5.00 |
| Feed cost | 395.90 | | 343.75 |
| Interest on investment in feed @ 8% over 24 months. | +31.67 | | 27.50 |
|  | **$427.57** | | **$371.25** |
| Labor: 50 hr. @ $2.50/hr. | 125.00 | 40 hr. | 100.00 |
| Buildings & Equipment costs | 14.50 | | 10.00 |
| Breeding | 20.00 | | 20.00 |
| Bedding | 11.25 | | 7.00 |
| Veterinary and medical | 10.00 | | 10.00 |
| Insurance | 3.25 | | 3.25 |
| Utilities | 1.50 | | 1.50 |
| Miscellaneous | 5.68 | | 5.68 |
| Mortality cost (90% survival of live births) | 15.25 | | 15.25 |
| Value of the calf | 150.00 | | 150.00 |
| Interest on investment other than feed @ 8% over 24 months | 28.51 | | 23.41 |
| **Total other than feed costs** | **$384.94** | | **$346.09** |
| **Total cost of replacement at 24 months of age** | **$812.51** | | **$717.34** |

[a] Based on 1975 Virginia average values.
[b] Whole milk and dry-lot feeding of good-quality mixed grass, legume hay, and corn silage.
[c] Colostrum, dry-lot feeding for five months per year, and pasturing for seven months per year.

supply of high-genetic-potential heifers to replace cows culled for low production or other reasons. Assurance of this supply is dependent on the breeding (both genetic and reproductive) program and the dry-cow program as well as the replacement-rearing program. Rearing or buying replacements is expensive and is a major cost on most farms. Economies can be realized, without sacrificing quality or quantity of the supply of replacements, with careful analysis and other good management practices.

## REFERENCES FOR FURTHER STUDY

Appleman, R. D., and Owen, F. G., Breeding, Housing, and Feeding Management, Symposium, *J. Dairy Sci.*, **58:** 447 (1975).

Collins, W. H., Dairy Calf-Raising Facilities, *V.P.I. & S.U. Ext. Pub.* 572, 1973.

Hartman, D. A. et al., Calf Mortality, *J. Dairy Sci.*, **57:**576 (1974).

Hillman, D. et al., Raising Calves, *Mich. St. Ext. Bull.* 412 (1971).

Jones, G. M., Dairy Replacement Heifers, *V.P.I. & S.U. Dairy Guideline Series* 291 (1975).

La Salle, T., and Kalison, S. L., Dehorning, Vaccination and Removing Extra Teats, *V.P.I. & S.U. Dairy Guideline Series* 287 (1972).

Rakes, A. H., and Davenport, D. G., Complete Feeds can Simplify Your Heifer Raising, *Hoard's Dairyman* **119:**1341 (1974).

# 15

# Dairy Buildings: Types and Arrangements

**15**

The basic objective for dairy buildings and equipment is to provide a system that will give cattle the opportunity to be housed, fed, and milked properly. The system is interrelated with almost every other aspect of dairy management. It involves a major capital investment, and annual costs are normally the third-largest production cost in dairy operations. The system, to a very large degree, determines labor efficiency. Decisions in this area have long-term effects on the dairy operation, as the expected life of most buildings is 20 or more years and that of equipment 5 to 15 years. Major changes within these time periods are expensive. These factors stress the importance of thorough planning before new systems are built or existing systems are remodeled. Many existing systems are examples of poor planning and design. They are characterized by low labor efficiency, drainage problems because of poor site selection, high maintainance costs, and lack of flexibility with regard to possible expansion or changes in materials-handling management. These problems have been accentuated because of the changes that have occurred in recent years, that is, increasing herd size and new technical developments in feeding, milking, and waste-handling systems.

The concept of engineering management, as it applies to dairy systems, has not been well understood by many dairymen or many builders. Dairymen know cattle, builders know construction, and neither group is greatly knowledgeable about both.

Engineering management can be defined as the process of designing and constructing dairy production facilities that will function or perform according to a desired scheme of control. The dairyman who installs a group of gates to control cow movement in a production facility or who changes the length of his free stalls has engineered management into the system. In other words, the act of managing a dairy system is determined to a large degree by the physical facilities that make up the system.*

## PURPOSES OF DAIRY BUILDINGS AND EQUIPMENT

The basic purposes of dairy systems must be met if the system is to function properly. Some of these basic purposes are listed below.

1.  Protect the farm animals from inclement weather. The amount and type of inclement weather largely determines the type and amount of protection needed. This may vary from only shade in warm areas such as the

* W. H. Collins, Engineering Management, *V.P.I. & S.U. Ext. Ser.*, 1974.

southern United States to completely enclosed structures and systems in northern regions.

2. Minimize risk of injury. Many cases of bruised hocks, lameness, stepped-on teats and udders, capped hips, and other injuries occur because this purpose is not engineered into the system. Narrow openings, slippery floors, bare concrete resting areas, etc., must be avoided if a high incidence of such injuries is to be avoided.

3. Provide the cattle with ample opportunity to be well fed. Sufficient, accessible feeding and watering areas must be a part of the system.

4. Provide an area for special handling of cattle, such as treating for various diseases and conditions, breeding, calving, clipping, diagnostic work, etc.

5. Meet all legal requirements for the production and sale of milk. These codes vary among states and it is essential that applicable requirements and regulations be met. Perhaps equally as important, the system should be conducive to the production and temporary storage of quality milk. The system should be such that cattle are kept reasonably clean and buildings and equipment used in milking and milk handling are easily cleaned and kept clean.

6. Afford labor protection from long exposure to inclement weather. Providing reasonably comfortable working conditions for labor is an important factor in hiring and retaining good labor. The importance of this item increases as herd size increases and more nonfamily labor is used.

7. Be conducive to a high degree of labor efficiency in materials handling. In the average dairy operation labor is the second-largest production cost. The buildings and equipment system affects both the type and amount of labor needed. Labor efficiency can be measured in terms of milk per man, per year, or cows per man. The former is preferred as it is an indication of quality as well as quantity. Laborers' jobs in the dairy operation are largely confined to moving feed, manure, milk, and cows. The volume of each is considerable. Moving the first three of these by mechanical means such as augers, scrapers, flushing systems, and pipelines greatly increases labor efficiency. Avoiding corners, constraints, bottlenecks, steep grades, and slippery floors enhances smooth cattle flow.

8. Be within economic limitations for which cattle can pay during the expected life of the system. They must be economical to build, yet durable, and have low maintainance costs. They should be conducive to

a high return per dollar invested. If the system results in high production per cow and high labor efficiency, the amount that can safely be invested can be sizable. Conversely, if the system results in low per-cow production and/or labor inefficiency, the amount one can afford to invest in buildings and equipment is very low. The limitations of the investment in a system should be determined after careful financial analysis (Chapter 3), including calculation of current equity (net worth), projected budget (to determine additional debt-paying capacity), and cash flow. It has been relatively easy in recent years for many established dairymen to overinvest in buildings and equipment because of the appreciated value of owned land. Often overlooked is the fact that, although inflated land values provide excellent collateral, they cannot pay for buildings and equipment unless the land is sold. Capital expenses involved in dairy buildings and equipment must be paid for by income generated by the cattle, not the appreciated land value, if the dairy operation is to continue operations. Not understanding this basic principle has forced many dairymen to sell out.

9. Be legally and conveniently located. Many areas have zoning, environmental, and other regulations and ordinances controlling new construction and environmental pollution from existing operations. These must be thoroughly checked before final site selection for new construction or major renovations are made. Other potential problems such as complaints from urban neighbors, possible runoff into streams, etc., should also be ascertained and avoided. New construction sites should be conducive to good drainage and allow for future expansion. They also should be located as near to forage production areas as possible and be accessible to large trucks in all types of weather, as the loss of a couple of tanks of milk because of a snowstorm can seriously reduce profits.

10. Be flexible enough to allow future expansion if desired and to adapt to new technology as it is discovered. This is easy to say, but difficult to do, for who knows what the future holds? Nevertheless a certain degree of flexibility can be engineered into dairy systems with very little extra cost at the time of construction. Milk houses can be of sufficient size to accommodate a larger tank; building close to roads, creeks, or steep hills can be avoided; boxing in of the layout with silos, machine sheds, etc. can be avoided; and free-span buildings can be constructed.

11. Meet the personal preferences of the owner or manager. Many alternatives are available that can meet the functional purposes of a system. Within these alternatives, personal preference is an important factor and should be given strong consideration, as most dairymen spend a major portion of their awake hours working with the system.

Any system that meets these purposes can be defined as a good one. It will be conducive to the health and productivity of the cattle and labor and, when accompanied by the other essentials of a profitable herd, it will contribute to productivity and profitability of the entire operation. There are many good systems, though the authors have never seen any two exactly alike. This is because dairy buildings and equipment systems, like so many others aspects of dairy management, must be tailor made to fit the resources and goals of each dairy operation and meet the personal preferences of the owner.

## COMPONENTS OF THE DAIRY SYSTEM

Regardless of the type of system, certain essential areas or component systems must be included. These are the following.

1. A milking, milk storage, and milking equipment cleaning system. Harvesting a quality crop (milk) efficiently and maintaining the quality of that crop is the single most important job on a dairy farm. Fifty or more percent of the total labor in most dairies is concerned with the milking operation. Milking center design and equipment are a major factor in determining both the labor efficiency of this job and the quality of the product.

2. A feeding system. This includes one or more feeding areas, feed storage facilities, and a system of moving the feeds from storage to the feeding area. On most dairy farms the feeding operation is the second most time consuming job.

3. A waste-handling system, including facilities for both collection and disposal of animal and other wastes. Problems in this area have greatly increased in recent years because of increasing herd size, more intensive cropping, and environmental protection regulations.

4. A longing or resting area.

5. A special area or areas for confining and restraining cattle when they are sick or for breeding, treatment, calving, etc. This is an essential but often overlooked area in loose or free-stall systems.

6. Service areas, including roads, lanes, and alleys for the movement of cattle and materials.

7. An area or areas for replacement herd if replacements are raised rather than purchased.

Each of these components is essential to the operation of the system. The challenge is not only to have each component conducive to high performance of cattle and labor, but to arrange the components into an efficient production unit that not only satisfies current needs but may also be easily adapted to future needs, either anticipated or unforseen.

## BASIC ALTERNATIVES

Most dairymen or prospective dairymen have three basic alternatives in regard to building and equipment decisions. These are: (1) continue operations in the existing system with few or no alterations; (2) renovate or alter the existing system by adding on or replacing component parts of the system; and (3) building a completely new system. The decision among these three alternatives will depend on the type and condition of existing facilities, the nature and extent of existing problems, and the effect of each alternative on probable costs and returns.

The first step should be to identify the problems within the existing system in view of the basic purposes of dairy building and equipment systems. For example, if cow and labor performance is high, with resulting high profitability, in the current system, and if the buildings and equipment are in good condition, alternative (1) is a logical choice. If, however, the existing system does not meet legal requirements for the production and sale of milk of the desired grade, or cow performance is low because of deficiencies in the system such as poor milking equipment, lack of adequate feeding space, etc., or labor efficiency is low and costs high because of inefficient materials handling, then either alternative (2) or (3) is indicated. If no system exists, or the existing one is completely worn out, then the only alternative is to build a new system.

If problems exist that must be corrected, compare alternatives (2) and (3) in terms of their probable effects on solving existing problems, achieving the purposes of dairy building and equipment systems, and on projected costs and returns. If renovation, replacement, or expansion of one or two components of the system will solve the existing problems at nominal cost and result in a system that achieves the purposes of the total system, then this is a logical way to proceed. If, however, extensive and expensive renovation is required to correct existing problems, and there is considerable question whether the system will achieve the purposes of a good system even after renovation, then rebuilding is a logical choice.

If basic alternative (2), renovation, is indicated after this thorough evaluation, the extent and type of renovation and/or equipment replacement must be designed to solve the specific problems of the existing system. The only recommendation that is valid is that when the renovation is completed, the system should meet the basic purposes of a dairy buildings and equipment system.

If, however, building a new system is indicated, then careful evaluation of the two basic types of systems (conventional and loose) and evaluation and planning of various designs and layouts for each system, various designs for component parts of each system, and equipment needs are required. The decisions regarding which system should be built and the degree of automation installed for any individual dairy operation depend primarily on: (1) the climate, (2) the size of the herd, (3) the feeding system, (4) the quantity and quality of available labor, (5) the availability and cost of capital, and (6) the personal preferences and goals of the owner or manager. The alternatives should be evaluated in terms of these factors and their ability to satisfy the purposes of a dairy building and equipment system.

## TYPES OF DAIRY FACILITIES

In 1974 about 50% of the dairy cattle in the United States were housed in conventional (stanchion or tie stall) barns.* This type of housing was the predominant type in the Lake States and the Northeast, while some type of loose housing was the most popular type in the remaining areas (Table 15-1). The effects of climate, herd size, and personal preference or custom are apparent from these statistics.

### Conventional Facilities

In conventional facilities most of the system components are contained in the same structure. Most consist of two or more rows of stanchions or tie stalls with a manure gutter, feed manger, and service alley(s) for each row. Cows are fed, milked, and lounge in the stanchion or tie stall. Often box stalls, calf facilities, and hay and grain storage areas are contained in the same structure. Silos and milk house are usually located nearby or attached to the dairy barn. The barn is normally enclosed and has window and fan ventilation. Many barns are equipped with gutter cleaners and pipeline milkers to automate handling of manure and milk and increase labor efficiency. A typical layout for a stanchion barn is presented in Figure 15-1.

**Advantages.** Some of the major advantages of conventional barns are the following.

* C. R. Hoglund, *Res. Rept.* 275, Mich. St. Univ., 1975.

**Table 15-1** Percentage Distribution of Dairy Farms and Milk Cows by Type of Housing, 10 Areas of United States, 1974 Estimate

| | | Percentage distribution | | | | |
|---|---|---|---|---|---|---|
| | Stanch-ion | Loose housing | | Corral or dry lot | Other | Totals |
| | | Loaf-ing | Free stall | | | |
| **Number by type** | | | | | | |
| Northeast | 73 | 7 | 16 | — | 4 | 100 |
| Lake States | 80 | 6 | 11 | — | 3 | 100 |
| Corn Belt | 34 | 38 | 24 | — | 4 | 100 |
| Appalachian | 14 | 47 | 34 | — | 5 | 100 |
| Delta | 10 | 50 | 17 | 20 | 3 | 100 |
| Southeast | 15 | 36 | 22 | 24 | 3 | 100 |
| Northern Plains | 35 | 40 | 17 | 3 | 5 | 100 |
| Southern Plains | 13 | 43 | 7 | 33 | 4 | 100 |
| Mountain | 5 | 33 | 32 | 28 | 2 | 100 |
| Pacific | 17 | 17 | 35 | 28 | 3 | 100 |
| **Number of cows** | | | | | | |
| Northeast | 67 | 6 | 23 | — | 4 | 100 |
| Lake States | 76 | 5 | 16 | — | 3 | 100 |
| Corn Belt | 28 | 33 | 34 | — | 4 | 100 |
| Appalachian | 9 | 40 | 44 | — | 5 | 100 |
| Delta | 8 | 48 | 20 | 22 | 2 | 100 |
| Southeast | 7 | 32 | 21 | 36 | 4 | 100 |
| Northern Plains | 27 | 40 | 25 | 3 | 5 | 100 |
| Southern Plains | 7 | 36 | 6 | 46 | 4 | 100 |
| Mountain | 3 | 21 | 32 | 32 | 2 | 100 |
| Pacific | 8 | 10 | 30 | 50 | 2 | 100 |

[a] From C. R. Hoglund, *Res. Rept.* 275, Mich. St. Univ., 1975.

1. Individual cow care. Cows are confined to one stanchion most of the day so it is easier to give a cow individual attention. Cows can be fed individually, abnormal conditions (off-feed, diarrhea, constipation, etc.) are more easily detected, and cows are readily available for insemination, treatment, clipping, or other special attention.

2. Maximum protection from inclement weather is afforded both cattle and dairymen. In cold weather animal heat can be utilized to maintain comfortable temperatures for the cows, labor, and calves, if they are housed in the same structure.

3. Timid cows are protected from abuse at the feed manger and are assured of receiving adequate feed.

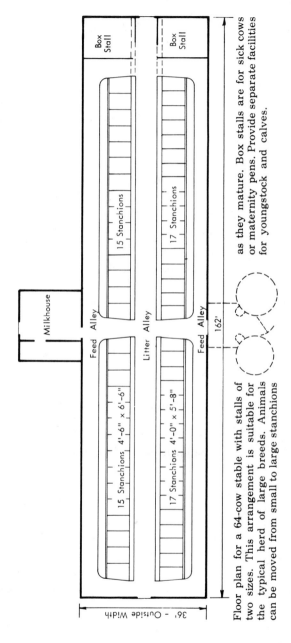

Floor plan for a 64-cow stable with stalls of two sizes. This arrangement is suitable for the typical herd of large breeds. Animals can be moved from small to large stanchions as they mature. Box stalls are for sick cows or maternity pens. Provide separate facilities for youngstock and calves.

**Fig. 15-1.** Typical stanchion barn arrangement. (U. of Minnesota Bul. M-13, 1968)

371

4.  Cattle can be displayed to advantage to visitors and prospective buyers. They can be arranged by sire groups, according to size, or by other arrangements that are advantageous. Platforms can be elevated and the front can be higher than the rear to help the cow look her best.

**Disadvantages.** Some of the major disadvantages of conventional housing are the following.

1.  A high degree of labor efficiency is difficult to achieve. Feeding requires moving the feed from the place of storage to individual cows. This can be mechanized, but it is expensive and difficult to do. Milk and manure removal can be mechanized more easily and has been done in most conventional barns. Considerable time is spent unfastening and fastening cows, brushing and cleaning cows, bedding cows, cleaning service alleys, cleaning mangers, etc.

2.  Many of the jobs that must be performed are undesirable to those who must do them. Milking involves stooping and bending on the part of the milker, feeds must be shoveled or forked by hand—and often the same feed must be handled several times before it reaches the cow—and cleaning cows and service alleys is tedious work for most.

3.  Construction and maintenance costs are usually higher than other systems, although this may vary depending on the degree of automation involved. More costly items such as plumbing, individual drinking cups, stanchions, ventilation equipment, and conventional rather than pole type of construction are included in conventional barns. Many of these items are also more costly to maintain in good operating condition.

4.  Cows are more susceptible to injuries to the hocks and teats.

5.  Lack of flexibility. It is often difficult to expand, even slightly, or to use the facility for any other enterprise.

Conventional housing can be described as an intensive system of dairying. Greater individual attention is possible while sacrificing labor efficiency. It is probably best suited for small herds, where most of the work is performed by family labor, for some purebred herds, and in colder climates. This dairy system is decreasing in popularity in most of the United States because of labor inefficiency, construction and maintainance cost, and lack of flexibility, although over 50% of the new dairy barns recently built in Wisconsin and nearly 50% of those in New York and Pennsylvania were stanchion barns.*

* C. R. Hoglund, *Res. Rept.* 275, Mich. St. Univ., 1975.

**Desirable Characteristics of Conventional Systems.** For most satisfactory results in conventional systems, the following features should be included.

1. Two-row design; this facilitates ventilation and the operation of gutter cleaners.

2. Adequate ventilation for all seasons that is thermostatically controlled. This requires fans, and, in cold climates, ceiling insulation (hay in two-story barns or commercial insulation).

3. Gutter cleaners.

4. Ample-sized stalls to afford cow comfort and avoid injuries.

5. Tie stalls to increase cow comfort and make it easier for the cow to rise.

6. Cow trainers to help keep cows cleaner.

7. Gutter grates if manure gutters are very deep or minimum bedding is used. This eases crossing the gutter and avoids injuries.

8. The bottom of the manger 4 to 5 in. above the stall floor to help reduce injuries and increase cow cleanliness.

9. Stall floors or platforms 2 to 3 in. higher than the litter alley and sloped toward the litter alley.

10. A properly installed milk pipeline.

11. Feed-storage structures such as vertical silos accessible from the stanchion barn, with sufficient capacity for herd needs; mechanical unloaders; hay storage overhead or in a nearby, readily accessible storage barn; and concentrate storage in a metal bin(s) with mechanical delivery to the barn.

## Loose Housing Facilities

Loose housing facilities include loafing, free-stall, corral, and dry-lot systems. These systems are characterized by separate areas for milking, feeding, and lounging. In principle, the cows move freely to lounging and group-feeding areas and are milked in a milking center.

In loafing systems the lounging area is a weather-protected (usually a three-sided pole barn), bedded area of 60 to 80 sq. ft. per cow (Figure 15-2). The area is well bedded and the manure pack is allowed to accumulate for several months. It is usually cleaned by use of a mechanical manure loader two to three times per year. Considerable bedding is added daily, and picking

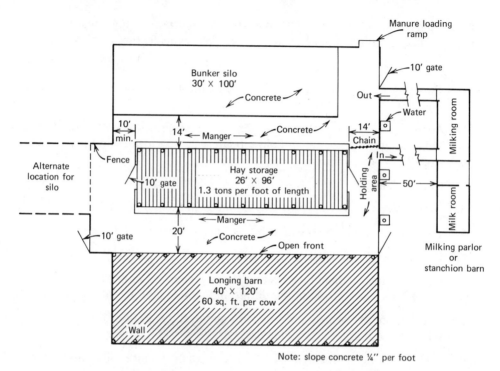

**Fig. 15-2.** Loose housing arrangement for 80 cows arranged for self-feeding from a horizontal silo [*V.P.I. & S.U. Ext. Pub.* 928 (1967)].

up droppings one or two times daily is required to maintain cleanliness of cows. This type of facility is used extensively in the southern, Corn Belt, and Plains states (Table 15-1).

Free-stall systems differ from loafing systems primarily in the lounging area. Many loafing systems have been changed to free-stall systems by installing free stalls in the lounging area. Free stalls reduce the bedding requirement and the risk of injury as cows rest in an unrestrained stall rather than an open area. This type of facility has gained in popularity in recent years in all but the southern Plains states.

Free-stall systems can be constructed in various arrangements. The basic designs include the following.

1. Warm or completely enclosed systems where all facilities except feed storage and in some cases replacement facilities are combined and enclosed in one insulated structure that is warm in cold weather. Maximum protection is afforded to cattle and dairymen. Mechanical

ventilation is required. This design is most popular in extremely cold climates. (Figure 15-3)

2. Cold covered or partially enclosed systems where most of the facilities are combined in one structure. However, the structure is not insulated, is often open on one side, and is designed to keep inside temperatures near outside temperatures through use of natural ventilation. These systems are popular in moderate climates (Figure 15-4).

3. Open-lot systems where the feeding and watering facilities and milking facility are separate from the lounging area. The lounging area may be cold enclosed or consist of rows of roofed free stalls. Feed bunks may be covered or uncovered.

Corral and dry lot systems vary from the other two loose systems in that a corral or lot, rather than a loafing or free-stall structure, is provided for each group of cows. Most of these corrals or lots are equipped with shade structures of various types. More recently, many corral-type systems have been equipped with roofed free stalls that provide a dry, shaded, resting area (Figure 15-5). The free stalls may be used continuously or only during rainy

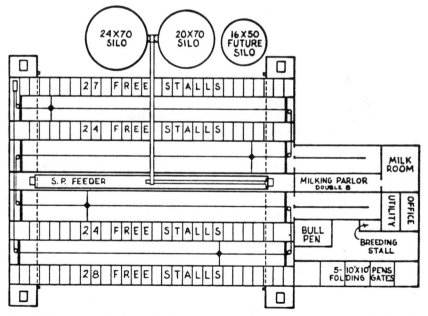

**Fig. 15-3.** Automated free-stall arrangement (Conference Papers—1973 National Dairy Housing Conference. ASAE-SP-01-73).

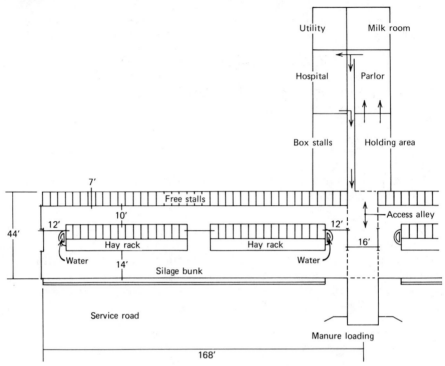

**Fig. 15-4.** A covered free-stall system for 70 cows (W. H. Collins, V.P.I. & S.U. Ext. Serv., 1974).

seasons when mud is a problem in the earth-floor corrals and lots. This system is most prevalent in warm climates and predominates in the southwestern and southern states (Table 15-1).

**Advantages.** Some of the major advantages of loose housing systems are listed below.

1.  They are conducive to a higher degree of labor efficiency because of the ease of mechanizing materials handling. In milking, the cows move to the man and milking machine rather than the man and machine moving to the cow. They are group-fed forages and, in many cases, complete feeds, the handling of which is easily automated. Manure handling is also easily mechanized.

2.  They decrease the number of undesirable jobs that need to be performed by hand and require less bending, stooping, and lifting.

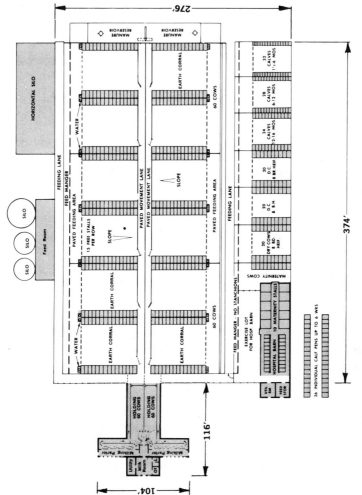

**Fig. 15-5.** Corral system with fence line feeding and roofed free stalls for 360 cows (Babson Bros. Dairy Research Service, 1976).

377

3. They are usually more economical to construct and maintain, as less expensive types of construction, less plumbing, less mechanical ventilation, etc., are required. However, depending on the degree of automation and type of facility, they can be as or more expensive than conventional facilities.

4. Greater opportunity for cow comfort and reduced risk of injury can be afforded if stalls and other parts of the system are properly designed. The cow is free to seek a place to lie down that is comfortable, and in the case of free stalls, is protected, rather than being restrained in a stanchion or tie stall.

5. Less bedding is required in free stalls; however, in loafing systems bedding requirements are very high.

6. If properly designed, they enable cows to be kept clean more easily. This contributes to milk quality and milking efficiencies, since milking is in a separate area away from feed, manure, and dust. This facilitates the production of quality milk.

7. They afford greater flexibility. Herd size can usually be increased moderately with little or no additional construction. Major increases in herd size are often possible with minimal expense as often only certain areas of the system must be expanded. In cases of decreased herd size or elimination of the milking enterprise the lounging and feeding areas are often suitable for other enterprises such as beef feeding or heifer rearing.

**Disadvantages.** Some of the major disadvantages of loose housing systems are the following.

1. There is less opportunity for individual cow attention. Cows are housed, fed, and often milked in groups. Individual feeding and care is difficult to achieve. Off-feed conditions, abnormal bowel movements, etc., often go undetected until the condition is acute.

2. Spread of contagious diseases is enhanced because of the common use of feeding and watering facilities.

3. Labor often is less protected from inclement weather. In many facilities, much of the time spent feeding, handling manure, and moving cows is spent outside or in a partially open structure, regardless of weather conditions.

4. Manure is dropped on a variety of surfaces rather than in a gutter. This

detracts from the neatness of the appearance of the facilities and necessitates the wearing of boots by dairy workers. Because the manure contains little bedding, it is often sloppy and difficult to handle.

5. It is more difficult to display cows to advantage to visitors or prospective buyers. Often, finding the particular animal in a group is difficult. When she is located, typically she wishes to stand with her front feet lower than her rear feet (this doesn't help her appearance), or another cow obstructs the vision of the visitor, or she moves away.

6. Special areas must be provided for milking and for restraint of cattle for breeding, treating, clipping, diagnostic work, etc. In the latter situation, cattle must be separated from the rest of the cattle and driven or led to and from these areas. This is often difficult, especially with cows in heat or other particularily nervous or cantankerous cows.

Loose housing systems can be described as a less intensive, or more extensive, type of dairying than the conventional systems. Group handling of cattle and mechanized materials handling replace individualized handling of cows and materials. Labor efficiency is higher. These systems are best suited to large herds where the costs of the more expensive parts of the system (milking parlor, automatic devices, etc.) can be distributed over more production units, thus decreasing the cost per producing unit, and where labor efficiency is more important because of the cost of that labor. These are also well adapted to smaller operations, providing an operation is large enough to justify the cost of milking facilities or can utilize existing milking facilities. Many smaller dairies with conventional barns have constructed new lounging and feeding facilities and use the conventional barn for milking. This is often an interim step in the change from a small, conventionally housed operation to a larger, loose-housed operation. Loose systems, especially free-stall systems, are increasing in popularity in nearly all parts of the United States, primarily because of increasing herd size, labor costs, bedding costs, and the desire to reduce time spent in bending, stooping, lifting, and cleaning and brushing cows. Reducing these tasks also increases the ability to hire and keep labor.

**Desirable Characteristics of Loose Systems.** For most satisfactory results in loose systems, the system chosen must be suitable for the climatic and resource conditions of the individual operation and personal preference of the owner or manager. Each component must be suited to the service it provides—the feeding system to the type of feed(s) fed, the milking system to the number of cows milked, etc.—yet well adapted to the other components of the system so that the result is an efficient production unit. Available alternatives, from general type, design, and layout of the total system to type, design, equipment, and degree of automation involved in individual components of the system, are numerous. When one considers all the possible combinations, the possibilities approach infinity. This is why the authors have not seen any two systems exactly alike and why planning the system must be carried out on an individual farm basis. It is beyond the scope of this text to discuss all the alternatives in loose systems. We will, however, reemphasize the importance of ensuring that any system fulfill the 11 basic purposes mentioned earlier in this chapter, and the need for thorough planning, including financial planning as well as design and equipment planning, before decisions are made. Visiting existing operations is also advised. We will briefly discuss some of the basic principles involved in selection of the system and components of the system.

**The System as a Whole.** Some of the basic principles that apply to most loose systems that enhance efficiency and productivity are the following.

1. Straight-line design. The moving of materials, including cattle, is usually facilitated by movement in straight, horizontal lines. Corners, turns, severe slopes, and gates that must be opened and closed should be avoided wherever materials must be moved. Bottlenecks to cattle movement, such as narrow openings, right-angle turns into milking parlors, steep grades, and slippery surfaces must be avoided if cow flow is to be smooth. Manure scraping is also facilitated by moving it in straight lines rather than around corners. In addition, construction costs are usually less per square foot for this type of design.

2. Group cattle. In larger herds, grouping cattle reduces stress, makes more precise feeding possible, especially if the cattle are fed a complete ration, facilitates heat detection, and allows more uniform milkout. The maximum size of groups is not well defined; however, groups in excess of 100 cows appear to be more susceptible to stress problems in confinement systems.* Methods of grouping are variable, but most dairymen group according to level of production. If complete rations are fed, three to four production groups plus a dry cow and springing heifer group are recommended. Grouping by production level also facilitates heat detection, as most or all of the open cows are located in the same group. Recent research indicates that disruption caused by shifting cows does not significantly adversely affect performance.** Major changes in

---

* B. Harris, Florida Large Herd Mgt. Symposium, 1976. (To be published by The University Presses of Florida.)
** T. Friend, M. Sowerby, and C. P. Polan, V.P.I. & S.U. unpublished data, 1976.

nutrient composition of the ration, however, may cause significant production decreases. Group size should be such that a group of cows is not required to remain in the milking parlor holding area more than two hours unless the cattle in the area are well protected from adverse climatic conditions.

3. Special areas for breeding, treatment, and other routine individual care should be located such that they are easily accessible from milking, feeding, and lounging areas. Individual animals that require attention will be identified in each of these areas and at times will need to be returned to each of these areas. Treatments and other practices, such as obtaining blood samples, vaccinations, etc., that are unpleasant for the cow should not be performed in the milking parlor lest the experience cause future reluctance of the animal to enter the parlor. Cow experiences in the parlor should be pleasant ones associated with milking if efficient movement through the parlor and maximum milk-ejection response is to be obtained.

4. Free stalls should be properly sized for the cows that use them, constructed so they afford minimum risk of injury, and durable. Free stalls of excessive length result in manure being dropped in the stall rather than the litter alley. This results in either extra labor to clean the stalls, or soiled cows, or both. Chest boards or cables can be used to help avoid this problem if stalls are too long. If stalls are too wide, cows may turn around in them and defecate and urinate in the front of the stall. If they are too short, the rear end and udder of the cow may lie on the curb or in the litter alley. If wood construction is used, the stalls should be bolted rather than nailed together.

5. The system should be flexible to accommodate future changes, foreseen and unforeseen. Dairymen who built new facilities some years ago but left room for additional feeding and lounging units did not need to rebuild the system completely when they increased herd size. Those who allowed sufficient space and openings for passage of large equipment in the feeding areas did not have to rebuild or add feeding areas when they switched to complete feeds. Those who allowed proper slope and avoided corners in litter and traffic lanes did not have to rebuild in order to convert to liquid manure systems. These are a few examples of what is meant by flexibility of the system. Boxing in of the system by structures or natural barriers such as streams or steep hills should be avoided, as should creating areas not readily accessible by good-sized equipment.

6. Allow access to well-drained, earthen exercise areas. There is considera-

ble difference of opinion concerning whether this is necessary. The authors believe that the availability and use of such an area reduces risk of feet and leg problems, facilitates heat detection (cows are more likely to mount other cows if the footing is firm but not slippery as paved areas often are), and promotes the general health and well-being of the cows.

7.  Provide adequate ventilation. In enclosed or partially enclosed systems, adequate ventilation is essential if excessive condensation and warming are to be avoided. Ridge vents, openings under the eaves, and wall openings or hinged walls will facilitate ventilation. Fans may also be needed in some types of systems.

**Milking Systems.** The milking system consists of milking facility and holding area, the milking and milk-transfer equipment, the milk storage tank, and milking-equipment cleaning facilities. As this system occupies over 50% of the labor on most farms, its potential effect on overall labor efficiency is considerable. As its function is harvesting the crop, the need for high-quality performance of this system should be apparent; poor performance adversely affects milk quality and udder health (see Chapter 12).

**Milking Facilities.**   There are five basic types of milking arrangements used in loose systems. They can be equipped to various degrees with automatic devices. Their cost and labor efficiency varies with the type of parlor and degree of automation. The five basic types are: (1) the stall or stanchion barn (Figure 15-1), (2) side-opening parlors (Figure 15-6), (3) herringbone parlors (Figure 15-7), (4) polygon parlors (Figure 15-8), and (5) rotary parlors. Rotary parlors may be either the tandem (Figure 15-9), herringbone (Figure 15-10), or turnstyle type (Figure 15-11). A holding area to retain cows near milking area is needed in all these types except the stall barn. The use of some type of automatic crowd gate usually facilitates movement into the milking facility.

Distribution of dairy farms and cows milked by type of milking system in 1974 are given in Table 15-2. From these data it can be seen that type of milking system (stanchion or some type of parlor) closely parallels type of housing system. Use of pipelines in stanchion systems is popular in all areas of the United States except the Northeast and Lake States areas. It can also

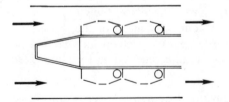

Fig. 15-6.  A double-2 side opening parlor (Michigan State Univ. Ext. Bul. E-1034, 1976).

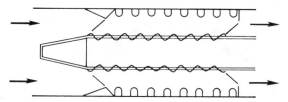

**Fig. 15-7.** A double-8 herringbone parlor (Michigan State Univ. Ext. Bul. E-1034, 1976).

be seen that herringbone parlors are the most popular type of parlor in all areas.

The decision regarding the type and size of parlor and the type and degree of automation of the milking equipment to choose depends on the herd size, initial and annual costs per unit of production (cows or milk), availability of reliable service, and the personal preferences of the owner or operator in regard to individual cow attention and labor efficiency. All types can perform in a satisfactory manner in regard to quality milking performance if they are properly installed, maintained, and operated (Chapter 12).

HERD SIZE. Size of milking herd affects choice of milking facilities in two ways, through economic and time aspects (Table 15-3). Smaller herds are limited to facilities that are less expensive, as cost becomes prohibitive on a

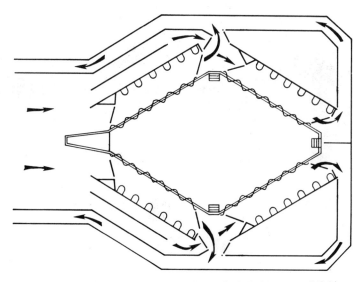

**Fig. 15-8.** A 24-stall polygon parlor with a single holding pen (Michigan State Univ. Ext. Bul. E-1035, 1976).

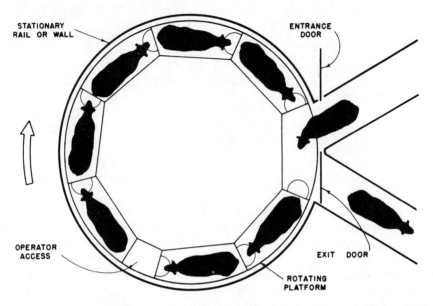

**Fig. 15-9.** An 8-stall rotary tandem parlor (Michigan State Univ. Ext. Bul. E-1047, 1977).

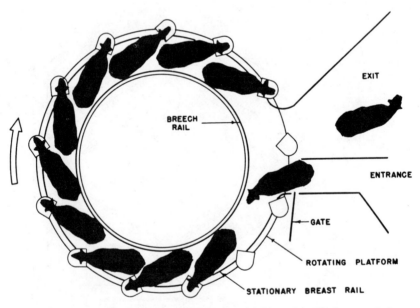

**Fig. 15-10.** A rotary herringbone milking parlor (Michigan State Univ. Ext. Bul. E-1047, 1977).

384

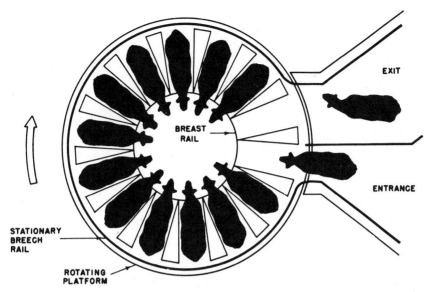

**Fig. 15-11.** A rotary turn-style milking parlor (Michigan State Univ. Ext. Bul. E-1047, 1977).

per-cow or per-unit-of-milk basis with more expensive systems. Herds of 100 cows or less are economically best suited to single-operator parlors such as double 3 or 4 herringbones or three- or four-stall side-opening parlors, or to stall barn milking facilities. Some dairymen with herds of less than 100 cows have installed multiple operator parlors, primarily because they prefer not to spend more than three to four hours per day in the milking operation.

Milking facilities can be used to milk cows a maximum of 18 to 20 hours per day. They are rarely used for milking more than 14 to 16 hours per day and usually much less time than this. If milking is performed in two eight-hour shifts, the milking arrangement must be large enough to milk the herd in one eight-hour shift. This is a limiting factor in large herds; the facility must have enough capacity to get this job done in the time available to do it. Because of the large number of cows, initial costs are distributed over more cows, and costs per cow can be reasonable even though total costs are high. Large herringbones (double 8, 10, 12s or more), polygons, rotaries, and side-opening parlors with a high degree of automation are suitable for larger herds, especially if the system is used at or near capacity. Some dairymen feel that efficiency and quality of milking are higher in single-operator parlors, and consequently have installed two or more single-operator parlors rather than one multiple-operator parlor.

COST OF MILKING FACILITIES AND MILKING. Investment in milking facilities var-

**Table 15-2**  Percentage Distribution of Dairy Farms and Cows Milked by Type of Milking System, 10 Areas of Unites States, 1974 Estimate[a]

| Area | Stanchion milking | | Parlor milking | | | |
|---|---|---|---|---|---|---|
| | Bucket | Pipe-line | Herring-bone | Side-opening | Other[b] | Total parlor |
| **Dairy farms** | | | | | | |
| Northeast | 65 | 17 | 12 | 4 | 2 | 18 |
| Lake States | 71 | 15 | 9 | 3 | 2 | 14 |
| Corn Belt | 40 | 24 | 17 | 11 | 8 | 36 |
| Appalachian | 23 | 24 | 19 | 19 | 15 | 53 |
| Delta | 10 | 49 | 17 | 14 | 10 | 41 |
| Southeast | 12 | 36 | 28 | 14 | 10 | 52 |
| Northern Plains | 40 | 23 | 18 | 10 | 9 | 37 |
| Southern Plains | 11 | 44 | 22 | 13 | 10 | 45 |
| Mountain | 18 | 29 | 27 | 16 | 10 | 53 |
| Pacific | 12 | 40 | 22 | 17 | 9 | 48 |
| **Milk cows** | | | | | | |
| Northeast | 50 | 20 | 24 | 4 | 2 | 30 |
| Lake States | 60 | 18 | 17 | 3 | 2 | 22 |
| Corn Belt | 27 | 24 | 27 | 14 | 8 | 49 |
| Appalachian | 14 | 25 | 31 | 20 | 10 | 61 |
| Delta | 7 | 46 | 24 | 15 | 8 | 47 |
| Southeast | 4 | 38 | 42 | 11 | 5 | 58 |
| Northern Plains | 25 | 24 | 30 | 12 | 9 | 51 |
| Southern Plains | 9 | 38 | 32 | 13 | 8 | 53 |
| Mountain | 7 | 26 | 40 | 18 | 9 | 67 |
| Pacific | 4 | 44 | 30 | 14 | 8 | 52 |

[a] C. R. Hoglund, *Res. Rept.* 275, Mich. St. Univ., 1975.
[b] Mainly walk-through systems.

ies with type and size of facility, degree of automation, brand of equipment, and location. A comparison of 1975 estimated total and per cow investment of various types of parlors and herd sizes is presented in Table 15-3. Comparisons of characteristics, investments, and annual costs of four alternative milking and grain-feeding systems for a 120-cow operation are given in Table 15-4 and for a 240-cow operation in Table 15-5.

Labor is the major annual cost of milking in most systems. It may, however, vary from as little as 40 to 50% in highly automated systems where facilities are used relatively few hours per day and wages are low, to 80 to 90% in cases where milking facilities are used 10 to 12 hours per day or more and wage rates are high.

Annual parlor and equipment costs, however, can also significantly increase annual per-cow milking costs in cases of large investments in smaller

**Table 15-3**  Estimated Cows Milked Per Man Hour and Per Hour, Per-Day Capacity and Cost of Various Milking Systems

| System[a] | No. operators | No. units | Cows per hour | Cows per man hour | Cows per day, 3 hr./ milking | Cows per day, 6 hr./ milking | Cost of system[c] ($) | Herd size | Investment per cow[d] ($) |
|---|---|---|---|---|---|---|---|---|---|
| Stall barn (buckets) | 1 | 2 | 15–20 | 15–20 | 45–60 | 90–120 | | | |
| Stall barn (pipeline) | 1 | 3 | 20–25 | 20–25 | 60–75 | 120–150 | | | |
| Herringbone-double 3 | 1 | 3 | 30–35 | 30–35 | 90–105 | 180–210 | 40,000 | 100 | 400 |
| Herringbone-double 4 | 1 | 4 | 35–40 | 35–40 | 105–120 | 210–240 | 44,000 | 100 | 440 |
| Herringbone-double 6 | 2 | 6 | 60–70 | 30–35 | 180–210 | 360–420 | 50,000 | 150 | 333 |
| Herringbone-double 8 | 2 | 8 | 70–80 | 35–40 | 210–240 | 420–480 | 56,000 | 150 | 373 |
| Herringbond-double 10 | 3 | 10 | 90–105 | 30–35 | 270–315 | 540–630 | 64,000 | 200 | 420 |
| Side opening-in line 3 | 1 | 3 | 25–30 | 25–30 | 75–90 | 150–180 | 42,000 | 100 | 460 |
| Side opening-in line 4 | 1 | 4 | 26–32 | 26–32 | 78–96 | 156–192 | 46,000 | 100 | 400 |
| Side opening-double 3 | 2 | 6 | 50–60 | 25–30 | 150–180 | 300–360 | 60,000 | 150 | 453 |
| Side opening-double 4 | 2 | 8 | 52–64 | 26–32 | 156–192 | 312–384 | 68,000 | 150 | 375 |
| Polygon-24-stall[b] | 2 | 24 | 130–140 | 67–70 | 390–420 | 780–840 | 150,000 | 400 | 480 |
| Rotary-8-stall tandem[b] | 1 | 8 | 50–55 | 50–55 | 120–150 | 240–300 | 120,000 | 250 | 300 |
| Rotary-13-stall herringbone[b] | 2 | 13 | 130–140 | 65–70 | 390–420 | 780–840 | 120,000 | 400 | 300 |
| Rotary-17-stall turnstyle[b] | 2 | 17 | 90–100 | 45–50 | 270–300 | 540–600 | 110,000 | 400 | 275 |

[a] Base system with crowd gate.
[b] Some automation.
[c] Estimated 1975 prices.
[d] Based on quoted herd size.

387

**Table 15-4** Investments and Annual Costs for Four Milking and Grain-Feeding Systems for a 120-Cow Free-Stall, Herringbone Milking System[a]

| Item | Milking and grain feeding system | | | |
| --- | --- | --- | --- | --- |
| | Feed all grain in parlor | | Feed grain in parlor and bunk | Feed all grain in bunk |
| System | 1 | 2 | 3 | 4 |
| Characteristics | | | | |
| Number of milker units | 4 | 8 | 4 | 4 |
| Number of men milking | 1 | 1 | 1 | 1 |
| Annual hours per cow[b] | 28.0 | 25.6 | 29.5 | 25.6 |
| Cows milked per man hour | 40 | 44 | 40 | 44 |
| Total investments ($)[c] | 30,700 | 32,300 | 31,700 | 29,700 |
| Annual costs ($) | | | | |
| Parlor and stalls | 1,840 | 1,840 | 1,840 | 1,840 |
| Milking system | 1,520 | 1,840 | 1,520 | 1,520 |
| Grain storage and feeding | 1,000 | 1,000 | 1,200 | 800 |
| Labor | | | | |
| $2/hr. | 6,720 | 6,144 | 7,080 | 6,144 |
| $3/hr. | 10,080 | 9.216 | 10,620 | 9,216 |
| $4/hr. | 13,440 | 12,288 | 14,160 | 12,288 |
| Total costs | | | | |
| Labor/$2 hr. | 11,080 | 10,824 | 11,640 | 10,304 |
| Labor/$3 hr. | 14,440 | 13,896 | 15,180 | 13,376 |
| Labor/$4 hr. | 17,800 | 16,968 | 18,720 | 16,448 |

[a] C. R. Hoglund, Dairy Facility Investment and Labor Economics. Reprinted from *J. of Dairy Sci.*, **56**:491 (1973). Published by the American Dairy Science Association.
[b] Includes time to set up and clean equipment and parlor, collect cows into holding area, feed grain, and do milking.
[c] Includes investment in parlor, stalls, milking system and grain-feeding equipment, and sotrage, but not a bulk tank.

herds. Overinvestment in milking facilities in proportion to herd size can also create cash-flow problems if debt repayment obligations must be met.

OTHER CHARACTERISTICS OF MILKING SYSTEMS. Stall milking barns vary in size from 8 to 10 to 50 or more stalls usually arranged in one or two rows. Cows may be milked by single or multiple operators. Bucket-type milkers, buckets with a transfer system, or milking units and a pipeline may be used. In some operations a unit is provided for each two stalls. Cost and efficiency of these systems varies with the size of barn, degree of automation, and number of units per operator. When bucket milkers are used and milk is carried by hand

**Table 15-5** Characteristics, Investments, and Annual Costs for Milking Parlors, Equipment, Grain-Storage and Feeding, and Labor to Milk 240 Cows, Seven Systems.[a]

| Item | Double 4 herringbone milking system | | | Double 8 herringbone milking system | | | |
|---|---|---|---|---|---|---|---|
| System | 1 | 2 | 3 | 4 | 5 | 6 | 7 |
| **Characteristics** | | | | | | | |
| 4 milker units | ✓ | | | | | | |
| 8 milker units | | ✓ | ✓ | ✓ | ✓ | | |
| 16 milker units | | | | | | ✓ | ✓ |
| Automated milking equipment | | | | | ✓ | | ✓ |
| Feed grain in parlor only | ✓ | ✓ | | ✓ | ✓ | | ✓ |
| Feed grain in bunk only | | | ✓ | | | ✓ | |
| Number of men milk at one time | 1 | 1 | 1 | 2 | 2 | 2 | 1 |
| Total number of men milking | 2 | 2 | 2 | 2 | 2 | 2 | 1 |
| Cows milked per man hour | 40 | 44 | 48 | 38 | 42 | 46 | 60 |
| Annual hours per cow to milk, etc. | 28.0 | 25.6 | 23.2 | 29.2 | 26.8 | 24.5 | 19.0 |
| **Investments** | | | | | | | |
| Parlor and stalls | 14,600 | 14,600 | 14,600 | 20,600 | 20,600 | 20,600 | 22,700 |
| Equipment | 7,600 | 9,200 | 9,200 | 12,100 | 15,300 | 15,300 | 27,000 |
| Grain storage and feeding equipment | 12,000 | 12,000 | 10,800 | 12,000 | 12,000 | 10,800 | 10,800 |
| Totals | 34,200 | 35,800 | 34,600 | 44,700 | 47,900 | 46,700 | 60,500 |
| **Annual costs** | | | | | | | |
| Parlor and stalls | 2,208 | 2,208 | 2,208 | 2,596 | 2,596 | 2,596 | 2,860 |
| Equipment | 1,976 | 2,592 | 2,592 | 2,420 | 3,060 | 3,060 | 5,400 |
| Grain storage and feeding | 1,400 | 1,400 | 1,200 | 1,400 | 1,400 | 1,200 | 1,200 |
| Subtotals | 5,584 | 6,200 | 6,000 | 6,416 | 7,056 | 6,856 | 9,460 |
| **Labor** | | | | | | | |
| $2/hr. | 13,440 | 12,288 | 11,136 | 14,016 | 12,864 | 11,760 | 9,120 |
| $3/hr. | 20,160 | 18,432 | 16,704 | 21,024 | 19,296 | 17,640 | 13,680 |
| $4/hr. | 26,880 | 24,576 | 22,272 | 28,032 | 25,728 | 23,520 | 18,240 |
| **Total annual costs[b]** | | | | | | | |
| Labor/$2 hr. | 19,024 | 18,485 | 17,136 | 20,432 | 19,920 | 18,616 | 18,690 |
| Labor/$3 hr. | 25,744 | 24,632 | 22,704 | 27,440 | 26,352 | 24,496 | 23,250 |
| Labor/$4 hr. | 32,464 | 30,776 | 28,272 | 34,448 | 32,784 | 30,376 | 27,810 |

[a] C. R. Hoglund, Dairy Investments and Labor Economics. Reprinted from *J. of Dairy Sci.*, **56:**492 (1973). Published by the American Dairy Science Association.

[b] Costs do not include relief milkers when regular milkers are off duty. Investments do not include bulk tank.

389

or transported with a transfer system the operator usually uses two units (although the range may be one to six) and averages 15 to 20 cows per man per hour; if a pipeline and three units are used, 20 to 25; and if pipeline and four units, 25 to 35. The major advantages of stall barns are low initial cost, especially when existing stanchion facilities are used; cows can easily be given individual attention; and slower milkers do not normally hold up other cows. The primary disadvantages are labor inefficiency (although automation can relieve much of this) and the type of labor involved (bending, stooping, and, if bucket milkers are used, carrying milk).

In side-opening parlors cows can be handled individually, and a slow milker in one stall does not hold up the use of other stalls, thus allowing individual attention. Such parlors are easily adapted to use of automatic prep stalls, feeders, and detachers. The major disadvantage is diminished labor efficiency, when compared to comparably equipped herringbone parlors, because of the longer distance between udders. They are also somewhat more expensive to construct.

Herringbone parlors are conducive to higher labor efficiency than side-opening parlors because of the shorter distance between udders. They are well adapted to mechanized feeding and automatic detachers. Initial costs are lower than for side-opening parlors because a smaller structure and less stall equipment is needed. The primary disadvantage is that they are less conducive to individual attention because cows are milked in groups, and a slow milker in a group holds up the entire group.

In rotary parlors of the tandem or herringbone types the cows rotate on a circular platform and are milked by an operator in the middle. In the turnstyle type the operator works on the outside of the platform. The most prevalent rotary types in the United States are the eight-stall tandem and 13-stall herringbone. These types of parlor are conducive to high labor efficiency and adaptable to automatic prep stalls and detachers. The major disadvantage is the initial cost. Because of the high degree of mechanization involved, potential for breakdown and costly repairs is also high.

Polygon parlors are usually four-sided parlors with the operator working in the middle of the polygon. They may be constructed with more or less sides to the parlor and with four or more cows per side, depending on degree of automation. Because of the design, distances between udders are minimal and labor efficiency is high. They are well suited for automatic prep stalls, feeders, and detachers. The primary disadvantage is initial cost.

MECHANIZATION. Various degrees of mechanization can be installed in milking systems. Mechanization in stall barns is limited to pipelines for milk transport and group udder-washing devices in holding areas. There are many more possibilities in milking parlors. Some of these are listed below.

*1. Mechanized udder washing and stimulation.* Three types currently in use are: (a) group washing, with cool water in holding areas, (b) individual washing, with cool water followed by stimulation with warm water in a prep stall or stalls separate from the milking stall, and (c) washing and stimulation with warm water in the milking stall.

*2. Mechanized feeding.* Types commonly used include: (a) dial-controlled devices on which the operator turns a knob to the desired amount of grain to be fed, (b) automatic devices that deliver concentrate in proportion to milk produced, and (c) dribble feeders that deliver a steady amount of feed to the cow when the stall is occupied.

*3. Mechanized detaching units.* Types in general use, from most to least automated, include: (a) those that remove the machine from the teats and from under the cow when milking ceases, (b) those that remove individual teat cups when milk flow ceases but do not remove the machine from under the cow, and (c) those with variable vacuum level and pulsation ratio to protect the cow from injuries of overmilking and allow the operator a greater degree of flexibility of time to remove the machine.

*4. Mechanized crowd gates.* These typically involve a false back in a rectangular or circular holding area. They are usually driven by a power unit or weighting device capable of moving the gate and thus the cows toward the milking parlor. The control of the gate may be automatic through the use of a sensing device or timing mechanism or controlled by the operator.

*5. Power-operated gates.* Parlors are usually equipped with power-operated entry and exit gates that are controlled by the operator from the pit. Some are also equipped with power-operated stall gates.

*6. Automatic feed gates.* This device is designed to facilitate cow flow through herringbone systems. The system is composed of power-operated covers for the feeders, which are designed to cover the grain when all cows in a group have been milked and released. When the first cow reaches this feeder, she triggers a switch that uncovers the second feeder. This process continues until all feeders are uncovered and the group is in place and ready to be milked.

A summary of the effect on milking efficiency and relative costs of various degrees of automation in herringbone and side-opening parlors is presented in Tables 15-6, 15-7, 15-8 and 15-9. Prices are 1976 estimates; however, their relative position should be similiar.

**Milk Houses.** The purposes of milk houses and their equipment are to cool the milk and provide temporary storage for milk and areas for cleaning milking equipment. In some newer arrangements most of the milking equip-

**Table 15-6** Throughput for Herringbone Parlors with Various Mechanization[a]

| Mechanization[b] | Parlor size | | | |
| --- | --- | --- | --- | --- |
| | d-4 | d-6 | d-8 | d-10 |
| | (cows per hour) | | | |
| None[c] | 37[1] | 60[2] | 74[2] | 86[2] |
| Crowd gate | 42[1] | 65[2] | 81[2] | 94[2] |
| Crowd gate and stimulating sprays | 42[1] | 68[2] | 84[2] | 97[2] |
| Crowd gate and feed gates | 42[1] | 68[2] | 84[2] | 98[2] |
| Crowd gate, stimulating sprays and feed gates | 44[1] | 71[2] | 87[2] | 101[2] |
| Detaching units | 41[1] | 59[1] | 72[1] | 78[1] |
| Detaching units and crowd gate | 45[1] | 64[1] | 78[1] | 85[1] |
| Detaching units, crowd gate and stimulating sprays | 47[1] | 67[1] | 81[1] | 89[1] |
| Detaching units, crowd gate and feed gates | 47[1] | 67[1] | 82[1] | 89[1] |
| Detaching units, crowd gate, feed gates and stimulating sprays | 49[1] | 70[1] | 85[1] | 93[1] |

[a] W. G. Bickert, and D. V. Armstrong, reproduced by permission from February 25, 1977 issue of *Hoard's Dairyman*. Copyright © 1977 by W. D. Hoard and Sons Co., Fort Atkinson, Wisc. 53538. Superscript ([1], [2]) denotes number of people milking.
[b] Steady-state throughputs. Time spent with setup, cleanup and changing groups is not included. Good milking management is assumed.
[c] None denotes a parlor with base equipment including stalls, feeders, feed distribution and storage, pipeline milking system, ventilation, plumbing, hot water, electrical and other essentials.

ment is cleaned in place (CIP). The milk house should be located in a well-drained area on a clean side of the barn or parlor, with easy access for the milk truck. It should provide adequate space for the bulk tank, wash vats, or other washing equipment, and a lavatory. Inclusion of an adjoining utility room, toilet, and office is desirable. Construction details regarding size, materials, drainage, etc., are beyond the scope of this text, but thorough checking of applicable health department codes in regard to these details is essential before construction.

Often observed problems in milk houses are inadequate space and milk-storage capacity. Many milk houses have had to be enlarged after very few years of operation in order to accommodate larger bulk tanks. Allowing extra space, a large door or easily enlarged opening, and purchasing a somewhat larger-than-necessary bulk tank initially can often result in considerable future savings.

Bulk tank size needed is affected by the number of cows, level of production, and pickup schedule. Accurate knowledge or estimation of these factors is essential. A suggested procedure is to multiply the number of cows in the herd (include dry cows for a bit of added flexibility) times the highest expected average daily per-cow production of the herd times 2½ days. Most dairymen are on an alternate-day pickup schedule, so allowing room for another milking allows for moderate expansion, increased production level, or delayed pickup because of weather conditions. An example of this calculation for an 80-cow (Holstein herd with a high per-day average production during the flush period of 52 lb. (approximately 6 gal.) follows:

$$80 \text{ cows} \times 6 \text{ gal./day} \times 2.5 \text{ days} = 1{,}200\text{-gal. tank needed}$$

With conservative calculations, 90% in milk, and two-day storage ($72 \times 6 \times 2 = 864$ gal.), a 1,000-gal. tank could probably meet herd needs. However, a 20% increase in herd size or level of production or a 12-hour pickup delay would create problems. The difference in price (1975 estimates) between a 1,000-gal. and 1,200-gal. tank was approximately $900. Replacement of the 1,000-gal. tank with a 1,200-gal. tank in two to four years can easily cost five

**Table 15-7** Throughput for Side-Opening Parlors with Various Mechanization[a]

| Mechanization[b] | Parlor size | | |
| --- | --- | --- | --- |
| | d-2 | d-3 | d-4 |
| | (cows per hour) | | |
| None[c] | 33[1] | 49[1] | 56[2] |
| Crowd gate | 36[1] | 51[1] | 60[2] |
| Prep stalls | 42[1] | 51[1] | 63[2] |
| Crows gate and prep stalls | 46[1] | 55[1] | 66[2] |
| Detaching units | 44[1] | 54[1] | 59[1] |
| Detaching units and prep stalls | 48[1] | 60[1] | 65[1] |

[a] W. G. Bickert, and D. V. Armstrong, reproduced by permission from February 25, 1977 issue of *Hoard's Dairyman*. Copyright © 1977 by W. D. Hoard and Sons Co., Fort Atkinson, Wisc. 53538.

[b] Steady-state throughputs. Parlor setup and cleanup and changing groups not included. Good milking procedure is assumed.

[c] Base equipment as in footnote [a], Table 15-6. Superscript ([1], [2]) denotes people milking.

**Table 15-8** Investments in Herringbone and side-Opening Parlors, 1976 Prices

| | Construction[b] | + | Base equipment[c] | + | Mechanization Partial[d] | All[e] |
|---|---|---|---|---|---|---|
| | | | (in dollars) | | | |
| Herringbone parlors | | | | | | |
| Dbl.-4 | 12,500 | | 24,700 | | 15,500 | 17,100 |
| Dbl.-6 | 16,200 | | 29,700 | | 21,100 | 23,600 |
| Dbl.-8 | 19,800 | | 34,800 | | 26,700 | 29,800 |
| Dbl.-10 | 24,000 | | 39,900 | | 32,300 | 36,000 |
| Side-opening parlors | | | | | | |
| Dbl.-2 | 18,800 | | 20,000 | | 10,100 | 14,100 |
| Dbl.-3 | 24,300 | | 29,600 | | 10,700 | 14,700 |
| Dbl.-4 | 32,300 | | 37,100 | | 13,400 | 21,400 |

[a] W. G. Bickert, and D. V. Armstrong, reproduced by permission from February 25, 1977 issue of *Hoard's Dairyman*. Copyright © 1977 by W. D. Hoard and Sons Company, Fort Atkinson, Wisc. 53538.
[b] Building for parlor ($15 per square foot) and holding pen ($8 per square foot).
[c] See footnote c, Table 15-6, for base equipment.
[d] Partial mechanization includes crowd gate, power-operated gates and automatic detaching units.
[e] All mechanization includes crowd gate, power-operated gates and automatic detaching units plus stimulating sprays and feed gates for herringbone or prep stalls for side-opening.

times this amount. The difference in cost (1975 estimates) between a 1,000-gal. and a 1,500-gal. tank is about $1,800. In this example the authors would recommend purchase of the 1,500-gal. tank because: (1) 50% more storage capacity can be purchased for 25% added cost, (2) this added flexibility allows for increased herd size and level of production (both are rising in the United States and should continue to do so); (3) the larger tank has greater cooling capacity, which can allow more rapid milk cooling and more easily avoid temperature increases of stored milk during milking, and (4) it provides a reserve if needed in case of inclement weather that prevents regularly scheduled pickup.

**Feed-Storage and Feeding Equipment.** Type and capacity of feed-storage structures is dictated by the cropping and feeding program. Comparisons of types of silos in relation to nutrient preservation and costs and methods of determining needed capacity are discussed in Chapter 6.

Feeding-equipment needs are influenced by the type of housing arrangement, type of feed-storage structures, the feeding program, and the method of feeding (Chapter 8).

Mechanized feeding of silages is facilitated by tower silos equipped with mechanized unloaders or by horizontal silos that can be emptied with front-end loaders or other similar equipment. If silage is stored in tower silos, they usually discharge into adjacent bunk feeders equipped with various types of distributors (augers, shuttles, etc.), as shown in Figures 15-3, 15-13, 15-14, and 15-15. When the silage is stored in horizontal silos, it is usually delivered in some type of self-unloading wagon to a fence line or other accessible type of feeder as shown in Figures 15-4 and 15-12.

**Table 15-9**  Annual Milking Cost per Cow at 1976 Prices[a]

| Parlor | Mechanization[b] | No. of milkers | No. of milking cows | | | | |
|---|---|---|---|---|---|---|---|
| | | | 50 | 100 | 200 | 400 | 600 |
| | | | Dollars per cow (hours per milking[c]) | | | | |
| Herringbone, double-4 | None | 1 | 223 (1.4) | 145 (2.7) | 107 (5.4) | | |
| Side-opening, double-2 | None | 1 | 228 (1.5) | 152 (.30) | 114 (6.1) | | |
| Herringbone, double-6 | None | 2 | 274 (0.8) | 179 (1.7) | 131 (3.3) | 107 (6.7) | |
| Herringbone, double-6 | Partial | 1 | | 184 (1.6) | 112 (3.1) | 75 (6.3) | |
| Side-opening, double-3 | Partial | 1 | | 179 (1.9) | 113 (3.7) | 80 (7.4) | |
| Herringbone, double-8 | None | 2 | | 180 (1.4) | 124 (2.7) | 96 (5.4) | 86 (8.1) |
| Herringbone, double-8 | Partial | 1 | | 208 (1.3) | 120 (2.6) | 76 (5.1) | 61 (7.7) |
| Herringbone, double-8 | All | 1 | | 213 (1.2) | 121 (2.4) | 75 (4.7) | 60 (7.1) |
| Side-opening, double-4 | All | 1 | | 225 (1.5) | 131 (2.9) | 84 (5.9) | 68 (8.8) |
| Herringbone, double-10 | None | 2 | | 190 (1.2) | 124 (2.3) | 91 (4.7) | 80 (7.0) |
| Herringbone, double-10 | All | 1 | | 244 (1.1) | 135 (2.2) | 81 (4.3) | 63 (6.5) |

[a] W. G. Bickert, and D. V. Armstrong, reproduced by permission from February 25, 1977 issue of *Hoard's Dairyman*. Copyright © 1977 by W. D. Hoard and Sons Company, Fort Atkinson, Wisc. 53538. Annual cost was based upon: Depreciation of 12 years for parlor and holding pen and 7 years on all equipment, interest at 8% on unpaid balance, insurance at $4.65 per $1,000 of original investment, repairs at 2.5% for parlor and holding pen and 5% on equipment, labor charged at $10,000 per man.

[b] None denotes base equipment only as in footnote c, Table 15-6. Partial and all mechanization are defined in footnotes d and e, Table 15-8.

[c] Parlor setup and cleanup and changing groups not included.

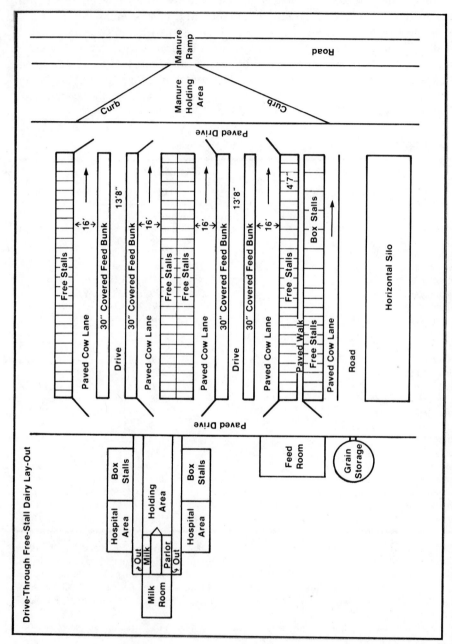

**Fig. 15-12.** Drive-through free-stall dairy layout [*V.P.I. & S.U. Ext. Pub.* 644 (1975)].

396

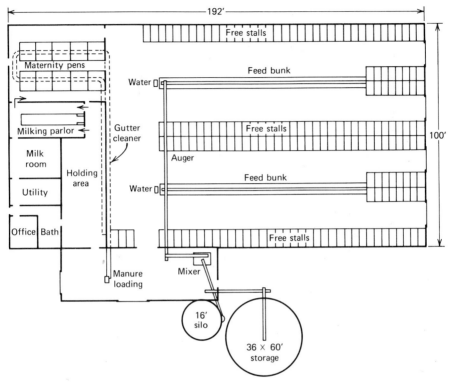

**Fig. 15-13.** An enclosed free-stall arrangement with mechanized complete feeding facilities (Reproduced by permission from Jan. 10, 1972 issue of Hoards Dairyman, Copyright 1972 by W. D. Hoard and Sons Co., Ft. Atkinson, Wisconsin 53538).

Self-feeding of silages can be achieved from horizontal silos, providing the depth of settled silage does not exceed 8 to 10 ft. (Figure 15-2). Hay may also be self-fed from certain types of storage structures.

The use of a complete ration system is facilitated by a drive-through or fence-line feeder arrangement, a mobile mixer wagon, or a mixing box mounted on a truck, and storage facilities for various ration components located near each other and easily accessible to the feeding areas. Figures 15-4, 15-5, and 15-12 represent layouts that are well adapted to this feeding system. Silage storage may be in either tower or horizontal silos. Concentrate storage may be in individual bins for various ingredients or in a single bin if a concentrate mixture is used. A mixer wagon or truck is filled with the specific quantity of each ration ingredient and mixes and delivers the ingredients to the feed bunks. Cows can be conveniently grouped according to level of production and different rations fed each group. Complete rations can also be used in arrangements with auger feeding systems as in Figures 15-3, 15-13, or

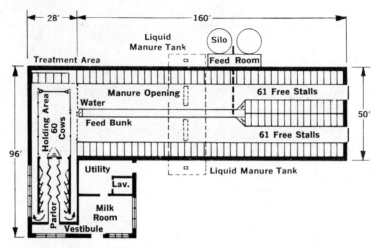

**Fig. 15-14.** Environmental free-stall system for 120 cows with automated feeding and liquid manure handling (Babson Bros. Dairy Research Service. 1976).

15-14. Concentrate storage facilities and a metered mixing box would need to be added to adapt these layouts for complete ration feeding.

The essentials of feed storage and equipment facilities for a complete ration feeding system are: (1) sufficient storage space for forages (usually silage), (2) a mechanized system of moving the forage from storage to the mixing box (mechanical unloaders in tower silos and bucket- or conveyor-type unloaders for horizontal silos), (3) sufficient storage space for other ration ingredients (single or multiple bins, depending on the type and variety of ingredients used), (4) a mixing box capable of accurately measuring and thoroughly mixing the ingredients, (5) a method of transporting and feeding the ration (conveyor system or portable mixing box), and (6) sufficient feed bunk space (12 in./cow is sufficient if fed on an ad-libitum basis).

**Manure Handling Systems.** The basic purpose of the manure handling system is to provide an efficient and economical method of collecting and disposing of animal wastes while avoiding air and water pollution and retaining a high proportion of the nutrient value of the wastes for recycling. The system should also be conducive to cow cleanliness and fly control. Other dairy wastes, including milking parlor waste and lot runoff, must be disposed of.

Dairy cows produce urine and feces at a rate of about 8% of their body weight daily. This amounts to about 19 tons/1,300-lb. cow annually. The nutrient content of these wastes as excreted varies considerably, but on the

average it contains about 10 lb. nitrogen, 4 lb. phosphorus, and
per ton. Based on prices of $.22/lb. for nitrogen and phosphorus
of potash, the annual value of 19 tons of such animal waste has
value of $77.52 if none of the nutrients are lost through evaporatior

Even with this potential value, the costs of disposing of the
often greater than the potential or actual fertilizing value. In many systems a
significant part of the nutrient value is lost in evaporation or runoff. However,
if part or all of the costs of the collecting and disposing of the wastes can be
recovered through recycling of its nutrients, this can contribute significantly
to dairy farm profitability. The major challenge facing dairymen, however, is
designing a waste-disposal system that is labor efficient, economical, meets
antipollution requirements, and provides a clean, fly-controlled environment
for cattle and labor. Increased herd size, more extensive use of year-round
confinement of dairy cattle, increased awareness of environmental problems,
and, in many cases, urbanization of areas adjoining dairy farms have
increased the number and severity of waste disposal problems.

There are two basic manure-handling systems, solid and liquid, and many
variations of each are available to dairymen. Parts of the system and
equipment needs vary considerably with the choice of a basic system and
variations of each, so these items will be covered as each system is discussed.

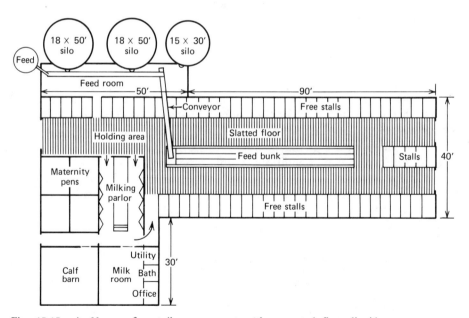

**Fig. 15-15.** A 66 cow free-stall arrangement with a slatted floor liquid manure system.
(Reproduced by permission from Jan. 10, 1972 issue of Hoards Dairyman. Copyright 1972 W. D.
Hoard and Sons Co., Ft. Atkinson, Wisconsin 53538).

**Table 15-10** Percentage Distribution of Manure-Handling Systems Used, 10 Areas of United States, 1974 Estimate[a]

| Area | Spread as: Solid | Spread as: Liquid | Lagoon system | Apply on or sell to other farms | Other methods | totals[b] |
|---|---|---|---|---|---|---|
| Northeast | 93 | 8 | 1 | 2 | 1 | 105 |
| Lake States | 94 | 7 | — | 2 | 1 | 104 |
| Corn Belt | 92 | 8 | 2 | 2 | 2 | 106 |
| Appalachian | 90 | 10 | 3 | 3 | 6 | 112 |
| Delta | 66 | 8 | 15 | 6 | 11 | 106 |
| Southeast | 68 | 11 | 16 | 7 | 10 | 112 |
| Northern Plains | 92 | 7 | 2 | 2 | 5 | 108 |
| Southern Plains | 72 | 10 | 8 | 9 | 10 | 109 |
| Mountain | 88 | 11 | 2 | 7 | 5 | 113 |
| Pacific | 73 | 28 | 9 | 14 | 6 | 130 |

[a] C. R. Hoglund, *Res. Rept.* 275, Mich. St. Univ., 1975.
[b] Adds to more than 100% because some dairymen use a acombination of manure systems.

There is no one best system for all loose housing arrangements, but rather there are several systems that can achieve the basic purposes of a manure-handling system. The major factors that should be evaluated in choosing a system are: (1) environmental control compliance, both current and anticipated, (2) location of the dairy in relation to residences, (3) size of the dairy, (4) availability of land, (5) labor efficiency, (6) cost, (7) climatic conditions, and (8) type of loose housing arrangement.

Solid systems of manure handling are easily the most popular method used in the U.S. (Table 15-10). These systems are likely to continue to be popular in the northern half of the country, because of climatic conditions, and in smaller herds. Liquid, or lagoon, systems are used on a significant number of dairy farms in the Pacific and southern areas, as they are well adapted to those climatic conditions and to larger herd sizes.*

SOLID MANURE SYSTEMS. In lounging arrangements the manure contains ample amounts of bedding and is allowed to accumulate for several months before loading and spreading. This is a low-cost system, but bedding requirements are high and can be costly. In free-stall arrangements daily scraping of paved areas, either by permanently installed (Figure 15-3) or mobile (usually tractor mounted) scrapers (Figure 15-12) is the usual practice. Hauling may be on a daily basis, or the manure may be stored and hauled periodically (Figure 15-16). The major advantage of solid systems is the low investment in equipment and facilities. The major disadvantage is the high

* C. R. Hoglund, *Res. Rept.* 275, Mich. St. Univ., 1975.

labor requirement associated with daily scraping of paved areas. This can be offset by installing mechanized scraping devices that operate similarily to gutter cleaners in a stanchion barn, but this may increase equipment costs.

Major advantages of daily hauling, in comparison to storage and periodic hauling of solid manure include minimum odor and fly problems, even distribution of labor needs for manure disposal (although this may be a disadvantage during planting, harvesting, or other busy seasons), and lower investment in manure-handling equipment. Disadvantages include lack of availability of land on which to spread the manure during the growing season (this becomes more critical as cropping systems intensify) and the need to spread manure daily regardless of adverse weather and soil conditions. This can result in cutting ruts during wet weather and runoff problems when the soil is snow-covered or frozen, which can result in loss of nutrients and pollution of nearby streams.

These disadvantages of daily spreading have created increased interest in storing solid manure for periodic hauling. In comparison to daily hauling the major advantage is the elimination of the need for daily spreading and, if sufficient storage capacity is available, manure can be spread when most convenient for labor, more beneficial for soil, and runoff problems are minimal. Limitations of solid manure storage or stacking include increased investment in the manure-handling system, fly and odor control problems,

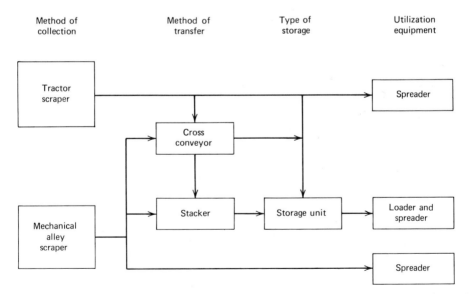

**Fig. 15-16.** Alternatives for solid manure handling systems for free-stall arrangements [*V.P.I. & S.U. Agr. Eng. Ext. Pub.* ME-10 (1977)].

which limit length of storage time during warm weather and cause concern for dairy inspectors and dairy regulatory personnel, and controlling runoff from the stack. These latter two limitations, in conjunction with the advantages of the system, have led to the development of a variety of semisolid manure-storing systems (Figure 15-17). These include systems (1) in which liquid is allowed to run off and solids only are retained for spreading, (2) in which solids and liquids are retained in one structure, and (3) in which liquids and solids are separated and the liquid is drained into a second structure for spreading. Thorough study of available systems and consultation with Agricultural Engineers and Dairy Regulatory Personnel should be made before storage systems are constructed.

LIQUID SYSTEMS. There are two general types of liquid systems, the conventional type, in which the manure is stored in a holding tank and periodically hauled with a liquid spreader or pumped out daily through an irrigation system (Figure 15-14), and the lagoon system, in which the manure is stored in a lagoon or pond. Generally these systems are characterized by higher investment costs than solid manure systems, and increased labor efficiency. This makes them better adapted to larger herd sizes, where the larger investment can be spread over a larger number of cows.

CONVENTIONAL LIQUID SYSTEMS. In conventional systems manure is usually stored in an underground tank, although more recently, above-ground

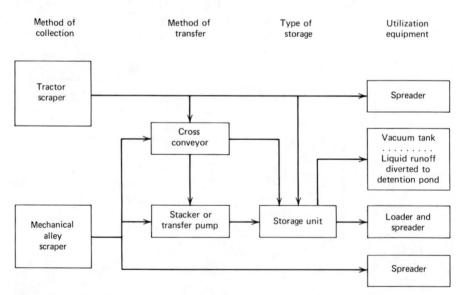

**Fig. 15-17.** Alternatives for semi-solid manure handling systems for free-stall arrangements [*V.P.I. & S.U. Agr. Eng. Ext. Pub.* ME-10 (1977)].

storage tanks are being used because of relative cost advantage. Manure may be scraped into the tank with mobile or permanently installed scrapers or, for above-ground tanks, scraped into a small collection area and pumped into the tank. The manure is dropped into the tank when slatted floors are used (Figure 15-15). Water must be added to the manure to increase the moisture content from 85 to 87% (the average water content of mixed urine and feces from dairy cattle) to 90 to 92% so that the material can be handled by pumps. Often, lot runoff and milking-parlor wastes are collected in the tank, thus providing a method of handling these wastes and avoiding pollution runoff problems, as well as adding the necessary water to the storage tank. The material is periodically agitated and pumped into a liquid manure spreader for application to the land.

The essential components of a conventional liquid manure system are: (1) a building arrangement that allows efficient collection of manure and other wastes; (2) an efficient method of adding water if needed; (3) a water-tight storage structure with sufficient capacity for the desired storage period (1,300- to 1,400-lb. dairy cows will produce about 1.8 to 2.0 cu. ft., or 13 to 15 gal., of urine and manure daily; capacity should be allowed for three to five months storage); (4) a means of agitating the material, as settling will occur; (5) a means of emptying the storage facility; and (6) a means of hauling and applying the waste to the land.*

Advantages of this system include: (1) elimination of the need for daily hauling, as manure may be hauled when convenient for labor and when weather and soil conditions are favorable; (2) water pollution problems caused by lot runoff, parlor wastes, or spreading on frozen ground can be avoided if the system is properly designed; (3) fly and odor problems are reduced during storage periods; and (4) labor savings may be realized, especially in slatted-floor systems.

A major disadvantage of the system is the initial cost of the storage tank, agitating and pumping equipment, and the liquid spreader. This cost varies considerably with herd size and the size and type of storage facility, but can amount to two to three times as much per cow as solid systems (Table 15-11). The per-cow investment and annual costs are considerably less in larger herds than in smaller herds (120- vs. 60-cow herds).**

This system is best suited for larger herds, although it may be necessary for some smaller herds because of runoff pollution problems. Storage capacity for a minimum of three months is needed and four to five months is desirable in cases where the ground is frozen or snow covered for long periods or where limited land or intensive cropping conditions prevent spreading during

* E. R. Collins, Liquid Manure Handling, *V.P.I. & S.U. Ext. Pub. No.* 598, 1974.
** C. R. Hoglund et al., *Res. Rept.* 91, Michigan St. Univ. 1969.

**Table 15-11** Investments and Annual Costs for Alternative Manure-Handling Systems for a 120-Cow Holstein Herd, Covered Free-Stall Housing, and Labor at $3 per Hour.[a]

| | | Manure system | | |
| | | Liquid manure | | |
| Item | Spreader, scraper, and loader | Tractor scraper | Mechanical scraper | Slatted floors |
|---|---|---|---|---|
| Investments | | | | |
| Conventional storage | $ 1,000 | | | |
| Liquid storage tanks | | | | |
| 3 mo. | | $14,400 | $14,400 | $14,400 |
| 6 mo. | | 28,800 | 28,800 | 28,800 |
| Tractor, scraper, loader | 6,000 | | | |
| Tractor and blade | | 4,400 | 4,400 | 4,400 |
| Manure spreader | 3,300 | 1,000 | 1,000 | 1,000 |
| Liquid spreader | | 1,700 | 1,700 | 1,700 |
| Pump and agitator | | 1,600 | 1,600 | 1,600 |
| Mechanical scraper | | | 2,500 | |
| Slatted floors | | | | 7,000 |
| Totals | | | | |
| 3 mo. storage | $10,300[b] | $23,100 | $25,000 | $30,100 |
| 6 mo. storage | 16,300 | 37,500 | 40,000 | 44,500 |
| Annual costs | | | | |
| Storage and equipment | 2,146 | 3,180 | 3,680 | 3,980 |
| Labor—handle and haul | 3,240 | 1,980 | 1,188 | 1,008 |
| Power—handle and spread | 1,500 | 1,200 | 870 | 870 |
| Total annual costs | | | | |
| 3 mo. storage | $ 6,886[b] | $ 6,360 | $ 5,738 | $ 5,858 |
| 6 mo. storage | 7,486 | 7,800 | 7,178 | 7,298 |

[a] C. R. Hoglund, Dairy Investments and Labor Economics. Reprinted from *J. of Dairy Sci.*, **56:**493 (1973). Published by the American Dairy Science Association.
[b] Minimum storage based on spreading manure every suitable day.

the growing season. Storage capacity needed can be calculated by multiplying the number of animals in the facility times the daily manure production times the desired day of storage plus needed dilution water (15–25%). An example of this calculation for a 100-cow Holstein herd with a desired storage period of 150 days follows:

100 cows × 15 gal. manure/day × 150 days + 25% = 281,250 gal.

This can be converted to cubic feet by dividing by 7.5. Thus,

281,250 ÷ 7.5 = 37,500 cu. ft. of storage capacity

Building design should be such that moving manure to the storage area is efficient. Lot runoff and parlor wastes should be collected in the storage tank. Spreading should be avoided at times when odors may reach other residences. In slatted-floor arrangements ample ventilation is required, or, better still, cattle should be removed from the barn when agitating the manure. Gases are released when the stored material is agitated and spread, and these can cause severe air pollution problems. They can also be lethal, as indicated in a report of 25 bred heifers dying in a slatted-floor barn when the manure in the storage facility under the barn was agitated on a humid day. The ventilating system of the barn was inadequate to remove the gases rapidly under existing weather conditions.* A per-cow comparison of investment and annual costs of various alternative manure-handling systems is presented in Tables 15-12 and 15-13.

LAGOON SYSTEMS. In these systems dairy wastes and runoff are moved from the facilities and collected in a lagoon or pond. In some operations lagoons handle only the runoff and parlor wastes and in others manure is also collected in the lagoon. These lagoons may be either aerobic (populated by organisms that must have oxygen to survive) or anaerobic (populated by organisms that do not need oxygen to survive), as shown in Figure 15-18. Some systems may include both types, as shown in Figure 15-19. Aerobic lagoons, when functioning properly, are relatively odor free. To maintain aerobic status, however, large amounts of oxygen are needed, especially when large amounts of organic material are collected. Large-surface-area, shallow lagoons facilitate natural transfer of oxygen from the air to the lagoon. Additional mechanically provided aeration may be needed when larger quantities of organic material are collected in the lagoon. Because of the large surface area needed, the need to provide additional aeration, and the relative inactivity of aerobic organisms in colder temperatures, aerobic lagoons are most suitable for disposal of milking parlor wastes and collecting lot runoff. Costs of pond construction and aerating costs make them prohibitive for handling all wastes from most operations.

The use of anaerobic lagoons can eliminate some of the problems of aerobic lagoons; however, some other problems may be present. As oxygen is not needed for proper function, the need for aeration equipment is eliminated and surface area can be reduced, allowing the construction of deeper but smaller-surface-area lagoons. Anaerobic bacteria act faster and generate heat, thus reducing length and severity of bacterial inactivity in colder environmental temperatures. Anaerobic reactions produce gases that can cause objectionable odors. This problem, however, is minimal when a proper biochemical balance is maintained. Anaerobic digestion of waste materials is less complete

* A. Mackiewicz, Manure Gases Kill 25, *Hoard's Dairyman,* October 10, 1974, p. 116.

**Table 15-12** Investment per Cow, Alternative Manure-Handling Systems[a]

| System | Cleaner | Spreader | Tractor scraper/loader | Mechanical scraper | Stacker | Agitator/pump | Storage | Underground pump transfer | Total investment per cow |
|---|---|---|---|---|---|---|---|---|---|
| Stanchion barn; barn cleaner-daily hauling (40 cows) | $2,500 | $1,600 | — | — | — | — | — | — | $103 |
| Stanchion barn; barn cleaner-stacker-storage-spreader (40 cows) | $2,500 | $1,600 | — | — | $3,000 | — | $ 3,000 | — | $253 |
| Open-lot free-stall; solid manure, tractor scraper-loader-spreader (80 cows) | — | $4,100 | $3,000[b] | — | — | — | — | — | $ 89 |
| Open-lot free-stall; solid manure, tractor scraper-stacker-storage-spreader (80 cows) | — | $4,100 | $1,800 | — | $3,000 | — | $ 9,450 | — | $229 |
| Covered free-stall; semisolid manure, tractor scraper-loader spreader (160 cows) | — | $4,100 | $3,600[b] | — | — | — | — | — | $ 48 |
| Covered free-stall; semisolid manure, tractor scraper-storage spreader (160 cows) | — | $4,100 | $2,400 | — | — | — | $17,800 | — | $152 |
| Covered free-stall; tractor scraper-silo liquid storage-spreader (80 cows) | — | $5,000 | $2,400 | — | — | $3,000 | $13,000 | $4,000 | $343 |

[a] E. H. Collins, *V.P.I. & S.U. Agr. Eng. Pub. ME*-101 (1977). Based on data from J. B. Johnson et al., *J. Dairy Sci.*, **56:**1354 (1973), and O. I. Berge et al., *Agr. Eng. Ext. Pub. AEN*-7, Univ. of Wisc. (1975).
[b] Includes $1,200 for pushoff ramp.

**Table 15-13**  Annual Costs Per Cow, Alternative Manure-Handling Systems[a]

| System | Equipment | Storage (or ramp) | Labor | Tractor use | Electricity | Total cost per cow |
|---|---|---|---|---|---|---|
| 1. Stanchion barn; barn cleaner—daily Hauling (40 cows) | $ 852 | — | $ 912 | $ 353 | $ 5 | $53.05 |
| 2. Stanchion barn; Barn cleaner-stacker-storage-spreader (40 cows) | $1420 | $ 420 | $ 822 | $ 331 | $ 20 | $75.33 |
| 3. Open-lot free-stall; solid manure, tractor scraper-loader-spreader (80 cows) | $1262 | $ 168 | $1824 | $ 997 | — | $53.14 |
| 4. Open-lot free-stall; solid manure, tractor scraper-stacker-storage-spreader (80 cows) | $1780 | $1323 | $1611 | $ 912 | $ 30 | $70.70 |
| 5. Covered free-stall; semi-solid manure, tractor-scraper-loader-spreader (160 cows) | $1382 | $ 168 | $3831 | $2678 | — | $50.37 |
| 6. Covered free-stall; semisolid manure, tractor scraper-storage-spreader (160 cows) | $1300 | $2492 | $3174 | $1940 | — | $55.66 |
| 7. Covered free-stall; tractor scraper-silo liquid storage-spreader (80 cows) | $2940 | $1820 | $1611 | $ 912 | $162 | $93.06 |

[a] E. H. Collins, *V.P.I. & S.U. Agr. Eng. Pub. ME*-101 (1977). Based on data from J. B. Johnson et al. *J. Dairy Sci.*, **56**: 1354 (1973), and O. I. Berg, et al., *Agr. Eng. Ext. AEN*-7, Univ. of Wisc. (1975).

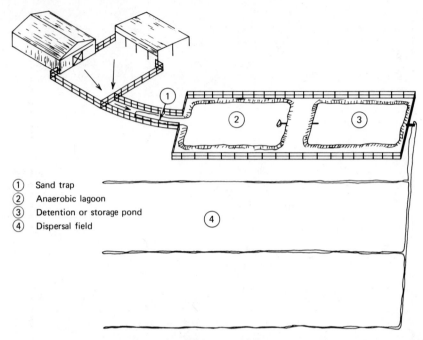

① Sand trap
② Anaerobic lagoon
③ Detention or storage pond
④ Dispersal field

**Fig. 15-18.** An anaerobic lagoon liquid manure system. (Conference Papers—1973 National Dairy Housing Conference. ASAE-SP-01-73).

than aerobic digestion. This results in more rapid accumulation of sludge and necessitates more frequent cleaning.

The use of anaerobic lagoons, usually in combination with detention ponds, is becoming more prevalent for larger herds in the warmer climates of the southern and southeastern United States. Many are used in combination with flush systems of manure removal from paved areas, while others scrape paved areas and flush the material to the lagoon. These systems typically are highly labor efficient and have lower per-cow investment costs than conventional liquid manure systems, especially in larger herds. However, the fertilizer value of the manure is largely lost and flies and odors can be a problem if not properly controlled. If flush removal systems are used, large quantities of water are needed and this may be a problem in some areas.

For satisfactory results, lagoons should be constructed a minimum of 300 ft. from other buildings; they should be fenced and identified as a sewage disposal area; drainage ditches should be constructed to keep surface water out; and an overflow pipe should be installed at least 6 in. below maximum water level.

OTHER SYSTEMS. Flush systems of waste removal can be used with lagoon systems or with irrigation systems. In the latter, storage capacity may be small and daily removal through the irrigation system may take place. These systems are moderate in cost; however, labor required to move irrigation pipes can be considerable and availability of land on which to pump the material can be a limiting factor. Waste from flush or from scrape-and-

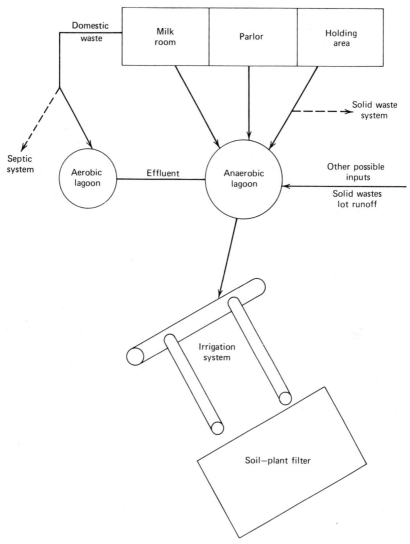

**Fig. 15-19.** A milking center waste-handling system with alternatives (aerobic and anaerobic) (Conference Papers—1973 National Dairy Housing Conference. ASAE-SP-01-73).

flush systems is also collected in aerated storage ponds on some farms. Liquid from these ponds is used for irrigation. Use of these types of pond increases flexibility of disposal by the use of irrigation equipment.

Water disposal is a growing concern for many dairymen. We have briefly discussed some of the alternatives. The selected reference list at the end of this chapter contains several publications that cover the subject in a more detailed fashion. The final decisions in regard to choice of system should be made after evaluation of the alternatives in view of the purposes of the system and the resources of the individual farm. Manure-handling technology is changing rapidly, so dairymen need to continue to be alert for the development of improved systems.

**Lounging or Resting Area.** In conventional loose housing arrangements, 60 to 80 sq. ft./cow should be allowed. In free-stall arrangements, the traditional recommendation has been to allow one free stall per cow plus an additional 5 to 10% for flexibility. Recent information indicates that under conditions of ad-libitum feeding of complete feeds, free-stall needs are .71 free stalls per cow. At this rate, occupancy time of free stalls was not different than when one stall per cow was provided.* Similar observations have been reported from a Wisconsin study.** Cow density in relation to number of stalls needed should be further investigated, as building costs can be significantly decreased if density can safely be increased.

Free-stall size, partition design, stall surface, and bedding material have been the subject of considerable research during the last 10 to 15 years. In spite of this there is no general agreement on these details. Indiana workers reported poor cow acceptance and high incidence of leg and udder problems on uncovered concrete free stalls.† In a Wisconsin study** cows showed a decided preference for clay-filled stalls, followed in order of preference by concrete-covered with rubber mats, indoor-outdoor carpeting, and chopped bedding. These reports also indicate an investment cost advantage for clay-filled stalls. The major problem with clay-filled stalls appears to be potholing, or digging out of the clay at the rear of the stalls and beneath the partitions. This can largely be eliminated by the use of a 2 × 6 bedding board at the bottom of the partition.

The combined use of smaller stalls (6 ft. 6 in. to 6 ft. 10 in. in length including the curb), a partition consisting of a 2-in. × 6-in. or 2-in. × 8-in. bedding board at the base, and a 2-in. by 6-in. or 2-in. by 8-in. partition board at the top or 3 ft. 4 in. above the curb has given satisfactory results (Figure

* T. H. Friend, unpublished data, V.P.I. & S.U., 1976.
** H. S. Larson et al., Chore Reduction *Res. Progress Rept. Msh.* 5105, Univ. of Wisc., Madison, 1976.
† *Hoard's Dairyman*, June 25, 1974.

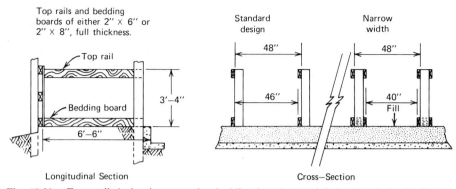

**Fig. 15-20.** Free-stall design incorporating bedding boards to minimize "potholing" of earth-filled stall platforms (W. H. Collins, et al. Paper presented Sou. A.D.S.A. Meeting, 1976).

15-20). Stalls of this type take less upkeep, reduce bedding waste, and reduce injuries.* Bedding boards can be installed in many existing free stalls, but care must be taken to ensure that there is a minimum of 12 in. of space between the top of the bedding board and the bottom of the first partition board or rail. Narrower openings can contribute to a high incidence of leg injuries.

Because of the lack of ample, conclusive data regarding cow density and surfaces for free stalls, the authors recommend visiting a variety of layouts, and considering personal observations strongly, in addition to recommendations and cost factors, before reaching a decision in this regard.

**Special Areas.** Facilities for calving, cow treatment, breeding, routine cow care such as diagnostic work, and clipping must be a part of loose housing arrangements. All too often these areas are forgotten in the planning of a new facility. They should be readily accessible from other parts of the layout, and convenient to use. Often old, stanchion-type barns that have been replaced by new facilities for the milking herd serve this purpose adequately. When designing such facilities into a new layout the following areas should be included.

1.  A temporary treatment area where cows can be confined for short periods for artificial insemination, diagnostic work, treatments, etc. This area should be accessible from the exit lane(s) of the milking parlor so that cows can be retained as they leave the parlor. It should be equipped with at least as many stanchions or similar restraint stalls as the

* W. H. Collins, *Hoard's Dairyman*, 120, June 10, 1975.

capacity of one row of milking-parlor stalls, or one stanchion per 15 to 20 cows in the herd.

2. A treatment area where cows with long-term problems such as downer syndrome, feet and leg injuries, etc., can be separated from the rest of the herd for longer periods. This facilitates needed treatment and protects them from abuse by other animals. This area should be equipped with a head restraint and feeding and watering facilities. It can consist of a large box stall, preferably a rectangular one so that an individual area can be made if needed by the use of a temporary gate. Capacity for two to four cows per 100 milking animals is usually sufficient. To facilitate milking, it should be near the milking parlor. It should also have a wide access to the outside to facilitate removal of dead animals.

3. Maternity pens for use in inclement weather should also be included. Usually four to five of these pens per 100 cows is sufficient. These stalls should be a minimum of 10 × 12 ft. for the larger breeds and 10 × 10 ft. for smaller breeds, and equipped with feeding and watering facilities. Proximity to the milking parlor facilitates milking. Dirt floors provide better traction for cows but are difficult to disinfect, so roughened concrete with ample bedding is recommended. The use of maternity stalls as long-term treatment stalls or vice-versa is not advisable because of the possibility of spreading diseases to fresh cows or calves.

4. A loading chute should be provided for safe (for cattle and labor), efficient movement of cattle to and from the facilities. The chute should be attached to a box stall or small corral or holding pen to facilitate separation of cattle to be moved and loaded.

The precise arrangement and location of these special areas is largely a matter of personal preference. Designing use-convenience into these facilities is very important. If they are located and arranged so that moving cows in and out is convenient, cows can easily be moved to and from the milking parlor, they are easy to clean, and can be easily observed, they will be used more and more effectively. Arrangements such as shown in Figs. 15-3, 15-4, 15-12 and 15-21 should work very well.

**Replacement Rearing Facilities.** Facilities for the replacement herd may be located in, attached to, or separate from the cow facility. Many dairymen include facilities for calves in or near the cow facility and utilize other facilities for postweaning-age calves and heifers. Details of replacement housing were discussed in Chapter 14.

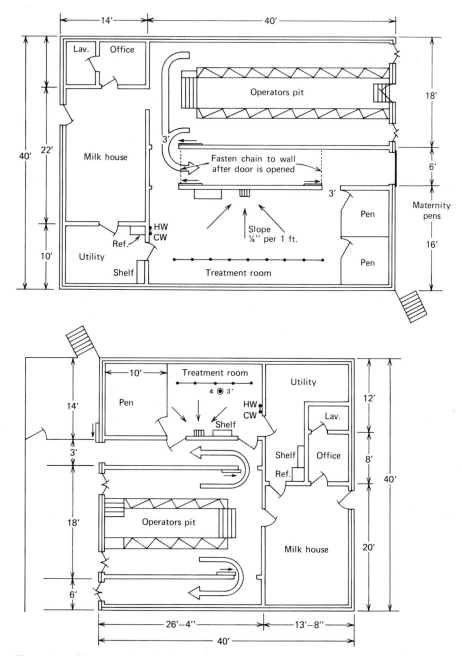

**Fig. 15-21.** Treatment areas adjacent to milking area (Conference Papers—1973 National Dairy Housing Conference. ASAE-SP-01-73).

**Table 15-14**  Investments Needed for Building[a]

| | Size | Price |
|---|---|---|
| Stanchion barn—warm | 40 cows | $25,815 |
| Building | (36 × 134) | 18,090 |
| Plumbing, wiring fand | | 2,490 |
| Metal stalls | | 2,200 |
| Gutter cleaner | | 3,035 |
| Stanchion barn—warm | 60 cows | $31,545 |
| Building | (36 × 160) | 21,600 |
| Plumbing, wiring fans | | 3,360 |
| Metal stalls | | 3,300 |
| Gutter cleaner | | 3,285 |
| Free-stall barn—warm | 60 cows | $20,890 |
| Building | (44 × 120) | 14,520 |
| Plumbing, wiring fans | | 2,200 |
| Concrete alleys | | 2,640 |
| Metal stalls | | 1,530 |
| Free-stall barn—warm | 80 cows | $27,620 |
| Building | (44 × 160) | 19,360 |
| Plumbing, wiring fans | | 2,700 |
| Concrete alleys | | 3,520 |
| Metal stalls | | 2,040 |
| Free-stall barn—cold | 60 cows | $17,970 |
| Building | (44 × 120) | 13,200 |
| Plumbing, wiring | | 600 |
| Concrete alleys | | 2,640 |
| Metal stalls | | 1,530 |
| Free-stall—cold | 80 cows | $23,890 |
| Building | (44 × 160) | 17,600 |
| Plumbing, wiring | | 730 |
| Concrete alleys | | 3,520 |
| Metal stalls | | 2,040 |
| Milk house | (22 × 20) | $ 8,800 |
| Milking parlor | | |
| Herringbone | | |
| Double-3 | (36 × 19) | $13,680 |
| Double-4 | (40 × 19) | 15,200 |
| Double-6 | (48 × 19) | 18,240 |
| Side opening | | |
| Double-2 | (39 × 19) | 14,820 |
| Double-3 | (47 × 19) | 17,860 |

[a] M. Boehljie, and R. Herr, What Will a New Modern Dairy Barn Cost You? reproduced by permission from May 25, 1976, issue of *Hoard's Dairyman*. Copyright © 1976 by W. D. Hoard and Sons Company, Fort Atkinson, Wisc.

[b] Based on a 20-year life, salvage value of the buildings would be estimated at 10% of initial cost. Lifetime repairs would run 5%.

414

**Planning** Thorough planning of the total system and its component parts, accurate investment-cost estimates, and financial analysis are essential when renovation or new construction is anticipated. Consultation with engineers, extension personnel, regulatory people, contractors, bankers, and other dairymen, and thorough study of printed materials, are highly recommended before proceeding. Errors on paper are much easier to correct than errors in concrete and steel. Ensure, as far as possible, that the system will meet the basic purposes of a good system and meet personal goals, resources, and preferences of the owner or managers. Costs are extremely variable between and within areas for buildings and equipment, so accurate cost estimates are often difficult to obtain for new construction and even more difficult for remodeling. Initial and annual maintainance costs are greatly affected by labor efficiency; thus labor efficiency of the system can be very important. Of total annual operating costs on an average dairy farm, labor and annual building and equipment costs rank second and third. The total of both of these is much more important than either alone. Relatively high labor costs, if accompanied by relatively low annual building and equipment costs, can be conducive to profitable dairying, especially in smaller, family-type operations. Relatively high annual building and equipment costs, if accompanied by relatively low labor costs, can be conducive to profitable dairying, especially in larger commercial operations. Problems arise, however, when large investments in buildings and equipment do not result in high labor efficiency and low labor

**Table 15-15**  Costs of Metal Milking Stalls[a]

|  | Price[b] | Salvage value[b] | Lifetime repair costs[c] |
|---|---|---|---|
| Side-opening stalls |  |  |  |
| Double-2 | $2,060 | $412 | $103 |
| Double-3 | 2,900 | 580 | 145 |
| Herringbone |  |  |  |
| Double-3 | $1,935 | $387 | $ 97 |
| Double-4 | 2,345 | 469 | 117 |
| Double-6 | 3,020 | 604 | 151 |

[a] M. Boehljie, and R. Herr, What Will a New Modern Dairy Barn Cost You? Reproduced by permission from May 25, 1976, issue of *Hoard's Dairyman*. Copyright © 1976 by W. D. Hoard and Sons Company, Fort Atkinson, Wisc. 53538.
[b] Includes cost of stalls, drains, grates, other essential items, and installation. Does not include cost of concrete work.
[c] Salvage value would be computed as 20% of price; repair costs are estimated as 5% of price, based on 10-year life.

**Table 15-16**  Investment in Milking Equipment[a]

| | Price[b] | Salvage value[c] | Yearly repair costs[d] |
|---|---|---|---|
| 2-unit milker for a double-2 side-opening | $4,200 | $ 840 | $300 |
| 3-unit milker for a double-3 herringbone | 4,970 | 994 | 360 |
| 4-unit milker for a double-4 herringbone | 5,450 | $1,090 | 480 |
| 6-unit milker for a double-6 herringbone | 6,500 | 1,300 | 620 |
| 3-unit milker, Permanent pipeline in 40-cow stanchion barn | 6,035 | 1,207 | 360 |
| 2-unit bucket carry | 1,500 | 300 | 60 |
| 2-unit milker, portable pipeline for 40-cow stanchion barn | 2,470 | 494 | 100[e] |

[a] M. Boehljie, and R. Herr, What Will a New Modern Dairy Barn
Cost You? Reproduced by permission from May 25, 1976, issue of
*Hoard's Dairyman*. Copyright © 1976 by W. D. Hoard and Sons
Co., Fort Atkinson, Wisc. 53538.
[b] Installed price for milker units, pipelines, pumps, motors, wash
vats, and other essential equipment; does not include the bulk milk
cooler and the metal parlor stalls.
[c] Salvage value is computed as 20%, based on 10-year life.
[d] Includes cost of inflations.
[e] Includes cost of inflations and replacement milk lines.

costs. This, again, emphasizes the importance of thorough planning and accurate budgeting before large investments are made. The combination of building and equipment annual costs plus labor costs should not exceed 30 to 35% of the total costs of producing milk.

Tables 15-14 through 15-19 include a summary of investments in buildings and equipment needed for milk production in the midwest in 1975.* The information in these tables may be useful for comparative purposes or for classroom or laboratory exercises. They should not be used for estimating costs for actual farms because of wide variation in prices in various areas. In

* M. Boehljie and R. Herr, What a New Dairy Barn Would Cost You, *Hoard's Dairyman*, p. 639, May 25, 1976.

**Table 15-17** Free-Stall Parlor System Costs[a]

| | Double-2 side opening 60 cow | | Double-3 herringbone 60-cow | | Double-3 side opening 80-cow | | Double-4 herringbone 80-cow | |
|---|---|---|---|---|---|---|---|---|
| | Warm | Cold | Warm | Cold | Warm | Cold | Warm | Cold |
| Free-stall barn | $20,890 | $17,970 | $20,890 | $17,970 | $27,620 | $23,890 | $27,620 | $23,890 |
| Milkroom-parlor | 14,820 | 14,820 | 13,680 | 13,680 | 17,860 | 17,860 | 15,200 | 15,200 |
| Milking equipment | 4,200 | 4,200 | 4,970 | 4,970 | 4,970 | 4,970 | 5,450 | 5,450 |
| Milking stalls | 2,060 | 2,060 | 1,935 | 1,935 | 2,890 | 2,890 | 2,345 | 2,345 |
| Mechanical feeder | 3,316 | 3,316 | 3,316 | 3,316 | 4,163 | 4,163 | 4,163 | 4,163 |
| Stock waterers (2) | 300 | 300 | 300 | 300 | 300 | 300 | 300 | 300 |
| Bulk milk cooler | 5,750 | 5,750 | 5,750 | 5,750 | 7,200 | 7,200 | 7,200 | 7,200 |
| Total | $51,336 | $48,416 | $50,841 | $47,921 | $65,003 | $61,273 | $62,278 | $58,548 |

[a] M. Boehlje, and R. Herr, What Will a New Modern Dairy Barn Cost You? Reproduced by permission from May 25, 1976, issue of *Hoard's Dairyman*. Copyright © 1976 by W. D. Hoard and Sons Co., Fort Atkinson, Wisc. 53538.

**Table 15-18** Investments for Bulk Milk Coolers[a]

| Size, gal. | Price[b] | Salvage value[c] | Repair costs during lifetime[c] |
|---|---|---|---|
| 400 | $3,870 | $ 400 | $200 |
| 500 | 4,450 | 500 | 200 |
| 600 | 4,900 | 600 | 200 |
| 800 | 5,750 | 800 | 300 |
| 1,000 | 7,200 | 1,000 | 350 |
| 1,500 | 8,970 | 1,500 | 425 |

[a] M. Borhljie, and R. Herr, What Will a New Modern Dairy Barn Cost You? Reproduced by permission from May 25, 1976, issue of *Hoard's Dairyman*. Copyright © 1976 by W. D. Hoard and Sons Co., Fort Atkinson, Wisc. 53538.
[b] $500 to $650 would be added to the price for an automatic washer unit.
[c] Salvage value and repair costs are for 20-year life.

these summaries, feed-storage areas are not included. Stanchion barns are equipped with pipelines and gutter cleaners, but no automated feeding equipment. Free-stall systems include a silage bunk and mechanical feeder, and stalls are filled with dirt or crushed stone. Milking systems contained a holding area but no automated equipment.

**Table 15-19** Stanchion Barn System Costs[a]

| | 40-Cow pipeline warm | 40-Cow transfer unit warm | 60-Cow pipeline warm |
|---|---|---|---|
| Stanchion barn | $25,815 | $25,815 | $31,545 |
| Milk house | 8,800 | 8,800 | 8,800 |
| Milking equipment | 6,035 | 2,470 | 7,200 |
| Bulk milk cooler | 4,450 | 4,450 | 5,750 |
| Total | $45,100 | $41,535 | $53,295 |

[a] M. Boehljie, and R. Herr, What Will a New Modern Dairy Barn Cost You? Reproduced by permission from May 25, 1976, issue of *Hoard's Dairyman*. Copyright © 1976 by W. D. Hoard and Sons Co., Fort Atkinson, Wisc. 53538.

## REFERENCES FOR FURTHER STUDY

Johnson, J. B. et al., Animal Waste Management: Symposium, *J. Dairy Sci.,* **56**:1354 (1973).

Arave, C. W., and Albright, J. L., How Much Space Do Your Cows Need? *Hoard's Dairyman,* **120**:1116 (1975).

Babson Bros. Co. The Way Your Cows Will be Milked Tomorrow, Babson Bros. Dairy Res. Serv., Oak Brook, Ill. 8*th* edition, 1976.

Bates, D. W., Dairy Waste Management Systems, *J. Dairy Sci.,* **56**:495 (1973).

Bickert, W. G., and Armstrong, D. V., How Herringbone and Side Opening Parlors Compare, *Hoard's Dairyman,* 208, February 25, 1977.

Bickert, W. G. et al., Milking Systems for Large Herds, *J. Dairy Sci.,* **57**:369 (1974).

Boehljie, M., and Herr, R., What a New Dairy Barn Would Cost You, *Hoard's Dairyman,* May 25, 1976.

Collins, W. H., *Dairy Systems Development: Engineering Enterprise Management,* Multilith, V.P.I. & S.U. Agric. Engineering Dept. 1975.

Collins, E. R., *Solid and Semi Solid Manure Handling Systems, V.P.I. & S.U. Ext. Pub. ME*-101, 1977.

Coppock, C. E., Feeding Methods and Grouping Systems, paper presented at the 1976, ADSA meeting, Raleigh, N.C.

*Dairy Housing and Equipment Handbook,* Midwest Plan Service, Iowa State University, Ames, Iowa, 1971 and 1976.

Hoglund, C. R., Dairy Facilities Investments and Labor Economics, *J. Dairy Sci.,* **56**:488 (1973).

Hoglund, C. R., *Mich. St. Res. Rept.* 275 (1975).

*Livestock Waste Facilities Handbook,* Midwest Plan Service, MWPS-18, Iowa State Univ. 1975.

*Proceedings of the National Dairy Housing Conference,* Am. Soc. of Agr. Engineers, 1973.

*A Successful Farming Book—Livestock,* 9*th* edition, Meredith Corp., 1976.

# 16

# Herdsmanship and Labor Management

Herdsmanship, or cowmanship, may be defined as **interest in** and **concern for** cattle; **awareness** or **perception**, that is *seeing, hearing,* and *sensing* when an animal is normal or abnormal; and **action**, or taking *prompt, meaningful action* to correct abnormal conditions of animals. Cowmanship, or the lack of it, can and does determine to a large extent the profitability or lack of profitability of a dairy herd of any size. Having one or more cowmen, people with livestock sense, working with the cattle and performing the daily tasks of caring for the animals is essential to success in dairying. The dairy cow is a marvelous creature, a highly specialized but very complex biological system capable of growing, reproducing, and efficiently synthesizing and yielding large quantities of milk. She is also susceptible to many diseases and abnormal conditions that can prevent normal growth, reproduction, synthesis, and let-down of milk and that can cause her premature death. People largely determine whether she will be healthy and productive or unhealthy and unproductive for, in spite of all her marvelous qualities, she is still a mute animal and is dependent on people to perceive whether she is normal and abnormal, and what the abnormality may be; whether she is well or ill, and what illness is present; whether she is contented or discontented, and what is causing the discontentment; whether she is in heat; when she is finished milking; etc. Some people can readily perceive or even anticipate abnormal or unusual conditions, and are interested enough to take prompt, meaningful action. These are the cowmen. Others cannot perceive abnormal or unusual conditions until the animal is so abnormal or the condition so serious that performance is likely to be permanently impaired or even until death is imminent.

## SYMPTOMS

The authors have visited many herds, large and small, over the years and have developed a list of symptoms that are typically associated with herds that are handled by people who lack cowmanship. Some of these symptoms are listed below.

*1. Calf and heifer problems.* Calf mortality rates are high and a large percentage of those that survive are unthrifty and stunted in appearance. We refer to the latter as "big head" disease, as the head seems to continue growing even if the heifer is stunted. Average age at first calving is high, largely because the heifers that do survive aren't large enough to breed or calve at a normal age.

*2. Low genetic (low production) and high nongenetic culling rate.* The latter is caused by a high number of cows culled for disease, injury, and reproductive problems.

*3. Low reproductive efficiency.* Average herd calving intervals of 15 months or even as high as 20 or more months are common. Often, services per conception is average or lower than average. This indicates lack of heat detection. Often these managers, after trying artificial insemination for a few years, return to using a bull. This, in turn, decreases genetic progress for improved production.

*4. Periodic high incidence of various diseases such as mastitis, foot rot, metritis, etc.* Incidence of cows with unbalanced udders, three-quartered cows, lame cows, and cows with vaginal discharges is often high.

*5. Periodic milk-quality problems.* This may be caused by abnormal milk from undetected cases of mastitis, unclean cows, dirty or malfunctioning milking equipment, or other careless milking procedures. On many of these operations, milking is considered the lowest-priority job on the farm. It is often performed by the least skilled, least interested person(s) on the farm.

*6. Wide fluctuations in daily production.* Variations in total daily production often exceed 10%. Irregular milking hours and inconsistent feeding programs are two of the major causes.

*7. Nervous cows.* Often the mere sight of people terrifies the cattle. As one walks through lots containing cattle, they scatter in all directions. They are often nervous and constantly move their feet during milking. They have learned to associate the presence of a person, or the milking process, with pain or discomfort. Some years ago, one of the authors visited a stanchion-housed herd that had scheduled a dispersal sale in a few weeks. On approaching the barn, he looked through a window and observed all but two of the 50 cows lying down. As he opened the barn door and entered, every cow in the barn was on her feet immediately with head turned and eyes fixed on him. As he walked down the aisle the cows pranced nervously from side to side and, when approached, crouched in their stalls. Small wonder the herd average was less than 7,000 lb. of milk.

*8. Low production and associated low profitability.* The owner or other herd personnel are usually discouraged with the dairy business because it isn't profitable. They blame the government, weather, taxes, etc., and threaten to sell out if things don't improve. They can't identify individual cows (except the ones that kick or jump fences), can't tell you how much they produced, who their sire is or the bull they are bred to, or even if they are bred.

Characteristics of herds that indicate the presence of cowmanship are the opposite of the previously listed symptoms. The cattle are healthy and contented. They produce large quantities of a high-quality product and reproduce regularly. They are profitable operations and those associated with these herds are optimistic about dairying and enthusiastic about their profession.

## CHARACTERISTICS OF GOOD COWMEN

A good cowman realizes his importance to his cattle as well as their importance to him. He is aware that his success is dependent on their high performance and their high performance is dependent on his ability to keep them alive, healthy, well fed, and contented. Most have highly visible characteristics that identify cowmanship ability. Some of these follow.

1. He knows his cows. He can identify individuals by sight, he knows their performance credentials, their sire and dam, and the bull they are bred to. He enjoys visitors and is proud to show them and tell them about his cattle.

2. He typically visits the barn or lots at night, usually just before retiring, for he knows that one can hear, see, and sense unusual conditions more easily at a time when cows are not being fed or milked, lots are not being scraped, etc. He knows that a high percentage of cows exhibit heat symptoms at night, so he checks cows that are expected in heat or those nearing calving. He listens for coughing or abnormal breathing in the calf barn.

3. He moves quietly and easily when near cattle. He avoids sudden rapid movements and making unusual loud noises. He and the animals appear at ease with each other. As he moves among cattle, he is constantly checking droppings (for symptoms of digestive upsets), cow's feet and legs and cow movements (for symptoms of lameness), udders (for unbalanced quarters), tailheads and vulvas (for symptoms of heat), and overall condition of the cow. Is she bright and alert or dull and listless?

4. He takes prompt, meaningful action when unusual conditions exist, rather than waiting for a convenient time to do so. He knows what to do and he does it. Doing it may be something as routine as assisting a newborn calf to breathe, separating a cow in heat from the rest of the herd, or closing a door to prevent a draft in the calf barn, but he realizes it is useless to assist a newborn calf to breathe after the calf is dead, meaningless to separate a cow in heat from the herd after she has caused injury to two other cows, and much easier to close the door than to treat calves for pneumonia.

The first two characteristics indicate interest, the next one awareness or perception, and the last one ability to take proper corrective action. Combined, they indicate cowmanship ability, an essential ingredient in maintaining a profitable herd of any size.

## LABOR MANAGEMENT*

Labor management can be defined as the planning, organizing, and directing of the operative functions of personnel. The purpose of the labor management program on a dairy farm, or in any business, is to get the job(s) done right and on time. When most dairy farms were one-man or small, family operations, and for those that still are, labor management problems are usually minimal as management and labor responsibilities are performed by the same person or within the same family. As herd size increases, more time is needed for management per se, and less time is available to milk, feed, haul manure, and perform other routine tasks. Increasing amounts of the daily work are performed by hired personnel, and the importance of the labor management program increases. Continued success of operation depends not only on the dairyman's ability to manage land, capital, cows, etc., but also on his ability to get the job done well through others. Many dairymen who, through hard work and good cowmanship, developed successful family-size dairy operations, failed when they expanded to larger operations. For many, a major reason for failure was inability to get the job done as well through others as they could do themselves.

Labor management increases in importance as herd size, level of production, and degree of mechanization increases. More cows means increasing number of hired personnel, and this creates more opportunity for disagreement or friction between employees. Higher production increases susceptibility of cows to a variety of problems and requires more precise feeding, milking, etc., to avoid a high incidence of problems. Increased mechanization increases cows per man and decreases individual cow observation and care. To be a successful manager of any dairy operation that is large enough to require labor in addition to his or her own, a dairy owner must learn how to get the job done right and on time through others—to manage people as well as cows, land, capital, etc. This involves learning to plan manpower needs, hiring the needed manpower, training them to do the job(s) correctly and

* Many of the ideas and principles discussed in this section are based on presentations made at the Large Herd Management Symposium, University of Florida, Gainesville on January 19–21, 1976, by J. A. Speicher, of Michigan State University, R. C. Wells of North Carolina State University, and R. D. Appleman of the University of Minnesota.

efficiently, motivating them, providing relatively safe and healthy working conditions, and maintaining good employer-employee relations. Many of the principles involved in these functions of personnel management apply to family as well as nonfamily labor, and many dairymen would do well to heed them in regard to their children who help on the dairy farm.

## Manpower Planning

The objective or purpose of the manpower planning program is to determine the number and kind of personnel that need to be hired. This varies tremendously, depending primarily on the size of the operation and the jobs the owner or manager expects to perform himself. For example, on large operations the need may be for many highly specialized laborers such as milkers, feeders, etc., while on smaller units the need may be one person who is a generalist and is expected to do many different jobs. Some usual problems are the following.

1. Hiring people who are capable of accepting considerable responsibility and expect to be involved in the thinking, planning, and meaningful work on the dairy but are assigned only to the routine, menial, repetitious jobs that require little judgment or initiative. This situation soon results in boredom or dissatisfaction of the employee, who leaves at the first opportunity for a more challenging position.

2. Hiring people who feel comfortable, secure, and happy doing routine, menial, repetitious, tasks and then assigning them responsibilities that require initiative and decision making. This situation soon becomes frustrating for the employee who cannot or does not want to cope with mental challenges. He will also leave at the first opportunity.

3. Hiring too many or too few people. Hired labor is a cash expense like feed, fuel, or fertilizer; however, if excessive amounts are purchased they can't be stored for future use, and it is difficult to purchase labor in small units as needed. It must be used at a scheduled time or it is lost. This loss can significantly affect profitability. Overuse because of insufficient amount of labor often results in poor quality of performance and dissatisfied employees who seek other jobs.

There are several basic principles that may be helpful in planning manpower needs. Some of these are listed below.

1. Identify the total work requirement of the operation and decide which

jobs management will perform and which ones will be delegated to others. Following are some guidelines which may be helpful in deciding this. Jobs that should not be delegated include those (a) that are of a confidential nature, (b) that others cannot do satisfactorily, and (c) that take less time to do yourself than to delegate to others, unless they are repetitive jobs. Jobs that should usually be delegated include those: (a) that others can do as well as you, (b) that you do poorly because of lack of time to do them properly, (c) that are repetitive or routine, (d) that are low-risk jobs that may improve employee morale or development if they are delegated, and (e) that significantly interfere with the time available for managing. Occasionally, managers need to do certain normally delegated jobs to boost morale—to demonstrate a better way to do them, or that the manager can do them, or to get the feel of a job and its problems so that the manager is in a better position to evaluate it.

2. Divide the jobs to be performed by others into work units such as milking, calf care, etc., and define the number of persons needed to get the total job done right and on time.

3. Write a job description for each labor unit (person). This should include job title (herdsman, milker, etc.), duties and responsibilities, supervisory responsibility (whom the person is responsible to and whom he must supervise), working conditions (hours, vacation, sick leave, etc.), and any other information that may help explain or clarify the duties and responsibilities of the job, or the relationship of the job holder to other employees or management. Especially in larger operations the line of authority and delegation of responsibility must be well defined and clearly understood by all those affected. An organizational chart clearly indicating vertical and horizontal relationships is often helpful (Figure 16-1). The job description, when accurately completed, enables management to better match job needs and employee capability, thus helping to avoid problems. Job descriptions, especially for smaller operations with one or two employees, should also be flexible enough to accommodate enlarging the job with more meaningful and challenging responsibilities; those for large operations should provide for advancement to higher-level jobs or rotation of jobs or shifts.

In completing job descriptions, dairy managers should keep the need for and importance of cowmanship uppermost in their mind. Descriptions of jobs concerned with working with the cattle should include evidence that being a good cowman is an important part of the job responsibility.

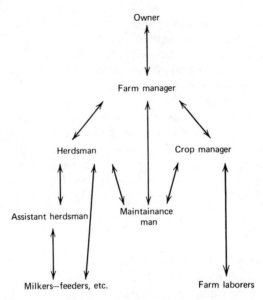

**Fig. 16-1.** Example of a dairy farm organizational chart that defines vertical and parallel relationships.

## Hiring

The hiring process consists of finding or recruiting, selecting, and hiring. Locating and hiring well-qualified dairy farm labor is not an easy job, nor is it likely to be so in the future. Traditional concepts that dairy labor must be farm-reared and have a lifetime of experience may have to be laid aside as the supply of farm-reared people dwindles. Hiring energetic and enthusiastic people without experience but with the desire and ability to learn, and training them, can be an effective method of meeting dairy labor needs for many dairymen.

There is no one best source of labor or best way to find labor. An accurate job description can be helpful in finding labor. It can be used in advertising to accurately describe the available position and attract the type of person needed. Some methods of finding labor are the following.

1.  Ads in local and area newspapers and area and national farm magazines. In writing ads it is important to remember the purpose of the ad—that is, to sell your job—to stimulate the interest of qualified persons sufficiently that they apply for the position. To do this, the ad must accurately describe the position in an imaginative, positive, and desirable fashion. Ads that reflect owner optimism and desirable opportunity for employees

are more likely to attract optimistic, energetic people than those that reflect owner pessimism and undesirable work conditions. Brief but accurate description of job responsibilities should be included. Three examples follow.

a. Dairy labor needed. Reply Box 270, this magazine.

b. Hard-working man needed for milking and general farm work. Wages plus house provided. No smoking, drinking, or pets allowed. Call 111-2222 if interested.

c. Assistant herdsman needed. 300-cow purebred Holstein herd handled in a modern, free-stall, herringbone arrangement. Current herd average 16,000 lb. milk and improving. Responsibilities include calf rearing, relief milking, heat detection, and AI, 5½-day work week with 2-weeks annual vacation. Experience desirable, but not necessary. Will train ambitious, energetic person. Competitive salary plus incentive plan. Fringe benefits include modern 3-bedroom home, garden space, paid health insurance. Good fishing on the farm. Located 6 miles from a small town in northwestern Virginia, with excellent schools, a variety of churches, and an abundance of community activities. Hospital located 16 miles from farm. Send resume to John Doe, Box 101, Brown, Virginia or call 703-111-2222, 8–10 p.m.

Ad (a) is too brief. It contains no job description, farm location, benefits, etc. Ad (b) is negative. It reflects owner dissatisfaction with dairying (hard work), pessimism (call *if* interested), and intolerance of people who are different from himself (he apparently doesn't smoke, drink, or like pets, or he has had a bad experience with someone who did). Ad (c) reflects owner optimism, confidence, and success in dairying (16,000-lb. herd average and improving), gives a brief description of what is expected of the person in this position, working conditions, indicates extra reward for a good job (incentive plan), evidences concern for employee and family well-being (by describing available school, health, and recreational facilities), and gives the location of the farm. It appeals to a broad audience by indicating experience desirable, while leaving the door open for energetic persons who want to learn.

2. Current employees. This may be an excellent source of prospects if the current labor relations program is good. If current employees like their jobs, their employers, and working conditions, they can often identify and help sell others on the desirability of working on a farm.

3. Personal contact with those associated with agriculture or service industries. Some of these are: county or state extension personnel, vocational agriculture teachers, AI technicians, DHI supervisors, fieldmen for milk cooperatives, feed industry fieldmen or salespeople, employment offices,

high school or community college placement personnel, dairy science faculty members, ministers, veterans' employee organizations, etc.

4. Personal contact with special-interest groups. These include parole officers, foreign exchange student coordinators, and minority groups. Parolees often experience great difficulty obtaining jobs, and many can be excellent dairy workers. Careful screening is critical, however. One of the authors has had personal experience with parolees and has known several dairymen who also have hired parolees with excellent results. Many foreign exchange students desire work on U.S. dairy farms to gain experience in U.S. methods of dairying. Again, careful screening is important, but results are often excellent. The same is true for minorities. This type of personnel is often unexperienced, so more thorough training is needed. However, these people represent a large labor pool that is often overlooked.

5. Unions. In areas of the country where farm labor is organized and unionized, this can be an excellent source.

Information acquired by the employer and the applicant during the interview, in most cases, is the primary basis for selection by both. The purposes of the interview are to learn as much as possible about the applicant, expose him to the working conditions of the operation, and give him the opportunity to interview the employer. To achieve these purposes, the interview should be held on the farm to allow the prospect to see working and in many cases living conditions. Prospects should be given the opportunity to ask questions and employers should ask questions and listen carefully as well as talk about the operation. Some important points concerning the interview are the following.

1. Give the interview top priority. Hold it on the farm, but not in the milking parlor while milking, or anywhere else where either party can be easily distracted. Show the prospect the farm, cattle, facilities, housing, etc., and explain the operation as a part of the interview, but also reserve time and a place for uninterrupted conversation.

2. When the applicant is married, invite him or her to bring the spouse or family along. If living quarters are provided, allow ample opportunity for the family to inspect the quarters.

3. The employer should sell himself and the job by conveying an enthusiastic, positive, successful, optimistic attitude about dairying and its future, rather than being pessimistic about the future of his operation and dairying in general. He or she should stress the attractive parts of the job such as outdoor living, local community activities, opportunity to go

home for lunch with the family, opportunity for advancement, pleasures of working with good cattle, etc., instead of complaining about long hours, hauling manure, equipment breakdowns, etc. People would much rather work in a positive than a negative environment.

4. Learn about the prospect. The employer hopes to gain the information needed to make a decision. To do this, he needs to systematically collect the needed information. An interview form listing the pertinent questions can be very helpful in avoiding omission of important points. In asking questions it is necessary to avoid biases that put the applicant on the defensive. Questions such as "Do you always let your hair grow that long?" "How far could you get in school?" etc., put the applicant on the defensive. They evidence employer bias concerning long hair and imply that the applicant wasn't intelligent enough to stay in school. A better procedure is to explain to the applicant that certain information is necessary to determine suitability for the job and proceed to ask questions in an objective manner. Much can be learned about a person's stability by asking about past employment record; about his or her loyalty by the manner of talking about the last employer, about his or her confidence by the directness of answers, etc.

5. Give the prospect the opportunity to learn about you. Explain that you realize the need for certain information to evaluate the suitability of the job. Answer questions in a direct, honest manner rather than hedging or being inaccurate. This can only lead to problems later on the job.

6. Ending the interview. Give the prospect a copy of the details of the job (a written job description is excellent), including responsibilities, work schedule, salary scale, and fringe benefits (it often helps to indicate a dollar value for these). If a written contract is used, and it should be, give the person a copy to study. Arrange a return conference in a few days to allow time for the employer to check references and the employee to decide whether he or she desires the job. A definite answer should be given by both at that time.

A general verbal agreement is often the case in hiring dairy farm employees, whereas nonfarm employees often have detailed written contracts with employers or labor organizations. Many dairymen are reluctant to offer employees labor contracts or wage and benefit agreements because employee requests for such contracts are often regarded as indicating a lack of trust. Both employers and employees should insist on such an agreement for the mutual benefit of both. Misinterpretation or inaccurate recollection of verbal agreements can be a major cause of discontent and dissatisfaction. Written agreements can help avoid many of these problems and should be considered an expression of mutual trust and understanding rather than mistrust. Agreements should clearly indicate the wage rate, fringe benefits, working hours, time off, vacation and sick leave, and job responsibilities. Other items such as period of agreement and renewal and termination procedures, incentive plans, care of employer-provided housing, grievance procedures, etc. can also be included. These agreements may vary considerably depending on the degree of specialization or lack of it required for the job, whether housing is provided, incentive plans, etc.; thus they should be tailored to fit individual farm needs. An example of a workable agreement is presented in Appendix H.

## Training

Dairy farm work, especially smaller operations with one or a few employees, requires a wide range of biological and mechanical skills. Good cows and good equipment are expensive and easily damaged or completely ruined. Proper training is important in preventing these problems. Most dairymen are well aware of the need for proper training; however, most either wish to rely on someone else to do it (by hiring people who work for a neighboring dairyman, or those who have been trained in college or vocational schools) or expect the employee to teach himself (often by starting in a lower level job and learning by observing other employees). These methods can be helpful, but personal training and instruction are also needed, and many dairymen do a poor job of it. The result is often wasted time and material and misunderstandings.

In training new employees, several points may be helpful.

1. Encourage them to ask questions about things they don't understand. Avoid embarassing them or implying that they are stupid by the way you answer questions or explain things.

2. Recognize that each new employee is different, with different skills and experience and that some understand and follow instructions more easily than others.

3. Don't assume they know how to do a job, but rather be sure by asking and observing.

4. In actual training of employees in specific jobs, first determine their degree of knowledge by friendly conversation. Next, explain the importance of the job and of doing it correctly. Follow this by illustrating how to do the job. Then ask new employees to demonstrate to you how to do

it and tactfully correct them if necessary. Have them repeat it, and if they do it satisfactorily, express your confidence in their ability to do it, then leave them on their own but tell them where you'll be if they experience difficulty. It does take time and patience to properly train employees, especially inexperienced ones; however, the benefits can be the development of good employee work habits, increased quantity and quality of employee output, improved morale, reduced labor turnover, and reduced cow and equipment losses. These result in increased profits.

## Motivation

Motivation may be defined as wanting to do the job correctly and on time. Motivation of dairy labor is a major factor in obtaining both quantity and quality of performance. Employees who want to do the job correctly and on time usually do so, while the performance of those who are not highly motivated is usually low in quantity and quality. A major responsibility of management or employers, then, is to provide an environment that does instill employee desire to do the job right and on time.

Motivation is based on human needs. Some of these are: (1) physical needs such as food, housing, etc.; (2) safety and security; (3) social needs such as friendship, love, and a sense of belonging; (4) esteem and self-respect; and (5) self-actualization or the need to become what one is capable of becoming.* To motivate dairy labor, employers must then provide the opportunity for employees to fulfill these needs to as large an extent as possible—to earn reasonable wages, to work in reasonably safe working conditions with a measure of job security, to feel they are a valuable part of the operation and not "just the hired man," to earn personal recognition for achievements, and to have the opportunity for personal growth and advancement. In short, employees must be treated with dignity, respect, and understanding—as fellow human beings with personal goals and needs, and not "just the hired man." Failure to recognize human needs of employees and to provide an environment conducive to fulfilling these needs is a major reason for low employee performance, dissatisfaction of both employer and employee, and continuing labor problems. Providing a proper environment, conversely, can result in satisfaction for the employer from increased productivity and profit and for the employee in additional financial reward, achievement, recognition, advancement, and personal growth.

There is no one best way to motivate all employees, as individual human

* A. H. Maslow, *Motivation and Personality*, Harper, New York, 1954.

needs and goals vary considerably. There are, however, some basic principles that are applicable in most situations.

1. Determine the personal needs and goals of each employee. Does the person wish to learn more about all aspects of dairying, advance to more responsible jobs, acquire a capital interest in dairying, etc., or would he or she rather perform less demanding, more repetitive tasks?

2. Pay a decent base wage and provide the opportunity for employees to receive additional remuneration for high performance. This additional remuneration may be in the form of salary increases, promotions, or bonuses based on incentive plans. Criteria for all of these need to be defined. Are raises and/or promotions automatic after a specified length of time, or are they based on certain criteria, and if so, what criteria? Incentives can be defined as something designed to enhance production and/or profitability. Good incentive plans should: (a) be mutually beneficial for employer and employee—that is, the employee should be rewarded for working for the benefit of the employer; (b) be impartially measured by a third party; (c) be based on performance within the control of the employee; and (d) be designed to improve an existing condition in the herd that needs improvement. Incentive plans for specialized personnel such as milkers can be based on milk quality (bacteria count, sediment content, leucocyte count, etc.), which is under the control of the milkers to a large extent and measured by marketing or processing personnel, or for calf rearing employees on the basis of calf mortality or percent of calves born alive that are still alive at weaning. For employees with more general jobs, bonuses based on milk production (daily, weekly, or monthly milk sold, or DHI averages), heat detection, calving interval, or a per-cow bonus based on pregnancy by 100 days after freshening, confirmed by pregnancy palpation by a veterinarian, can be effective in improved herd productivity. Incentives based on a "good job" of milking, a "good year," increased herd average (when the employee has little control over feeding, breeding, etc.), or increased profitability (profit can be defined in several ways) do not usually work well and should be avoided because they do not meet one or more of the criteria of a good incentive plan.

## Job Safety and Job Security

Most dairy farm accidents are caused by an unsafe condition, an unsafe action, or a combination of both. Unsafe mechanical and other conditions must be eliminated promptly and unsafe actions discouraged by good example and by prompt correction of employees performing them. This will emphasize concern of management for employee safety and discourage carelessness. Employee security is enhanced by the success of the business and the attitude of the manager. Highly profitable operations rarely sell out. A confident, optimistic manager commands the confidence of employees in his ability to do his job well, thus helping to ensure continued profitability of the operation and job security of employees.

## Employer–Employee Relations

Maintaining good employer–employee relations is an important factor in retaining good dairy labor. Application of the Golden Rule, "Do unto others as you would have them do unto you," is the key to success in the area. Periodically dairymen need to ask themselves "How would I like to work for me?" and employees need to ask themselves "How would I like to hire me?" Some factors that are important in maintaining good relations, whether they are between employer and employee or intrafamily, are the following.

1.  Develop mutual respect for each other and appreciation for the impor-
    tance of both in the success of the operation. Labor must realize that
    good management is essential for the profitability of the operation,
    which in turn is necessary for the continuation of their job. Conversely,
    management must realize that high-quality labor is essential for manage-
    ment success and continued profitability. Both good management and
    good labor are essential for the success and profitability of the total
    operation. If either fails, both lose, and if both do a good job,
    opportunity for success and advancement is enhanced. Mutual respect
    and appreciation cannot be bought, ordered, or requisitioned, but must
    be earned and developed. Employers can earn respect by being honest
    with employees, by telling them they are in business to make a profit and
    how they plan to do it, and that they are willing to share these profits
    with those who help create profits by doing a top job. Employees can
    earn respect by doing a top job, the kind of job that does contribute to
    increased profits. Both can earn the respect of the other by treating each
    other as fellow human beings with personal feelings, goals, and aspira-
    tions.

2.  Have a clear understanding of the job conditions, responsibilities,
    supervision, wages, hours, etc., before employment begins. Misunder-
    standings lead to dissatisfaction and can be largely avoided through the
    use of a written job description and work agreement.

3.  Properly train employees to do a good job. People generally like to do a

good job, like to be productive, and like to feel they are making a valuable contribution to the success of the operation. This is enhanced by good training.

4. Encourage suggestions and listen to ideas from employees. Carefully evaluate and discuss them with employees. If a suggestion is not a good one or is one that you had previously considered and discarded, make sure they understand why. If the idea is a good one and is adopted, give proper credit to the individual.

5. Compliment in public and criticize in private. Publicly recognize workers when they do a good job, make an excellent suggestion, or solve a problem. Give credit when credit is due. Equally important, constructively criticize when mistakes are made or for inferior performance, but do this in private and accompany the criticism with suggestions or demonstrations for improvement.

6. Provide opportunities for advancement. People like to feel there is opportunity to advance, to improve their position in regard to salary and responsibility. Position advancement is difficult in small operations, but additional responsibilities can be assigned to enlarge the current position. Wise managers look for new abilities acquired by current personnel and use these abilities to enlarge positions as a method of advancement.

7. Keep the lines of communication open. Periodic frank discussions of employee performance, suggestions for improvement of employee or herd performance, employee and management suggestions and ideas, current status and future plans for the herd and farm etc., clear the air, promote mutual understanding of the problems and goals of both employer and employee, and foster a sense of belonging.

8. Choose compatible people. This reduces interemployee friction. Compatibility among wives is also important when housing is included with the job.

9. Provide decent housing when housing is included. Unsatisfactory housing stimulates domestic problems and implies a lack of employer consideration for the employee and his family.

10. Have a successful operation. Success attracts and holds good people. They prefer to be a part of a successful operation managed by an optimistic, enthusiastic manager rather than be associated with a "sinking ship."

## Summary of Labor Management

The labor management program involves finding, hiring, and keeping the right kind of labor and avoiding or firing the wrong kind. Some dairymen rarely have labor problems, while others almost continuously have labor problems. The former usually are confident and optimistic about the dairy business; they have the ability to perceive and react to the human needs of their employees; they respect employees as fellow human beings with personal feelings, goals, abilities, and problems, and treat them with dignity and respect. They expect high performance from employees and reward them for this high performance by providing recognition, advancement, and decent wages. They realize that their success is, to a large degree, dependent on their ability to surround themselves with capable people and to motivate these people to get the job done right and on time.

## OPERATING AGREEMENTS, PARTNERSHIPS, CORPORATIONS

As dairy operations become larger, degree of mechanization increases and cows produce increasing amounts of milk on a per-cow basis; the competency and cowmanship ability of those working with the cattle must also increase. Attracting and retaining the type of people needed—those with cowmanship ability and willingness to accept responsibility—is difficult on a salary basis. People who have the needed ability also have the ambition and desire to become junior partners, part owners, or owners themselves. Written agreements, profit-sharing plans, and the potential for acquiring an interest in the business are often necessary to attract and retain this caliber of person. The results can be mutually beneficial. The enthusiasm and ideas of youth tempered with the wisdom of age and experience can be and have been a vital reason for the success of many dairy operations.

Acquiring experience and sufficient capital resources is a major problem for many young people who aspire to become dairymen. Many established, older dairymen wish to gradually transfer management and ownership responsibilities to one or more of their sons or daughters, or other competent young people, to provide for the orderly transfer of the operation from generation to generation while maintaining an adequate retirement income. A wide variety of methods can be utilized to help solve these problems. It is not the purpose of this discussion to examine all of the alternatives in detail, but rather to discuss some of the basic principles involved in planning and implementing several of the more common ones.

The basic purposes or objectives of most operating agreements, partnerships, and corporations are: (1) to provide reliable, competent, highly motivated labor and management assistance for the operation; (2) to provide the opportunity for competent, highly motivated young people with inadequate experience or capital resources to acquire the needed experience and capital resources to assume major dairy management and ownership responsibilities; (3) to provide present and future security for all parties; and (4) to provide a mechanism for keeping the operation intact as it passes from generation to generation. There are also tax (income and estate) considerations, which may be listed as a purpose or objective of some types of agreements. Many agreements are within family (such as father and son, brother and brother, brother and brother-in-law, etc.) and some involve nonfamily members. The authors have worked with many families and nonfamilies on various agreements and arrangements. Some of these have worked very well and were successful in achieving their purposes. Others failed. Certain characteristics are apparent in most successful arrangements and lacking in those that have failed. Some of the characteristics that are essential ingredients for success are the following.

1. The operation must be large enough to support the families of those involved at their minimum acceptable standard of living. If this condition is not met, friction within and between families soon develops and the chances for success are poor.

2. All people involved in the agreement and their families must be compatible. Those directly involved must have mutual respect for each other and treat each other with dignity and understanding. They must be able to discuss differences of opinion objectively and rationally, resolve differences, and reach mutually agreeable decisions. Older people must accept some of the ideas of youth, while young people must recognize the value of experience. The enthusiasm and ideas of youth, tempered by the wisdom of age and experience, can be an excellent situation or a very poor situation depending on the personalities of those involved. Disagreements and problems between spouses and families must also be avoided. Separate housing is a must in most arrangements.

3. There must be a written, legally acceptable document that specifies the details of the agreement, including financial arrangements and responsibilities, management and labor responsibilities, renewal and dissolution procedures, mediation provisions, and, in cases where capital resources are shared, provision for dissolution or acquisition of assets in case of death, disability, or retirement of one or more of those involved. Preparation of such a document requires careful thinking, planning, and

competent assistance. It will also help avoid many potential misunderstandings and problems. It must be tailor made to fit the individual situation. When joint ownership of assets is involved, more thorough planning and completeness of the document is necessary. Experience has shown, however, that all possible situations cannot be anticipated and covered in this document; therefore mutual trust and respect and a certain amount of give and take between associates is essential.

Agreements that include these three ingredients have a good chance of success. If one or more is lacking, the chances for success are low.

**Types of Agreements.** Agreements may vary from an employer–employee agreement (Appendix H) to a corporation, depending on the situation and people involved. However, agreements generally fall into one of three types: (1) wage or wage and bonus, (2) partnership, and (3) incorporation. These three basic alternatives and the many possible variations of each should be evaluated in relation the basic purposes of agreements and the specific purposes and objectives of those involved in each individual situation.

WAGE OR WAGE AND BONUS AGREEMENTS. In these agreements, initially, the owner retains essentially full ownership and management responsibility and receives the profit from the operation. Labor receives a specified wage and in some cases a bonus, either in cash or personal property such as heifers or cows. These types of agreement give owners and employees time to test their relationship—to determine whether they are compatible and whether the employee is really interested in becoming a dairyman. It also gives him the opportunity to gain experience. If these agreements continue for several years, the employee is generally delegated increasing amounts of management responsibility and granted increasing amounts of personal property, thus enabling him to acquire additional experience, assets, and security. These agreements often continue until the employee owns one-third or more of the personal property such as cattle and machinery. Such agreements are common among parents and children who have recently graduated from high school or college or between older dairymen and nonrelated young people aspiring to become dairymen. They are much preferred to oral wage agreements and the promise that "some day it will all be yours," which some parents prefer.

PARTNERSHIPS. A partnership is defined by the Uniform Partnership Act as "an association of two or more persons to carry on as co-owners of a business for profit." A partnership implies sharing of management, ownership, profits, and losses of the business. In addition each partner may be held responsible for business debts contracted for and wrongful acts committed by the other partner(s) in the ordinary course of conducting the business.* Because of the potential legal implications and problems with partnerships, the advice of experienced persons such as farm management extension specialists, attorneys, and/or bankers should be solicited. All parties involved should review sample partnership agreements such as Appendix Table I or others available from state extension services, decide on details peculiar to their situation, and inventory assets all partners plan to bring into the business. Then a written agreement should be drafted by those involved and given to an attorney who will write the legal document. Partnerships are among the best and worst agreements the authors have seen. Partnerships can successfully achieve the four basic objectives of farm agreements if the three key ingredients for success discussed earlier are included. They are likely to fail if one or more of these ingredients is missing.

CORPORATIONS. A corporation is a legal entity that can hold and transfer property and carry on other business transactions much as an individual person can. The corporation is owned by stockholders, who elect officers, delegate management responsibility, and share in profits in proportion to the extent of their owned stock. State laws specify procedures for incorporating a business and for properly maintaining the corporation. Farm corporations may be either "Subchapter C," which are regularly taxed corporations, or "Subchapter S," in which income is distributed to stockholders and each is individually liable for income tax payments.

Incorporating a dairy operation involves considerable expense in legal fees and fees for filing articles of incorporation. Maintaining it involves considerably more paper work than do sole proprietorships or partnerships; thus incorporation is usually limited to larger operations. Incorporation, however, can offer many tax, operational, and estate transfer advantages. Some tax advantages are the lower tax brackets for higher earnings and the fact that costs of retirement plans, insurance, etc., can be deducted as ordinary business expenses. Significant savings in estate taxes may also be realized. Operational advantages include the facts that stockholder liability is limited to holdings in the corporation and that the corporation is not automatically terminated at the death of one of the stockholders. Estate transfer is facilitated by transfer of stock rather than of personal and other property.

As dairy operations become larger, incorporation should be investigated. The advice of a competent attorney and tax accountant is essential in comparing types of incorporation as well as whether to incorporate. Decisions concerning whether to include assets such as land in the corporation or whether to retain individual ownership and rent these or other assets to the

---

* H. W. Hannah, *Beuscher's Law and the Farmer,* Springer, New York, 1975.

corporation, when to incorporate, choice of fiscal year, method of accounting, etc. can greatly affect taxation and should be thoroughly investigated.

## REFERENCES FOR FURTHER STUDY

Brown, L. H., and Speicher, J. A., How Can We Compete for Good Labor? *Hoard's Dairyman,* p. 693, June 10, 1972.

Brown, L. H., and Speicher, J. A. How to Keep a Good Man after He Has Been Hired, *Hoard's Dairyman,* p. 744, June 25, 1972.

Brown, L. H., and Speicher, J. A., Short of Help—You May Need a Partner, *Hoard's Dairyman,* May 25, 1971.

Hannah, H. W., *Beuscher's Law and the Farmer,* Springer, New York, 1975.

Nixon, J. W., F. C. White, and B. R. Miller, Elements of Successful Labor Management among Georgia Farmers, mimeo, Dept. Agr. Econ., Univ. of Georgia, 1975.

Reiss, F. J. *Father–Son Farm Operating Agreements, University of Ill. Ext. Cir.* 969 (1967).

Speicher, J. A., Personnel Management and the Dairy Manager, paper presented at the Large Herd Management Symposium, Univ. of Fla., January 1976.

Wells, R. C., Have You Considered A Father–Son Partnership? *Hoard's Dairyman,* September 10, 1971.

The Large Dairy Farm Management Seminar, Univ. of Fla., 1976. Talks to be published by The University presses of Florida, Gainesville, include: Holt, J., Conditions for Manager and Employee Motivation; Wells, R. C., Recruiting and Training New Employees; Bishop, S. E., Incentive Programs for Dairy Employees.

# 17

# PRODUCTION OF HIGH-QUALITY MILK

The objective of the milk quality-control program on a dairy farm is to produce high-quality milk; that is, milk characterized by: (1) composition that meets or exceeds minimum legal standards (3.25% fat and 8.25% solids-not-fat), (2) good flavor and odor (a bland but slightly sweet taste and no odor), (3) freedom from all pathogenic bacteria, (4) a low bacteria content (less than 10- to 15,000/ml. in raw milk), (5) a low content of somatic cells or leucocytes (less than 0.5 million/ml.), (6) freedom from all drug and pesticide residues and other adulterants, and (7) freedom from foreign or extraneous materials. The production of high-quality milk requires healthy, normal cows; clean cows and sanitary milking practices; clean milking equipment; rapid cooling; discarding milk from cows that have been treated with drugs or that have consumed feeds that contain pesticides; and avoiding exposure of the milk to daylight, artificial light, or excessive agitation. Failure to meet any one of these requirements can result in the production of milk of inferior quality. As the sale of milk accounts for about 90% of the income of most dairy farms, the production of high-quality milk is essential to the success of the operation. Less than high quality jeopardizes both the immediate and future markets.

Milk and milk products are tested for quality (all seven of the characteristics mentioned earlier) regularly, in fact more often than nearly any other food, because of their perishability and widespread use. Standards of minimum acceptable quality have been established for the production, processing, and sale of the various grades of milk and milk products. Failure to meet these standards can result in temporary or, in severe cases, prolonged loss of a market for milk. This is as it should be, for if low-quality milk reaches consumers, the long-term demand for milk and dairy products can be adversely affected and the future of the dairy industry jeopardized.

## PRODUCING HIGH QUALITY MILK

A goal of every dairyman should be to produce milk that meets all standards and far exceeds most minimum or maximum standards. For example, the maximum bacterial limit for grade A raw milk is 100,000/ml. and the maximum somatic cell limit is 1,500,000/ml. Reasonable goals are maximums of 10 to 15,000 and 3 to 500,000 respectively.

The basic principles of a high-quality milk-producing program are meeting the requirements for the production of high-quality milk and avoiding practices that result in an inferior product. The following discussion is concerned with these items and with identifying some of the more commonly occurring problems.

## Healthy, Normal Cows

Milk from unhealthy or abnormal cows can have pathogens, high bacteria and somatic cell counts, and off flavors and odors. The organisms that cause brucellosis, tuberculosis, and Q fever are transmissible to man through milk. Fortunately, all can be effectively destroyed by pasteurization so there is little danger to consumers of pasteurized milk. However, animals that have contracted these diseases must be eliminated from the herd because of the danger of spreading the disease to other cattle or to people who may consume unpasteurized milk. Periodic testing of milk (ring test) for brucellosis and animals for both tuberculosis and brucellosis has largely eliminated problems of transmission of these diseases to man.

A larger current problem is mastitis. Common mastitis-causing organisms do not normally cause disease in man, with the possible exception of some of the streptococci. Mastitis-causing organisms are susceptible to pasteurization temperatures. Milk from mastitic cows, however, does contain large amounts of bacteria and somatic cells and is often abnormal in other characteristics such as appearance, flavor, and chemical characteristics. Therefore milk from cows with mastitis must be discarded. Other diseases, such as acetonemia, can cause an off flavor and odor in milk, and the milk from cows so affected should be discarded.

Milk produced by cows before freshening and for the first five to six milkings after freshening (colostrum) is also abnormal in composition and should be fed to calves (Chapter 14) or discarded.

## Clean Cows

Milk from dirty cows or cows subjected to unsanitary milking practices often contains dirt and other extraneous materials and excessive amounts of bacteria. It may fail to meet bacteria count and sediment standards. Most of the dirt and dust that gets into milk comes from the cow's flanks, udder, or belly during milking time. Keeping cows clean and thoroughly washing and drying teats and udder just before milking can avoid these problems. Keeping cows clean is facilitated by proper stall platform or free-stall length, by use of sufficient bedding, by keeping lots and lanes clean, and by keeping the hair clipped from the udder, flanks, tail, and belly. Research has shown that clipping udders, tails, and flanks will reduce raw bacteria counts by 45 to 75%. In temperate climates the udder, flanks, and belly should be clipped early in the fall, before the hair grows long, and periodically (usually one to two more times) as needed during the colder season. In other climates clipping should be performed as needed. Figure 17-1 illustrates the areas that may be clipped. Clipping can be effectively done by clipping against the

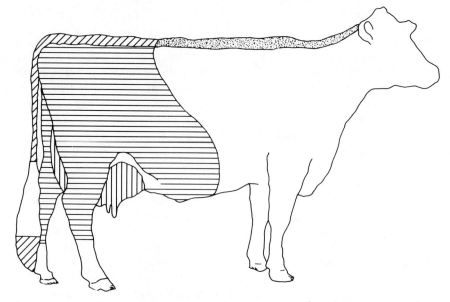

**Fig. 17-1.**   Areas which may be clipped [*V.P.I. & S.U. Dairy Guideline* 475 (1976)].

natural lay of the hair. The following procedure is suggested if this extensive
clipping is used.

1.   Clip the switch about 1 ft. from the ground and clip the tail starting 1 to 2
     in. above the long hairs on the switch and continuing over the tailhead
     and rump.

2.   Clip the udder. Stretching the skin facilitates closer clipping and avoids
     nicking.

3.   Clip the legs, hocks, flanks, thighs, and rump.

4.   Clip a strip 6 to 8 in. or more wide along the back to the poll. This has
     little to do with quality milk production but is effective in reducing
     irritation caused by lice and dandruff.

Some dairymen do not like to clip as extensively as indicated in Figure
17-1 for various reasons. Clipping need not be this extensive in all areas of the
United States. Practical minimum areas to clip in cows housed in stanchion
barns (solid line) and free stalls or well-bedded loose housing (dotted line) are
shown in Figure 17-2.

Some animals become frightened or nervous when clipping starts. When

**Fig. 17-2.** Minimum areas to clip [*V.P.I. & S.U. Dairy Guideline* 475 (1976)].

this occurs, rest the clippers, with the motor running, against the animal's shoulder to accustom her to the sound and feel of the clippers.

Clipper malfunctions and expensive repairs can be greatly reduced by proper preventive maintenance, care, and use. Preventive maintenance includes: (1) removing and cleaning the screens that allow air circulation to the motor whenever they become clogged with dirt or hair, (2) periodic lubrication of the motor armature shaft and clipper head, and (3) cleaning accumulated dirt and hair from the motor and surrounding area as needed.

Proper care and use include: (1) using sharp blades, as using dull blades results in a poor job of clipping, overloading the motor, and overheating the motor and blades; (2) adjusting the tension to the minimum needed for effective clipping as excessive tension overloads the motor and may cause overheating, causes excessive blade wear, and dulls blades more rapidly; (3) oiling the clipper head through the oil holes provided; (4) dipping the clipper head occasionally, while running, in fuel oil or a similar light lubricant to flush out dirt and hair and to lubricate and cool the blades (always hold the clippers with the head down after flushing to allow excess lubricant to drip off rather than seep into the motor); (5) occasionally remove blades and clean dirt and hair from the clipper head; and (6) grasp the clippers firmly when using them so as not to drop the clippers or strike the blades against hard objects. Clipper blades are very hard and brittle and striking them against concrete or steel often results in broken blade teeth. Extra sets of sharp blades should always be kept on hand. Clippers should be cleaned, blades oiled and stored with the head lower than the motor in a clean, dry place when not in use.

## Sanitary Milking Practices

These enhance milk quality by preventing contamination of milk during milking and by preventing spread of mastitis. The udder and teats should be thoroughly washed with water containing a disinfectant solution. Spray nozzles or individual towels (preferably disposable paper towels made especially for this purpose) should be used to remove all dirt and dust from the teats and udder. The teats and udder should also be thoroughly dried, again preferably with a disposable paper towel, to prevent dirty wash water from entering the teat cup. A few squirts of milk should be removed from each teat before attaching the machine. This helps detect clinical mastitis and removes the foremilk, which contains a high bacteria population. After milking, teats should be dipped in a disinfectant solution to prevent incidence of new cases of mastitis. Cows with mastitis, those producing colostrum, and those treated with drugs should be milked last, and their milk must be kept from the saleable milk supply. If milking last is not feasible, their milk should be diverted into a separate container and the milking claw assembly and milk lines thoroughly cleaned before they are used on another cow.

Persons who do the milking should be healthy, free of communicable diseases, clean, and wear clean clothing in addition to following sanitary milking practices.

## Clean Milking Equipment

All milking equipment, lines, and utensil surfaces that come into contact with milk must be thoroughly cleaned, and after each milking sanitized before the next milking. Storage tanks must also be cleaned and sanitized after each milk pickup. The purpose of cleaning is to remove milk soils, organic and mineral solids of milk that remain on surfaces after the milk is removed. The purpose of sanitizing is to kill residual microorganisms. Inadequate or improper cleaning and/or sanitizing allows bacteria to remain on equipment surfaces and to multiply. This results in elevated bacteria content of the milk.

Organic soils are composed of the major organic constituents of milk: sugars, fats, and proteins. It is important to remove these soils from surfaces as quickly as possible after milking because their adhesion to surfaces increases with time, dryness of the soils, and heating. After they harden, they are difficult to remove.

Mineral soils are composed of precipitated salts (by alkaline conditions or heat) of various minerals, usually calcium, magnesium, iron, and manganese, in water or milk. Cleaning agents can actually enhance precipitation of minerals if they are not compatible with water hardness conditions or are used in amounts or at temperatures contrary to manufacturer's recommendations. Precipitated minerals on surfaces of milking and milk storage equipment can combine with organic soils to form milkstone.

Effective cleaning of milking equipment begins with analysis of the water used in washing for mineral content or hardness and choosing a cleaning compound that is compatible with the water. In very hard water (30 or more grains/gal.) a water softener should be used. The compatible cleaner should then be used according to manufacturer's directions in relation to amount and concentration of the cleaner, temperature of the cleaning solution, and contact time of the cleaning solution and the surface to be cleaned.

Usually an alkaline or chlorinated cleaner (alkaline cleaners with added chlorine) followed by an acid cleaner is recommended. Chlorinated alkaline cleaners and acid cleaners should never be mixed, as a poisonous chlorine gas is released. Alakline cleaners are effective in dissolving organic soils; the acid cleaners are designed to remove or prevent an accumulation of mineral soils.

Cleaning reduces bacterial concentration on surfaces but does not eliminate all types of bacteria. Sanitizing surfaces before next use destroys nearly all remaining organisms if: (1) the sanitizing solution used is of proper strength, and (2) a thorough cleaning precedes the sanitizing. Some sanitizing compounds lose strength with time in storage (chlorine compounds) or increasing pH (chlorine and iodine compounds). Some are also unstable at temperatures above 120°F, or 49°C (iodine compounds), while others are not compatible with hard water (quaternary ammonium compounds). Improper cleaning results in residual soils that can protect bacteria from the sanitizing action.

Equipment and bulk tank cleaning procedures should be posted on the milk-house wall and rigidly followed. An example of equipment cleaning procedures is presented in Table 17-1. The precise procedure, compounds used, and water temperatures will vary. In general, equipment should be rinsed with lukewarm 100 to 110°F, or 38 to 43°C water immediately after milking to prevent drying of milk solids on surfaces. Water that is too hot can cause denaturization of proteins and a protein film on surfaces, while water that is too cold can cause fat crystallization and the formation of a greasy film on surfaces. Washing and rinsing should follow. In clean-in-place (CIP) systems, velocity and air in the system are also essential. A minimum velocity of 5 ft./sec. is necessary to ensure proper cleaning action. Introducing air into the system provides turbulence and increases scouring action.

**Table 17-1** Example of Cleaning Procedures for Milking Equipment[a]

| | | |
|---|---|---|
| **For hand washing:** | | |
| 1. | **Prerinse** | Rinse all equipment and utensils and flush pipeline with lukewarm (110–120°F) water immediately after use. |
| | | Disassemble all parts that must be hand washed. |
| 2. | **Wash** | Mix cleaning solution. |
| | | _____gallons hot water (160°F). |
| | | _____ounces alkaline cleaner. |
| | | Soak all parts at 120–135°F for at least 5 minutes. |
| | | Brush all parts thoroughly. |
| | | Drain. |
| 3. | **Rinse** | Final rinse with acidified solution of: |
| | | _____gallons clean water (100–120°F). |
| | | _____ounces acid cleaner. |
| | | Drain. |
| **For pipelines:** | | |
| 1. | **Wash** | Mix cleaning solution. |
| | | _____gallons hot water (160°F). |
| | | _____ounces alkaline cleaner. |
| | | Circulate cleaning solution for 10 minutes. |
| | | Brush all parts not designed for cleaning by circulating solution. |
| | | Drain. |
| 2. | **Rinse** | Rinse with lukewarm acidified water. Do not recirculate rinse solution. |
| | | Drain. |
| | | Visually inspect line, receiver jars, etc., for proper cleaning. |
| **Immediately before milking:** | | |
| 1. | **Sanitize** | Flush pipeline with sanitizer immediately before milking, using: |
| | | _____gallons clean water. |
| | | _____ounces sanitizer. |
| | | Let drain. |
| | | Sanitize hand-washed parts. |
| | | Let drain. |

[a] G. M. Jones, Managed Milking Guidelines, *V.P.I. & S.U. Ext. Pub.* 633, revised September 1975. Example for posting in the milk house. Consult dairy fieldman or inspector for assistance.

All equipment and utensils should be stored in a manner that permits water to drain and equipment to air dry. In CIP systems a drain should be located at the lowest point in the system.

Inflations and other rubber parts that come into contact with the milk must also be thoroughly cleaned and sanitized after each milking. They should be replaced when they become soft, cracked, or rough, as pores and cracks in rubber parts protect soil and microorganisms from effects of cleaning and sanitizing.

Bulk tanks also must be properly cleaned and sanitized, or psychrophillic bacteria (microorganisms capable of rapid growth at temperatures of 35 to

50°F, or 2 to 10°C) multiply rapidly. Tanks are cleaned by following essentially the same procedures as recommended for milking equipment. The milk hauler is normally responsible for rinsing the tank immediately after the milk is removed. Following this the tank must be washed, rinsed, and sanitized. Sanitizing should take place just before the next milking. Tanks may be cleaned and sanitized manually or with CIP or mechanical systems.

Improper or careless cleaning and sanitizing of equipment and tanks is a major cause of inferior milk quality. It need not be if cleaning water and cleaning compounds are compatible and a precise procedure is formulated and followed.

## Rapid Cooling

Milk should be cooled to 40°F, or 4°C, as rapidly as possible after completion of milking (within two hours) and maintained at that temperature to inhibit bacterial multiplication. As all raw milk contains some microorganisms, improper cooling allows rapid growth of these bacteria with resulting high bacteria counts and possible off flavors and odors. Table 17-2 illustrates the importance of keeping milk cool.

Nearly all milk produced in the United States is cooled and stored in bulk milk tanks. Most are designed for every-other-day pickup of the milk. Such tanks must have sufficient cooling capacity to enable them to cool milk equivalent to one-fourth of the volume of the tank from the temperature of milk entering the tank (about 90°F, or 32°C) to 40°F, or 4°C, within two hours, and to cool milk from succeeding milkings rapidly enough to prevent the temperature of the milk in the tank from rising above 50°F, or 10°C. If milk is

**Table 17-2** Effects of Varying Temperatures upon the Bacteria Growth in Milk

| Temperature maintained for 12 hours, °F | Bacteria per ml. at end of 12 hours |
|---|---|
| 40 | 4,000 |
| 47 | 9,000 |
| 50 | 18,000 |
| 55 | 38,000 |
| 60 | 453,000 |
| 70 | 8,800,000 |
| 80 | 55,300,000 |

picked up daily, the cooling capacity must be for one-half instead of one-fourth of the volume of the tank. Many larger tanks are equipped with dual compressors, one for each half of the tank. This can increase cooling capacity and allow a margin of safety in case of compressor breakdown, as one compressor is usually sufficient to hold stored milk at safe temperatures. More recently, in-line heat exchangers or cooling units have been installed in many systems. These are installed between the milk pump and the bulk tank and reduce the temperature of the milk to 40 to 60°F, or 4 to 16°C (depending on rate of milk flow and precooler capacity) before it reaches the bulk tank. This further increases rate of cooling and in some cases may result in more efficient cooling.

## Preventing Residue Contamination

Milk must not contain chemical residues from antibiotics, pesticides, or radionucleides because of the potential harmful effect on consumers. Antibiotics may also interfere with manufacture of cultured milk products by inhibiting desired bacterial growth.

Public health laws specify "zero tolerance" for antibiotic residues in milk. "Zero tolerance" is defined as any quantity that can be detected. This quantity is currently very small and is getting increasingly smaller as testing techniques improve. Most drugs used for treating lactating cows are eliminated from the milk within 48 to 72 hours, and, by regulation, all drugs must be certified by manufacturers to be eliminated within 96 hours. Antibiotic residues in milk can be prevented by precisely following the manufacturer's directions concerning withholding or discarding of milk following treatment of the cow. As mentioned previously, treated cows should also be milked after normal cows, or equipment must be thoroughly cleaned and sanitized before reuse.

Residues of pesticides used to control flies or insects that damage crops can contaminate milk. Various residues may be ingested with the feed, inhaled, or absorbed through the skin. Some are stored in body tissues for long periods and secreted in the milk for long periods of time. Federal standards have been established concerning the registration and use of pesticides. Use of pesticides is important in controlling insects and pests in feed crops and dairy barns. It is also quite safe, and residue contamination can effectively be avoided, so long as approved materials are used properly. New materials and methods of use are being continually developed. It is, therefore, recommended that dairymen check with extension personnel annually to determine which pesticides are safe and effective in a given area and ascertain the latest recommendations on usage.

## Preventing Off Flavors and Odors

Most consumers evaluate quality of milk primarily by the flavor of the milk. Therefore continued consumer demand and the future of the dairy industry is dependent not only on providing a wholesome product, but also one that is free of off flavors and odors. Milk flavor is determined by its odor, texture, and taste. Ideally, it should possess no detectable odor, have a smooth texture and a bland, slightly sweet taste that leaves a pleasing sensation in the mouth after swallowing. The flavor and odor of milk as it is secreted from healthy normal cows is ideal. However, milk is very susceptible to a wide variety of off flavors and odors, either from the cow prior to milking or during handling of the milk from cow to consumer. Preventing these off flavors and odors is an important part of the milk-quality-control program. Off flavors may be absorbed or caused by bacteria, chemicals, or improper handling. Table 17-3 indicates possible causes and prevention of some of the most common off flavors.

Absorbed off flavors can be caused by the cow consuming feeds containing objectionable flavors, by the cow inhaling objectionable odors, or by exposing the milk to objectionable odors. Some of the more common absorbed off flavors are feed, weed, and barn flavors (Table 17-3).

Bacterial off flavors are usually the result of poor sanitation or cooling or both. As milk is often held in the raw state two to four days before pasteurization, psychrophyllic bacteria have ample opportunity to multiply. These bacteria pose little potential danger to consumers, as they are destroyed by pasteurization; however, they can change the physical appearance of milk and cause the development of off flavors such as acid, bitter, unclean, etc., and coarse texture.

Various chemicals produced or acquired by the cow or acquired from handling equipment and other compounds used in the cleaning and sanitizing of equipment can cause a variety of off flavors. Oxidized, salty, rancid, medicinal, or cowy (ketosis) flavors are some of the more common chemical off flavors.

Improper handling on the farm or during processing can also cause a variety of off flavors. Rancid, bitter, and cooked off flavors are some of the more common ones.

Various other compounds and conditions can cause inferior flavor and odor in milk and dairy products. Most occur on the farm and therefore prevention is largely the responsibility of dairymen.

## SUMMARY

Milk, as ejected from normal, healthy cows is wholesome, nutritious and has a pleasant taste and no odor. The objective of the milk-quality-control

**Table 17-3** Raw Milk Flavor Chart[a]

| Off flavor | Possible causes | Suggested preventive measures |
|---|---|---|
| Feed | Silage—eating or breathing<br>Pasture or alfalfa hay<br>Distillers' or brewers' grains<br>Vegetables—cabbages, turnips, etc. | Withhold objectionable feed or remove cows from pasture 4 hr prior to milking. Feed silage *after* milking. Remove old silage from mangers and feeding bunks and store carts of silage outside of the milking barn. |
| Weed | Wild onion or garlic<br>Others: ragweed, bitterweed, honeysuckle, skunk cabbage, mustard, dog fennel | Renovate pastures when possible. Eradicate weeds through a well-planned program of spraying. Clip pastures or allow only dry cows and heifers to graze. Remove milking herd from infested pastures 4 hr prior to milking and feed cows good quality hay. |
| Cowy, barny, or unclean | Dirty cows<br>Rank air in housing and milking areas<br>Dirty milking equipment<br>Unclean cloths or sponges for washing udders<br>Ketosis | Clip cows, and clean them regularly.<br>Maintain clean environment and ventilate where necessary.<br>Clean and sanitize all milking equipment twice daily.<br>Clean and sanitize them between milkings or use single-service paper towels.<br>Manage dry cows properly. Withhold milk from affected cow. |
| Salty | Mastitis<br><br>Late-lactation cows | Remove infected cows from milking herd—discard milk. Milk disease-free cows properly.<br>Discard milk from late-lactation cows. |
| Rancid, bitter | Excessive agitation or air admission in pipeline milking systems. Breakdown of the fat globule membrane permitting lipase enzyme action and free fatty acids.<br>Cooling too rapidly or at too low a temperature (must freeze to be a problem)<br>Late-lactation cows; cows with mastitis or other illness<br>Protein-deficient ration | Eliminate risers in pipelines, keep fittings tight and air admission at the claw to a minimum; don't run milk pumps in starved condition or agitate milk too rapidly. Avoid filtering under vertical lift conditions.<br>Cool milk to below 40°F within 2 hr. after milking and hold; avoid extremely low temperatures. Do not freeze milk.<br>Discard milk from low-producing or late-lactation cows or cows with mastitis.<br>Have forages tested for protein content and have concentrate ration reformulated if necessary. |
| Flat | Low fat and solids content | Cull these cows from the herd; evaluate feeding program. |

454

| Oxidized | Copper-bearing metals, rust | Use only stainless steel, glass, plastic, or rubber for milk contact surfaces. |
|---|---|---|
| | Dry-lot or winter-feeding conditions | Provide green feed such as green chop or silage; feed 1,000 I.U. vitamin E per cow daily; after 2 weeks reduce to 500 I.U. |
| | Sunlight or artificial light | Keep milk covered. Avoid placing glass pipeline under fluorescent lights. |
| Malty or highly acid | Dirty milking equipment | Properly clean and sanitize all milking equipment. |
| | Slow or inadequate cooling | Cool milk quickly to temperatures below 40°F within 2 hr. after milking and hold. |
| Unnatural | Medications, insecticides, or disinfectants | Use colorless and odorless medications when possible. Use all materials according to directions. |
| | Certain sanitizing agents | Prevent introduction of sanitizing solutions to milk. Drain all equipment well. |

[a] Adapted from G. M. Jones, Managed Milking Guidelines, *V.P.I. & S.U. Ext. Bul.* 633, revised 1975.

program is to ensure that the cows are healthy and to preserve the wholesomeness and flavor of the milk. Success of the program is essential to maintaining current and future markets.

Because of its widespread use and its perishability, milk is regularly evaluated, and dairy farms are regularly inspected. Many dairymen regard these examinations of milk and milk-producing facilities as unnecessary nuisances. Inspection personnel should, however, be regarded as public servants whose primary duty is to ensure the wholesomeness of the public milk supply. By performing this function, they actually become allies, not adversaries, of the dairy industry as their service helps ensure immediate and future consumer acceptance of milk and other dairy products. Their advice to producers concerning various aspects of producing high-quality milk should be solicited and heeded. Because of their wide experience, they can often identify problem areas or potential problems and thus offer valuable assistance to producers. By and large, if milk inspectors or quality-control agencies reprimand dairymen, the dairymen are careless about various practices and deserve the reprimand.

## REFERENCES FOR FURTHER STUDY

Campbell, J. R., and Marshall, R. T., *The Science of Providing Milk for Mankind*, McGraw Hill, New York, 1975.

Jones, G. M., Managed Milking Guidelines, *V.P.I. & S.U. Ext. Pub.* 633, revised 1975.

Wagner, P., Jones, G. M., and Brown, C. A., Cow Clipping Is Important for Quality Milk Production, *V.P.I. & S.U. Dairy Guideline* 475, 1976.

# 18

# Marketing Milk

**18**

Dairy farmers receive 80 to 90% of their dairy income from the sale of milk. Successful marketing of milk is as essential as is volume and economy of production. In some localities there are markets for both fluid milk and milk for manufacturing. In other cases there is only one outlet.

## COMPOSITION OF MILK

Milk, on the average, has a specific gravity of 1.032 at 68°F, or 20°C. One quart weighs 2.15 lb., or 0.98 kg.; 1 gal. weighs 8.6 lb., or 3.91 kg.

The average percentage composition of milk is approximately 3.8% butterfat, 4.6% lactose (milk sugar), 3.2% protein (casein and albumin), 0.7% minerals, and 87.7% water.

There are variations caused by breed of cow, stage of lactation, season of the year, unusual rations, the individual cow and other factors.

Market milk is a blend of milk from many herds that tend to maintain the composition quite constant. The percentage of nonfat solids varies approximately 0.4% for each 1.0% change in butterfat percentage.

The trend toward less fat in the human diet has reflected in a similar trend in the butterfat content of market milk.

The United States Public Health Service defines milk as containing not less than 3.25% butterfat and not less than 8.25% solids-not-fat. The minimum legal standards vary from state to state (3.0 to 3.8% butterfat and 11 to 12.3% total solids).

## FLUID MILK

Fluid milk, or market milk, is milk eligible for class I usage, that is, for consumption as fluid milks and creams. This is Grade A milk, which is produced under very rigid sanitary requirements. Cows must be free of disease (by veterinary or other tests) and clean. Buildings and equipment must meet construction and sanitary requirements. Methods of milking and cooling and storage of the milk is also under supervision of the inspection of sanitarians of the state department of health or the state department of agriculture. The dairy industry is one of the most highly regulated of all industries.

During 1976 Grade A milk represented 80.8% of all milk marketed in the United States. In 21 states all milk sold was Grade A. The majority of these were in the northeast, west coast, and extreme south. In 10 states the percent of milk which is Grade A ranges from 23 to 73% of the total milk produced in those states.

All Grade A milk does not go into fluid consumption, only about 60% does. The remainder is used in the manufacture of dairy products. These may be labeled Grade A or made from Grade A milk.

## MANUFACTURING GRADE MILK

Milk for manufacturing is milk that is used only for the manufacture of processed dairy products such as cheese, ice cream, dry milk powder, evaporated and condensed milk, and some other minor products (usually referred to as class II usage). It is not eligible for use in class I products. It is also produced under sanitary regulations. These are not as stringent as are those for Grade A milk production; however, the milk must be clean and sweet when delivered.

Requirements vary in different states. In some states the manufacturing milk requirements are about as rigid as are those for Grade A in some other states. For several years some people have thought that all milk would be one grade in the near future. Everyone does not believe that this will be the case soon.

The percentages of use of milk for various purposes in 1976 is shown below.*

| | |
|---|---|
| Fluid milk and cream | 42.5% |
| Cheese | 23.8% |
| Butter | 16.6% |
| Ice cream, etc. | 9.5% |
| Evaporated and condensed | 2.2% |
| Other products | 2.8% |
| Used on farms where produced for family use and to feed calves | 2.6% |

## PURPOSE OF THE MILK MARKETING PROGRAM

The availability of a good market is critical to the success of any dairy farm. A good market may be defined as one where prices paid to dairymen are high enough to provide the opportunity for a reasonable level of profit and secure enough to provide assurance of a continuing outlet for the milk. These two

* From *Milk Facts,* 1977, Milk Industry Foundation.

factors are essential, as the opportunity for a reasonable profit is necessary to continue the business and, as milk is highly perishable, assurance of a continuing daily market is necessary.

The objective of the marketing program for any individual milk producer is to receive a fair price for his milk on a continuing basis or with market security. Collectively, the objective of the U.S. milk marketing program can be stated as that of supplying consumers with all of the high-quality dairy products desired, when they are desired, as efficiently, economically, and profitably as possible. To effectively achieve these collective objectives, an adequate supply of milk as needed must be assured, it must be of high quality, fair prices must be assured for both producers and consumers, and the milk must be efficiently handled from producer to consumer. In an effort to achieve these market conditions in the United States, a milk marketing system has evolved with pricing based on usage, supply and demand, and milk composition. Market stability is enhanced by milk marketing cooperatives, federal orders, and state milk commissions.

Alternative marketing routes available to many dairymen producing Grade A milk include individual marketing agreements with processors, collective marketing thorough milk marketing cooperatives, and individual or collective marketing in a federal order or milk commission market. In some markets higher prices are enhanced by base plans, which are designed to ensure an adequate, but not excessive, supply of milk for that market. This can result in higher class I usage and a higher price received by producers. The choice of market, if a choice is available, should be made on the basis of price and market security.

## GRADE A MILK PRICING

Most dairymen who produce Grade A milk receive a blend price for that milk. The blend price is established on a 100-lb. (cwt.) basis and is affected by (1) the class usage and price of each of the classes of usage, and (2) the butterfat content and differential price.

### Usage

This refers to the proportion of the milk that is used for (1) fluid milk products, usually referred to as class I milk, and (2) manufactured milk products, usually referred to as class II milk. The milk used for class I products is paid for at a class I price, which is the highest price. The milk used for class II products is paid for at a class II price, which is lower.

Under this system the same quality milk is paid for at different levels of prices. The system is a fair one, however, for both the distributor and producer, if the prices are arrived at fairly and correctly for that market. The producer is assured of receiving the highest price for all the milk that is sold as fluid milk or cream. The distributor, on the other hand, can afford to pay the higher price for all the milk sold as fluid milk because he gets the remainder of the milk at a price consistent with the use that he can make of it.

## Butterfat

Milk contains varying amounts of butterfat and other components that can affect its value for use in various milk products. A butterfat differential system of pricing has evolved to more equitably price milk of varying butterfat content. The common practice is to establish a price per 100 lb. of milk with a specified butterfat test. The usual standard for pricing is 3.5% butterfat. There is a tendency to have this standard as near as possible to the general average of the milk sold in the market. A price differential is set up for milk testing above or below this percentage. For milk testing above the specified butterfat test (in this case 3.5%) the price is a certain number of cents above the established price per 100 lb. of milk for each 0.1% butterfat increase. Likewise the price is decreased by the same amount for milk testing below the butterfat test. The differential price varies at different times and in different markets.

## Blend Price

The blend price can be determined for a given market if the class I utilization and price, the class II utilization and price, and the butterfat standard and differential are known. The following example illustrates the calculation of the blend price per 100 lb. of milk containing 3.5% butterfat.

**Example:**
1. Market conditions:
   a. Class I utilization = 50%
   b. Class I price = $11/cwt.
   c. Class II utilization = 50%
   d. Class II price = $8/cwt.
   e. Butterfat differential = $.10/point (both class I and II), with a standard of 3.5%.

2. Blend price calculation:

$$\% \text{ class I} \times \text{class I price} = \$5.50$$

$$(50\% \times \$11)$$

$$\% \text{ class II} \times \text{class II price} = \$4.00$$

$$(50\% \times \$8)$$

Blend price for 3.5% milk = $9.50/cwt.

3. At varying butterfat contents, the blend price would vary:
   a. Blend price for 4.0% milk = $10/cwt.

   $$\$9.50 + (5 \times \$.10)$$

   b. Blend price for 3.3% milk = $9.30/cwt.

   $$\$9.50 - (2 \times \$.10)$$

4. Blend price would change if usage changed.
   a. Blend price for 3.5% milk with 90%
      class I and 10% class II usage = $10.70/cwt.

      $$(90\% \times \$11) + (10\% + \$8)$$

   b. Blend price for 3.5% milk with 30% class I and 70%
      class II usage = $8.90/cwt.

      $$(30\% \times \$11) + (70\% \times \$8)$$

## Component Pricing

The emphasis on the food value of milk is shifting away from the butterfat content. More emphasis is being placed on other solids, especially protein. If nutritional labeling is required on the milk carton, or whatever the container of the future is, all components may be listed.

There is much interest in pricing milk on the solids-not-fat (SNF) or protein content as well as on the fat content. The other components of milk vary with the fat content. On the average the SNF content varies about 0.4% for each 1% variation of fat. However, this is quite variable, and milk of the same fat content from different herds or from individual cows varies in the SNF content including the protein content.

At the present time, producers of high-testing milk are the main proponents of component pricing. Producers of low-testing milk may have as much to gain or lose, since the ratio of SNF to fat is greater in milk of the low-testing breeds.

Component pricing does not mean that dairymen would be paid a higher price for milk, but that the price would reflect the value of more components of the milk rather than one component. Each producer would be paid on the actual composition of his milk. The gain or loss over the conventional pricing system would depend on the differential value placed on each of the components used in price base.

For milk used for manufacturing milk products, the component value depends on the product. For butter making only the butterfat is used. For cheese the main components are protein and fat. In the manufacture of nonfat dry milk all the solids-not-fat are used.

The problem of pricing milk on all solid components has been considered for a long time. Some form of component pricing is in effect in some markets in California, New England, the Mississippi Valley, Associated Milk Producers Incorporated, and Canada. It is anticipated that some system will extend to other milk markets.

## Base-Surplus Plans

The amount of milk produced in most milk sheds is greater during the spring months and lower during the fall months. In order to compensate the dairyman who maintains a high fall production, when more milk is needed, many markets operate on the base-surplus plan. The base period is established during the lower-production months, using a period of three to six months. A dairyman's base is determined by the average amount of milk delivered during the base period. He normally receives a blend price for his base amount of milk and a lower, or class II, price for milk delivered in excess of his base. The base may be modified in future years if production during the base-making months is changed. This encourages the dairyman.to produce a more uniform supply throughout the year. A dairyman's base, figured as a certain percentage of all bases on a specific market, indicates his portion of the total market.

Base plans have also been implemented in some areas to ensure an adequate supply of fluid milk and to help maintain balance between the supply and the demand for milk, especially fluid milk. The lower price paid for excess milk can discourage overproduction and maintain a higher market class I usage thus avoiding decreased blend prices due to lowered class I usage.

A combination of the milk-use and base-surplus systems appears to be a rather complicated method of payment for milk, but it is workable and used in many markets.

## Special Milks

"Golden Guernsey" milk is produced by purebred Guernsey herds that comply with the regulations of the American Guernsey Cattle Club and is sold under the "Golden Guernsey" trademark at a premium above regular milk.

"All Jersey" milk is produced by purebred Jersey herds that comply with the regulations of the American Jersey Cattle Club and is sold at an advanced price under the trademark of "All Jersey."

## METHODS OF SELLING MILK

### Milk Producers' Cooperatives

The trend is for dairy farmers to sell their milk through milk producers' cooperatives. Earlier these cooperatives were organized around one market area. Today a milk producers' cooperative may have producers in several states. A large number of them are regional and largely control the majority of milk in that region. These are federated cooperatives.

The cooperative management will contract to supply milk to plants in the area with all the milk they need, but do not require them to take any surplus milk. The management will then take the additional milk and place it where it will return the most money for the producers. It may be shipped to other areas needing fluid milk, to commercial manufacturing plants, or to manufacturing plants owned by the cooperative. Some cooperatives process and distribute a part or all of their members' milk.

**Services of the Cooperative.** The dairy farmer historically sold his milk individually to a dairy processor. He had only a very limited voice in the market place. As farmers organized, they controlled larger volumes of milk and moved into a stronger bargaining position.

The cooperative keeps control of the milk until it is delivered to the plant. It owns tank trucks and hauls the members' milk, or contracts with haulers. Most of the cooperatives have large refrigerated storage tanks that aid in the distribution of milk to the best advantage.

**Fieldmen and Servicemen.** The cooperatives have their own fieldmen. One of their main responsibilities is working with producers to see that the milk meets their quality standards. Also, they serve as liaison between the member and management.

**Service Charge to Processor.** A negotiated service charge, above the regulated price, may be arranged between the cooperative and the processor. The basis for this is the service that the cooperative is providing. The processors formerly included as a part of their responsibility the fieldmen, quality control, and milk hauling. This service charge is a bargaining deal in addition to the price set by a state milk commission or by a federal marketing order.

## Milk Control Boards and State Milk Commissions

Several states have milk control boards or milk commissions that have the power to establish the price of milk to the producer and in some cases to the consumer. They may also manage a base plan. Each producer is assigned a milk base. Milk base may be bought and sold. At times, 1,000 lb. of monthly milk base is valued at about as much as a cow.

The commission or board licenses producers and distributors. The license is required to sell milk on a certain market. The purpose of the commission is to assure fair prices to the producer and in turn to provide good assurance to the consumer that an adequate supply of milk will be furnished.

The method of establishing the commission's price varies. One method is to have hearings at which both distributors and producers present data and arguments for changes, if any, that they desire. Later, after consideration of all information presented and of other basic data, the price is determined.

Another method used is to use an economic formula that accounts for several factors, including the Minnesota-Wisconsin (M-W) price, cost of feed index, labor price index, general price index, price in surrounding markets, etc.

The established price is for that portion of milk which is used for fluid milk. The remainder is paid for at a lower price.

## Federal Milk Marketing Orders

In addition to state milk commissions, there are federal milk marketing orders. Like state regulations, their purpose is to set fair prices that the processor pays to the producer. This should assure an adquate supply of milk for the consumer at a fair price. Also, it eliminates, in a manner similar to that of the state milk commission, much of the chaos on a market through unfair competition. Federal orders cross state lines, which state regulations cannot do.

The fluid milk price is currently based on the Minnesota-Wisconsin price, with an additional amount added to cover extra costs involved in the production of grade A milk. Also, transportation, base production, and other factors may be considered.

The M-W price is derived by averaging the monthly price paid by more than 500 manufacturing plants in Minnesota and Wisconsin for manufacturing grade milk. The competitive price paid by these plants reflects the demand for manufactured products, especially cheese and butter. As many of the fluid milk markets are so far from these two states, and many states produce little or no manufactured milk products, the use of the M-W price is being questioned and probably will have a lesser input into the price of fluid milk in the future.

## SELLING MILK FOR MANUFACTURING PURPOSES

Milk for manufacturing purposes is sold for a lower price than market milk. It can be produced somewhat more cheaply because regulations of milk ordinances are not quite so strict, and there is usually no surplus problem. Neither is there the same necessity for uniform production throughout the year, even though this would be desirable.

Plants using milk for the manufacture of dairy products are usually located away from large centers of population so that they will not be competing with market milk distributors.

The majority of the milk produced primarily for this purpose comes from farmers with fewer cows and more limited facilities than Grade-A producers. Many of these farmers, if properly located, are building up their herds and facilities so that ultimately they may sell their milk for use as market milk.

### Condenseries and Powder Plants

Milk plants that condense or powder the milk utilize all of the solids of the milk. The price is usually based on a formula that takes into account the price of butter and of the nonfat milk solids. The price is quoted to the farmer as a certain price per 100 lb. for 3.5% milk (or some other standard) with a stated fat differential.

### Cheese Factories

In the manufacture of cheese some of the milk solids remain in the whey. Whey contains most of the lactose, albumin, minerals, and some of the fat. Whey has some value commercially. There are many researchers looking for commercial uses for whey. Some possibilities are for food, medicinal products, animal feed, and others. It also has some feeding value, either in the liquid form or as dried whey. The whey may be returned to the producer for livestock feeding.

The price of milk for cheesemaking is based on the price of cheese. The price is usually stated on 100 lb. of standard test milk with a specified fat differential. On the average, about 10 lb. of cheese can be made from 100 lb. of 4% milk.

## Selling Sweet Cream

Some dairymen are so situated that they can sell sweet cream to ice-cream manufacturers or for making sweet-cream butter. Sweet cream may be sold on the basis of the pounds of fat that it contains or at a stated price per gallon of cream of a certain test. Single cream, also called coffee cream, contains 18% fat. Whipping cream, or double cream, usually contains 35% or more fat.

## YOUR MILK CHECK

There are several factors that determine the amount of your milk check. These include the total pounds of milk, the butterfat test, the price for class I milk, the price of class II milk, your milk base, the percentage of all milk on this market that is used in class I, in some cases quality payments; and deductions, which may include your cooperative's operations and management costs, a checkoff for advertising, promotion, and consumer education, hauling costs, and probably others.

### Advertising, Promotion, and Consumer Education

The dairy industry has for many years promoted the use of milk and milk products through advertising, promotional activities, and consumer education. This is financed by money checkoff from the producers' milk checks. This money supports educational programs of the National Dairy Council (NDC), the advertising and marketing programs of the American Dairy Association (ADA), and the milk product development and research of Dairy Research, Inc. (DRINC).

In recent years the work of these three organizations is coordinated under the overall organization of the United Dairy Industry Association (UDIA).

The checkoff is handled largely by the milk-producer cooperatives. The amount varies in different associations. The range is usually from 1 to 4 cents per 100 lb. of milk used for fluid milk. Some cooperatives support UDIA with 100% of members contributing, others are on a voluntary basis. Some make contributions from manufacturing grade milk; the plan varies. This is the dairy farmer's means of supporting programs that are vital to the sale of milk and milk products.

## CONCLUSION

Available milk marketing alternatives for U.S. dairymen include producer processing, individual contracts with processors, producer milk marketing cooperatives, federal order markets and state milk commission markets. All of these alternatives are not available in all areas or states. Some of these alternatives overlap; for example, marketing cooperatives may operate in federal order or state milk commission markets.

When more than one alternative is available to a dairyman, the decision concerning choice of market should be based on net price per 100 lb. of milk, after marketing charges, including hauling, service charges, cost of base, etc. are deducted, and market security (assurance of a continuing market for the milk). Current and historical marketing conditions, such as class I utilization, stability of base plans, etc., should be carefully analyzed before the decision is made.

Milk pricing methods may change in the near future; however, the basic factors that determine the wholesale price of milk are not likely to change. Important factors include supply and demand, product quality and variety, competitive economic position with other foods within the United States, and competitive economic position with dairy exporting nations.

Historically, relative supply and demand for dairy products has significantly affected price. Nationally, increasing demand through effective advertising offers the opportunity to increase consumption and thus demand. Another alternative is effective supply control. The former or a combination of the former and the latter seem more desirable for long-term growth of the U.S. dairy industry.

Product quality and variety have increased in recent years. Product quality is largely under the control of producers. It is every dairyman's responsibility to maintain and improve milk quality to ensure continued consumer confidence and thus demand for dairy products (Chapter 17).

Competitive position in the U.S. food market can be achieved through efficient production, processing, and distribution methods.

The economic position of the U.S. dairy industry in the world market must remain competitive if excessive imports are to be avoided. Currently, the U.S. dairy industry is economically competitive with all countries except Australia and New Zealand. However, the supply from these two countries is small relative to world production and is not likely to be sufficient to reduce world prices to their low production costs in the long run.*

---

* B. M. Burton and G. E. Frick, Can the U.S. Compete with Dairy Exporting Nations? *J. Dairy Sci.* **59**:1184 (1976).

## REFERENCES FOR FURTHER STUDY

Bartlett, R. W. Government Control of the Dairy Industry, *J. Dairy Sci.* **39**:892. (1956).

Buxton, B. M., and Frick, G. E., Can the United States Compete with Dairy Exporting Nations? *J. Dairy Sci.* **59**:1184. (1976).

Dairy Marketing Facts. *Ext. Bul.* AE-4348, University of Illinois, 1974.

Henderson, J. L., *The Fluid Milk Industry*, AVI Publishing Co., 1971.

Hillers, J. K., et al., Effect of Seventeen Milk Pricing Systems on Producers' Price of Milk, *J. Dairy Sci.*, **54**:187 (1971).

Milk Industry Foundation, *Milk Facts*, Washington, D.C., 1977.

USDA, Milk: Production, Disposition, Income, 1974–76, *Crop Rept. Bd.*, Statistical Reporting Services, May 1977.

USDA, Federal Milk Order Market Statistics—Annual Summary for 1975, *Sta. Bul.* 554, 1976.

*USDA Econ. Res. Serv.*, Dairy Situation, published five times a year.

*USDA, Agr. Econ. Rept.* 278, The Impact of Dairy Imports on the U.S. Dairy Industry, 1975.

# 19

# Establishing A Dairy Herd

The objectives or purposes of establishing a dairy herd are: (1) to earn a reasonable profit for the owner, one that will enable him to meet the financial needs of his family and eventually provide for a satisfactory retirement income and (2) gain personal satisfaction and enjoyment from his occupation. Achieving these purposes requires acquisition of needed resources (capital, cattle, land, etc.) to establish a herd that is of sufficient size and quality to have a good chance of generating sufficient income to meet financial needs. It is also necessary to provide the management skills needed to maximize probability of success in meeting the objectives. Acquiring the needed resources and skills and achieving the objectives is a tremendous challenge, but there is also tremendous opportunity in the dairy business. This latter statement was true 20, 30, and 40 years ago; it is true today and it is likely to be true for many years in the future. To be sure, capital and management skill requirements have increased because of larger herd size, larger investments per cow, and higher production levels. But opportunity has also increased because of increased genetic ability of dairy cows, the availability of better methods of growing feed, feeding, breeding, milking, etc. The availability of vaccines has reduced the risk of heavy disease losses. More stable market conditions have also developed. Opportunity can be defined as a favorable set of circumstances and the authors sincerely believe establishing a dairy herd does present a good opportunity for those who possess or can acquire the needed personal characteristics of successful herd managers and acquire the needed resources.

For those who are successful in meeting the challenges of establishing a herd, the rewards can be a reasonable profit margin, sizable long-term financial growth, personal pride, and satisfaction of ownership of productive land and cattle. There can also be achievement of short, intermediate, and long-term personal goals and a way of life for the family that is difficult to equal in most occupations. Concerning this latter point, what is the real value of giving children the opportunity to learn to work and to accept responsibility, to learn first hand of the wonders of nature, to learn to accept challenges with subsequent defeats as well as victories, and of working and sharing rewards of hard work and disappointments of misfortune together as a family unit? Each person must answer this question for himself, but the authors have had the pleasure of working with hundreds of youngsters from such families and we sincerely believe there is no better environment than a dairy farm for rearing youngsters. Unfortunately, many successful dairymen discourage interested offspring and other young people from entering the dairy business, primarily because they do not want their youngsters to face the type of problems, hard work, and disappointments they encountered on their road to success.

The decision to establish a dairy enterprise is a long-term one; that is, once it is established it is difficult to change for many years without serious consequences. Because of this, the decision to establish a dairy enterprise should be made only after: (1) careful identification of individual and family personal interests, characteristics, and goals (both financial and personal), (2) thorough evaluation of various alternatives in regard to type of dairy enterprise to establish (size of herd, breed, purebred or grade, location, etc.) in view of personal preferences and goals, (3) realistic appraisal of availability of needed resources and market, and (4) careful analysis of the probability for success in goal achievement. Doing this requires extensive self-examination, extensive knowledge of all aspects of dairying and dairy herd management, first-hand experience in dairying, and extensive planning and budgeting. Careful analysis of successful and unsuccessful dairying operations to determine reasons for success or failure can also be helpful, as can advice from experienced dairymen, extension personnel, farm credit agency personnel, etc.

This chapter is devoted to a discussion of the challenges of establishing a dairy herd, personal characteristics needed to meet the challenges, characteristics of various types of dairy enterprise, determining resources needed for the various types, planning for establishing a dairy enterprise, and identifying and evaluating some of the ways to establish a dairy herd.

## CHALLENGES

The challenges facing potential new dairymen can be categorized into two general areas: (1) the challenge of developing the personal characteristics needed for success in the dairy business, and (2) the challenge of acquiring the resources needed to establish a dairy enterprise. These challenges existed for many years, but acquire added significance as level of production increases, herd size increases, and capital needed to establish a dairy enterprise increases. Which is more important, personal characteristics or resources? Much more has been written about problems of acquiring needed resources than of acquiring needed personal characteristics; however, the authors sincerely believe that if the personal characteristics are present, the resources can be acquired. We have annually seen and are still seeing young persons, many with little or no capital assets, who possess necessary personal characteristics, find ways to acquire the needed resources to successfully establish dairy enterprises.

**Personal Challenges.** Many people who have indicated a desire to establish a dairy enterprise but have been unsuccessful in doing so, or who have established one and subsequently failed, did so because they did not possess or develop the personal characteristics necessary for success in a highly competitive, high-risk, high-opportunity business. Some of the major personal challenges that must be met if failure is to be avoided are the following.

1. Acquiring thorough, up-to-date knowledge of all aspects of dairying and dairy herd management.

2. Gaining practical experience in dairying.

3. Possessing interest, motivation, and willingness to work hard, both physically and mentally, and to make personal sacrifices, especially during beginning years.

4. Avoiding overconfidence accompanied by an unwillingness to accept advice from more experienced persons.

5. Avoiding unrealistic expectations. Many beginners observe other established dairymen with top-quality cattle, facilities, and equipment and a high income, and expect similar resources and income level the first year of operation. Trying to emulate successful dairymen is certainly a good long-range goal, but it is unrealistic to expect in beginning years. Beginners must realize that most well-established, successful operations started with more modest resources and income level and reached their current status only after years of wise management and hard work.

6. Avoiding discouragement when temporary setbacks occur. These will happen in any high-risk business.

7. Avoiding careless or unrealistic planning.

8. Accepting dairying as a business as well as a way of life.

**Resource Challenges.**  Availability and/or cost of needed resources or capital to purchase these resources may be a major challenge in establishing a dairy herd. Major resources needed that require major capital outlay include: (1) dairy cattle, (2) facilities and equipment to house and handle the cattle, (3) land for growing all or part of the forage for the cattle in most areas of the United States, and (4) marketing privileges (marketing base) in some areas. Cows and equipment usually can be purchased and facilities can be built if capital is available; however, lack of availability of land to raise feed and/or lack of availability of market base can be major problems. Lack of availability of an adequate amount of suitable land at reasonable cost to raise all or part of needed feed can result in high feed costs in many areas and contribute significantly to failure. Lack of a good market for milk reduces income and can also contribute significantly to failure.

Obtaining sufficient capital to purchase needed resources has long been a major challenge for aspiring dairymen. Depending on the location, type of facilities, type of cattle, and many other factors, capital needed to purchase all the needed resources may vary from $2,000 to $4,000 or more per cow. The average per-cow investment of 59 dairy farms who participated in the VAMA* business record program in Virginia in 1976 was $3,350. This was divided between investments in land (40%), dairy cattle and milk base (25%), buildings and equipment (25%), feed and seed supplies (9%), and other (1%).

Some basic principles of money management for beginners in the dairy business are the following.

a. The amount of borrowed capital and the interest rate on that capital are important, but not so important as the actual or potential earning power of that capital. For example, borrowed capital totalling $1,000 or less per cow can be excessive if net profit per cow per year is low, whereas borrowed capital totalling $2,000 per cow might be considered very reasonable if net profit per cow per year is high (Table 19-1). In regard to interest rates, 5% is too much to pay for capital that returns 2% net income on investment, whereas 10% is a reasonable interest rate if the capital returns 20% net income on investment. This is not to imply that there is no upper limit on the amount of capital that should be borrowed or that one should not obtain the most reasonable interest rates available. It is discussed to emphasize the point that the actual, or a realistic estimation of, potential earning power of that capital is more important than the amount and interest rate. Earning power determines debt repayment capacity (the amount of money available to service debt after all operating and family living expenses have been paid). In general, high earning power of capital invested in the dairy business is associated with high production per cow, a good market, and production cost control. These factors increase profit margin on any dairy farm. This in turn increases earning power of invested capital and increases debt-servicing capacity. Table 19-1 contains sample summary budgets for two 100-cow dairy farms (Herd A and Herd B). The money available for debt service in Herd A is almost three times as much as for Herd B, primarily because of a higher level of production. Variable costs such as feed, livestock

* Virginia Agribusiness Management Association (VAMA), analysis compiled by the V.P.I. & S.U. Agr. Econ. Dept., 1976.

**Table 19-1** Summary Cash Budgets for Two 100-Cow Holstein Herds

|  | Herd A | Herd B |
|---|---|---|
| Cash farm receipts |  |  |
| Milk (15,000 lb./cow @ $10.20/cwt.) | $153,000 |  |
| (12,000 lb./cow @ $10.20/cwt.) |  | $122,400 |
| Cattle (cull cows, heifers, bull calves) | 7,000 | 7,000 |
| Other | 1,000 | 1,000 |
| **Total cash farm receipts** | **$161,000** | **$130,400** |
| Cash farm operating expenses |  |  |
| Purchased feed | $ 39,000 | $ 34,000 |
| Crop expenses | 15,000 | 15,000 |
| Hired labor | 15,000 | 15,000 |
| Livestock services | 10,000 | 9,400 |
| Machinery & facility maintenance & repair | 6,000 | 6,000 |
| Fuel and supplies | 8,000 | 8,000 |
| Taxes, insurance, and utilities | 8,000 | 8,000 |
| Miscellaneous | 3,000 | 3,000 |
| **Total cash farm operating expenses** | **$104,000** | **$ 98,400** |
| Other cash expenses |  |  |
| Family living | $ 7,000 | $ 7,000 |
| Income taxes | 4,000 | 2,000 |
| Capital expenses (machinery replacement) | 10,000 | 10,000 |
| **Total other cash expenses** | **$ 21,000** | **$ 19,000** |
| **Total cash expenses** | **$125,000** | **$117,400** |
| **Balance available for debt service** | **$ 36,000** | **$ 13,000** |

services (milk hauling), etc., increase with increasing level of production; however, fixed costs remain essentially the same regardless of production level. The result is increased earning power of invested capital, profitability, and debt-servicing capacity.

The amount of capital that can be serviced by the cash available for that purpose is dependent on the interest rate and length of the loan. Table 19-2 is an abbreviated annuity table that can be used to determine the amount of debt that a given amount of cash can service at various interest rates and time periods. For example, the $36,000 available for debt service from Herd A can service a loan of $353,288 at 8% for 20 years ($36,000 ÷ .1019). Under the same interest and time conditions, the $13,000 from Herd B can only service a $127,576 loan ($13,000 ÷ .1019). If the interest rate in both cases were 10% instead of 8%, the loan limit would be reduced to $306,383 ($36,000 ÷ .1175) for Herd A and $110,638 ($13,000 ÷ .1175) for Herd B. If the time period were reduced to seven years at 10%, these amounts would be $175,268 for Herd A and $63,291 for Herd B. These examples illustrate the effect of earning power of invested capital, interest rate, and loan time period on debt repayment capacity of a dairy business.

In Herd A a total debt load of $200,000 ($2,000 per cow) composed of an intermediate loan of $90,000 at 10% for seven years and a long-term loan of $110,000 at 8% for 20 years could be handled with a reasonable safety margin, but a $100,000 total debt load ($1,000 per cow) composed of $40,000 for seven years at 10% and $60,000 for 20 years at 8% could be excessive for Herd B. Debt service needs for Herds A and B under these conditions can be calculated by multiplying the amount of the loan by the annuity factor (from Table 19-2) for the appropriate interest rate and time period. These computations for Herd A and Herd B under the described conditions are as follows

Herd A = $ 90,000 × .2054 =        $18,486
           110,000 × .1019 =        11,209

Total dollars needed for debt service    $29,695
Total dollars available for debt service   $36,000

Herd B = $ 40,000 × .2054 =        $ 8,216
           60,000 × .1019 =        6,114

Total dollars needed for debt service    $14,330
Total dollars available for debt service   $13,000

b. When capital is limited, proper investment priorities must be established and followed. Investment in income-producing assets, such as high-production cows, and in milk base in areas where acquiring base is

**Table 19-2** Annuity Factors

| No. years | Interest Rate | | |
|---|---|---|---|
| | 8% | 10% | 12% |
| 3 | 0.3880 | 0.4021 | 0.4164 |
| 5 | 0.2505 | 0.2638 | 0.2774 |
| 7 | 0.1912 | 0.2054 | 0.2191 |
| 15 | 0.1168 | 0.1315 | 0.1468 |
| 20 | 0.1019 | 0.1175 | 0.1339 |
| 25 | 0.0937 | 0.1102 | 0.1275 |

essential for a good market, must take top priority. Investment in rapidly depreciable and convenience items (new buildings, equipment, etc.), which may be prestigious to own and ease the work load but do not increase the productivity of the operation, must receive low priority during beginning years. Concentrating investments in productive assets can help minimize capital needed to get started, thus enhancing the probability of obtaining the needed capital and helping to minimize debt load. Beginners with good cows, a good market, productive land (owned or rented), and adequate used buildings and equipment are more likely to be successful than those with inferior cows, market, and land, new facilities, and new automated equipment. The former are also much more likely to obtain needed capital than the latter. In general, money lenders look much more favorably on applications for capital to purchase cows and land than they do on applications to purchase automated equipment and new facilities.

c. Negotiate for favorable interest rates and repayment conditions. Beginners often overestimate production level and underestimate expenses or fail to allow enough financial flexibility to accommodate a bad crop year, lowered milk prices, or other temporary adversities that can occur. Based on optimistic estimates, they often borrow considerable short- and intermediate-term capital with accompanying high repayment demands. It is better to estimate repayment capacity conservatively and negotiate for longer terms with lower annual repayment demands, thereby providing a safety margin in case of temporary reversals. Shopping for the most favorable terms and interest rates is just as important as shopping for other resources.

Proper management of cattle, land, and people as well as other aspects of money management are also important, but are discussed elsewhere in this book. All are critical to the success of any dairy operation, but even more critical to the beginner because of his typically heavy debt load.

## PERSONAL CHARACTERISTICS

Before accepting the challenge of establishing and operating a dairy herd, individuals should identify those personal characteristics needed for success by thorough, careful self-examination to determine their presence or absence. If the needed personal characteristics are not present, perhaps the individual should seek another occupation, one that is better suited to his interests, abilities, and goals. Some of these needed characteristics are discussed below.

1.  Interest in dairying, accompanied by a positive and confident but realistic attitude. Interest in dairying as a way of life as well as a means of earning a living is important in achieving personal satisfaction and financial success. A positive and confident attitude is a trait of most successful people. It is associated with people who look for ways to solve problems and are confident they can do so. People who lack this characteristic often look for reasons why the problem can't be solved and convince themselves that a solution is impossible. Problems will occur in establishing a dairy herd; positive, confident persons will become a part of the solution to these problems, whereas those who lack this attitude will become a part of the problem. Realism, however, is also needed to avoid errors resulting from overconfidence. Unrealistic initial expectations concerning level of production, price of milk, profitability, etc., must be avoided.

2.  A high degree of motivation, perseverance, and willingness to make personal sacrifices is essential, especially during the first few years. Motivation is a powerful force in achieving success. It is difficult to define, but relatively easily identified. Highly motivated people usually have well-defined goals. They know where they are going, and this provides the incentive for them to identify what is necessary to get there and to do those things that are necessary to achieve their goals. Perseverance might be defined as the ability to persist in any enterprise undertaken, in spite of setbacks and reversals, until success is achieved. Because of the many risks involved in dairying, temporary setbacks will occur, and these can discourage people who lack perseverance. New dairymen and their families need to be willing to sacrifice personal luxuries during beginning years. Wise money management during these years, that is, using generated capital to improve equity position or increase productivity of the operation instead of purchasing personal luxury items is a good business practice and often essential. Substituting hard work for automation may also be necessary until capital position is improved. This does not imply that establishing a herd should be considered as buying a lifetime of hard work, long hours, and substandard family living, but rather that these personal sacrifices may be necessary for some years until a favorable capital position is established.

3.  Thorough, up-to-date knowledge of all aspects of dairying and the desire and ability to continue learning. Knowledge of good business practices as well as up-to-date knowledge of feeding, breeding, etc., is essential in any competitive business, as is keeping up to date as new knowledge is discovered and new techniques developed.

4. Practical experience. Knowledge is important, but equally important is the ability to apply this knowledge. Practical experience can also help determine an individual's degree of interest in dairying and add realism to his ideas and expectations. There is no substitute for practical experience for persons who wish to establish a dairy herd.

5. The ability and willingness to work hard, both physically and mentally, is essential to success. As mentioned previously, hard work may need to be substituted for mechanization in beginning years. Some hard, physical work is continually a part of all dairy operations and must be accepted as part of the occupation. Mental work is just as much a part of dairying as physical work. This responsibility becomes increasingly important as herd size, level of production, and capital investment increase.

6. The ability to manage cattle, land, money, and people. As indicated earlier in this chapter, capital requirements to establish a herd are high. In most cases, a high proportion of the capital needed to establish a dairy enterprise is necessarily borrowed capital. The additional cash costs incurred as the result of using large amounts of borrowed capital can be substantial and this fact makes wise money management essential if failure is to be avoided.

7. The desire, determination, courage (guts), physical and mental ability, and disposition to accept, even welcome, the challenges and risks involved in establishing and operating a dairy enterprise.

Thorough self-examination in regard to these seven characteristics is recommended before a decision is made to establish a dairy herd. Persons whose personal goals and interests tend more toward security and regular hours, or who are adversely affected by pressures of debt load and other risks (uncertainity of weather, diseases, etc.), would do well to avoid ownership of a dairy enterprise. Conversely, those who thrive on challenge and possess the other characteristics previously discussed can derive a great deal of personal satisfaction and achieve a high degree of financial success from establishing and operating a dairy enterprise.

## TYPES OF DAIRY ENTERPRISE

Selecting the type of dairy enterprise to establish is largely a matter of personal preference, within certain limitations. There are highly successful dairy enterprises of all breeds, herd sizes varying from 25 to several thousand, grades and purebreds, and in nearly all parts of the United States. There are also unsuccessful dairy operations of nearly all of these types and sizes. This opportunity to achieve success in a wide variety of types of dairy enterprise allows individuals the flexibility to choose the type of operation best suited to available resources and to personal interests and goals.

For comparative purposes, we will classify types of dairy enterprise as grade or purebred, family-size (100 cows or less) or large herds, farm-raised feed (forage or forage and part of the concentrate) or purchased feed, and total herds (cows and replacements) or milking herds only. The factors that should be evaluated in deciding among these and additional alternatives in dairying are: (1) personal preference, (2) availability of resources and (3) market conditions.

## Grades or Purebreds

In general, grade cattle can be purchased more economically per unit than purebreds of comparable genetic ability. This results in lower initial capital needs for cattle purchases but less opportunity for higher income from later cattle sales. Conversely, potential returns from sale of breeding stock can be significantly higher from purebred cattle if the owner is interested in and has the ability to merchandise cattle and there is an available market for those cattle. Personal preference is a strong factor in this decision. It is questionable whether those interested only in producing milk should invest the additional capital needed to purchase purebreds. If personal preference is for purebreds, this may need to be tempered with availability of extra needed capital for initial purchases and for merchandising programs during beginning years, availability of purebred cattle at reasonable cost, and purebred market availability. These statements are not meant to imply a preference for grade cattle, but rather to stress the importance of evaluating all three factors—personal preference, resource availability, and market—before the decision is made. For many beginners a combination of the two is often preferable to either one. Starting with a combination of grades and purebreds can help reduce capital requirements for establishing the herd and operating it during beginning years, yet allow for eventual significant income from purebred cattle sales.

In regard to breed of cattle, there are good cows and profitable herds and inferior cows and unprofitable herds of all the major U.S. dairy breeds. Choice of breed is largely a matter of personal preference. The choice of breed is usually based on one's association with it. One is often influenced by the breed kept by his neighbors. The availability of foundation animals for starting the herd is a factor. The type of market for milk may play a part in the selection.

The selection of the breed is not as important as getting the right kind of foundation animals with which to begin. The major breeds vary little in real efficiency. There is much more difference among cows in the same breed than there is among breeds. A person who does well with one breed can be expected to do well with another.

## Size of Herd

In general, larger herds require a greater capital investment and a higher degree of management skill than smaller ones. They also offer a greater opportunity for total profit if profit margin per cow can be maintained at the same level as in smaller herds. Smaller, family-size dairy operations usually have the advantages of lower total initial investment, a greater degree of financial flexibility (because labor expenses are chiefly noncash expenses), and they can offer the owner the opportunity to gain experience, management skills, and equity growth with less financial risk.

Decisions regarding initial herd size, then, are primarily determined by the extent of available resources and management ability. A certain minimum size, however, is necessary to ensure opportunity for adequate income to meet family living expenses and debt-servicing obligations. Personal preferences, goals, and interests are also important.

## Type of Feeding Program

Climate, cost and availability of land, and cost and availability of feed for purchase are the primary determining factors in regard to choice of feeding program. This, in turn, greatly influences initial capital requirements.

The primary advantages of raising all or most of the forage and part of the concentrates needed are better assurance of high-quality forage in most areas of the United States and more economical feed costs if good farming practices are followed (see Chapter 6). The disadvantages are larger initial investments needed to purchase land and cropping machinery; larger initial operating capital needs; additional management skills and labor required; and needed land resources may not be available at reasonable cost. Depending on the area of the United States and the proximity of urban developments, land purchases for feed production may double or triple the amount of initial capital required to establish a herd, or may not be available. Renting land resources may provide temporary relief, but is risky unless long-term lease conditions can be arranged. Machinery investments are sizable, but may be minimized by careful shopping for used equipment and/or hiring someone to perform the operations that require very expensive machinery (custom harvesting, etc.), in areas where reliable service is available at reasonable cost.

The major advantages of purchased feed, conversely, are lower initial investment and more time available to devote to acquiring and using herd-management skills. Major problems include providing an adequate supply of high-quality feed (forage especially) at reasonable cost, and loss of the potential value of land appreciation.

Alternatives in regard to feed production or purchase vary considerably in various areas of the United States and within each area depending on the degree of urbanization. Decisions in this area should be made only after thorough evaluation of the alternatives available in a given location. In some areas a readily available supply of by-product feeds, combined with high land cost and/or unsuitable climatic conditions, may dictate a purchased feed program. In others, more favorable climatic conditions combined with reasonable land costs and lack of available high-quality feeds for purchase may necessitate raising nearly all of the feed supply.

### Cows Only or Total Herds

The advantages of rearing versus purchasing replacements were discussed in detail in Chapter 14. We will mention here only that the major advantage of purchasing replacements is that this can significantly reduce initial capital needs for cattle, land, and facilities. The major disadvantages can be lack of availability of and high cost of high-quality replacements when needed. Again, the decision in this area should be based on careful evaluation of available alternatives in a given location.

As mentioned initially in this section, a wide variety of types of dairy enterprise can be successful in most parts of the United States. Market conditions and availability and cost of needed resources vary widely in different parts of the country. Before choosing a location to establish a herd, these factors should be determined and carefully evaluated.

### PLANNING A DAIRY ENTERPRISE

The plan should include: (1) realistic estimates of the amount and cost of resources needed to establish the desired type of dairy enterprise, (2) a herd management plan that is conducive to a continuing high and profitable level of production, and (3) financial planning, including realistic operating budgets to determine the financial feasibility of the plan. Such planning can help avoid many serious errors and enhance the chance for success in achieving both profit and personal goals.

Planning for establishing any dairy herd must be carried out on an individual basis because of variations in personal goals, abilities, and re-sources and is therefore, beyond the scope of this text. Herd management practices needed for continuing high and profitable levels of production have been discussed elsewhere in this book. Certain basic principles of planning the establishment of a dairy enterprise, however, can be applied to most situations, and these will be discussed.

## Amount and Cost of Needed Resources

Kind, amount, and cost of resources vary considerably with many factors; however, the following basic principles should be followed in acquiring resources.

1. Establish proper investment priorities. The major source of income in the dairy enterprise will be the sale of milk. Acquisition of high-production-potential cows and replacements must therefore be the top priority investment. Their productivity will largely determine the amount of income generated to pay operating expenses and family living expenses, to service debt, and to acquire additional resources. In areas where market base is essential to assure a good market, purchase of this resource should also receive top priority. High production per cow, a good market for milk, and a high percentage of the total investment in cows (and base where applicable) have consistently been reported as characteristics of profitable herds.

   The major production expense is feed, and most dairy farms produce most or a significant part of the needed feed. On these farms acquisition of a sufficient amount of productive land to produce this feed economically should be the next priority item. When capital is severely limited, renting this resource may be a suitable alternative. Enough land should be acquired to allow a reasonable degree of flexibility. For example, if average yield data indicate 1.5 acres/cow are needed to provide sufficient forage, an additional flexibility allowance of 15 to 25% additional land should be acquired (a total of 1.7 to 1.9 acres/cow) to allow for poor crop years.

   Facilities and equipment should receive third investment priority. Overinvestment in elaborate or unnecessary facilities and equipment can greatly increase debt load and the amount of money needed to service this debt. This reduces profits and should be avoided.

2. Shop carefully so as to minimize initial capital requirements. Purchase only what is needed to get the job done right with a reasonable margin of flexibility. This does not imply that one should buy cheap, inferior cows,

land, and equipment, but rather that the earning power of each resource investment in relation to its cost should be carefully considered, and that purchasing unneeded or unprofitable resources should be avoided. One high-producing cow may return more net profit than three low producers; one acre of good land may produce more feed at less expense per unit of feed than two acres of marginal land; and undependable equipment may break down at inopportune times and cause severe losses.

3. Allow for acquisition of sufficient operating capital to obtain needed supplies and services until these expenses can be met from generated income. In establishing a dairy herd, as in most other businesses, expenses are incurred before any or significant amounts of income are realized. In many cases, expenses for crop supplies, equipment and facilities, labor, etc., precede income by several months. In addition, income level may not reach full potential for several more months or even a year or more after the enterprise is established. Proper planning in this regard can avoid serious problems.

An example of resource planning for a 100-cow grade Holstein herd is presented in Table 19-3. The investment estimates were derived from 1975 Virginia investment summary data,* prevailing costs of good grade Holstein cattle, and used facilities and equipment in good condition. This information is presented as an example of resource planning for discussion purposes. It should not be used to plan any individual dairy enterprise, but rather as a format for estimates based on actual conditions prevalent in your specific area and time.

From this example it can be seen how capital investment requirements could be significantly increased if purebred rather than grade cattle were purchased (estimated at $150 more per cow, or an increase of $15,000), if new facilities were built (at least $500 more per cow, or an increase of $50,000), if new equipment were purchased (estimated at $350 more per cow, or an increase of $35,000). If all three were included, the total investment could easily reach over $400,000, or exceed $4000 per cow. Purchasing marketing base in some areas could add another $400 to $800 per cow and increase per-cow investment to over $4,500. The effect on capital needs of renting part of the resources such as land and facilities, purchasing replacements, and/or having part of the cropping work performed on a custom basis rather than purchasing expensive equipment can also be clearly seen. Again, there is no one best answer for all, but individual planning to estimate needs and costs is necessary to determine resource and capital needs.

---

* VAMA analysis compiled by the V.P.I. & S.U. Agri. Econ. Dept. 1975.

**Table 19-3** Sample Resource Amount and Cost Planning for Establishing a 100-Cow Dairy Grade Holstein Herd (1975 Virginia Estimate)

| Item | Number | Total cost | Cost per cow |
|---|---|---|---|
| Cattle | | | |
| Cows, milking age | 100 | $ 75,000 | $  750 |
| Heifers, bred | 20 | 12,000 | 120 |
| Heifers, yearlings | 20 | 8,000 | 80 |
| Heifers, birth-yearlings | 30 | 10,000 | 100 |
| **Total cattle** | **170** | **$105,000** | **$1,050** |
| Land for forage production | | | |
| Crop land | 150 | $ 90.000 | $  900 |
| Other pasture, lots, etc. | 50 | 15,000 | 150 |
| **Total land** | **200 acres** | **$105,000** | **$1,050** |
| Facilities, existing, used | | $ 50,000 | $  500 |
| Machinery & equipment, used | | $ 50,000 | $  500 |
| Miscellaneous | | $  5,000 | $   50 |
| **Total facilities, machinery, equipment, etc.** | | **$105,000** | **$1,050** |
| **Total resource investment** | | **$315,000** | **$3,150** |
| Operating capital allowance | | $ 30,000 | $  300 |
| **Total capital needs** | | **$345,000** | **$3,450** |

## Financial Feasibility

The potential of the newly established dairy enterprise to meet financial goals depends on the earning power of the invested capital. This can be estimated by budgeting, that is, calculating an estimated income-cost statement using realistic (conservative) income estimates and realistic (liberal) expense estimates. Sample summary budgets for two herds were presented in Table 19-1. To estimate financial feasibility accurately, the following procedure should be followed.

1. Plan the total herd management program, including feeding program, feed production program, labor program, breeding program, etc., including amount and estimated cost of each operating expense item.

2. Prepare detailed cash enterprise budgets for the dairy operation and each

crop enterprise on the basis of the information included in the herd management program and the best available current cost estimates (similar to the profit-and-loss statements presented in Chapter 3).

3. Prepare a cash budget for the total dairy farm operation, including family living, income taxes, and capital expenses (Table 19-4) for the initial year and summary budgets for several succeeding years (Table 19-5).

If realistic estimates are used in calculating these budgets, they can reliably evaluate the potential profitability and financial feasibility of the operation. They can also be used effectively to determine debt-servicing capacity, to support loan applications, to estimate net worth growth potential, and to calculate projected cash flow statements. They form the plan for the new dairy enterprise.

## WAYS TO GET STARTED

Persons who possess the personal characteristics of successful dairymen and a significant amount of capital can establish a dairy herd without great difficulty. Those who possess neither should not try to establish a herd. Therefore the following discussion pertains primarily to those persons who possess the needed interest, determination, courage, and other personal characteristics, but who lack sufficient capital to establish a dairy herd. As mentioned earlier, it can be done; it is being done annually, as evidenced by success stories in farm magazines and the personal observations of the authors for many years.

For those with a limited amount or no capital resources, three basic alternative methods, and many variations of each, of getting started should be evaluated. These are:

1. Work for someone else to obtain needed capital.

2. Present a sound plan to a credit source to obtain use of borrowed capital or low-equity financing.

3. Stretch limited capital by minimizing purchased resources and renting the rest.

All three methods involve making maximum use of abilities and talents that aspiring dairymen possess (the ability to work hard and think clearly, ambition, energy, dependability, up-to-date knowledge of dairying, etc.) to obtain needed capital. All three methods and various combinations and variations of these methods have been successful. For example, many have

**Table 19-4** Sample Cash Budget for a 100-Cow Dairy Operation for the Initial Year of Operation

| Item | Price/unit | Quantity | Total |
|------|-----------|----------|-------|
| Cash receipts | | | |
|   Milk (13,500 lb. sold/cow) | $10.20/cwt. | 13500.0 | $137,700 |
|   Cull cows and heifers | $240.00/head | 25.0 | 6,000 |
|   Bull calves | $20.00/head | 50.0 | 1,000 |
|   Miscellaneous | | | 1,000 |
|     **Total cash receipts** | | | **$145,700** |
| Cash farm operating expenses | | | |
|   Purchased feed (2.5T./cow) | $130.00/T. | 250.0 | $ 32,500 |
|                 (.5T./heifer) | $140.00/T. | 35.0 | 4,900 |
|   Crop expenses (from enterprise budget) | | | |
|     Corn silage | $100.00/A. | 130.0 | 13,000 |
|     Hay | $50.00/A. | 20.0 | 1,000 |
|     Pasture | $25.00/A. | 40.0 | 1,000 |
|   Hired labor | $10,000.00/man | 1.5 | 15,000 |
|   Milk hauling | .40/cwt. | 13500.0 | 5,400 |
|   Veterinary and medicine | $20.00/cow | 100.0 | 2,000 |
|   DHI | $10.00/cow | 100.0 | 1,000 |
|   Breeding (130 breeding age females) | $15.00/ | 130 | 1,950 |
|   Equipment maintenance and repair | — | — | 3,000 |
|   Facility maintenance and repair | — | — | 2,000 |
|   Fuel | — | — | 4,000 |
|   Supplies | — | — | 4,000 |
|   Taxes | — | — | 3,000 |
|   Insurance | — | — | 2,000 |
|   Utilities | — | — | 3,000 |
|   Miscellaneous | — | — | 3,000 |
|     **Total cash farm operating expenses** | | | **$101,750** |
| Other cash expenses | | | |
|   Family living | — | — | $ 7,200 |
|   Personal income tax | — | — | 3,000 |
|   Capital expense (machinery and equipment replacement) | — | — | 10,000 |
|     **Total other cash expenses** | | | **$ 20,200** |
| Summary | | | |
|   **Total cash receipts** | | | **$145,700** |
|   **Total cash expenses** | | | **$121,950** |
|   **Total available for debt service** | | | **$ 23,750** |

**Table 19-5** Sample Summary Cash Budgets for a 100-Cow Dairy Operation for Four Years of Operation

| Item | Year 1 | Year 2 | Year 3 | Year 4 |
|---|---|---|---|---|
| Cash receipts | | | | |
| Milk (13,500 lb./cow sold @ $10.20/cwt.) | $137,700 | | | |
| (14,200 lb./cow sold @ $10.20/cwt.) | | $144,840 | | |
| (14,600 lb./cow sold @ $10.20/cwt.) | | | $148,920 | |
| (15,000 lb./cow sold @ $10.20/cwt.) | | | | $153,000 |
| Cull cows and heifers (35/yr. after 1st yr.) | 6,000 | 8,400 | 8,400 | 8,400 |
| Bull calves | 1,000 | 1,000 | 1,000 | 1,000 |
| Misc. (surplus heifers after 2 yr.) | 1,000 | 1,000 | 2,500 | 4,000 |
| **Total cash receipts** | | | | |
| | **$145,700** | **$155,200** | **$160,820** | **$165,400** |
| Cash farm operating expenses | | | | |
| Purchased feed | $ 37,400 | $ 39,000 | $ 40,000 | $ 41,000 |
| Crop expenses | 15,000 | 15,000 | 15,000 | 15,000 |
| Hired labor | 15,000 | 16,000 | 17,000 | 18,000 |
| Livestock services | 10,350 | 10,700 | 11,000 | 11,300 |
| Machinery, equipment, and facilities maintenance and repair | 5,000 | 5,000 | 5,000 | 5,000 |
| Fuel and supplies | 8,000 | 8,000 | 8,000 | 8,000 |
| Taxes, insurance and utilities | 8,000 | 8,000 | 8,000 | 8,000 |
| Miscellaneous | 3,000 | 3,000 | 3,000 | 3,000 |
| **Total cash farm operating expenses** | **$101,750** | **$104,700** | **$107,000** | **$109,300** |
| **Other cash expenses** | | | | |
| **Family living** | $ 7,200 | $ 8,000 | $ 9,000 | $ 10,000 |
| **Personal income tax** | 3,000 | 4,000 | 5,000 | 6,000 |
| **Capital expenses** | 10,000 | 10,000 | 10,000 | 10,000 |
| **Total other cash expenses** | **$ 20,200** | **$ 22,000** | **$ 24,000** | **$ 26,000** |
| Summary | | | | |
| **Total cash receipts** | **$145,700** | **$155,200** | **$160,820** | **$165,400** |
| **Total cash expenses** | **$121,950** | **$126,700** | **$131,000** | **$135,300** |
| **Total available for debt service** | **$ 23,750** | **$ 28,500** | **$ 29,820** | **$ 31,100** |

worked for wages until a limited amount of capital was acquired and then borrowed additional capital or purchased minimum resources and rented the rest.

## Exchanging Service for Capital

This involves working for wages in on- or off-farm jobs until sufficient capital is accumulated and credit established to purchase needed resources. Persons who possess the characteristics of aspiring dairymen rarely have difficulty obtaining good jobs either on the farm or in agri- or other businesses. It is difficult, however, and requires great self-discipline, to save significant amounts of capital from wages because of the financial pressure of living expenses and taxes. A more desirable method for many is to work on a successful dairy farm, initially with a wage agreement and then with a wage and bonus agreement, and eventually in some type of partnership or corporate structure, or in lease or purchase arrangements (Chapter 16).

## Low-Equity Financing

Low-equity financing is possible, but not easy to obtain. If equity or owned capital is minimal (20% or less than the amount needed), the two major sources are F.H.A. or individuals (parents, relatives, etc.). If equity reaches 25% or more, many more credit agencies are available such as commercial banks and banks of the Farm Credit System, Federal Land Banks (FLB's), Federal Intermediate Credit Banks (FICBs), which supervise the local Production Credit Associations (PCAs), and the Bank for Cooperatives. Lending agencies have variable policies in regard to equity and/or collateral requirements, loan size, interest rate, and time period; therefore thorough investigation of the alternative sources in regard to these items is important. Often a combination of two or more credit agencies can be used to advantage (one for long-term capital and another for intermediate or operating capital).

The four things that are important in obtaining financing are collateral, a plan, evidence of ability and experience to make the plan work, and evidence of personal reliability. As collateral is limited in low-equity financing, the latter factors become increasingly important. A sound plan, one that includes financial as well as herd management planning, developed for the specific proposed operation, is essential. Realistic estimates of capital needs, uses of capital, expected income and expenses, cash flow, and evidence of profitability or ability of the proposed operation to generate sufficient income to pay the bills, meet family living expenses while maintaining productivity, and meet loan service obligations are essential (discussed earlier in this chapter and in Chapter 3). Ability to implement the plan is best evidenced by experience and demonstrated ability to work with cows, land, etc. Several years of experience working with a successful dairy operation, evidence of up-to-date knowledge of dairying, and ability to use this knowledge to obtain high performance from cattle, land, etc., is usually necessary to inspire confidence in potential lenders. Evidence of personal reliability can be established by proper handling of personal finances while gaining experience, that is, by steadily improving net worth and paying bills promptly during the years spent gaining experience.

All too often aspiring dairymen approach lenders with ideas and enthusiasm, but without sound plans and evidence of ability to implement the plans. Many do not appear to understand that lenders are not necessarily gamblers, but rather they are looking for investments that show evidence of minimum risk and maximum opportunity for success. A sound plan, experience, and evidence of personal reliability minimize risk and maximize opportunity for success for both the dairyman and the lender.

## Minimizing Capital Needs

Various ways of minimizing capital requirements—renting various resources and services, starting small (but large enough to be an economically feasible unit), establishing proper investment priorities, and substituting personal labor for expensive automated equipment—were discussed earlier in this chapter. Careful shopping for all resources can also be very helpful. In regard to the latter, proper planning greatly enhances ability to more economically purchase resources. If one knows what is needed and when it is needed, this allows ample time to compare alternative sources of the needed resource and purchase at the most favorable price. Often lower-cost heifers or calves, rather than cows, can be purchased, as can good used equipment rather than new, etc.

## SUMMARY

Ways to get started are almost as numerous as dairymen who have established dairy enterprises. Common to most are possession of the personal characteristics discussed earlier in this chapter, as these enable one to overcome capital acquisition challenges. Key factors in establishing a herd are making maximum use of personal ability to acquire needed capital, minimizing initial capital needs, and developing a sound plan. For the beginner it will not be easy, and it will require a great deal of motivation, perseverance, hard work, and personal sacrifice. All who try will not be successful, but then all who started 20, 30, or 40 years ago were not successful either. For those who are, the financial and personal satisfaction rewards are high. Key factors in success for the new operation are adequate volume, high production, and cost control. These factors increase the probability of high earning power of invested capital and continued profitability and financial growth. This, in turn, improves asset-liability ratio and enables one to obtain financing more easily for enterprise growth, should this be desired.

## REFERENCES FOR FURTHER STUDY

Brown, L. H., and Speicher, J., What Is Your Debt Paying Capacity? *Hoard's Dairyman,* August 25, 1972.

Brown, L. H., and Speicher, J., How Much Debt Can a Cow Carry? *Hoard's Dairyman,* November 1973.

Harris, H. M., Jr., and Loope, K. E., Getting Started in Dairying—Problems and Potentials, *Dairyman Inc. News,* June 1971.

Hoglund, C. R., Five Ways to Get Started in Dairying, *Hoard's Dairyman,* August 1971.

Longo, L. P., The "Beginner" in the Dairy Business, *Hoard's Dairyman,* November 1970.

Smith, R. S., Is It Possible to Finance a Start in Dairy Farming? *Hoard's Dairyman,* January 1975.

# 20

# Expanding the Dairy Herd

**20**

The trend toward increasing size of dairy herds in the United States is well established. The primary reason for expansion by most dairymen is that expansion can afford the opportunity to increase total profitability of the dairy enterprise. The major objective or purpose of expansion is, therefore, to increase total profit to meet the desire for higher income level. Other objectives may also be important in some situations. Some of these are: (1) to justify purchases of automatic materials handling equipment or modernization of facilities in order to reduce physical labor requirements; (2) to overcome dissatisfaction with current conditions such as lack of time off in one-man operations or loss of interest because of lack of challenge; (3) to make fuller use of available resources such as labor as children grow, land that may currently be used for other purposes or becomes available, etc.; and (4) to enable owners to meet current debt obligations.

Expanding herd size can offer the potential to achieve these objectives in the following ways.

1. Generate more total gross income from the sale of more milk, breeding stock etc.

2. Justify the cost of increased automation of materials handling equipment by spreading fixed costs over more producing units. This can also result in increased labor efficiency.

3. Generate greater output per dollar invested by increased utilization of some equipment and facilities.

4. Result in lower investment costs on a per-cow basis through increased utilization of some equipment and facilities.

5. Increase the opportunity for discount purchasing due to quantity buying of supplies and services.

Most dairymen are fully aware of these advantages of larger herd size. Fewer are aware of the limitations and potential problems of larger herd size. Just as size can increase profit potential, it can also increase potential for loss, because as opportunity increases, so does risk. Below are some conditions that often accompany expansion that can increase risk.

1. Decreased production per cow, from a variety of factors. Some of the more prevalent ones are lowered reproductive efficiency, less rigid culling, an inadequate supply of high-quality feed, and increased incidence of herd health problems.

2. Higher animal mortality, especially calves, primarily because of less individual care or neglected cowmanship. Contagious diseases also tend to spread more quickly and extensively when cattle are kept in close proximity to each other, which is more usual in larger herds.

3. Additional expenses of borrowed capital. Expansion usually requires borrowing additional capital to purchase additional cattle, buildings, equipment, etc., and this increases debt service expenses.

4. Increased hired labor expenses, as a greater proportion of the labor is often hired, rather than upaid personal and family labor.

5. Decreased financial flexibility, primarily from items three and four. As the proportion of owned to indebted assets decreases (a greater proportion and amount of borrowed capital), the financial obligation to meet interest and principal payments increases and must be met on a regular basis. If most of the capital is owned, the obligation of the business to generate sufficient income to pay a reasonable interest on the investment is present, but it is not a cash obligation that must be met regularly, and an occasional bad year or two can be tolerated. A similar condition exists in regard to labor expenses. It is desirable for the business to generate adequate profit to pay the owner and his family a good wage; however, in lean periods the amount and regularity of payments can be flexible. In cases of hired labor, meeting the payroll is a rigid, regular obligation. Often, in large herds, a greater proportion of the feed, other supplies, and services are purchased rather than home grown or provided by the owner. These items can also increase actual financial obligations and decrease financial flexibility.

6. Inadequate resource and/or financial planning. Inadequate resource planning can result in shortages of feed, land on which to grow feed, cattle, labor, etc. Expansion of herd size without adequate supporting resources must be avoided. Inadequate financial planning can result in major errors in initial costs of expansion, operational cash needs of the expanded operation, and profit margin on the expansion only as well as on the total operation after expansion.

Decisions regarding whether to expand, when to expand, and how to expand should be made only after thorough evaluation of: (1) current status, (2) personal goals, preferences, and abilities, (3) availability and cost of needed resources, and (4) availability of a good market. If the decision is made to expand, then thorough resource, operational, and financial planning must take place to maximize probability of success and minimize probability of failure. This chapter is devoted to a discussion of the principles and factors involved in making these decisions.

## CONSIDERATIONS IN EXPANSION

The basic decision of whether to expand depends primarily on current status, personal factors, availability of needed resources, and market availability.

### Current Status

Expansion is not a substitute or a cure for poor management; rather it is likely to increase management problems. Therefore only dairymen who have demonstrated the ability to manage smaller herds for a high and profitable level of production should consider increasing herd size. If current performance of the herd is low because of poor management of the feeding, breeding, health care, etc., programs, these dairymen should improve the current operation before expanding—that is, get better before they get bigger.

Conversely, if current productivity and profitability per cow is high and a favorable asset-to-liability ratio has been achieved, but total income is below desired level because of lack of an adequate number of cattle or operating inefficiency of the current operation, then expanding herd size should be considered.

### Personal Goals, Preferences, and Abilities

If personal goals include retiring debt free in 5 to 10 years, personal preferences include a desire to give personal attention to cows and to work alone, and abilities are strong in land and cattle management areas but weak in money and people management areas, expansion is likely to cause considerable personal dissatisfaction and create many unwanted problems. Significant expansion is likely to require increased debt load and change the responsibilities of the owner from primarily working with cattle and land to managing people and money.

Conversely, if individuals have long-term goals of high net worth achievement, welcome the challenge of increasing responsibility, and have the ability to manage people and large amounts of capital, expanding herd size should be considered.

### Availability and Cost of Needed Resources

Additional cattle, land, buildings, equipment, and labor are normally needed for expansion. Most of these additional resources usually need to be purchased. Availability of needed land at reasonable cost is usually the major problem, so should receive first attention. Availability of good labor is also a problem in many areas. It is very important to determine if all the needed resources are available at reasonable cost before expanding, as a shortage or very high cost of any one can result in serious consequences. If an adequate amount of productive land is owned or available at reasonable cost, and competent, reliable labor can be assured by family members or others through the use of various wage and profit-sharing agreements, partnerships, etc. (Chapter 16), the other resources can usually be acquired at reasonable cost.

## Market Availability

The availability, and in some cases, the cost of a good market (favorable price with market security) is essential if expansion is considered. Increased production that must be sold at class II prices of $1 to $3 less than prevailing blend prices is not likely to be profitable. In areas where market base must be purchased, the price of the base in relation to earning power and the stability of the base program must be carefully evaluated, as major capital outlays are often involved.

In summary, expansion is most likely to be successful in meeting the major purpose of increasing total profitability when the herd is currently producing at a high and profitable level; the owner is at least 10 to 15 years from retirement, desires a higher income level, welcomes and has the ability to meet the challenges that accompany expansion; adequate additional resources are available at reasonable cost; and a good market for the additional milk is available. When this set of circumstances is present, that is the time to expand. The expected or unexpected availability of additional land resources nearby is often a prime factor in stimulating expansion. In other cases, the desire of a son or daughter who recently has completed or will soon complete high school or college to go into the dairy business may stimulate the expansion. In either case additional income is needed either to pay for the land purchase or to support the offspring joining the home farm, and this precipitates expanding herd size.

## PLANNING FOR EXPANSION

The purposes of thorough resource, financial, and operational planning are to maximize potential for achievement of personal and financial goals while minimizing risk of failure by avoiding problems that can and often do accompany expansion. This might be stated as taking advantage of the economics of increased herd size while maintaining or increasing level of production and profit margin per cow. It is easy to state the purposes, but more difficult to achieve them, as indicated by numerous published reports. Planning may be facilitated by developing and following a step-by-step procedure. Some of the more important steps include the following.

## Inventory Current Assets

A detailed and complete inventory of cattle by age (cows, bred heifers, open heifers over 12 months of age, etc.), land, facilities, equipment, etc., and value of all assets and extent of liabilities on each should be compiled. This inventory is recommended to determine current resource and financial status. It defines where one is. If good herd and business records are maintained, performing this step will be relatively easy. If they are not, the advisability of expansion is questionable.

## Define Desired Expansion Size

Tentative definition of proposed size or acceptable size range of the expanded operation establishes a goal that can be used in further planning. In combination with the inventory, it can help identify additional resource needs. Personal needs, goals, and abilities; current financial status in regard to asset-liability ratio and net worth; probable availability and costs of major land and capital needs; and market availability should be carefully considered when defining size goals.

## Determine Amount and Type of Additional Resources Needed

Detailed planning in this regard is essential as the amount and type of resources defined will largely determine financial planning for the expansion as well as affect future profitability of the operation. If the current inventory is complete and the desired size goal is reasonably accurately defined, the amount and type of resources should be fairly easy to determine.

Major resource needs for the expanded operation usually lie in four areas: (1) land and equipment for feed production; (2) feed storage, feeding, housing, milking, and milk storage facilities and equipment; (3) cattle and, in some areas, milk base; and (4) labor. The first three comprise the major capital outlay costs of expansion.

Feed needs of the expanded herd will be determined largely by the size of the herd, the type of feeding program, and the expected level of production, these, in turn, will determine land and feed production equipment needs. Accurate estimates of projected cattle inventory (cows, heifers, and calves) after expansion and feed requirements for expected level of production are essential. Type of feeding program, definition of amounts of forages and concentrates, and evaluation of alternative sources of each should be made. In most of the Unites States forages will have to be primarily home grown. Allow a 15 to 25% margin of flexibility in estimating forage requirements to accommodate storage losses, increased level of production, or a bad crop year. Evaluate purchasing versus home growing of concentrates. In many cases, changing from home-grown to purchased concentrates and utilizing current land resources for forage production can allow considerable expansion without need of additional land resources.

Expansion may require only minor alterations or additions to existing facilities, or it may require an extensive building program. Alternatives of renovating existing facilities or building new ones, and various types of facilities, should be evaluated and a detailed plan of the expanded facilities should be made. More usual problems in this area include: inadequate facilities for dry cows, heifers, calves, forage storage, and milk storage (Chapter 15).

In regard to cattle needs, problems often arise because needs for replacement heifers is often underestimated or overlooked entirely, or unrealistic estimates are made concerning ability to expand from within the existing herd. Assuming a culling rate of 25 to 30% per year (it may need to exceed this during expansion), average calf mortality of 10 to 15% (it may be much higher during expansion), and average freshening age of 26 to 30 months, 70 to 75% of all heifer calves born alive will be needed to maintain herd size. That leaves only about 25% of all heifers born alive available to increase herd size. This amounts to about 12 to 13 extra heifers per 100 cows per year. Doubling herd size from within the herd will require 8 to 10 years under average conditions. Expanding herd size from within, therefore, is a suitable method if size increases are limited to 10 to 15% per year, but unsuitable for more rapid expansion programs. If additional cows are purchased to increase herd size, additional heifers should also be purchased or additional replacement cows anticipated and budgeted for. In regard to heifers needed, assuming average culling rates, etc., about 15 to 20% of the number of cows in the milking herd should be on hand as pregnant heifers (due in three to 9 months), 15 to 20% as year-old through recently bred heifers (12 to 18 months of age), and 35 to 40% as calves and heifers under one year of age, to maintain herd size.

## Estimate Costs and Identify Sources of Additional Resources

Realistic estimates of costs of expansion must be made for accurate financial planning. This can be difficult, especially during inflationary periods. It is greatly facilited, however, by completion of the previous step. If a detailed list of needed resources has been compiled, realistic estimates of construction costs can be obtained from contractors or building materials suppliers, equipment costs from dealers, and cattle costs from attending auctions or pricing cattle privately. Alternative sources of all these items should be identified and evaluated. Additional land costs usually can be determined accurately only after the particular acreage has been found and priced.

After the estimated costs have been compiled, financial analysis of individual expansion items, the total expansion costs, and investment in the expanded operation is made on a per-cow basis. This analysis can help identify problem areas such as excessive investment on a per-expansion-cow basis, and total investment per cow, or in individual areas such as facilities or equipment investment per cow. If such analysis reveals overinvestment, either on a total-herd or expansion-herd-only basis, other alternatives should be evaluated. If costs estimates exceed owned capital, alternative sources of capital should also be investigated to ascertain availability and approximate costs of borrowed capital (see Chapter 3).

## Budget

Three types of budgets should be completed before expansion begins. These budgets are: (1) an annual operating budget for the expansion year and at least the ensuing three years; (2) a monthly or quarterly cash flow budget for the expansion year (monthly is preferable) and for the ensuing two to three years; and (3) an annual net worth analysis for the expansion year and ensuing two to three years (see Chapter 3).

Detailed planning will take considerable time and effort. It should be further supplemented with visits to other dairymen who have recently expanded to get ideas, and with advice from extension personnel, money lenders, and other knowledgeable persons, to help avoid problems. When detailed planning is completed before expansion begins, the probability of success of goal achievement (increased profit) is greatly enhanced. It is much easier to correct errors made on paper than those made in concrete, steel, and dollars. This justifies the time and effort involved in careful planning. More recently, computerized expansion planning programs have been developed that can greatly increase accuracy and efficiency of planning.

Operational aspects of the expanded operations should also be included in the various parts of the plan as it is developed. Development of the feeding program, milking operation, etc., should accompany planning of the additional resource needs. This will identify labor needs, both amount and type, and also help avoid omissions such as lack of adequate facilities for replacements, dry cows, cow treatment, manure storage, etc.

## IMPLEMENTING THE EXPANSION PLAN

Proper implementation of the plan is equal in importance to developing the plan. However, if planning is complete, implementation is greatly facilitated as the route has been identified. Problems will occur, however. Delivery of needed materials may be delayed, some purchased cows may not adapt to the system, actual costs of some resources may exceed estimates, etc. For these reasons, some flexibility in scheduling of expansion events and capital needs is advisable.

## REFERENCES FOR FURTHER STUDY

Anonymous, Computer Advises Farmers on Their Proposed Expansion Plans, *Hoard's Dairyman,* December 1975.

Brown, C. A., and White, J. M., Immediate Effects of Changing and Herd Size upon Milk Production and Other Dairy Herd Improvement Measures of Management, *J. Dairy Sci.,* **56** 799 (1973).

Brown, L. H., and Speicher, J., Getting Big—Can Be a Headache, *Hoard's Dairyman,* February 10, 1971.

Tinsley, W. A., and Hoglund, C. R. Financial Planning for Dairy Herd Expansion *J. Dairy Sci.* **56.** 830. (1973).

# 21

# The Purebred Dairy Cattle Business

**21**

The breeder of purebred dairy cattle is engaged in two separate enterprises: first, the production and selling of milk; and second, the developing and selling of purebred cattle. To many, the challenge of developing a herd of superior animals, with the associated opportunity for recognition and increased profitability from sales of breeding stock, is the spark that gives them keen interest in the business. The development of an outstanding purebred herd of any of the dairy breeds can be profitable and fascinating for those who will take the time and trouble to carry out the exacting details that are necessary.

A purebred, or registered, animal is one whose lineage is traceable through its breed association registry, whose name has been registered with its respective breed association, and for which a certificate of registry has been issued by the association. Only about 10% of the dairy cattle in this country are purebreds. Most of the remainder are grades (animals that possess the characteristics of a breed, but whose lineage is not traceable) that have acquired their good qualities through purebred sires.

## THE PURPOSE OF PUREBREDS

The purpose of purebred breeds is to perpetuate the desirable characteristics of their breed, thereby furnishing superior seed stock for use in both purebred and grade herds. If they are to continue to achieve this purpose, they must be of superior breeding and managed in such a way that they are given the opportunity to perform in a superior fashion. It is necessary that they meet certain requirements. Some of these requirements are the following.

1. High production. The first requisite of a purebred is that it be a profitable producer of milk of acceptable composition. Purebred breeders must give top priority both to breeding for high production performance and to providing feeding and management conditions that will give the animal the opportunity to produce at high levels.

2. Good type. Good type, as well as high production, is one of the characteristics that distinguish purebred herds from many grade herds. Special emphasis should be given to developing a herd with superior utility type traits such as well-attached and shapely udders, correct feet and legs, and dairy character. The purebred breeder whose plans include showing should also try to breed and develop superior show animals.

Good type is not an acceptable substitute for high production in purebreds, it is a good supplement.

3. Good health. A purebred herd should be healthy. It should be accredited free from tuberculosis, accredited free from and vaccinated for brucellosis, and vaccinated for other diseases for which there is an effective vaccine (Chapter 13).

## THE PUREBRED BREEDER

Achieving these requirements for a purebred herd is the responsibility of the breeder. The characteristics a man or woman needs to achieve this are many and varied. Some of these characteristics are necessary; others are desirable, yet their absence may be overcome. These characteristics are as follows.

2. Love for the dairy cow. It is rare that a person becomes a great breeder unless he has a sincere love for the dairy cow. If the love is there, the breeder will see that the cow is fed and managed in such a way that it will be given the opportunity to perform at or near its maximum genetic potential.

2. Knowledge of the basic principles of feeding, breeding, milking, etc., and use of this knowledge to ensure that the various programs are designed to give the cattle the opportunity to perform at high levels.

3. Knowledge of type. A good breeder should have a good working knowledge of good type. He need not necessarily be an expert judge, but he should know how to evaluate type strengths and weaknesses of his cattle and of daughters of sires he plans to include in his breeding program. If showing is included in his program, he must be knowledgeable about the type of cattle that are competitive so that he may show his best animals.

4. Knowledge of pedigrees. A successful breeder is often a student of pedigrees. He should know popular sires, females and lines important in his breed and herd.

5. Integrity. The purebred business is based on the honesty of the breeder and the accuracy of his records. Persons who have a reputation for honest representation and dealing, and whose word is as good as their bond, are the ones who are most likely to be successful purebred breeders. The breeder who honestly helps those who buy from him, does

not misrepresent the animals, or who will not sell an animal to a buyer if he knows it is not in the best interest of the buyer, is more likely to be a successful breeder.

6. Attention to details. Many details are involved in maintaining a superior purebred herd and unless a breeder pays attention to these, he has a minimum probability of becoming successful. He must be prompt and careful in the identification and registration of animals and in transferring them when they are sold, prompt and courteous in answering inquiries, logical yet imaginative in developing a naming system and preparing ad copy, careful in presenting his animals to best advantage on the farm and in sales and shows, and in keeping all his records accurately and current.

7. Participation in and help with local, state and national programs. All the breed associations have breed-improvement and breed-promotion programs. Ultimately, the success or failure of these programs and the breed itself depends on participation and support of the programs by the breeders.

The breeder of purebred cattle must invest more money and time in his business than the commercial dairyman. He should expect a greater return for this additional investment. If the herd and farm are operated on a sound basis and attention is given to all the details, the breeder should be rewarded with the following results.

1. Having a high and profitable level of milk production.

2. Getting a high price for cattle.

3. Breeding outstanding cattle.

4. The opportunity for leadership.

5. The pride of ownership of outstanding animals.

6. The possibility of building up a herd of high value, which will be used as an estate, an annuity, etc.

7. The personal pleasure of working with, competing with, and enjoying the fellowship and friendship of other breeders.

## PROGRAMS OF THE BREED REGISTRY ASSOCIATIONS

There are currently six recognized breeds of dairy cattle in the United States. Each of these recognized breeds has a breed registry association that looks after the interest of the breed it represents. The members of the associations are owners of the cattle of the breed who have joined the association and paid membership fees. To become a member a breeder must have his application for membership approved by other members. This membership gives the member a voice in the operation of the association, and permits him to register and transfer animals at a reduced fee.

Each association holds an annual meeting where the members have the opportunity and responsibility of helping to determine the policies of the association, either directly by personal attendance or through representation by delegates. Each association has an elected board of directors who operate the association through an executive secretary and staff. The associations must have an adequate building to house the staff and keep the records. The building must be equipped with fireproof vaults to preserve the valuable records. Each association also publishes or is associated with the publication of a breed journal, which serves as a means of communication with breeders and is an excellent medium for advertising, promotion, and education. The following are the official names, locations, breed journals, and executive secretaries of the various dairy breed associations as of 1977.

The Ayshire Breeders Association; Brandon, Vermont
    05733; *The Ayrshire Digest*; David Gibson Jr.

The Brown Swiss Cattle Breeders Association;
    800 Pleasant St., Beloit, Wisconsin 53511;
    *The Brown Swiss Bulletin*; Marvin L. Kruse.

The American Guernsey Cattle Club; 70 Main St.,
    Peterborough, New Hampshire 03458; *The Guernsey
    Breeders Journal*; Max L. Dawdy.

The Holstein-Friesian Association of America;
    P.O. Box 808, Brattleboro, Vermont 05301;
    *Holstein-Friesian World*. Sandy Creek, N.Y.
    Charles Larson.

The American Jersey Cattle Club; 2105-J S. Hamilton Rd.,
Columbus, Ohio 43205; *The Jersey Journal*:
James F. Cavanaugh.

The American Milking Shorthorn and Illarawa Society;
    313 South Glenstone Ave., Springfield, Missouri,
    Harry Clampitt.

Table 21-1 presents membership, registration, and transfer information for each of these associations.

**Table 21-1** Breed Registry Association Data: Membership, Registrations, and Transfers, 1976

| | Active members, no. | Junior members, no. | Membership fee | | Registrations no. | Identification enrollment | Transfers, no. |
|---|---|---|---|---|---|---|---|
| | | | Life | Limited | | | |
| Ayrshire | 998 | 100 | $100 | $40–10 yr. | 11,237 | Yes | 7,981 |
| Brown Swiss | 1,024 | 957 | $25 (ind.) | $25–10 yr. (others) | 15,919 | Yes | 10,403 |
| Guernsey | 3,526 | 1,307 | $50 | $50–20 yr. (corp. & part.) | 26,210 | Yes | 13,529 |
| Holstein | 27,286 | 10,760 | — | $20–5 yr. renewal $10–5 yr. | 279,146 | Yes | 148,825 |
| Jersey | 1,721 | 2,009 | $50 | None | 40,616 | Yes | 22,077 |
| Milking Shorthorn | 900 | 1,028 | — | $15 + $18 annual dues | 4,797 | Yes | 2,919 |

## Purposes of Breed Associations

The purposes of the dairy breed associations are (1) to preserve the purity of the breed, (2) to improve the breed, and (3) to advertise and promote the breed. Although these purposes are fairly specific, there is some overlapping. For example, if breed improvement occurs in production and type, these factors in themselves will go a long way toward advertising and promoting the breed.

**Preserving the Purity of the Breed** The original purpose and prime objective in the organization of the breed associations was to preserve the purity of the breed by recording (registering) the identity and lineage of animals of that breed. Rules and regulations on registration of purebred animals were formulated, and, currently, the six dairy breed associations have similar registration requirements. In fact, application for registry of animals of all six breeds may be made by the use of a Dairy Breeds Unified Application for Registry (Figure 2-11 in Chapter 2) included in the DHI program. The essential parts of all breed registry applications are, (1) permanent identification of the animal to be registered (tattoo, or sketch or photo that clearly shows the color markings); (2) registration name and number of the sire and dam; (3) date of service, or a breeding receipt if the animal is sired by artificial insemination; and (4) date of birth. The application for registry must be completed by the owner or his representative and mailed to the association with the appropriate fee included.

The association then issues a registration certificate for each animal accepted. These registration certificates are sent to the owner of the animal and include the name of the animal, its registration number, birthdate, sire and dam, identification, and the name of the breeder and owner. Some of the associations' registry certificates also include additional information.

TRANSFER OF OWNERSHIP. When a registered animal is sold, the change of ownership must be recorded with the breed association in order to show proper title to the new owner and enable him to register future offspring from that animal. The transfer application form must be completed by the seller and submitted, along with the registration certificate and appropriate fee, to the breed association office. The change of ownership is recorded on the registration certificate and it is returned to the seller, who sends it to the buyer.

HERD REGISTRY BOOKS. Initially the dairy breed associations published herd registry books, which listed every animal that had been registered, with its name and number and the name and number of its sire and dam. Because of the expense involved in publishing these books, most of the associations have discontinued publication of herd registry books. However, the records are kept on file in the association office, and whenever a detailed pedigree is required it can be obtained from the association.

When it is realized that the total number of animals registered to date with the dairy breed associations runs well into millions in some of the breeds, the magnitude of the job of keeping these records can be seen. However, the life of the purebred dairy breeds depends on the keeping of these records. This, then, is the first and probably the most important duty of the breed association—preserving the purity of the breed.

**Improving the Breed**   Dairy breed association members have long been the leaders in dairy cattle improvement. The fact that their cattle were registered and thus considered purebred was not enough; continued improvement in performance of their cattle was also necessary to maintain their position of leadership. It still is.

The associations have breed improvement committees, whose responsibilities include the development of programs that will result in continued improvement. These programs are primarily in the areas of genetic improvement for production and type characteristics. Responsibility for breed improvement for each breed rests with each breeder. Personal success of each breeder in improving his herd is primarily dependent on his use of up-to-date information generated by researchers and by participation in breed improvement programs. Breed improvement is limited to the cumulative success of the breeders. Breed associations can plan programs to improve their breed, but only breeders can implement these plans.

PRODUCTING TESTING.   Early breeders conceived the idea of a system of testing for production, wherein the animals would be put in advanced registry if they produced the required amount of milk and fat. The breed associations started a program to certify the authenticity of the records. It is now conducted under the Unified Rules for Official Testing, as adopted by the Purebred Dairy Cattle Association and the American Dairy Science Association, and is called Dairy Herd Improvement Registry (DHIR). Production testing procedures are similar to DHI (Chapter 2); however, participation in the DHIR production testing program is limited to purebreds. Records of animals enrolled in the DHIR program are forwarded to and published by the respective breed associations. These publications are available and can be used in many ways, such as the writing of pedigrees, determining the eligibility of cows and bulls for performance awards, and selecting breeding stock.

TYPE CLASSIFICATION.   Each breed association has developed a program to evaluate the type or physical conformation of the animals of its breed. The purposes of these type classification programs are to improve the physical conformation and consequently the utility value of cattle of the breed and to preserve the characteristics peculiar to the breed.

Each association describes the ideal type for its breed, a standard of perfection to which animals of the breed can be compared. Official type classifiers, trained and supervised by the various breed associations, evaluate the conformation of animals of the breed by comparing them to the breed ideal. Each breed has a scorecard, or a scale of points, so that the classifier can assign a point score on the overall type of the animal as well as the various components; dairy character, mammary system, etc. The breed ideal is scored 100 points. Each animal classified receives a point score, or classification rating, of less than 100, depending on how much its conformation deviates from that of the ideal. There is some breed variation, but most are similar. Animals that score 90 or more points are referred to as "Excellent"; 85 to 89 "Very Good"; 80 to 84, "Good Plus" or "Desirable"; 75 to 79 "Good" or "Acceptable"; 65 to 74, "Fair," and below 65, "Poor."

More recently, most breeds type classification programs also have included a descriptive trait system of evaluating particular characteristics of an animal. Descriptive terms are used to identify specific characteristics concerning each of the type components. Examples include stature (tall, medium, low set), feet and legs (correct set to hock, sickled, cowhocked etc.). These are designed to describe more specifically certain characteristics of the animals and therefore make classification summary data more useful to breeders.

Breed classification programs have evolved because of the desire of breeders to measure and thereby improve the physical conformation of their animals, just as production testing programs evolved to measure and improve the producing ability of their animals. There are, however, two important factors that should be considered when comparing the two kinds of programs.

First, production can be objectively measured; that is, milk can be weighed, percentage of butterfat or other components can be measured by chemical or other analysis and milk yield or yield of other components can be accurately calculated. Physical conformation, however, is measured subjectively; that is, the conformation of an animal is measured by an individual assigning a score based on his comparison of an animal to the breed ideal. Because of this, type classification data are subject to more error and less accuracy than are production data.

Second, the breed standard or ideal is not a static one, but rather is subject to change. The breed ideal is a consensus of the ideal of the individual breeders and as such, changes with time. The ideal of most of the breeds has, in fact, changed considerably in recent years. Most of these changes reflect increasing importance of the utility type characteristics such as feet and legs, mammary system, and dairy character, in addition to increased stature, with decreased emphasis on the fancy requirements.

Type classification programs can be effective tools for breed improvement when classifier variation is minimized by thorough training, by frequent coordination among classifiers, and when utility traits receive major emphasis.

TRUE-TYPE PICTURES AND MODELS. The breed associations have attempted by painting or by model or both to prepare a representative of the true type of their breed. These true-type pictures or models, even better than the scorecard, give the breeders an ideal toward which they should strive to develop the animals in their herds.

DAIRY CATTLE SHOWS. The breed associations have always encouraged the showing of cattle at parish, county, district, state, and national shows. Each of the breed associations annually selects one or more dairy shows as their national or regional national breed shows. The associations also select a list of approved shows and of approved judges. Judges for their approved and national shows are selected from the list of approved judges. These shows are a help in maintaining the true type. Here, breeders have an opportunity to compare their animals with others of the same breed as well as advertise their animals and enjoy the fellowship of other breeders.

ALL-AMERICAN AWARDS. The breed associations conduct annual contests whereby the outstanding animals of the breed for each classification are named as All-American, Reserve All-American and Honorable Mention All-American. Individual breeders submit photographs of animals with superior show ring performance and a panel of judges, usually those individuals who have judged at one or more approved shows during the year, select the All-Americans by ballot. Many individual state breed associations have similar programs for All-State recognition.

PROVISIONAL REGISTRY PROGRAMS. Most of the breeds have programs by which unregistered animals that possess the physical characteristics of the breed, and sired by registered bulls, and have superior performance characteristics may be identified and recorded in a provisional registry. These programs have the purpose of identifying superior unregistered females with registration of subsequent generations, thereby, increasing the number of genetically superior animals of the breed. Highly restrictive qualifications for provisional registry and registry of subsequent generations are maintained to ensure that only superior animals that are essentially purebreds qualify.

SIRE DEVELOPMENT PROGRAMS. Most of the breed associations have programs designed to increase the number of highly selected young sires that are accurately progeny tested, thereby increasing the number of plus proved sires of their breed. Some of these programs are administered almost completely by the breed associations, while in others the program is limited to defining standards for participation in breed-sponsored sire development programs.

PERFORMANCE EVALUATION PROGRAMS. Some of the breed associations have programs by which a breeder can enroll his herd in a program that includes both production (DHIR testing) and type (classification) evaluation programs for a single fee. Another has a program whereby breeders can obtain an evaluation of the strengths and weaknesses of both their genetic and management program. In these programs qualified consultants visit the herd and evaluate both the genetic and the herd management programs. Based on their identification of weaknesses, they recommend sires for use in the herd and improved management practices.

SALES PROGRAMS. Many breed associations are directly or indirectly involved with consignment sales at the state, regional, and national levels and with direct sales both domestic and foreign. The primary purpose of the consignment sales is to make superior purebred animals available to other purebred breeders or to grade breeders who desire to establish purebred herds. When high selection standards are maintained, these sales can be successful in achieving their purpose. Foreign market demand for U.S. purebred cattle of several breeds is good, and breed association programs have developed to facilitate sales by U.S. breeders.

OTHER BREED IMPROVEMENT PROGRAMS. In addition to these specific breed improvement programs, all breed associations employ breed fieldmen as program directors whose responsibilities include both breed promotion and breed improvement. Fieldmen assist individual breeders, local, state, and regional breed associations. They attend meetings of the breed clubs and other dairy organizations in the interest of the breed they represent; they help organize state and local breeders clubs, sales, and shows; they help organize junior member activities; and give service and advice to the breeders whenever possible in matters of registration, transfers, and breed improvement programs such as testing, classification, sales, shows, and other programs. Breed fieldmen are an important link between breeders and the association.

Various breed associations are also involved in conducting or supporting research programs designed to identify improved methods of breed improvement, calculating and publishing production-type indices, and other breed improvement programs and services.

As indicated earlier in this discussion, the purpose of the breed improvement program is to improve the performance ability of their breed. The success of the breed will depend to a large degree on the design of the breed improvement programs and the degree of participation of the breeders in those programs. Ultimately, the responsibility for breed improvement lies in the hands of the individual breeders.

**Advertising and Promoting the Breed.** The breed association in the broad sense, that is the breeders and the association staff, have the responsibility of advertising and promoting their breed. The breeders and field staff should accept primary responsibility at the local and regional level while the main association office has primary responsibility at the national level. The breed improvement programs are often ideal for helping in this work. Special awards and programs have been implemented by the various associations to advertise and promote the breed through recognition of bulls, cows, and breeders.

BULL RECOGNITION PROGRAMS. The bull recognition awards are based on either the production performance or the type performance, or a combination of both, of the daughters of a bull. For example, for a Holstein sire to become type qualified (TQ) he must have a minimum of 10 classified daughter-dam pairs with a minimum predicted difference type (PDT) of +0.50 if repeatability is 20 to 29%; +0.40 if repeatability is 30 to 69%; and +0.30 if repeatability is 70% or more. To become production qualified (PQ), he must have at least 10 daughters with official (DHIR) records; the predicted difference milk (PDM) must be at least +300 if the repeatability is 20 to 29%, +200 if the repeatability is over 30%,; the predicted difference fat (PDF) must be equal to zero or a plus value; and at least 51% of his daughters must be registered or positively identified by blood typing if fewer than 30 daughters are included in his sire summary. The bull is awarded the Gold Medal Sire recognition if he meets both the TQ and PQ requirements. Similar recognition programs are used by other breeds and serve the purpose of recognizing superior sire performance as well as breed promotion.

COW RECOGNITION PROGRAMS. Cow recognition awards are based on the performance of the cow, as in the Jersey Ton of Gold Awards; the performance of the offspring of the cow, as in the Ayrshire Approved Dam Awards; or a combination of both the performance of the dam and her progeny, as in the Holstein Gold Medal Dam Award. In many of these programs, both production (DHIR) and type (classification score) performance are required. These programs also serve to recognize outstanding females as well as promoting the breed.

BREEDER RECOGNITION PROGRAMS. Breeder recognition awards are based on breeder herd performance. Most include a minimum percentage of the herd registered and bred by the breeder, as well as minimum production (DHIR) and type (classification) performance. These recognition programs provide incentive for breeders to enroll their herds in breed improvement programs, to maintain both high production and type standards in their herds, and to contribute generally to both herd and breed improvement. Some breed associations also give special recognition such as the Jersey Master Breeder Award to breeders in recognition of long-time breeding excellence.

EXTENSION WORK. Extension work is carried on by the various associations for advertising and promoting the breeds. It is done in several ways: by preparing and mailing out printed information on the breed, by assisting at shows featuring the breed, by demonstrations and special exhibits at fairs and expositions, by press advertising, by publishing production records and recognition awards, by conducting special sales, and by other forms of field work. Many of these programs are the responsibility of the field staff of the association.

MILK PROGRAMS. To help with the promotion of the breed as well as promote milk sales from their breeds, the Guernsey and Jersey Associations have developed a milk merchandizing program. The Guernsey association's program goes under the name of Golden Guernsey Milk, whereas the Jersey association's program is called All Jersey Milk. The association licenses the producers and distributors and requires that they maintain certain quality standards necessary for the production of high-quality milk.

## THE PUREBRED
## DAIRY CATTLE ASSOCIATION

The Purebred Dairy Cattle Association (PDCA) is an organization composed of the six dairy-breed associations and others interested in purebred dairy cattle. Its purposes are to promote purebred dairy cattle, coordinate certain activities of all the dairy breeds, and to cooperate with the American Dairy Science Association, the National Association of Artificial Breeders, the National Dairy Herd Improvement Association, and others, in coordinating many of the programs of common interest to all of them. Some of the work of the association is as follows:

1. The preparation of unified rules for official testing.

2. The Unified Dairy Cattle Scorecard.

3. Classes for dairy shows.

4. Rules and regulations governing artificial insemination of purebred dairy cattle.

5. Codes of ethical practices for showing and selling dairy cattle (See Appendixes K and L).

6. The promotion of research work in the various experiment stations dealing with dairy cattle improvement.

## REFERENCES FOR FURTHUR STUDY

American Guernsey Cattle Club, historical items on The Guernsey breed, in various issues (especially July) of *The Guernsey Breeders Jour.*, 1977.

Annual reports of the breed associations.

Bowling, G. A., *History of Ayrshire Cattle in the United States*, McClain Printing Co., 1975.

Breed journals of each breed association.

Herd books and performance books of each breed association.

Prescott, et al., The Hostein Friesian History, *The Hols. Fries. World,* 1960.

Unified Rules for Official Testing, Purebred Dairy Cattle Association, Peterborough, N. H.

# 22

# MARKETING DAIRY CATTLE

The dairy operation on a farm may represent only a portion of the total farm income or it may be the sole source of income. The major part of the dairy income may come from the sale of milk, but often on a farm where purebreds are kept a large portion of the dairy income will come from the sale of dairy cattle. Commercial dairy operations normally receive 80 to 90% of their total gross income from the sale of milk. The remainder comes largely from the sale of cull cows, bull calves, and, in some cases, a few heifer calves, bred heifers, or cows for dairy purposes.

Herds with a high breeding efficiency, a low calf death loss, and a low culling percentage may have a greater than usual number of females to sell for dairy purposes.

A dairyman may decide against increasing the number of milking cows to increase his income and choose to raise dairy heifers to sell for replacements. In addition to his own surplus animals, he may purchase calves and heifers to raise.

Those heifers not needed for herd replacements may be sold whenever there is a demand for them. It is the usual practice, however, to sell the majority of them as heavy springers or soon after calving. If the dairyman wishes to maintain a young herd he may sell the older cows or those still in their prime (five to six years of age) and retain the first-calf heifers. This practice will usually result in a somewhat lower herd production, but, on the other hand, it will lower the number of cows to be disposed of to the butcher because of old age, udder trouble, nonbreeding, and other troubles. If a progressive breeding program is followed, it may also result in increased genetic progress for milk production.

## WHEN TO SELL COWS

### Time of Freshening

Usually the greatest demand for dairy cattle is for fresh cows or heavy springers. The advantages of buying springers (cows or heifers before freshening) are: (1) they will stand transportation better than the fresh cow, with less danger of udder injury; (2) they reach their new home in time to become accustomed to the new surroundings before they calve and are to be milked; and (3) the buyer has a chance of getting a valuable calf. Of course, half of the calves will be bulls.

The fresh cow or heifer has her production interrupted somewhat in making the change, but the advantages of buying at that time are: (1) many buyers need the milk at once and are not willing to wait for the cow to freshen; (2) the buyer can be sure that there are no blind or weak quarters, and the shape and attachments of the udder can be evaluated more accurately; (3) the amount of immediate production is a matter of record; and (4) there is not the problem of difficult calving, retained placenta, or milk fever.

Although there are advantages both ways, the fresh cow or heifer usually will bring a slightly higher price than will the springer, especially in auction sales.

### Season of the Year

The season of the year has a strong effect on the demand for dairy cattle. Calves and yearlings are in heavier demand in the spring, when they can be turned out to pasture.

Milking cows and close springers are in greatest demand in the fall, as this is the time of greatest demand for milk and is the period when the milk base for the year is determined for many dairymen selling fluid milk.

Some differences among breeds exist as to the best season of the year to sell. On most markets there is a surplus of milk in the spring and early summer, which favors the higher-testing breeds. In the fall the demand is for volume of milk, so the high-testing cows do not hold their advantage.

### MARKETING GRADE CATTLE

The marketing of grade cattle is somewhat different from the marketing of purebreds. The market for grades is largely to commercial dairymen with limited land and feed resources who do not raise their own replacements in sufficient numbers. There is, of course, also a continual movement of dairymen into and out of the business; those going into the business must purchase cattle, and they often start with grades.

Good grade cows that have been bred up by the use of high PD sires of the same breed for a few generations will be high-quality cattle. They can be expected to have high production. Their conformation and general characteristics often are indistinguishable from purebreds.

Some dairymen with surplus cattle maintain a standing agreement with other commercial dairymen to supply them with fresh cows and springers. Others sell regularly to dairy cattle commission men and still others have no regular method of disposal of their surplus. A few grade dairy animals are still sold through livestock auction markets. This is certainly an undesirable place to secure cattle. These markets have been developed to furnish a market for surplus livestock of almost any kind. In most cases the animals that are sold are expected to go to slaughter. They come from all kinds of farms and too often little effort is made to control disease. This offers a market for cull cows that should go to the butcher. A dairyman should not buy animals of any kind from these stockyard markets. The authors do not know of a profitable herd where the owner has made a practice of purchasing replacements from the stockyard market, but do know dairymen who purchased a great deal of trouble with cattle from this source. Probably the most prevalent of these troubles have been cows with mastitis, brucellosis, chronic digestive troubles, or mean dispositions.

## MARKETING REGISTERED CATTLE

Registered cows must be good producers; however, their value extends beyond their milk-producing value. The demand for them is also based on their value as foundation animals on which to build or improve a registered herd. Top-flight herds look to other registered herds for their sales.

The small purebred breeders will expect most of their sales to go as foundation animals to new breeders or to breeders with only a few purebreds.

It may be advantageous for smaller purebred breeders who do not exhibit their cattle widely to sell an occasional outstanding cow or heifer to an established breeder who regularly exhibits cattle at state, regional, and national shows. Exceptional animals often bring a very high price. This extra income can significantly improve the financial position of small and beginning breeders. In addition, they can receive free advertising and promotion of their herds at regional and national levels as the animals carry their herd prefix. This identifies them as breeders of outstanding individuals which in turn can create future demand for their cattle.

### Bull Calves

The market for bull calves is much more limited than for females. The greatest demand is for young bulls of breeding age, usually in the fall and early winter.

The widespread growth of artificial breeding associations has reduced the number of bulls needed and at the same time has made the demand much more discriminating. Herds with high production, good type, and strong cow families may profit from growing some bull calves to breeding age from selected cows to sell for breeders. Also, certain cows may be contracted to an AI organization to be bred to a bull chosen by the organization. If the resulting calf is a bull the AI organization agrees to buy it at the price stated in the contract.

While the AI organizations rely largely on purchasing (or contracting for) selected young bulls and proving them, they do purchase a few bulls that have been proved by the individual breeder or syndicates.

## New Breeders

A new breeder of registered dairy cattle must not at first expect high prices or too great a demand for his surplus cattle. He must first build his herd on a sound financial operation from the sale of milk. After he has established a healthy herd that has satisfactory production records and desirable type, as indicated by classification or by showing at local shows, and after he has advertised by some method, he can expect some demand for his animals.

## Expected Qualities

Certain factors are essential for a breeder wishing to sell surplus cattle. His herd must have the following qualities.

**Production Records.** High producing ability is the first requisite of registered dairy cattle. The proof of producing ability is through one of the systems of production testing.

**Type.** Type is not a substitute for production, but is a valuable addition to it. The type components with utility value such as strong udder attachments, sound feet and legs and dairy character are especially important.

To secure some of the higher prices, purebred cattle must have desirable type, but it is not necessary that they be show-ring winners. A part of the desire for ownership of registered cattle is in the pride of having something better.

One of the top-flight sale managers stated that when he selected cattle he looked first at the animal, then at the pedigree, and that a splendid pedigree and a high record would not sell an off-type animal.

**Health.** The third requisite for a breeder wishing to sell his surplus stock is to have a healthy herd—a herd that is free of brucellosis, tuberculosis, mastitis, and other diseases. He should develop an accredited herd. Buyers prefer to purchase animals that have been calfhood-vaccinated for brucellosis. All animals to be sold should have a negative test for both brucellosis and tuberculosis within 30 days of the selling or delivery date. In most states, calves whose calfhood vaccinations have not cleared up, but are under 24 months of age and come from a clean herd, are acceptable animals from a health standpoint. An attempt should be made to sell cows free from mastitis and other diseases. It is the seller's obligation to sell only healthy animals.

## ADVERTISING

Besides developing a herd with high production, with good type, and free from disease, the breeder must advertise if he wishes to sell his surplus cattle to advantage.

The art of advertising is to let the prospective buyer know what one has to sell and to present its good points in such a way that the buyer will want what one has to offer.

Some advertising may be free and other kinds may cost money. Whatever the cost, it may be placed in the same category as the expense of raising the animal to sell. It is a necessary part of the sales program.

There are many ways to advertise, among which may be listed the following.

1. Production testing. Milk production records are the first consideration in advertising and promoting dairy cattle.

2. Showing animals at local or state fairs. If this is to be done, the right kinds of animal must be shown and they must be well fitted and shown in an attractive manner. Signs over the heads of the cattle and printed material of information about the herd can be distributed.

3. Furnishing the local papers with information on the outstanding accomplishments of the herd. The production records, type classification scores, or show-ring winnings are always news that the local paper will print without charge.

4. Consigning a few top animals to a local or state consignment sale is another means of getting a herd before the buying public. The sale, as well as the show, can be a breeder's show window through which he can display his merchandise.

5. Purchasing advertising space in breed magazines or local papers may at times be one of the wisest expenditures in selling cattle. It may reach the person in need of the specific offering. The type and extent of purchased advertising that is justified depends upon the number and kind of animals to be sold.

A beginner would not expect to advertise his surplus through a national medium. He could expect his best results with local papers or statewide farm papers. The use of continuing small advertisements is more effective than a single large one costing the same amount.

6.  A breeder's reputation for honesty and integrity is the greatest advertisement for a herd of cattle. A reputation for square dealing is a requisite for a breeder to sell cattle at the best prices. This is true for either private or public sales.

    The activity of the breeder in breed clubs, dairy, and other types of organization is effective in getting to know more people and to have more people know about his cattle. A man and his cattle are closely tied together in the thinking of his fellow breeders.

## METHOD OF SELLING

Surplus animals may be sold in several ways: privately, by mail, or through one of the various forms of auction sales.

### Private Sales

The simplest and most used method of selling is by private sale direct from the seller to the buyer on the seller's farm.

The buyer sees the cattle in their usual surroundings, in their "everyday clothes." The seller does not have any selling expense. Blemishes, if any, can be looked over with deliberation. Related animals in the herd can be seen. More time can be given to making decisions. This method offers advantages that the auction sale does not.

Many buyers do not like to buy under the excitement and pressure of the auction sale. At the farm they can see the cows milked. They can learn more about the feeding and care of the herd as compared to their own.

The seller should have readily available complete production records, breeding dates, and other information on his cattle. Some pedigree information should be available. Most buyers of purebred cattle today expect much more information than is on the registration papers.

### Selling by Mail

In selling by mail, the seller should give the interested party as complete information as possible. This should include the age, a complete two- or three-generation pedigree, a photograph of the animal, and, if possible, photographs of the sire and dam.

This type of information is splendid for answering inquiries about cattle, even though the prospect is expected to come to the farm to see the cattle; in fact, it may be the deciding factor as to whether or not he will visit the farm.

### The Auction Sale

The auction sale is a very popular method of selling. A number of animals can be sold in a shorter time. A buyer can purchase more animals in a shorter time and with less traveling than he can privately.

**Dispersal Sales.** The dispersal sale is one in which the breeder is selling an entire herd, with perhaps a few exceptions that are noted in the announcement of the sale. This is a popular type of sale. Breeders realize that they have an opportunity to purchase animals that might not otherwise be offered.

**Reduction Sale.** The reduction sale is one held to reduce the number of animals in the herd. The owner has the privilege of selecting the animals to sell. This may be an average cross-section of the herd or it may be above or below the average.

**Consignment Sale.** The consignment sale is usually conducted by some breed club or group of breeders. The breed club consignment sale is made up of cattle selected from the herds of the members. A sale may also be sponsored by a group of breeders using the same blood lines or other characteristics common to their herds.

Some consignment sales are managed by professional sale managers. Others are handled by a sale committee selected from among the group sponsoring the sale. When sales are held as annual affairs, they can build up good will and prestige that is a definite asset in selling cattle.

### THE EVALUATION OF CATTLE

### General Considerations

It is often said that an animal is worth what it will bring. Of course, in an auction sale this is the value placed on it. But on the farm, how is the price to be determined? Both the buyer and the seller may compare the animal with other animals that have been sold recently. But what is a cow worth? A number of factors and conditions enter into the worth of a dairy animal. Those discussed below are important.

**The Price of Milk.** The trend of the price of dairy cows follows rather closely the price of milk. This holds from year to year and from season to season. Often when milk bases are being increased or base allotments are being changed, the price of cows increase greatly just prior to the base-making period and in the early part of it. Following the base-making period cow prices drop sharply.

**The Price of Beef Cattle.** The price of dairy cows is affected by the price of cull dairy cows sold for butcher. Beef price puts a floor price on the less desirable cows. With high prices, more of the borderline cows are sold. When cull cows sell for a very low price, the tendency is to keep more of the questionable cows and milk them. More milk is produced and the extent of it may depress the milk price.

**Age.** Generally, an animal increases in value until it reaches its prime, after which its value will decrease. In the case of outstanding breeding animals, their value remains high to an older age because of the value of their offspring.

**Soundness.** A blemish or unsoundness decreases the worth of an animal. The blemish may be negligible or it may be so great that the animal has value only for beef. A slightly capped hip that does not interfere with the animal's walking might not cause a serious reduction in price. On the other hand, an injured stifle that causes the animal to walk with difficulty might render her almost totally useless.

**Health.** There are probably four diseases that are looked into first: brucellosis, other reproductive diseases, mastitis, and tuberculosis. The health of the individual and also of the entire herd must be considered. Numerous other diseases are important in pricing.

**Condition.** The flesh and the condition of the hair of an animal will materially affect its price. For an animal sold at home the condition will not affect its selling price as much as in a public sale.

Cows and heifers that are extremely thin at calving time will not produce their best during the following lactation. Yearlings that are ready to go on pasture in the spring might not be discounted much for lack of flesh if they are thrifty and have good skeletal growth. Overfatness, especially in young cattle, is not looked on with favor.

**Present Production.**  If a cow is being purchased primarily to help out the immediate milk supply, her present production is of utmost importance. Also, this may be moderately good guarantee of her ability to continue to produce.

**Past Production.**  When the present production is known, the added information of past production gives more assurance as to what to expect during the remainder of the present lactation and subsequent lactations. Previous records are always valuable.

**Bred or Open.**  A buyer should know when a heifer or a cow has been bred. The bred heifer is just that much nearer to the time when she will start to bring in an income.

**When Due to Calve.**  A person buying cattle for production at a specific time will pay well for those calving at the time the milk will be needed, but will not be so interested in those due much earlier or later.

**Service Sire.**  In the case of top-quality purebred or grade animals the service sire often influences value. He may be a bull whose calves are sought after, or he may be a mediocre or poor bull, which may detract from the value.

**Pedigree.**  A great deal of importance is placed on the pedigree of high-quality cattle. The value of average cattle is not influenced as much by the ancestors, but many breeders will pay a premium price for animals from certain sires, cow families, or blood lines.

Young cattle cannot have milk production records, classification ratings, or sire proofs. Therefore a pedigree will give an indication of what might be expected of an animal at a later time. Even for animals with milk records or sire proofs, the data in the pedigree on the ancestors do give some indication of continued performance.

**Type and Conformation.**  A poor-type animal, regardless of pedigree, usually will not command a high price. Occasionally, if the buyer is interested in showing, the animal that is a show prospect takes on added value. In the majority of cases the consideration is whether or not the animal is typical of the breed, is straight with plenty of strength, shows evidence of milkiness, has a satisfactory udder, and sound feet and legs.

**Disposition.**  All other factors could be favorable and the cow have a nervous or bad disposition. She might be a kicker, be difficult to handle, or be a slow or hard milker.

**Other Considerations.** Cattle with horns are often a nusiance and a hazard to the health of other cattle in loose housing arrangements; hence the market for horned cattle may be limited. Milking-age cattle that were previously housed in free-stall or loose housing arrangements often do not adapt well to stanchion systems, and vice-versa. This again may limit their marketability.

Other considerations may also help determine what a cow is worth, but these are the basic ones.

## A Guide for Pricing

To arrive at a reasonable price for an animal, a dairyman can consider the price paid for similar animals in auction sales, the prices that are being paid or offered by dairy cattle commission companies, and prices paid at private sales.

Animals of various ages may be evaluated by starting with cows, 3 to 6 years old as 100%. If these are considered to be worth, say, $1000, the younger animals can be evaluated as a certain percentage of the value of the 3–6 year-old group. A bred heifer might be considered to be worth 60% of the $1000. Other ages could be figured according to the following schedule.

|  | % |
| --- | --- |
| Cows, 3 to 6 years old | 100 |
| Cows, 2 to 3 years old | 75 |
| Heifers, bred | 60 |
| Heifers, 12 to 18 months old | 50 |
| Heifers, 6 to 12 months old | 40 |
| Heifers, under 6 months | 25–40 |
| Heifers, at birth | 20–25 |

These figures can be a guide for pricing the average animals in the herd. For older cows, deduct 20 to 25% yearly for cows over six years of age, but not lower than salvage value.

For various levels of milk production, add 1 to 2% to the value of cows for each 1% their production exceeds herd average. For example, if the average price of grade cattle three to six years of age is $800 in a herd whose average production is 15,000 lb. of milk, a 16,500-lb. producer ($+10\%$ over herd average) would be worth $880 to $960. Similar amounts would be deducted for those producing below herd average. Thus, a 13,500-lb. producer would be worth only $640 to $720. This production level should be based on 2 × 305 day, mature equivalent records.

When pricing purebred cattle, the value of above-herd-average produc-

**Table 22-1**  Useful Life Span of Dairy Cattle

| Age | % | Age at disposal | Anticipated years of usefulness |
|---|---|---|---|
| 2 | 100.0 | 6.7 | 4.7 |
| 3 | 93.9 | 7.0 | 4.0 |
| 4 | 81.3 | 7.5 | 3.5 |
| 5 | 67.7 | 8.2 | 3.2 |
| 6 | 53.5 | 8.9 | 2.9 |
| 7 | 42.5 | 9.5 | 2.5 |
| 8 | 31.8 | 10.2 | 2.2 |
| 9 | 21.2 | 11.1 | 2.1 |
| 10 | 14.5 | 11.9 | 1.9 |

tion may be worth 1 to 3% for each 1% increase over herd average because of the increased merchandising as well as milk value. Classification score also adds considerable value to purebreds. Using a base classification score of Good Plus or Desirable (80 to 84 points), cows that are classified Very Good (85 to 89) would be valued 20 to 50% higher, those classified Excellent (over 90), 50 to 100% or more higher. Conversely, purebred cattle classified below breed average are usually valued at about the same price as comparable quality grades.

A study of the useful life span of dairy cows was made by Becker and co-workers\*. The results are tabulated in Table 22-1.

Table 22-2 is a summary of 1972 auction sales that illustrates the importance of performance records on the value of cattle. A summary of 1976 Holstein auction sales is listed in Table 22-3. This indicates the relative value of age on value of dairy cattle.

## Bulls

The same generalizations cannot be made on the worth of bulls as for females. The demand is different and is more exacting. Normally, young bulls from one to two years of age are in greatest demand. Three- to five-year-old bulls are as a rule difficult to sell. They will not command any higher price than a yearling. At this age they are yet unproved, except in unusual cases. Dairymen hesitate to buy a mature bull. The cost, the difficulty, and the danger of keeping mature bulls on the farm are some of the determining factors in using artificial insemination in dairy herds. Most bulls are six years

*Fla. Agr. Exp. Sta. Bul. 540.*

**Table 22-2** 1972 Auction Sale Summary[a]

| Breed | Age | Average price | | |
|---|---|---|---|---|
| | | With production records | Records dams only | No records |
| Guernsey | 2 yr. and over | $ 802 | $647 | $565 |
| Holstein | 2 yr. and over | 1,000 | 819 | 626 |
| Jersey | 2 yr. and over | 558 | 448 | 420 |
| Guernsey | under 2 yr. | | 478 | 319 |
| Holstein | under 2 yr. | | 669 | 366 |
| Jersey | under 2 yr. | | 506 | 342 |

[a] Based on data from Putnam, D. P., Reproduced by permission from March 10, 1974 issue of *Hoard's Dairyman.* Copyright © 1974 by W. D. Hoard and Sons Co., Fort Atkinson, Wisc. 53538.

old or older before they are proved. The value at this age depends largely on two factors: how good is his proof and how sure a breeder is he? The use of proved bulls by artificial breeding associations is increasing the price on well-proved bulls, even of advanced years.

## DAIRY CATTLE
## A PART OF NATIONAL ECONOMY

The marketing of dairy cattle is a part of the national economy that relocates surplus cattle from one farm or area into a deficit area. The prevailing price will be reflected in the balance of the supply and demand.

**Table 22-3** 1976 Purebred Holstein Auction Sales Summary[a]

| Sex and age | Number | Average price |
|---|---|---|
| Females over 2 yr. | 13,430 | $1,297 |
| Females, bred yearlings | 1,030 | 1,232 |
| Females, open yearlings | 1,774 | 993 |
| Females, calves | 2,953 | 838 |
| Bulls, over 3 months | 372 | 1,095 |
| Both sexes, calves under 3 months sold with dams | 1,264 | 329 |

[a] Based on data from *The Holstein World*, p. 21, March 25, 1977.

Within a state there are regions where surplus cattle are grown and from which dairymen in another part of the state normally purchase some, or at times, all of their replacements. The same is true for the country. Nationwide, Wisconsin has, over many years, produced more dairy cattle to sell elsewhere than any other state. Traditionally, the southern States have purchased large numbers of replacements. More recently the latter area has been becoming more self sustaining. The general economy, the price index, and the geographic location influence the price paid for cattle.

## International Marketing

There is a demand for U.S. dairy cattle in foreign countries. A large number are exported. European countries, a few African countries, Mexico, and Japan have been good markets. The exports are handled through the U.S. Foreign Agricultural Service and the breed registry associations, especially the Holstein-Friesian Services, Inc., of the Holstein Friesian Association of America. Some states, through the International Trade Director of the State Departments of Agriculture, assist foreign buyers in locating animals and in handling preparations and shipping. Shipments are made by both air and boat. This market has helped to increase the income from agricultural exports and improved the balance of trade.

## VALUE TO THE BREEDER

The selling of registered cattle and the contact with other breeders is a part of the romance of the dairy business. Dairying and the breeding of good dairy cattle may be a vitally interesting and profitable business. There are values not measured entirely in monetary units.

## THE BUYER

Decisions regarding purchases of dairy cattle should be made after defining one's own needs and resources, identifying availability and cost of alternative types and sources of needed cattle, and determining probable effect of the available alternatives on profitability of a particular dairy operation.

Needs should be defined in terms of the number, type (purebred or grade), and age of cattle needed. In the case of milking-age cattle, stage of lactation may also be important. Needs may vary due to a variety of factors such as; milk needed to establish base during a certain period, planned increases in herd size; desire to fill a show string or to improve the genetic value of the herd; or for many other reasons. The important factor is to accurately define the number and type of cattle needed for a particular operation.

Identifying current market conditions such as availability and costs of cattle can often be achieved by attending auctions or consulting other dairymen or marketing agents. Comparisons should be made in regard to auction and private sales prices. There may also be significant capital or other resource limitations to consider. The purchase of a heifer with great potential (based on pedigree and conformation) at a very high price as a speculative investment may be desirable for an established breeder. It is less likely to be a wise choice for a beginning dairyman who is heavily indebted, and needs milk to establish base. Inferior dairy cattle, those that are unhealthy or unsound or lack the genetic ability for high production, are generally sold for butchering.

Registered dairy cattle bring the best prices if they are high producers, conform to type, and are in excellent health. A successful breeder keeps careful records and, most importantly, becomes known for his integrity.

## REFERENCES FOR FURTHER STUDY

Bartlett, Prescott, and Crissey, *Selling Purebreds for Profit,* published by *The Hol.-Fries. World,* (1960).

Breed journals published by each of the dairy breeds associations.

Briggs, S., Merchandising Dairy Cattle, *The Hol.-Fries. World,* p. 50, August 25, 1975.

McKitrick, Dr. J. L., and McKitrick, John W., Herdsman's Diary, *Hoard's Dairyman,* February 10, March 10, May 25, and October 25, 1972.

McKitrick, J. W., Herdsman's Diary, *Hoard's Dairyman,* January 10, March 25, 1973; February 10 and August 10, 1974; and April 10, 1975.

Perry, H. M., (1976). What We Learned When Selling Our Herd. *Hoard's Dairyman,* **121** p. 1297, (1976).

Putnam, D. N., Production Tested Cows Brought $358 Premium. *Hoard's Dairyman,* p. 356, March 10, 1974.

Willet, G. S., How Much Is That Cow Worth? *Hoard's Dairyman.* p. 205, February 25, 1977.

# 23

# Fitting Dairy Animals For Show And Sale

**23** Participation in dairy cattle shows and in consignment sales is considered by many purebred breeders as a part of their total purebred program. Showing is an excellent means of advertising cattle and contacting prospective buyers. It can also be very educational, as attendance at dairy shows can create the opportunity to exchange ideas with fellow dairymen, to become acquainted with the desired type of a particular breed, and to participate in one of the more glamorous aspects of dairying. Showing dairy cattle has been responsible for maintaining the dairy interest of many youngsters.

## SHOWING CATTLE

The basic purpose of showing cattle is advertisement and promotion. To achieve this purpose it is necessary: (1) to exhibit only those animals that are of desirable type and competitive for the level of the show, and (2) are presented to best advantage. For many years the judges at our dairy shows have been selected on their ability to give unbiased decisions, and are compelled to be well informed about the desired type of the particular breed. The result is that, within minor degrees of difference, the animals with the most desirable type, as compared with the breed ideal, are the winners in the show ring. No one can hope to win with cattle of inferior type, even though they have been well fitted and presented. However, one may experience difficulty in winning or placing well even with animals of desirable type if they are poorly fitted and presented. It is difficult for the judge to properly evaluate cattle that are not properly fitted and are handling poorly at the time of showing.

### Selecting the Show Herd

In general, select only those animals that will be competitive or have a chance to win in their respective age classes at the level of competition of the show. Doing this requires knowledge of the age classes of that show, of desirable type of animals of various ages, and of the quality level of the competition at a particular show.

Age classes are fairly standard, although there is some variation among shows. The more usual female classes (based on the 1977 show season) are the following.

1.  Junior heifer calf—born between January 1 and April 30, 1977

2. Intermediate heifer calf—born between October 1 and December 31, 1976

3. Senior heifer calf—born between July 1 and September 30, 1976

4. Junior yearling heifer—born between January 1 and June 30, 1976

5. Senior yearling heifer—born between July 1 and December 31, 1975

6. Two-year-old cow—born between July 1, 1974 and June 30, 1975

7. Three-year-old cow—born between July 1, 1973 and June 30, 1974

8. Four-year-old cow—born between July 1, 1972 and June 30, 1973

9. Aged cow—born before July 1, 1972

Many shows also include dry cow, senior yearling-in-milk, and group classes such as get-of-sire (three to four animals sired by the same bull), best three females (three animals bred and owned by exhibitor), produce-of-dam (two animals from the same dam), etc. Information concerning classes and age specifications can influence selection of animals to show, hence it should be obtained before final selection takes place.

In nonmilking classes (junior calf through senior yearling), larger animals are generally given preference over smaller ones. Selecting animals born nearer the upper rather than the lower age limit of the class is advised, other things being equal.

In milking-age classes, special attention should be given to the stage of lactation of the cows at the time of the show. Cows in full flow of milk, showing bloom or fullness of udder and considerable dairy character, are generally given preference over stale, thicker cows with less bloom of udder. The most important type component of milking-age dairy cows is the mammary system; therefore only cows with good udders should be shown.

In group classes, mature animals are generally given preference over younger animals. In addition, uniformity of desirable type and size is evaluated favorably. It is therefore important to evaluate maturity and uniformity as well as individual quality in selecting animals for group classes.

It is often advisable to solicit the opinion of another breeder or other qualified person whose judgment you respect before final selections are made. Many times, owners' opinions are less objective than they should be, or, in the case of beginners, owners may not be thoroughly knowledgeable about desired type or level of competition at a particular show.

Availability of time to work with the show herd should also be carefully evaluated. If time is limited, it is often advisable to limit selections to the few very best and have them properly fitted, rather than trying to fill every class and doing a mediocre job of preparing all of the animals.

Selections of the show herd should be made well in advance of the time of the show so as to allow ample time to properly prepare the animals. It may be ncesssary to adjust feed intake on some animals, or it may be desirable to breed some cows to freshen at a certain time to look their best on show day.

## FITTING THE SHOW HERD

The purpose of fitting cattle is to have them looking their best at the time of exhibition. Achieving this purpose requires: (1) thorough analysis of the animal to identify her strong points and deficiencies, and (2) developing and implementing a plan to emphasize the strong points and correct the deficiencies so far as possible by the time of exhibition. This normally involves feeding, training, clipping, hoof trimming, brushing, washing, and often, blanketing.

### Feeding

Problems that often can be corrected by proper feeding include underconditioning or thinness, overconditioning or fatness, and lack of body depth and width. Depending on the degree of underconditioning, it may take several months for the animal to acquire the proper conditioning. Feeding underconditioned animals a higher proportion of high-energy concentrate feeds will usually correct underconditioning problems. It also may be advisable to determine if internal parasites are the cause of the thinness. Overconditioned animals should be fed a restricted diet consisting primarily of high-fiber forages. Care must be taken to ensure adequate intake of protein, minerals, and vitamins, while restricting energy to cause the animal to lose overconditioning. Animals, especially heifers, that have desired size and conditioning but lack width and depth of body should be fed bulky rations. A combination of hay and beet pulp often obtains the desired effect.

Toward the end of the fitting period, the animals should be acclimated to both the feeds and the feeding methods that will be used at the show. If silage is being fed during the fitting period, beet pulp and/or hay should be substituted, as it is almost impossible to obtain silage on the show circut. It is also advisable to tie the cattle and feed and water them from buckets. Doing this will minimize changes while at the show, and this can help prevent off-feed problems and help ensure that the animal will fill properly on show day.

## Training

For the animal to be presented at its best, it must be trained to lead and stand properly. It should be trained to walk slowly, taking short steps, with head up and alert; to stand with its weight evenly distributed on all four feet and with the feet in proper position; and to respond to a light touch from the leader. Proper leading and posing can overcome many minor weaknesses of dairy cattle, such as slightly weak loins, crooked legs, prominant shoulders, etc.; improper leading and posing can accentuate these faults. Efforts expended in properly fitting animals may go for naught if an animal is not properly trained. If she fights the halter or leader, takes long steps, or refuses to place her feet properly when being posed, this detracts from her overall balance and symmetry and makes it difficult for the judge to properly evaluate her conformation.

This training should take place well in advance of the beginning of the show season. In general, the younger the animal, the more easily it can be taught to lead and stand properly.

A rope halter is usually used in the initial training. It is advisable to tie an animal for a day or two before leading training begins to enable her to become accustomed to the halter and to being secured. Leading training should be achieved by coaxing and gentleness, rather than with physical abuse. Leading animals back and forth to water is helpful. Firm, yet gentle, control of the animal during training is important. A few minutes daily practice walking slowly with the head up and practice in posing are worth much more than occasional several-hour sessions of pulling, tugging, and pushing or allowing the animal to graze while being led. Toward the end of the training period, a leather halter of the type used while showing should be used in training sessions.

## Clipping

Proper clipping can also enhance the appearance of the animal. Areas normally clipped include the head and ears, neck, tail, and, for cows, the udder and mammary veins. The purpose of clipping is to improve the appearance of the animal; therefore, she should be clipped only after careful analysis of how her appearance can be improved by clipping. Often, leaving some hair in low spots on the rump, loin, or neck, shaping the hair to a point over the withers, or clipping some of the hair from the hocks, legs, and brisket will enhance the animal's appearance. The hair should always be blended at the place where clipped and unclipped areas meet. It is advisable to clip show animals about a month before the show to help determine the best way to clip each animal and to acquire experience in clipping. These animals should, however, be reclipped within a few days of the show.

## Hoof Trimming

The hooves of all show animals should be carefully trimmed and properly shaped. Long toes and misshaped hooves detract from the appearance of the animal and make it difficult for her to walk and pose properly. Hoof trimming should take place well in advance (at least two to four weeks) of the show season as often an animal can become temporarily lame after trimming. This is especially true for those who require extensive trimming.

## Brushing

Brushing stimulates the circulation of the blood and helps make a glossy coat of hair. It also removes dirt, dust, and loose hair and helps train hair to lie in place. A rubber currycomb is effective in removing dirt and loose hair, and is easy on the hide. A clean, soft, brush should be used for routine brushing intended to remove dust and increase the glossiness of the hair.

## Washing

The animals should be thoroughly washed with a mild soap such as castile or ivory, or with a soap especially designed for washing livestock (Orvus). The animal should be thoroughly wetted, then scrubbed thoroughly with the soap accompanied by considerable rubbing. After washing, all the soap must be rinsed from the hair. It is advisable to keep the animals lightly blanketed and to avoid exposing them to drafts and cold until they dry. Animals with considerable long hair should be washed and blanketed well ahead of the show season in order to loosen the long hair. All animals should be washed and allowed to dry before clipping. They should be rewashed within 12 to 18 hours of show time.

## Blanketing

Blanketing helps keep the coat clean, makes the coat smooth, and helps give the animal a smooth, clean appearance. It is especially helpful on longer-haired animals and is widely practiced in this country.

## Hauling

The transporting of the animals to the show in such a manner that they arrive uninjured and healthy is an important part of successfully showing cattle. Overcrowding, slippery floors, faulty loading chutes, or poor loading conditions enhance the risk of bruising and injury, and must be avoided. Squeezing 10 animals in a truck only large enough for eight is a poor practice. It is better to make two trips. Surfaces should be of the nonskid type and/or covered with sand, sawdust, or shavings. Inside surfaces should be free of projections and sharp edges. Loading chutes should be of strong construction, have a skidproof surface, and fit properly to the truck entrance. The truck should be in good condition and carefully operated. Sudden starts, stops, and speeding around curves must be avoided.

Transporting cattle long distances and/or during inclement weather increases the risk of disease and often requires special precautions. Periodic stops for unloading, milking, feeding, watering, and resting the cattle may be required if the trip takes more than 12 to 14 hours. During inclement weather, the sides of the truck may need to be covered. It is advisable to give the animals hemorrhagic-septicemia vaccine 10 days to three weeks before moving them, or to give them hemorrhagic serum three to four days before transporting. Antibiotics are also often used to increase resistance to infection that may result from stress and lowered resistance because of exposure.

Many experienced dairy cattle exhibitors prefer to transport cattle when they are relatively empty and avoid feeding them large amounts of concentrates and succulent feeds within 12 hours of transporting. Cattle handled in this manner defecate less during hauling, and so are cleaner when they arrive. This practice may increase their appetite on arrival and ease problems with getting the cattle on feed in the new surroundings.

Arrangements should be made to arrive at the show a minimum of one to one and a half days, and preferably two to three days, before the show. This allows time for the animals to adjust to the new surroundings and get back on feed, and allows ample time to get the animals in the best shape before showing.

## At the Show

The comfort of the animals should be the first concern after arriving at the show. They should be tied in a freshly bedded stall. Immediate needs such as feeding, watering, and milking, depending on the distance traveled, should be taken care of and a regular routine established for these necessities. Cattle usually prefer to lie down and perhaps eat a little hay after unloading. The animals should be washed as needed, or brushed to enhance their appearance.

Supplies and tack should be stored neatly and conveniently and the display area made and kept neat, orderly, and attractive. One of the purposes of showing is advertising and promotion of cattle. Another especially important promotional consideration at fairs attended by urban people, our milk and dairy products customers, is promotion of milk and dairy products. Both purposes are greatly enhanced by the neat, attractive appearance of the total display, by clean, contented, high-quality animals, and by courteous personnel caring for the animals. Conversely, dirty cattle, littered areas, and discourteous attendants detract significantly from the purpose of showing and must be avoided.

## Final Preparations

After the comfort of the animals has been cared for and the display area arranged, the entries should be rechecked to make sure everything is in proper order. The rules and methods of the particular show should be studied and every detail carefully watched. The showing schedule with approximate times should be learned, for this will determine when certain cattle-preparation tasks will need to be carried out. The purpose of all the prior preparation such as fitting, clipping, etc., and transporting the animals to the show is to have them at their best at the time of showing. The final preparations to make them look their best include washing, filling, and bagging. The final washing should take place several hours before show time so the cattle have ample time to dry and to eat. Many exhibitors prefer to wash their cattle the evening before the show. The timing of the last milking before the show should be determined by the milking interval at which the animal's udder looks best. Good showmen carefully observe the udders of show animals at various intervals after milking for several days before the show. Some may look best eight hours after milking, while others may look best with 14 or 16 hours milk in their udders. Individual quarters on some cows may not be completely milked out, in order to balance the udder. It is very important to determine the approximate time a particular class will be shown and then perform the last preshow milking at such a time that the correct amount of milk will be in the udder at show time. Overbagging to an extent that the udder and teats have a distended appearance should be discouraged.

Most animals look better when they are full. They give the appearance of possessing more depth and width of body and are often more contented. A good fill may also serve to strengthen a slightly weak loin. Most showmen prefer to feed liberal amounts of hay, but to withhold water, concentrates, and succulent feeds such as soaked beet pulp, for a period starting 12 to 15 hours before and extending to three to four hours before show time. This helps ensure a good appetite during the few hours just before the show when the cows are fed succulents and concentrates. Water may be limited or withheld completely until just before showing. Care should be taken, however, to avoid overfilling cattle, as this may cause them great discomfort accompanied by roaching and difficult handling during the show.

## Showing

The exhibitor should have final brushing, cleaning, and haltering taken care of and be nearby when the class is called. He should lead his animal around the ring in a clockwise direction, holding the halter strap in his left hand and walking backward. This enables him to see both his animal and the judge. From the time the animal enters the ring until it leaves it should be the business of the showman to see that the animal is being exhibited to its best advantage at all times. The judge's directions should be followed promptly and courteously. Most animals look better when walking than when standing, so the showman should keep his animal moving whenever possible. However, he should stop his animal and pose her whenever requested by the judge and when he is in the placing line. Cows in milk normally look best when standing with the rear leg nearest the judge slightly ahead of the other rear leg and with the front feet about even and spaced widely apart. Heifers are normally posed with the front legs in the same position as cows, but with rear leg nearest the judge slightly behind the other rear leg. The best position for any animal, however, is the one that presents her to best advantage. This should be determined prior to actual showing and the animal should be trained to pose in that position.

When the placings have been made the animals should be taken from the ring. Griping and criticism of the judge or fellow exhibitors must be avoided, even though the placement may not have been to the liking of the exhibitor. To be a good showman, a person must exhibit good sportsmanship.

## Showing Out of the Ring

Breeders are appreciating more and more the value of showing animals when they are not in the ring. There are often prospective buyers or other interested people who are attending the show and it is important, that such people have the opportunity to see the animals outside the ring. This is especially true of winning animals. It is usually profitable to have a courteous and knowledgeable attendant with the animals when visitors are about. Unless the animals are about to be shown, the attendant should be willing to remove the blanket and show them at any time.

## Showing Ethics

The Purebred Dairy Cattle Association (PDCA) Show Ring Code of Ethics was designed with the belief that it is in the best interests of all breeders of registered dairy cattle to maintain a reputation of integrity and to present a wholesome and progressive image in the show ring. It therefore published a list of practices and procedures that are considered unethical in the showring (Appendix K). These practices should be avoided by all dairy showmen.

## Learning To Fit and Show Dairy Cattle

Basic principles of fitting and showing dairy cattle can be learned from discussions such as in this chapter. Real proficiency can only be acquired through actual experience working with cattle and by carefully observing the techniques of superior showmen. When attending fairs and shows as an exhibitor or a spectator, careful observation of the real professionals can be a valuable learning experience and is highly recommended.

## PREPARING CATTLE FOR SALE

As the purposes of preparing cattle for sale are the same as for showing cattle, that is to present them to best advantage at the time of the sale, similar procedures should be followed in regard to clipping, training, hauling, etc. Consignors to sales should also be with their cattle while they are on display in the sales barn to promote their consignments and to answer questions from prospective buyers. All too often, consignors rely entirely on the sales team to not only sell their cattle, but to wash, clip, and train them to lead.

Sellers of dairy cattle should also be aware of the conditions of the PDCA Code of Ethical Sales practices and to strictly adhere to them (Appendix L).

## REFERENCES FOR FURTHER STUDY

Dickson, D. P., Clipping is a Big Part of Grooming *Hoard's Dairyman*, June 10 and June 25, 1972.

How to Buy Dairy Cattle, *Dairy Field Production*, **3**: 2(1962).

McKitrick, Dr. J. L., and McKitnick, Herdsman's Diary, *Hoard's Dairyman*, June 10, July 10, July 25, August 10, and August 25, 1972.

Moore and Gildow, *Developing a Profitable Dairy Herd,* Wood and Weber, Inc. Seattle, 1945.

Strohmeyer, Photography of Jersey Cattle, *Jer. Jour.,* **8:**17:20 (1961) and *The Hol.-Fries. World,* **58:**22:21 (1961).

# 24

# Some Details in Dairy Cattle Management

**24**

The person and the cow are the two most important factors in determining the degree of profitability and personal satisfaction derived from dairying. The person is responsible for doing many things, including a number of little things. Doing these things well and when they need to be done can make dairying a profitable business and an interesting and challenging way of life. Neglecting these details can make dairying a job of drudgery that returns little profit or satisfaction.

A good dairyman is not only a lover of good cattle but also one who has the knowledge and ability to get the job done. There are times when the comfort and well-being of his cattle are placed above his own desires.

## HANDLING THE HERD

### Regularity of Care

The dairy cow is a creature of habit. The same routine of feeding, milking, caring for her should be used each day. It is desirable that the cow have approximately the same amount of exercise, and that she be fed and milked at the same hour in the same manner daily. No system is stronger than its weakest link. A change of milkers or strange people in the barn at milking time can have an effect on some of the more sensitive animals. Major ration changes can cause cows, especially high producers, to go off feed. These should be avoided. Most cows can, however, become accustomed to a certain amount of change.

### Kindness in Handling

A cow must always be treated with kindness if she is to maintain production. The beating of a cow should never be tolerated under any circumstances. It is not only cruel, it reduces milk production. A man who cannot control his temper will never make a good dairyman. Many good herdsmen can go into the field or lot where their cattle are loose and pet or catch them without trouble, but when animals are handled roughly they will run away from their caretaker. When they are handled gently and quietly, they will move in and out of the barn or parlors slowly, with little disturbance or danger of injury. Dairy cows should never be hurried when moving from one place to another, such as to and from the pasture, milking parlor, exercise lots, etc. Seldom do dogs have a place in handling dairy cattle, as most dogs will move cattle too fast. Dogs around the barn or milking parlor at milking time may disturb cows to the extent that milk let-down is inhibited.

## Exercise

Dairy cows need only a limited amount of exercise. The work that they do in eating, chewing, and digesting their feed gives them considerable activity. Time studies on cows show that they spend 9 to 15 hours out of each 24 lying down. They do their eating and a part of their ruminating while standing, but most of their ruminating is done while lying down.

When cows are housed in some type of loose-housing arrangement, the freedom of movement may suffice. Some herds are kept in stanchion barns throughout the winter. In general, however, it seems desirable to give the animals at least some exercise, preferably on a sodded area. Allowing cows some time off concrete daily, whether housed in conventional or loose housing, helps avoid many feet and leg problems. It can also facilitate heat detection.

It is possible for a dairy cow to have so much exercise that she will use up energy unnecessarily. This is true when she is driven too far to pasture or when the pasture is so poor that she must cover too much ground to get enough feed.

## Keeping Cows Clean

Milking cows should be kept clean. In conventional barns, properly sized stalls, ample bedding, and regular grooming are necessary.

The equipment needed for grooming is a blunt currycomb for removing coarse material or manure from the cow and a heavy bristle brush for the main grooming job. The comb should be used gently, so as not to irritate the hide. Some herdsmen like a rubber currycomb. It is easy on the hide and is a good tool for removing loose dirt and hair. It is not satisfactory on soiled areas. There are electrically operated, revolving bristle brushes and cow vacuum cleaners on the market that do a satisfactory job of grooming cows, except on wet areas.

In loose-housing arrangements, little grooming is needed. However, ample bedding and frequent picking up of the droppings in the bedded area are helpful in keeping cows clean. In free-stall arrangements, proper-sized stalls and frequent raking of manure from the stalls are conducive to cow cleanliness. In all systems, muddy areas in lanes, around feed bunks, etc., should be avoided.

## Clipping

One of the greatest aids in keeping milk cows clean is to clip the belly, udder, and rear portion of the animal, including the entire area back of a line drawn from in front of the udder to the tail head (Chapter 17). Clipping this area removes the long hairs where the cow is most likely to be soiled when she lies down. Bedding and manure will not adhere to short hair to the same extent as to long hair. Any that does is easily removed.

### Clean Feeding Area

Profitable dairying is closely associated with a high level of production. This, in turn, is closely coupled with the cows' nutrient intake. Palatable feed is important in achieving high nutrient intake. So is the cleanliness of feed mangers.

When cattle are housed in conventional barns, feed mangers should be swept before each feeding. Concentrate or fresh silage should not be fed on top of soured feed.

In loose-housing arrangements, mangers should also be cleaned daily if feed is left over. Particular attention should be given to cleaning the feed trays or bowls in the milking parlor. Many parlor feeders are equipped with removable bowls. These should be washed regularly.

In general, cows that are milking the heaviest are the ones most likely to be adversely affected by unclean feeding areas. They have the need for greatest feed intake and are usually more easily thrown off feed than lower producers. They are also usually the most profitable cows, so overlooking this detail can severely affect profit.

### WATERING

Dairy cows must consume large quantities of water for the production of milk. The amount that a cow will drink depends largely upon the outside temperature, the kind of feed eaten, the amount of milk the cow is producing, and the temperature and cleanliness of the water. Most cattle normally consume 3 to 4 units of water for each unit of dry feed. A large cow producing 70 to 80 lb., or 30 to 35 kg., of milk daily will need as much as 15 to 25 gal. or 60 to 100 liters, of water daily.

A cow will not drink all the water she needs for the most profitable production of milk unless she can get it frequently, without discomfort, and at a temperature well above freezing. Many dairymen are using automatic drinking cups in their barns in order to supply these conditions.

If the weather is warm, more water is required than in moderate or cold weather. Cows will drink up to twice as much water in hot weather as in cold, winter weather. Since cows do not sweat, or, if so, relatively little, they eliminate more water as urine in hot weather in order to cool the body. If the feed is of a succulent nature, the cow will consume less water than if she were fed dry feed.

In experiments at the Beltsville Station, average-producing cows were watered once a day, twice a day, and at will from watering cups. The cows watered once a day drank less and produced less than those watered twice a day, and the cows watered twice a day drank as much but produced less than those watered at will. It was found that the higher the production, the greater the benefit derived from frequent watering. The cows consumed about 4 units of water for each unit of dry matter consumed.

The Iowa Station* found that dairy cows watered by means of water bowls in the barn consumed approximately 18% more water and yielded 3.5% more milk and 10.7% more butterfat than cows watered twice a day at an outside tank. Cows in the barn drank an average of 10 times per day, consuming two-thirds of their water in the daytime and the other third at night (5 p.m. to 5 a.m.). The cows watered outside twice a day in very cold weather drank only once a day about 30% of the time. The amount of water consumed was 3.0 to 3.5 units of water for each unit of milk produced.

In cold weather the cows will drink more water if it is slightly warmed. The economy of this practice depends on the climate. Ordinarily, if water is given in the barn at will, either in watering cups or in a tank, so that they can drink small quantities at frequent intervals, no advantage will result from warming it.

If the water tank is located where the water will freeze, a water heater is advisable. Milking cows should not be required to drink ice water. Water cups equipped with automatic heating units can be used outside or in any unprotected place to supply cows with water above the freezing temperature. One such cup will be sufficient to water 25 cows.

Some persons have believed that soft water is more desirable than hard water. Studies at the Virginia Station** showed no difference in water consumption or in milk production of cows from the use of either softened or hard water.

The size of the water tank needed depends on the number of cows to be watered, how often they have access to it, its rate of supply, and the number of times the tank is to be filled each day.

If the tank is to hold a day's supply, 15 gal. per cow should be allowed. When cows are watered twice a day, allowance should be made for 10 gal. capacity per cow, except where the water is replenished quickly. Where cows have free access to the water tank, a float valve can be used to keep the water

---

level constant. A small tank will suffice with this arrangement. One cubic foot of water is approximately 7½ gal. or 28.4 liter.

A water tank, no matter where it is located, needs a paved apron around it. Otherwise, a mudhole is almost sure to develop.

The availability of abundant water in summer is even more important than it is in winter. It is, however, more often taken care of automatically. Springs in pastures are valuable assets. The cow should have access to water continuously and at not too great a distance. In summer, algae growth is often prevalent in water tanks. Copper sulfate will prevent this growth. One ounce in 8,000 gal. or 1 tsp.-ful in 150 gal. is recommended.

Checking and cleaning watering facilities regularly, daily for water bowls in conventional barns, is an important detail in dairy cow management.

## TRIMMING HOOFS

In recent years the incidence of foot problems seems to be increasing. Part of this may be from the increase in confinement housing; some is unquestionably from inherent foot or leg problems; some because of disease or injury; and often a combination of these factors is involved. Regardless of the cause, the result is cattle that are lame, have sore feet, walk with difficulty, and lie down more than is usual. They often eat less and produce less milk.

In herds with a high incidence of foot injuries, the cause should be determined and eliminated.

In herds where there is a high incidence of shallow heels, spread toes, weak pasterns, post legs (legs that are too straight when viewed from the side), and legs that toe out (cow-hocked), some selection pressure should be applied when choosing bulls for use in the herd. This can help avoid problems in future generations.

Cattle that currently have long toes or misshaped hooves should have their feet trimmed to alleviate the current problems, regardless of the cause. Trimming should, after a period of adjustment during which the feet may be tender, help the animal feel better, move more easily, produce better, and extend her productive life.

To do a good job of hoof trimming it is necessary to (1) know what a normal foot looks like, (2) control the animal so as to avoid injury to the animal or to the person doing the trimming (various portable chutes and tilt tables are available commercially, or they can be constructed), (3) have the proper tools to get the job done (a portable box to place the animal's foot on, nippers or pincers, hoof knife, chisels, and rubber mallet), and (4) trim cautiously so as not to injure the animal.

The trimming should be done largely from the bottom of the hoof with a chisel, pincers, hoof knife, and rasp. Extremely long hoofs may be cut back on the front and side before the bottom is trimmed. This may be accomplished with pincers, or the foot may be set on a solid board and cut with a chisel.

Care must be taken not to cut too far back or too deep as there is danger of cutting into the quick, which may result in lameness and foot infection.

Figures 24-1 and 24-2 are photographs of a cow and the rear feet of that cow, which are badly in need of hoof trimming. The hooves are long and the outer halves are curled. Note also the shallowness of the heels. It is not unusual for cows with shallow heels such as these to have long hooves, as they typically walk back on their heels and the hooves do not wear off in front. They are also more succeptable to heel bruises, which leads to sore feet and lameness.

The cow in Figure 24-1 was secured to a tilt table and her feet and dew claws were trimmed (Figure 24-3). Note that her hooves were trimmed from both the front (to shorten her toes) and from the bottom (to provide her with a flat surface on which to stand and to correct the curled outer halves of her hooves). Figure 24-4 shows this same cow after hoof trimming was completed.

Considerable practice is required to become a proficient hoof trimmer. It

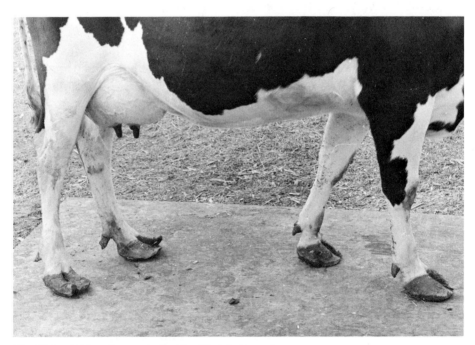

**Fig. 24-1.** All four feet need trimming (Carl A. Brown).

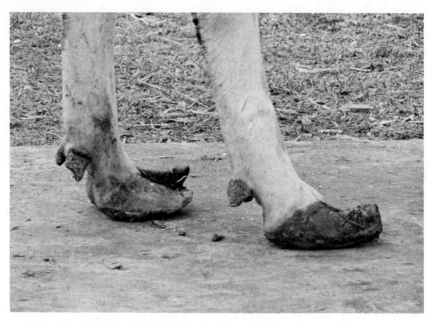

**Fig. 24-2.** Rear feet of cow in Fig. 24-1 (Carl A. Brown).

is highly unlikely that one can learn how to become proficient by reading about it. Better ways include practicing on smaller heifers, practicing under the supervision of a competent and experienced person, or enrolling in a hoof-trimming short course or similar training program.

## IDENTIFICATION OF CATTLE

It is necessary to have some means of designating and identifying the animals in a herd. Many dairymen name their cows but do not have identifications except as they know them. This may work very well in herds so small that one can keep in mind all the animals. In larger herds and in all purebred herds it is never safe to depend on memory. Some method of marking must be used.

Animals can be marked and identified by eartag, tattoo, number tags attached to straps or chains around the neck, leg straps, branding numbers on hips, ear notches, photographs, or color sketches.

### For Registration

Purebred animals to be registered must have a permanent identification. Color markings, which do not change with age, are used by some of the breeds. The spots on the broken-colored animal can be sketched on a cow outline, or a photograph may be taken. With solid-colored breeds, a tattoo in the ear is required.

## Eartag

The most common method of marking and identifying dairy cattle is by metal eartags placed in their ears. These tags have numbers or numbers and letters stamped on them. The standard Dairy Herd Improvement Association method of identification is the eartag. The calf may be eartagged by the DHIA supervisor when it is a month or less of age, and its identity recorded.

**Fig. 24-3.** Bottom view of hooves after trimming (Carl A. Brown).

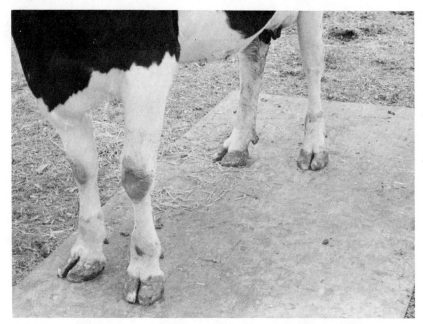

**Fig. 24-4.** Cow in Fig. 24-1 stands squarely on all four feet after trimming (Carl A. Brown).

Practically all cattle that are tested for Bang's disease or tuberculosis are eartagged or tattooed. The few exceptions are registered cattle that have registration papers available for identification. The standard DHIA eartag is also acceptable for health-test identification purposes and for identification of grade animals in artificial breeding.

## Tattoo

Animals with light-colored ears can be tatooed in the ear. The tattoo does not show too clearly in black colored ears. The method is to punch several small holes with a die in the form of numbers or letters through the skin on the inside of the ear and then fill them with tattoo ink. If done correctly, this is a permanent mark. Its disadvantage is that the animal must be caught and the inside of the ear cleaned to be able to read the identifying marks.

## Number Tags

Metal tags, large enough to be read at a distance, are quick means of identification. They may be fastened to a neck strap or chain. There is a chance that they may be lost.

## Branding

Branding numbers and letters on the hip or thigh can be done with either branding irons (heated or freeze-branding irons) or a branding fluid. Freeze branding has become increasingly popular in larger herds in recent years.

## Ear Notches

Notches cut in the ears make a rather easy method of identifying animals. A notch represents a number, depending on its location in the top, bottom, or end of the ear, and also, which ear it is in. Few dairymen are willing to disfigure the ears of their cattle in this way.

## VICES OF COWS

Cows may develop certain habits that range all the way from being a nuisance to rendering the animal almost useless as a dairy cow. For example, occasionally there is a cow that cannot be left in the barn with a drinking cup with a paddle valve because she will flood the barn.

## Sucking

Probably there is no more difficult problem to deal with than an animal that sucks another cow or herself. Often, there is much difficulty with calves sucking each other during the milk-feeding period. However, the problem is not limited to calves. Young calves can best be kept in individual pens or tied so that they cannot reach each other. With older heifers and cows, a more difficult problem prevails. Separating the animals concerned may be the easiest and most successful method if it can be done. Many patent devices are partially effective for overcoming this habit, such as muzzles with sharp prongs in them. Most of these require some attention. If they are used on cows that suck other cows there is some danger of injury to the udder. A simple method that is helpful in controlling sucking, but not always effective, is to put a bull ring in the cow's nose and attach two or three other rings to it. A special ring that has some sharp prongs soldered on to it is very effective. The ring needs no attention and does not interfere with the animal's eating but does interfere when she attempts to suck. An aerosol spray for use on udder or ears wherever animals are sucking offers some help. Many other devices have been tried, but sometimes nothing will work.

## Kicking

Some cows develop the habit of kicking through improper treatment, whereas others are characteristically vicious. Of the latter class, there are fortunately very few. Practically always, if the animals are carefully handled, especially during the first days of milking of their first freshening, they will be quiet milkers. Should they give difficulty, the teats and udder should be examined thoroughly for soreness and injuries. Patience and kindness for a few days will pay dividends in quiet, cooperative cows later. Some cows do not respond and remain vicious. If they are good producers and if it appears wise to keep them, they may be milked by securing them in various ways. One method is to tie the head high. Another is to tie a rope around the body of the cow just in front of the udder. With many cows, after this method has been used several times, it may be necessary only to throw the rope over the back.

Some men develop confidence in their cows and can get along in handling and milking them with little difficulty. Others seem to keep the cow afraid and on the defensive. In rare cases nervousness and viciousness may be inherited. This is a characteristic to consider in selecting cows for foundation animals.

## Fence Breaking

Usually cows break, crawl through, or jump fences because they have little to eat within their boundary. A few just seem to think the grass on the other side of the fence is greener. There is little that will stop a roguish cow except good fences.

## ARRANGEMENT OF
## COWS IN CONVENTIONAL BARNS

There should be, if possible, a systematic method of arranging cows in the conventional barns. Each dairyman can choose the arrangement that is best suited to his barn and the layout of his work, and the one that would be the greatest help to him in his work.

## By Size

It is advisable to construct the dairy barn with stalls of varying lengths. The cows can then be lined up in the row according to size. With this plan it is best to have the young cows or smaller ones at the entrance end of the barn. They will learn their stalls more quickly than if they must go the full length of the alley past all the other stalls.

## By Sires

Breeders of purebred cattle like to have all the daughters of a bull together. This is especially effective in showing animals to prospective buyers. When so grouped, they can be more carefully and thoroughly evaluated. Often the different qualities of first-calf heifers by a sire will be more easily seen if they are together than if they are scattered throughout the barn. In a specific case 10 daughters of each of two bulls freshened about the same time. They were scattered throughout the barn and produced about the same amount of milk. Later, however, when they were grouped according to sires, a specific characteristic became evident. The group by one sire had more quality in their udders and were easier to milk. This characteristic was much more evident when the young cows by different sires were milked by groups.

### By Health of Udder

When mastitis is an active problem in a herd, the cows should be arranged according to the health of the udders. The clean, first-calf heifers should be milked first; then the older cows that have no history of udder trouble; next, clean cows with a history of udder trouble; and last, the cows that, at the time, have mastitis in any form.

### By Ease of Milking

The easy-milking cows may be placed on one end and hard milkers on the other, with medium milkers in the middle. With this arrangement the easy milkers can be milked first with the more difficult ones left until last when more time is given to milking them out clean. Some dairymen, however, prefer to begin on the hard milkers and get them milked first while the men are fresh on the job. Either method gives satisfactory results if carried out properly.

## NAME PLATES

Good management of milking cows dictates that a name plate be placed over or in front of each stall, which should give the cow's name and her production. In purebred herds this is an advantage when showing the cattle to visitors and prospective buyers. The name cards should be covered with glossy isinglass, acetate paper, or some other transparent material to protect them and keep them clean.

## BEDDING

Bedding requirements vary considerably with the type of housing. Types of bedding for free-stall systems are discussed in Chapter 15.

In conventional and loose-housing arrangements, some sort of bedding is necessary to provide comfort for the animals, to keep them clean, and to absorb the liquid manure. Less bedding is needed if the cows are fastened in stanchions in stalls of proper length than if they are fastened with chains or left loose in a box stall or in loose housing.

The common materials used for bedding are straws of various kinds, corn stover, shavings, sawdust, and peanut hulls. The best kind depends upon local conditions. On some farms sufficient straw is available, so that the problem of the proper kind of bedding is easily solved. In many cases corn stover can be shredded and used for bedding. It makes very satisfactory bedding material.

If the bedding has to be purchased, the relative prices of the various bedding materials will usually be the deciding factor. Rubber mats can be used in stall barns where cows are kept in the barn for long periods and bedding is scarce and expensive. They appear to be very durable, and protect the cow well. There is not the problem of the cow's working all the bedding off the concrete platform into the gutter.

The absorptive properties of different bedding materials are given by Doane* in Table 24-1.

The fertilizing value of the straws and corn stover is much greater than that of either shavings or sawdust. Some dairymen are finding that cut straw is more economical than long straw. The absorptive power is not changed, but less short straw is wasted by undue amounts being worked back into the gutter.

The bedding cannot be used too sparingly if the cows' udders are to be protected and the cows kept clean. In trials at the Virginia Station,** large cows in stanchions and tie stalls required 6 to 8 lb. of long straw or about 30 lb. of sawdust per day.

In loose housing approximately 50% more bedding is required than in a conventional barn. This depends a great deal on the kind of bedding material and whether the cattle are fed and watered away from the bedded area. When these are all together, twice as much bedding may be required.

Workers at the Pennsylvania Station* studied conditions for minimum bedding needs in loose-housing systems. The building was completely open on the side adjoining a paved yard. There was no mud carried on to the bedding. The open shed gave complete ventilation, which allowed the manure pack to dry out somewhat on sunny days. The feeding area was away from the bedded area. No traffic lanes developed across the pack. The droppings

* *Md. Exp. Sta. Bul.* 104.
** J. W. Howe, M. S. Thesis, Virginia Polytechnic Institute, 1952.
* Pa. Agr. Exp. Sta. Prog. Rpt. 199, (1958); *PDCA Report* (1960).

**Table 24-1  Absorptive Properties of Bedding Material**[a]

| Material | Water-absorbing power of bedding | Pounds of bedding required to absorb for 24 hr. |
|---|---|---|
| Cut stover | 2.5 | 4.0 |
| Cut wheat straw | 2.0 | 5.0 |
| Uncut wheat straw | 2.0 | 5.0 |
| Sawdust | 0.8 | 12.5 |
| Shavings | 2.0 | 4.4 |

[a] Doane, *Md. Exp. Sta. Bul.* 104.

were picked up daily. Fresh bedding was added in the afternoon. The minimum acceptable amounts as determined under this system of management ranged from 5 to 8 lb. of straw per cow per day. Removing the droppings daily decreased the amount of straw by 2 lb. per cow per day. Reducing the bedded area per cow from 60 to 50 sq. ft. increased the requirement by 1 lb. per day. The 5 to 8 lb. of straw per day under the above conditions compared with 5 to 6 lb. in stanchion barns. In trials with sawdust, using regular management practices, they reported 29 lb. per cow per day in the pen barn and 43 lb. in the conventional barn as the minimum for reasonable cow comfort and cleanliness.

## FLIES

Flies are a great annoyance to cattle and persons working with the cattle. The appearance of flies is coincident with the appearance of warm weather. Their population increases rapidly and continues at high levels throughout the warm season unless a sound fly-control program is formulated and implemented.

### Houseflies and Stable Flies

The housefly and the stable fly are the two most common varieties found on dairy farms. The housefly is primarily a nuisance, but can spread disease. It is not thought to lower milk production. The stable fly is a blood-sucking insect that feeds on a variety of warm-blooded animals including dairy cows. Studies have shown that it can cause lowered milk production.

Both of these flies breed readily in manure piles or in moist rotting silage or grain found around feed bunks. If sanitation is poor and many such areas are available for fly breeding, the fly population will increase rapidly.

## Sanitation

A good fly-control program begins with good sanitation. Reducing the breeding areas reduces fly population and chances of developing a resistant strain of flies. Regular removal of accumulated manure from calf pens, loafing barns, and other areas, combined with regular cleanup of waste feeds and areas around feeders are essential if effective fly control is to be achieved.

## Chemical Control

Chemical control with insecticides approved for use in dairy facilities and on dairy cows is also a vital part of a good fly-control program. These may be residual or contact insecticides.

Residual insecticides, when properly applied, can give good control for six to eight weeks. These chemicals should be applied initially when flies first appear in the spring and then reapplied when they no longer appear effective.

Contact or knockdown chemicals can be effective in helping to control flies on cattle and in barns. Most are petroleum based and can be applied by sprayers or foggers.

Chemicals are also available in powder form to dust on cattle or in liquid form for use in a back rubber.

Generations of flies gradually build up resistance to many chemicals; therefore recommended insecticides change rapidly. It is wise to check current recommendations for both residual and knockdown chemicals annually. It is also essential to use only those chemicals approved for use in dairy facilities and on dairy cattle and to follow the manufacturer's directions when using them. Some may be approved for nonmilking animals, but not for milking cows. Others may be approved as residual chemicals for some areas of the dairy facility, but not in the milking area.

Chemical insecticides should never be used as a substitute for good sanitation.

## Deer Flies, Horse Flies, and Mosquitos

Other fly pests include face flies, horn flies, deer flies, and horse flies. Mosquitos can also be a nuisance.

The face fly does not bite, but rather feeds on tears and saliva. It can spread pathogenic organisms and can be responsible for spreading a disease such as pinkeye. The horn fly does bite and feeds on blood. Both horn and face flies are parasites of cattle and must live with cattle. Horn flies are relatively easy to control because they spend all their time in the adult stage on cattle. Face flies do not and thus are harder to control.

Deer flies, horse flies, and mosquitos are blood suckers and can inflict painful bites. Deer flies are particularly vicious. These three insects breed in swamps, lowland areas and, in the case of mosquitos, any pool of water that remains 10 days or more. Chemical insecticides are not practical in controlling these pests, so eliminating breeding areas and/or avoiding exposure of cattle to breeding areas is the only method of practical control.

## ENVIRONMENT

Dairy cattle performance can be adversely affected by climatological factors, especially by extremes in temperature and by excessive himidity at high temperatures. Fortunately, cattle adapt well to a wide range in temperatures. Although the ideal environment appears to be 30-60°F, or -1 to 15°C, little adverse effect on production is noted within a range of 5 to 80°F, or -15 to +27°C. Breed differences in heat tolerance have been reported, with Holsteins having the lowest (80.6°F, or 27°C) and Jerseys the highest (86°F, or 30°C) critical temperature. By comparison, the critical temperature for Braham cattle is 95°F, or 35°C.

At environmental temperatures below 5°F, or −15°C, feed consumption increases but milk production decreases. When temperatures rise above 70 to 75°F, or 2l to 24°C feed consumption decreases and may approach zero when temperatures climb above 100°F, or 38°C, especially when accompanied by high humidity. Milk depression usually begins at 80 to 85°F, or 27 to 29°C, and is severe at temperatures about 90 to 95°F, or 32 to 35°C, particularly when accompanied by high humidity.

The adverse effect of a combination of climatological factors such as environmental temperature, humidity, and solar radiation on conception rates is also well documented.

Decreases in conception rates during the summer months of 5 to 19% in most of the United States and 10 to 20% in the extreme southern areas are not unusual.

In view of these possible consequences of decreased dairy cow performance at temperature extremes, it becomes an important function of management to prevent exposure of cattle to such extremes. Preventing cold stress is usually not a major problem. Cattle housing systems have been developed for various parts of the United States that effectively avoid most problems of cold stress.

Heat stress is a more prevalent and difficult problem. In the northern two-thirds of the United States, most heat stress problems can be avoided by providing ample amounts of shade in pastures and lots and by having housing systems that can be opened for cross-ventilation. In severe periods, forced air ventilation with fans or cooling of structures by spraying metal roofs with water may be necessary.

The effects of various methods of preventing climatological stress have been widely studied in the southern part of the United States. These include shade, evaporative cooling with fans, circulating cooled air, and air conditioning. Most reports indicate that the use of air conditioning is not economically feasible in most cases, although the beneficial effect on production was apparent. Providing shade is helpful and highly recommended. A combination of shade and evaporative cooling, with air pulled across a cooling surface and directed across the shaded area is more effective. One study* reported consistent lowered temperatures of 20 to 24°F, or 11 to 13°C, when such a system was used. This report indicated an increase of 1,200 lb., or 540 kg., of milk per cow for 10-month lactations and increased breeding efficiency when such a system was used during the months of July, August, and September in Arizona, when compared to the control group of cows that were provided with shade only. The income-cost comparison indicated an increase in annual net income of $62 per cow, or $6,885 for the 110 cows in the study.

## APPEARANCE OF THE DAIRY FARM

Maintaining a neat, clean, and attractive appearance of the dairy farm buildings and the surrounding area can have a positive influence on public opinion in regard to the wholesomeness of milk and dairy products and on concern for farm ecology problems. Neglect of this detail can also have a negative effect.

Most of us would not be too enthused about stopping at a restaurant that was unpainted, with loosened siding and broken windows; surrounded by weeds, mudholes, piles of manure, rusting machinery, dirty cows, and broken-down fences; and was permeated by an unpleasant aroma. Yet many dairymen tolerate or ignore similar conditions on their farms while complaining about the lack of consumer acceptance of milk and dairy products.

Neglected appearance of the farm often accompanies neglected cows and general dissatisfaction of the owner with dairying. Conversely, a neat, attractive appearance often indicates the owner's personal pride, satisfaction, and optimism about dairying.

* G. H., Stott, and F. Wiersma, Now, Cooling Cows Can Be Profitable, *Hoard's Dairyman* 719, June 10, 1973.

## SUMMARY

Paying attention to the many details of dairy herd management mentioned in this chapter and throughout this book accounts for a major portion of the difference between successful dairymen and those who are not. The former have set realistic goals, planned and implemented programs by using up-to-date knowledge and applicable methods. They have carefully monitored and evaluated progress toward goal achievement and identified weak links, and strengthened those weak links. They are also satisfied with dairying as a business and a way of life and are optimistic about the future of dairying.

## REFERENCES FOR FURTHER STUDY

Hafez, E. S. E., *Adaptation of Domestic Animals,* Lea and Febiger, Philadelphia 1968.

Johnson, H. D. et al., Temperature-Humidity Effects Including Influence of Acclimation on Feed and Water Consumption of Holstein Cattle, *Mo. Agr. Expt. Sta. Bul.* 846 (1963).

Morris, J. L., and Hunt, G. E., This Chute Makes Hoof Trimming Easy, *Hoard's Dairyman,* p. 1388, October 10, 1974.

Shaw, J. E., *Hoard's Dairyman* p. 1174, October 25, 1972; p. 1232, November 10, 1972; and p. 1298, November 25, 1972.

Stott, G. H., and Wiersma, F., Now, Cooling Cows Can Be Profitable, *Hoard's Dairyman,* p. 719, June 10, 1973.

# Appendices

## DAIRY HERD MANAGEMENT PROBLEMS

The responsibility of the dairy herd manager is primarily decision making. To be successful in herd management, individuals must have knowledge in many areas and a well-developed capacity for decision making—the ability to identify problem areas and specific problems, to identify relevant alternatives and evaluate them in terms of probable costs and returns, and to choose and follow a prudent course of action. Managers must know how to apply the basic principles of cattle, land, feed, engineering, and other management to a particular set of resources or circumstances, keeping in view the goals of a particular dairy operation.

The following are example problems that are designed for teachers to use to give students the opportunity to learn to increase their ability to make prudent management decisions in various areas of dairy herd management. We have found that using these types of problem in some areas of herd management is an effective student learning tool. For most effective use, they should be altered to make them relevant to conditions and resources (herd size, land resources, etc.) prevalent in the area of the country in which they are used. The economic parameters (price of milk, cows, feed, labor, etc.) should realistically reflect current conditions.

**Problem 1.**   Use of DHI Performance Testing Information

**Reference:**   Chapter 2, Appendix Tables M, M-1, and M-2.
**Objective:**   To learn what information is contained in various DHI forms and how to use that information to evaluate cow and herd performance.
**Materials:**   DHI forms 200, 202, 203, 205, and 206. Example sheets from Chapter 2 or those from a herd or herds can be used.
**Part 1:**   The instructor prepares a list of questions pertaining to information contained on each of the DHI forms for students to complete. For example, from DHI form 200, ask the students to identify which cows calved or were turned dry, have completed or projected 2× 305 ME records, have been open over 100 days, etc.
**Part 2:**   Weak-link identification. The instructor prepares a list of herd parameters for several herds and the state average. From these data, the students are required to: (1) identify the weak links in each of the herds, and (2) make recommendations concerning

566

**Example, DHI Weak-Link Analysis Problem**

| | State average | Herd #5 | Herd #6 | Herd #7 | H |
|---|---|---|---|---|---|
| Breed | | 3 | 3 | 3 | 3 |
| Cow Years | 87.7 | 94.9 | 183.6 | 57.1 | 56.3 |
| Milk, lb. | 13,362 | 18,138 | 15,112 | 17,462 | 9,955 |
| Fat, % | 3.7 | 3.8 | 3.6 | 3.2 | 3.6 |
| Fat, lb. | 496 | 690 | 554 | 562 | 362 |
| Concentrate, cwt. | 49 | 53 | 36 | 53 | 30 |
| Silage, cwt. | 179 | 201 | 182 | 198 | 191 |
| Hay, cwt. | 13 | 28 | 35 | 24 | 3 |
| Pasturage, days | 131 | 300 | — | 239 | 122 |
| Rate forage feeding | 2.4 | 3.1 | 2.3 | 2.3 | 2.3 |
| Value product, $ | 1,259 | 1,726 | 1,376 | 1,604 | 934 |
| Cost concentrate, $ | 334 | 345 | 233 | 340 | 217 |
| Total feed cost, $ | 562 | 577 | 428 | 558 | 404 |
| Income over feed cost, $ | 697 | 1,149 | 947 | 1,046 | 530 |
| Feed cost/cwt., $ | 4.21 | 3.18 | 2.84 | 3.20 | 4.06 |
| 1st lactation age, mo. | 30 | 32 | 27 | 30 | 32 |
| Average age, mo. | 55 | 48 | 58 | 55 | 58 |
| Body weight, cwt. | 12 | 13 | 14 | 13 | 12 |
| % days milk | 86 | 93 | 84 | 91 | 87 |
| Calving interval, mo. | 13 | 14.1 | 12.9 | 12.7 | 15.2 |
| Days dry | 65 | 45 | 73 | 54 | 57 |
| Days open | 130 | 148 | 111 | 105 | 182 |
| No. services/conception | 1.7 | 1.53 | 1.76 | 1.26 | 1.00 |
| % sire identified | 60 | 98 | 44 | 100 | 69 |
| % left herd | 28 | 43.2 | 34.8 | 62.9 | 14.3 |

how these weak links might be strengthened (use Example table, p. 567).

**Problem 2.** Dairy Farm Business Analysis and Budgeting

**Reference:** Chapters 3 and 19.

**Objectives:** To learn to calculate dairy farm profitability, determine the effect of changes in various costs and levels of milk production on profit, and to budget a possible purchase.

**Materials:** A dairy farm annual profit and loss or operating statement (use Example on p. 568).

**Problem:** The following information is available about John Brown's dairy farm for last year. The farm includes only the milking herd and facilities for them (barn, parlor, feed storage, etc).

Replacements and feed are purchased from other enterprises or individuals.

Income and costs for the dairy farm are based on a herd of 100 milking cows, weighing 1,400 lb. each, and producing an average of 15,000 lb. of milk, 525 lb. butterfat with an average test of 3.5% (DHI). Cows are fed an average of 16 lb. of concentrate daily for 305 days, 4 lb. of concentrate daily for 60 days while dry, and 36 lb. of hay equivalent daily, allowing for feed wastage. The dairy is a modern facility (4 years old) equipped with a double 4 herringbone parlor, free stalls, bunk feeders, etc.

Herd replacements cost an average of $800 each, cull cows sold at an average of $250 each, heifer calves at $200 each, and

## Example, Profit-and-Loss or Operating Statement

|  | Annual herd total costs | Each cow per year |
|---|---|---|
| *Cash Receipts* | | |
| Milk: 1,440,000 lb. × $10.30/cwt. | $148,320.00 | $1,483.20 |
| Heifer calves: 55 @ $200.00 ea. | 11,000.00 | 110.00 |
| Bull calves: 55 @ $20.00 ea. | 1,100.00 | 11.00 |
| Cull cows: 28 @ $250.00 ea. | 7,000.00 | 70.00 |
| **Total cash receipts** | **$167,420.00** | **$1,674.20** |
| *Cash Expenses* | | |
| Feed | | |
| Corn silage | | |
| 12 T./cow/yr. @ $20.00/T. | $ 24,000.00 | $ 240.00 |
| Mixed alfalfa-grass hay | | |
| 1 T./cow/yr. @ $70.00/T. | 7,000.00 | 70.00 |
| Concentrates | | |
| 2.5 T. @ $130.00/T./cow/yr. | 32,500.00 | 325.00 |
| Labor | | |
| 2 men at $800.00/mo. ea. | 19,200.00 | 192.00 |
| Replacements (30 × $800 ea.) | 24,000.00 | 240.00 |
| Breeding costs | 1,500.00 | 15.00 |
| Taxes | 1,500.00 | 15.00 |
| DHI testing | 900.00 | 9.00 |
| Vetarinarian and medicine | 1,800.00 | 18.00 |
| Supplies | 2,500.00 | 25.00 |
| Milk hauling @ $.40/cwt. | 5,760.00 | 57.60 |
| Misc. (incl. repairs, maint. insurance, elect., etc.) | 5,000.00 | 50.00 |
| **Total cash expenses** | **$125,660.00** | **$1,256.60** |

**Example, Profit-and-Loss or Operating Statement**

|  | Annual herd total costs | Each cow per year |
|---|---|---|
| *Farm Net Cash Income* | $ 41,760.00 | $ 417,60 |
| *Depreciation (straight line)* | | |
| Bldgs., feed mangers, silos, etc. | | |
| $100,000 orig. value @ 25 yr. | $ 4,000.00 | $ 40.00 |
| $ 84,000 current value | | |
| Equipment | | |
| $40,000 orig. value @ 10 yr. | 4,000.00 | 40.00 |
| $24,000 current value | | |
| **Total depreciation** | **$8,000.00** | **$80.00** |
| *Change in inventory* | 0.00 | 0.00 |
| *Farm income* | $ 33,760.00 | $ 337.60 |
| *Interest on Investment @ 8%* | | |
| Land, 10 acres $1,000/A. = | | |
| $10,000 @ 8% | $ 800.00 | $ 8.00 |
| Bldgs., mangers, silos, etc., | | |
| $86,000 @ 8% | 6,880.00 | 68.00 |
| Equipment, $26,000 @ 8% | 2,080.00 | 20.80 |
| Cows @ $800 ea. @ 8% | 6,400.00 | 64.00 |
| **Total interest on investment** | **$ 16,160.00** | **$ 161.60** |
| *Total expenses* | $149,820.00 | $1,498.20 |
| Cost per cwt. milk (100% sales basis = 15,000 lb./cow) | $9.99 | |
| Cost per cwt. milk (96% sales basis or as sold = 14,400 lb./cow) | $10.40 | |

bull calves at $20 each. The herd was replaced on the basis of 28% culling and 2% death loss. Two men, required to do the milking, feeding, and manure disposal receive wages and benefits totaling $800 each per month.

## QUESTIONS

1. If Mr. Brown was not one of the laborers, and he was debt free, what was his:
   a. Total cash expense for the enterprise? _____
   b. Cash expense per hundredweight of milk sold? _____
   c. Farm net cash income? _____
   d. Farm income? _____

e. Total investment in the enterprise (average annual value)? _____

f. Labor income? _____
   (use 8% interest on invested capital)

g. Percent return on invested capital? _____
   (assume a value of $16,000 for Mr. Brown's labor and management)

2. What was Mr. Brown's feed cost per hundredweight of milk sold? _____

3. What was Mr. Brown's labor cost per hundredweight of milk sold? _____

4. If Mr. Brown was not one of the two laborers and this was his only source of income, how much money was available for family living expenses and debt service? _____

5. If Mr. Brown was not one of the two laborers, this was his only source of income, and his family living expenses, including income taxes, were $20,000 per year, how much money was available for debt service? _____

6. If Mr. Brown was debt free and he was one of the two laborers as well as the manager, what was his farm net cash income from the milking enterprise? _____

7. If Mr. Brown had an outstanding loan of $90,000 at 8% annual interest and he was one of the two laborers as well as the manager, what was his farm net cash income? _____

8. If Mr. Brown had a loan of $100,000 at 8% annual interest and another loan of $60,000 at 12% annual interest, and he was not one of the two laborers, what was his farm net cash income? _____

9. If Mr. Brown had a loan of $90,000 at 8% annual interest, and he was one of the two laborers as well as the manager, what would his farm net cash income have been if:
   a. Concentrate price increased to $160/T. with no concurrent increase in milk price? _____
   b. Milk price increased $0.40/cwt. (to $10.70) concurrently with the increase in concentrate price (to $160/T)? _____
   c. Concentrate price remained unchanged ($130/T.) and milk price decreased $0.50/cwt. (to $9.80)? _____
   d. Milk production decreased to 13,000 lb. sold/cow, with a concurrent decrease in annual feed cost/cow of $50? _____
   e. Milk production decreased to 11,000 lb. sold/cow, with a concurrent decrease in annual feed cost/cow of $100? _____
   f. Milk production increased to 16,000 lb. sold/cow, with a concurrent increase in annual feed cost/cow of $50? _____

10. Assuming you are interested in going into the dairy business, would you

buy this operation for $200,000 if Mr. Brown would finance purchase price at 8% interest due annually on the unpaid bal equal annual principal payments of $20,000 for 10 years?

a. Assumptions:
   (1) All buildings and equipment are in good condition and you can contract the forage on a long-term basis (5 yr. minimum).
   (2) Production of 14,400 lb. of milk sold/cow.
   (3) Prices (costs and income) will remain in the same relative position (if costs increase, milk price will also increase proportionately).
   (4) You are one of the laborers as well as the manager and an annual salary of $9,600 will cover your living expenses and the income tax on your salary.
   (5) 20% income tax on the taxable income (farm net cash income-depreciation) from the dairy farm.

b. Suggested procedure:
   Project an annual profit-and-loss or operating statement for the first year. From this calculate the expected farm net cash income, deduct income taxes, and determine if enough cash is available to meet debt service needs. Continue budget until the loan is repayed. Determine your net worth at the end of the repayment period.

**Problem 3:** Dairy Farm Business Analysis, Identification of Financial Weak Links and Alternatives.

**Reference:** Chapter 3

**Objectives:** To learn to calculate various profit measures, to use these to identify weak links, and to identify reasonable alternatives concerning strengthening these weak links.

**Materials:** An inventory and a receipts and expenses record for a dairy farm (see Example table, p. 573). A computer program to calculate various measures of profit and a terminal available for student use is very helpful, but not essential.

**Problem:** Ben Brown purchased a dairy farm 20 years ago with the help of a second mortgage from his father, which has been repaid. He lived in the existing house until 10 years ago, when he built a new one. He milked about 40 cows and farmed with used machinery until 5 years ago. At that time he built a new milking parlor (double 4 herringbone), a 100-cow free-stall barn, new upright concrete stave silos with unloaders and an automatic feeding system, and new calf and maternity quarters. He purchased new, modern equipment for both farm and dairy. The new buildings cost $70,000, and $60,000 of this was financed for 15 years at 8% (interest due annually on the unpaid balance and equal principal payments of $4,000 per year). The machinery and equipment cost $85,000, and $80,000 of this was financed at 9% for 10 years (interest due annually on the unpaid balance and equal principal payments of $8,000). He purchased another 40-cow herd and has since repaid the money borrowed for this purchase.

Ben is a bit discouraged. He owes more money now than he did 20 years ago. He and his family have worked hard and lived frugally for 20 years and they still don't have enough money to buy a new car or take a vacation. He anticipates a 20% increase in property tax next year and the hired man (a good one) has asked for a raise of $100 per month. After he paid all his bills, including interest and principal payments, family living expenses, etc., this year, Ben's checking and savings account balances were the same as one year ago. Ben tries to do things correctly, keeps a good set of business records (see Example table, p. 573), hires a good accountant to calculate his taxes, and takes good care of the land, cows, and machinery, but is discouraged at his inability to make enough money to enjoy life a bit more.

In order to evaluate business progress for the past year, and to help him decide whether to continue as is, make some adjustments, or sell out, he has asked you for assistance.

## QUESTIONS

1. For the past year, calculate Ben's:
   a. Farm net cash income.
   b. Farm income.
   c. Labor income (use 8% interest on investment).
   d. Labor earnings.
   e. Percent return on all invested capital.
   f. Increase or decrease in net worth from the previous January 1 to the current January 1.

   (To complete this part of the problem, it will be necessary to compile a profit-and-loss statement for the previous year and net worth statements for January 1 of the previous and the current years).

**Example, Summary of Business Records for Ben Brown (90-Cow Grade A Dairy Operation)**

| | Value at purchase | Value previous Jan. 1 | Value current Jan. 1 |
|---|---|---|---|
| I. **Inventory** | | | |
| *Land* | | | |
| 300 A. tillable | $75,000 | $180,000 | $190,000 |
| 40 A. waste | 5,000 | 10,000 | 10,000 |
| *Buildings* | | | |
| Home | $20,000 | $ 30,000 | $ 30,000 |
| Personal property (car, furniture, etc.) | 10,000 | 8,000 | 7,000 |
| Milking parlor (4 yr. old) | 20,000 | 16,000 | 15,000 |
| Free stall barn (4 yr. old) | 20,000 | 16,000 | 15,000 |
| Silos (4 yr. old) | 20,000 | 16,000 | 15,000 |
| Calf & mat. barn (4 yr. old) | 10,000 | 8,000 | 7,500 |
| Mach. storage & shop (old) | 6,000 | 2,000 | 1,500 |
| Hay storage | 4,000 | 3,200 | 3,000 |
| *Machinery and Equipment* | | | |
| Dairy milking, feeding, etc. (5 yr. old) | $20,000 | $ 12,000 | $ 10,000 |
| 2 large tractors (5 yr. old) | 22,000 | 14,000 | 12,000 |
| 2 small tractors (5 yr. old) | 12,000 | 8,000 | 7,000 |
| Scraper & spreader (5 yr. old) | 6,000 | 4,000 | 3,500 |
| Haying equipment—baler, mower, rake, wagons, etc. (5 yr. old) | 9,000 | 7,000 | 6,600 |
| Corn equipment—sod planter, chopper, sprayer, etc. (5 yr. old) | 11,000 | 7,000 | 6,000 |
| Truck (5 yr. old) | 5,000 | 3,400 | 3,000 |
| Miscellaneous | 1,000 | 1,000 | 1,000 |
| *Livestock* (Holsteins—original herd grades and purchased herd registered) | | | |
| Milking age cows | | $63,000(90) | $ 63,000(90) |
| Heifers (bred) | | 9,000(15) | 12,000(20) |
| Heifers (1 yr.—breeding) | | 8,000(20) | 10,000(25) |
| Heifers (under 1 yr.) | | 10,500(35) | 10,500(35) |
| *Feed and Supplies* | | | |
| Silage | | $ 30,000(2,000 T.) | $ 32,000(2,000 T.) |
| Hay | | 12,500(250 T.) | 12,500(250 T.) |
| Concentrate | | 1,400(10 T.) | 1,400(10 T.) |
| Miscellaneous (fuel, supplies, calf feed, etc.) | | 2,000 | 2,000 |
| *Cash* | | | |
| Checking account (farm) | | $ 2,000 | $ 2,000 |
| Checking account (personal) | | 1,000 | 1,000 |
| Savings account (personal) | | 4,000 | 4,000 |
| II. **Debts** | | | |
| Accounts payable (feed, fuel, etc.) | | $ 1,000 | $ 1,000 |
| Equipment (9% for 10 yr.) | | 48,000 | 40,000 |
| Buildings (8% for 16 yr.) | | 44,000 | 40,000 |
| Land (6% for 25 yr.) | | 15,000 | 12,000 |

|  | Receipts | Expenses |
|---|---|---|
| **III. Previous year cash receipts and expenses** | | |
| Sale of 1,170,000 lb. milk @ $9.60/cwt. | $112,320 | |
| Sale of 50 bull calves @ $10 ea. | 500 | |
| Sale of 30 cull cows @ $200 ea. | 6,000 | |
| Sale of 3 bred heifers @ $700 ea. | 2,100 | |
| Sale of 50 T. of hay @ $60/T. | 3,000 | |
| Miscellaneous cash receipts (gov't payments, coop. patronage refunds, etc.) | 580 | |
| | | |
| Purchased feed (cows) | | $ 25,200 |
| Purchased feed (heifers) | | 6,000 |
| Seed, fert., spray, lime, etc. for 150 A. corn | | 15,000 |
| Seed, fert., spray, lime, etc. for 100 A. hay | | 6,000 |
| Seed, fert., spray, lime, etc. for 50 A. pasture | | 1,000 |
| Hired labor costs (dairy) | | 8,800 |
| Hired labor costs (feed production) | | 4,000 |
| Machinery costs (fuel, maintenance, repairs, etc.) (dairy) | | 1,000 |
| Machinery costs (fuel, maintenance, repairs, etc.) (corn) | | 2,500 |
| Machinery costs (fuel, maintenance, repairs, etc.) (hay) | | 2,500 |
| Buliding repair & maintenance cost (dairy— parlor, barn, etc.) | | 760 |
| Building repair & maintenance cost (hay storage) | | 200 |
| Building repair & maintenance cost (silos) | | 800 |
| Freight (milk, cull cows, etc.) | | 6,500 |
| Fuel (dairy) | | 1,000 |
| Utilities (dairy) | | 2,500 |
| Taxes (land) | | 1,200 |
| Taxes (farm buildings) | | 800 |
| Taxes (house and personal property) | | 300 |
| Vet. & medicine | | 1,800 |
| Breeding | | 1,500 |
| DHI testing | | 700 |
| Miscellaneous farm expenses | | 1,000 |
| Cash family living expenses including income taxes | | 9,700 |
| Interest on equipment loan | | 4,320 |
| Interest on building loan | | 3,520 |
| Interest on land mortgage | | 900 |
| Debt principal payments | | 15,000 |
| | | |
| **IV. Other** | | |
| | | |
| Value of Ben's unpaid labor | | $ 7,000 |
| Value of his wife's and children's unpaid labor (farm only) | | 3,000 |
| Value of house, garden, meat, milk, etc. | $ 4,000 | |

2. Did he meet the criteria for a profitable business? Support your answer with factual evidence.

3. Does Ben have financial weak links in his dairy farm and herd operation? If so, what are they?

4. Would you recommend that he:
   a. Sell out?
   b. Continue as is?
   c. Make some financial adjustments? If so, what adjustments? Illustrate how your recommendations will help him financially.

**Problem 4.**   Evaluation of Alternatives—Budgeting

**Reference:**   Chapters 3, 19 and 20.

**Objective:**   To learn to evaluate the effect of various alternatives for solving a dairy herd problem(s).

**Materials:**   Information and answers from Problem 2. A computer program to project budgets and calculate profit measures from these budgets and a terminal available for student use are very helpful, but not essential.

Evaluate the effect of each of the following alternatives in terms of correcting Ben Brown's financial problems (Problem 3). Indicate which one(s) you would recommend for Ben. Support your answer with evidence that indicates improved cash flow, continued progress in net worth, and logical reasoning for your decision.

## Alternatives

1. Refinance the operation as of Jan. 1 of the current year.
   a. A first mortgage loan on the farm of $93,000 ($92,000 to pay off all current loans and $1,000 loan processing charges) for 20 years at 9% annual interest on the unpaid balance. The loan to be repaid in equal annual payments for 20 years.
   b. Assume no prepayment penalties on current loans or the refinancing mortgage.
2. Expand to a 130-cow herd.
   a. Additional facilities to be built in the spring and summer, cattle to be purchased in late summer and expansion to be completed by Sept. 1 of the current year.

b.  Expansion costs and financing.
   (1)  Costs

| | |
|---|---|
| 30 free stalls (added to present barn) | $10,000 |
| 40 cows @ $750 ea. | 30,000 |
| Additional feed storage | 10,000 |
| **Total** | **$50,000** |

   (1)  Financing
        A $50,000 loan beginning July 1 of the current year.
        Terms: 9% annual interest on the unpaid balance due Dec. 31 annually.
        Principal payment of $1,000, due Dec 31 of the current year and $7,000 due Dec. 31 annually thereafter.
c.  Additional operating expenses.
   (1)  Assume the same level of annual operating expenses (except interest) on a per-cow basis as the previous year (Problem 3). Charge one-third of the annual expenses for the current year (Sept.–Dec.) and the full annual amount thereafter for the additional cows.
d.  Additional income.
   (1)  Assume the same level of production for all cows and the same price for milk as the previous year (Problem 3). Allow one-third of the annual production for the current year (Sept.–Dec.) and the full annual amount thereafter for the additional cows.
   (2)  Assume no receipts from the sale of heifers or hay for the next 5 years.
   (3)  Assume the sale of 30 cull cows and 50 bull calves for the current year and 40 cull cows and 60 bull calves annually thereafter.

3.  Improve production by 1,000 lb. per cow (sold basis) with a concurrent increase in feed cost of $35/cow and concurrent increase in other operating costs of $11/cow.

4.  Implement alternatives 1 and 2 concurrently.

5.  Implement alternatives 1 and 3 concurrently.

6.  Implement alternatives 1 and 3 during the current year and 2 the next year.

**Problem 5.**   Analysis of a Current Feeding Program and Making Use of Current Feed Resources: Feeding Problem Part 1

**References:**   Chapters 4, 5, 6, 7, 8, and Appendix Tables A through E.

**Objectives:**   To learn to identify deficiencies in a feeding program and correct these deficiencies by making maximum use of feed currently available on the farm and from a local supplier.

**Materials:** A description of a current herd situation (preferably the same one as used in problems 3 and 4) and current listing of available feeds and prices common to the area (see Example table, p. 578). One or more computerized least-cost ration formulation programs and a terminal available for student use is helpful, but not essential.

## Problem

1. Identify the weak links in Ben Brown's current feeding program for the 90-cow milking herd (including dry cows) and springing heifers.

2. Design a feeding program for the milking herd, dry cows, and springing heifers for the current feeding year (Oct. 1 to Sept. 30).
   a. Restrictions:
      (1) Use any or all of the stored homegrown feeds and any of the locally available feeds from Group A—Commercial Mixed Feeds (Example table, p. 578).
      (2) Design a program that should result in an increase of 1,000 lb. of milk/cow for the period.
      (3) Include a minimum of 5 lb. of hay/milking cow/day.
      (4) Do not design a feeding program for the heifers (other than the springers), but do budget 10 lb./day of forage dry matter for each of the 70 heifers for the feeding period.
   b. Support your recommended program with:
      (1) Evidence that it should result in an increase of 1,000 lb. of milk/cow/yr. (a balanced ration to support the desired level of production).
      (2) Evidence that it is an economical feeding program. Specify:
        (a) Forage(s) intake and costs/cow/yr.
        (b) Concentrate intake and costs/cow/yr.
        (c) Feed cost—per cow per year and per cwt. of milk.
        (d) Income over feed cost—per cow per year and per cwt. of milk (milk value = $9.60/cwt.).
      (3) Any other pertinent information that will convince Ben to follow your recommendations.

## Current situation:

It is now Oct. 1 and Ben Brown started using his forages in his feeding program about 2 weeks ago. The forages are of similar quality to those he has

## Example, Locally Available Feeds

**Commercial mixed feeds[a]**

| Conc. | DM % | TP % (as fed) | CF % (as fed) | Cost/T. |
|---|---|---|---|---|
| 1 | 90 | 16 | 8 | $120 |
| 2 | 90 | 20 | 8 | 125 |
| 3 | 90 | 20 | 6 | 130 |
| 4 | 90 | 24 | 8 | 140 |
| 5 | 90 | 24 | 6 | 145 |
| 6 | 90 | 28 | 8 | 150 |
| 7 | 90 | 28 | 6 | 155 |
| 8 | 90 | 32 | 8 | 160 |
| 9 | 90 | 32 | 4 | 166 |
| 10 | 90 | 36 | 8 | 166 |
| 11 | 90 | 36 | 4 | 170 |
| 12 | 90 | 40 | 8 | 172 |
| 13 | 90 | 40 | 4 | 176 |
| 14 (Calf Starter | 90 | 20 | 6 | 180 |
| 15 (Milk Replacer) | 90 | 20 | .25 | 700 |

**Available ingredients[b]**

| Ingredient | Price/T. |
|---|---|
| Cane molasses | $106 |
| Dried beet pulp | 124 |
| Soybean meal (44%) | 250 |
| Peanut meal (45%) | 240 |
| Linseed meal | 190 |
| Corn gluten feed | 130 |
| Cottonseed meal | 240 |
| Corn, yellow dent No. 2 | 106 |
| Barley (grain) | 104 |
| Oats (grain) | 128 |
| Wheat (grain) | 132 |
| Wheat bran | 160 |
| Vitamin mix | 1,000[c] |
| Dicalcium phosphate | 200 |
| Ground limestone | 40 |
| Iodized salt | 62 |
| Trace mineral salt | 66 |
| Urea (42% N.) | 250 |
| Monosodium phosphate | 220 |
| Defluorinated phosphate | 220 |

[a] Fortified with sufficient vitamins and minerals.

[b] A local mill will mix and deliver a concentrate mixture according to your formula for a service (mixing and delivery) charge of $10/T. or $12/T. if pelleted. He will also store any homegrown grains for $.03/cwt./month.

[c] 2 lb./T. of concentrates needed to meet requirements for vitamins A and D.

578

been using for several years, so he is continuing on the same feeding program he has used for several years.

One of the alternatives suggested in the business problem was to increase production by 1,000 lb. of milk/cow during the current year. The current average PD of the sires of the milking herd is +500 lb. of milk. Ben has been using AI for several generations, so it is reasonable to assume the genetic potential of the herd for milk production is higher than the current average production (13,500 DHI, 3.6% BF, or 13,000 lb. of 3.6 milk sold). The average body weight of the cows is 1,350 lb.

Ben's corn crop for the current year was about average for his land. The 150 A. produced 2,250 T. as ensiled (150 A. at 15 T./A.). A small silo was filled (500 T.), one large one filled (1,000 T.), and one large one three-fourths filled (750 T.). As of Oct. 1, he has 2,200 T. left. Ben's hay crop for the current year was also about average. He harvested 275 T. (100 A. at 2.75 T./A) and this is stored in his hay barn. As of Oct 1, he has 270 T. left (160 T. 1st crop and 110 T. 2nd crop). His hay is a mixture of alfalfa and brome grass.

His feeding setup is arranged for complete ration feeding. He has the herd divided into 3 groups. One group contains the highest-producing half of the milking herd, another the lowest-producing half of the milking herd, and the third the dry cows and springing heifers. The milking groups are confined to dry lot except for small exercise lots. Each milking group area has 50 free stalls, a silage bunk, and a hay-feeding manger. The dry cows and springers are in a 5-acre shaded pasture lot with a fence line bunk that can be used to feed silage and/or hay and grain.

**Current feeding program:**

    (1)   Group 1, high producers—average production of the 40 cows in this group = 54 lb. of 3.5 milk/day.

|  |  |  |
|---|---|---|
| (a) Complete feed | | 3,920 lb./day |
| Corn silage | 3,200 lb. | |
| Concentrate | 720 lb. | |
| (b) Hay (2nd crop) | | 200 lb./day |

    (2)   Group 2, low producers—average production of the 40 cows in this group = 34 lb. of 3.8 milk/day.

|  |  |  |
|---|---|---|
| (a) Complete feed | | 3,600 lb./day |
| Corn silage | 3,200 lb. | |
| Concentrate | 400 lb. | |
| (b) Hay (1st crop) | | 200 lb./day |

    (3)   Group 3, dry cows (10) and bred heifers within 6 weeks of calving (10 now but varies from 5 to 10).

|  |  |
|---|---|
| (a) Complete feed (same as Group 2) | 1,000 lb./day |
| (b) Hay (1st crop) | 100 lb./day |

(4)  Heifers, the remaining 70 are divided according to age groups. They are housed inside and fed milk or milk replacer and grain to 6 to 8 weeks and then calf starter and hay until 5 to 6 months of age. After 6 months they are turned out to available pasture by age groups and supplemented with complete feed (same mixture as fed to Group 2 milking cows) as needed to keep they growing. Last year, Ben used an average of 30 lb. of corn silage and 3.75 lb. of concentrate/heifer/day for these heifers.

(5)  Analysis of feeds currently being fed:

| Feed | % D.M. | C.F. (D.M. basis) | C.P. (D.M. basis) |
|---|---|---|---|
| Corn silage | 27.9 | 26.3 | 8.4 |
| Alfalfa-brome hay (1st crop) | 88.0 | 34.0 | 12.5 |
| Alfalfa-brome hay (2nd crop) | 88.0 | 32.0 | 16.2 |
| Concentrate | 90.0 | 7.2 | 14.4 |

**Problem 6.**  Feeding Problem Part 2

**References:**  Same as for Problem 5.

**Objectives:**  To learn to use, efficiently and effectively, available land, storage, and other farm resources to design a herd feeding program that is likely to maximize income over feed cost.

**Materials:**  Information concerning land, feed-storage, and feeding facilities for the dairy herd situation described in Problem 5. Computer program(s) for evaluating land use alternatives in addition to least-cost ration-formulation programs are helpful, but not essential.

**Problem:**  In the fall of the current year, Ben is going to expand his milking herd to 130 cows (average weight = 1,350/cow) by purchasing an additional 40 cows. He plans to maintain this herd size in the future without additional purchases; i.e., by retaining all heifers born on the farm for the next 4 to 5 years. It is now February of the current year and Ben has asked you for your specific recommendations concerning a feeding program for the coming and future feeding years (Oct. 1 to Sept. 30). Include in your recommendations:

1.  Land use for the 250 tillable acres. Use any combination of the following crops:

| | Produc-tion cost/A. | Yield/A. | DM % | Oppor-tunity value/T. |
|---|---|---|---|---|
| Pasture | $ 70 | ? | ? | |
| Alfalfa-brome hay | 150 | 3 T. | 86 | $70 |
| Alfalfa-brome haylage | 150 | 6 T. | 50 | 35 |
| Alfalfa hay | 170 | 3.4 T. | 86 | 90 |
| Alfalfa haylage | 170 | 7.6 T. | 50 | 45 |
| Corn silage | 230 | 16 | 28 | 17 |
| Corn silage | 230 | 15 | 34 | 18 |
| Corn silage | 230 | 14.1 | 38 | 19 |
| Corn silage | 230 | 13.3 | 39–42 | 20 |
| Corn for grain | 220 | 90 bu./A. (2.52T.) | 86 | 92 |
| High-moisture corn | 220 | 3.2 T./A. | 72 | 75 |

2.  A feeding program for the milking cows, dry cows, and springing heifers, using any of the crops listed in the table above and any of the locally available feeds (group A and/or B). Budget 10 lb. of forage DM daily for each the heifers, (plus the 40 acres of permanent pasture) but do not design a feeding program for the heifers.

Justify your recommendations with evidence that the recommended ration should result in high production (over 15,000 lb. milk sold/cow), economical feed cost per unit of production, maximum income over feed cost (estimate $9.60 cwt. for milk) and efficient utilization of available land resources.

**Assumptions**

a.  Adequate feeding facilities to divide the herd into 2 milking groups and 1 dry cow and springing heifer group (for complete ration feeding).

b.  Adequate forage feeding facilities to group feed forages and individually feed concentrates in the double 4 herringbone parlor.

c.  300 T. of leftover silage from Problem 5.

d.  4 T. of 2nd crop and 10 T. of 1st crop hay left over from Problem 5.

**Problem 7.**  Breeding Problem

**References:**  Chapters 9 and 10.

**Objective:**  To learn to design an economical herd breeding program that should result in significant genetic improvement in performance of future generations and acceptable reproductive efficiency.

**Materials:**    A description of the current herd status and goals of several types of dairy herds typical of the area. Sire summaries, price lists, and within-stud bull reproductive efficiency data from AI organizations serving the area.

**Problem:** Design a breeding program for the herds below. Include a list of the bulls you would use, the number of cows bred to each, the type of cows bred to each (heifers, repeat breeders, etc.,) the projected annual cost of your program, the average predicted differences (dollars, milk, etc.) of the service sires, and any other information to justify your recommendations.

**Herd Situations**

**Herd A**

Commercial dairyman—180 grade Holsteins.
Average production—12,800, 3.9, 499.
Income—primarily milk sales.
Goal—net profit from milk sales.

**Herd B**

Breeder of registered cattle—70 registered Holsteins.
Average production—17,500, 3.2, 560.
Average type score—82.8.
Income—65% milk sales, 35% purebred sales.
Goal—net profit from milk and purebred sales.

**Herd C**

Farmer breeder—20 registered and 20 grade Holsteins.
Average production—14,000, 3.7, 524.
Average type score (on registered cattle only)—78.2.
Income—primarily milk sales.
Goal—develop a top purebred herd of 60 to 80 registered Holsteins.

**Problem 8.**    Building and Equipment Problem

**Reference:**    Chapter 15 and Appendix Tables F, F-1, and F-2.

**Objective:**    To learn to plan a dairy system that meets the basic purposes of a good dairy system.

**Materials:**    A description of the size and type of herd to be housed, the building site (a field trip to the proposed site), a current estimated cost listing for the various parts of the system and the goals of the operation.

**Problem:**    Because of current labor inefficiency, John Doe has decided to build a new setup for his Holstein herd. He has 190 cows with a 17,000-lb. DHI herd average and average body weight of 1,300 lb., and 180 replacement heifers of various ages. He has 450 A. of tillable land and 50 A. of pasture to use to supply feed for the herd. His goal is to maximize income and minimize production costs. He wants to build a system for 200 cows, but feels he may want to increase herd size to a maximum of 240 in the future.

You have been hired to design the new system for the herd, including replacements. Your recommendations should include a sketch of the layout with dimensions, cow and materials flow systems, etc., specifically:

1. Milking area—type and size of parlor, holding area, milking and milk storage, equipment, etc.

2. Feeding system—include method of feeding, group size and criteria for grouping and type, location, and dimensions of feeders, etc.

3. Waste handling system—include type, design, equipment, etc.

4. Resting area—include type and size of facilities, alley dimensions, type of bedding used, etc.

5. Hospital and special areas—size, layout, location, equipment, number of box stalls, stanchions, etc.

6. Calf housing area—type, number of pens, etc.

7. Estimated costs—based on the information included in this exercise, with adjustments for your design.

8. Labor requirements and utilization.

9. Any other facts that help justify your recommendations.

Students may utilize any printed source of information to assist in recommendations, including copies of layouts.

**Problem 9.** Milk Marketing

**Reference:** Chapter 18.

**Objective:** To learn to calculate Grade A blend price for milk under varying market conditions and to evaluate available marketing alternatives.

**Materials:** Class utilization rates and prices for several market situations prevalent in the area. Formulas and annuity tables for calculating time horizons and repayment schedules in areas where market base is purchased. In these areas, computerized programs to determine various economic parameters are also useful, but not essential.

**Problem:** Calculating milk prices

1. Current marketing conditions (based on a 3.5% fat standard).

| | Market N | Market O[a] | Market P[b] |
|---|---|---|---|
| Class I usage, % | 50 | 60 | 90 |
| Class I price/cwt. | $11.50 | $11.50 | $11.50 |
| Class II usage, % | 50 | 40 | 10 |
| Class II price/cwt. | $8.00 | $8.00 | $8.00 |
| Fat differential | $0.10 | $0.10 | $0.10 |

[a] Base-surplus marketing plan. Producers receive blend price for base milk and class II price for surplus milk. Base is established annually during a 4-month period (Sept. through Dec.)

[b] Allocated monthly base plan. Producers receive blend price for base milk and class II price for surplus milk. Additional base may be purchased from dairymen going out of business at current market value of base.

## QUESTIONS

1. What is the blend price per cwt. for 3.5 milk in each of these markets (all milk in market N and base milk in Markets O and P)?

2. What is the blend price per cwt. in each of these markets (base milk in O and P) for milk that contains 4.3% fat?

3. What is the blend price per cwt. in each of these markets (base milk in O and P) for milk that contains 3.2% fat?

4. How much would a market N producer receive for 50,000 lb. of 3.5 milk shipped this month?

5. a. If a producer in market O had 50,000 lb. of monthly base and he shipped 60,000 lb. of 3.5 this month, what is the blend price per cwt. and the total value of that milk?

   b. If, under the same base conditions as (5a), he shipped 100,000 lb. of 3.5 milk this month, what blend price per cwt. would he receive for that milk?

   c. If he were to continue to ship 100,000 lb. of 3.5 milk monthly during the base-setting period (4 months), what is the cost (in reduced income) of the additional 50,000 lb. of base?

6. a. If a producer in Market P had 50,000 lb. of monthly base and he shipped 60,000 lb. of 3.5 milk this month, what is the blend price per cwt. and the total value of that milk?

  b. If, under the same base conditions as (6a) he shipped 100,000 lb. of 3.5 milk this month, what blend price per cwt. would he receive for that milk?

  c. If he were to purchase an additional 50,000 lb. of base at $.60/lb., and borrowed all the money for this purchase at 9% annual interest on the unpaid balance, how long would it take him to repay the loan from the additional income generated by the investment?

  d. What is the maximum price he could pay for milk base and recover the cost of the base from the additional income within three years?

7. If the following conditions were true:

  a. Cost of resources (land, cattle, feed etc.) were similar in all 3 market areas,

  b. Cost of capital were 9% in all 3 market areas,

  c. Current marketing conditions (price, usage etc.), were relatively stable and were expected to continue to be stable,

  d. Milk base could be established by marketing at class II price for 4 months in Market O and purchased for $.60/lb. in Market N, would you rather start dairying in Market N, Market O or Market P if your projected average production was 15,000 lb. of 3.5 milk sold/cow/year for a herd of 100 cows? Why? Justify your answer with economic projections.

**Problem 10.** Overall Problem

**Reference:** Text, previous problems, bulletins, etc.

**Objective:** To learn to plan a feasible management program for a dairy farm.

**Materials:** A description of resources and prices of one or more dairy herd and farms typical of the area. Various computerized programs are helpful but not necessary.

**Problem:** Fran Trueblue, a graduating senior in dairy science who wants to go into the dairy business, has located a dairyman who would like to retire with a steady income. He has offered Fran two alternatives. He will (1) sell Fran an entire operation, or (2) sell him the cattle, machinery, equipment, base (optional) and stored feed and give him an 8.75-year lease on the land and buildings. Both alternatives and any of three options are available April 1, of the current year.

**Options:** 1. Purebred Jerseys with purchased base or no base.
2. Purebred Holsteins with purchased base or no base.
3. Grade Holsteins with purchased base or no base.
    Based on your preference, choose one of the alternatives. Define your goals for the operation and plan a management program to meet these goals. Include plans for breeding, land use, feeding, labor, health, marketing, etc., as well as financial projections that define profitability of the operation and financial growth.

## Alternative I Conditions

1. Fixed resources (assets 6, 7, 8 in Example table, p. 587) financed for 20 years at 8% annual interest on the remaining balance. Interest only due on Dec. 31 of the current year and interest plus equal principal payments due annually on Dec. 31 thereafter.
2. Current and intermediate resources [assets 1, 2 (optional), 3, 4, 5] financed for 8.75 years at 8% interest annual interest on the remaining balance. Interest only due on Dec. 31 of the current year and interest plus equal principal payments due annually on Dec. 31 thereafter.

## Alternative 2 Conditions

1. An 8.75-year lease on the fixed resources (assets 6, 7, 8) at 8% of their current value due annually on Dec. 31 (.75 year in the current year). Building and fence repair and maintenance, etc., are the responsibility of the owner.
2. Current and intermediate resources [assets 1, 2, (optional), 3, 4, 5] financed as in Alternative 1.

## Other Data

1. Short-term operating capital is available (up to $40,000) at 9% interest on the unpaid balance.

2. All buildings and equipment are in good condition.

3. The farm is currently being operated by the owner and two full-time men.

4. The average PD of the sires of the Jersey cows is +300, the Jersey

heifers +400, and the service sires +600. The average classification score is 84.5.

5. The average PD of the sires of both the grade and registered Holstein cows is +450, the heifers +600, and the service sires +800. The average classification score on the purebred Holstein cows is 80.5.

6. Available feeds for purchase are the same as used in the feed problem (Problem 5). Good alfalfa hay is also available at $100/T.

## Example, Assets

| | | Options | |
|---|---|---|---|
| Assets | 1 | 2 | 3 |
| 1. Cattle—90 milking age cows, 30 heifers over 1 year and 40 heifers under 1 year | | | |
|    a. All purebred Jerseys (11,000—550 sold) | $72,000 | | |
|    b. All purebred Holsteins (15,000—555 sold) | | $108,000 | |
|    c. All grade Holsteins (15,000—555 sold) | | | $90,000 |
| 2. Milk base | | | |
|    a. Purchased monthly base at $.50/lb. as needed with market of 90% class I at $11.50/cwt., 10% class II at $8.00/cwt. and $.10 BF diff. based on 3.5% (80,000 or 100,000 lb.) | $40,000 | $50,000 | $50,000 |
|    b. No base with market of 50% class I at $11.50/cwt., 50% class II at $8.60/cwt., and $.10 BF diff. based on 3.5% | | | |
| 3. Dairy equipment. | | | |
|    a. Milking & milk storage equipment; current value = $10,000, deprec. rate = $1,000/yr. | $10,000 | $10,000 | $10,000 |
|    b. Feeding and manure-handling equipment; current value = $12,000, deprec. rate = $1,500/yr. | $12,000 | $12,000 | $12,000 |
| 4. Farming equipment | | | |
|    a. Adequate equipment for silage, haylage, and hay planting and harvesting; current value = $30,000, deprec. rate = $3,000/yr. | $30,000 | $30,000 | $30,000 |
| 5. Stored feed | | | |
|    600 or 900 T. good corn silage @ $20/T. | $12,000 | $18,000 | $18,000 |
|    40 or 66 T. good mixed hay @ $76/T. | $ 3,000 | $ 5,000 | $ 5,000 |
| 6. Dairy buildings | | | |
|    a. 90-cow free stall barn, feed bunks, double 4 herringbone, holding area, maternity, calf, and special handling areas, etc.; current value = $30,000, deprec. rate = $3,000/yr. | $30,000 | $30,000 | $30,000 |
|    b. Silage storage—two or three 600 T. concrete stave uprights; current value = $5,000 ea., deprec. rate = $500 ea./yr. | 10,000 | 15,000 | 15,000 |

## Example, Assets

| | Options | | |
|---|---|---|---|
| Assets | 1 | 2 | 3 |
| c. Hay storage—150 T. capacity; current value = $3,000, deprec. rate = $300/yr. | 3,000 | 3,000 | 3,000 |
| d. Concentrate storage—two 10 T. bins; current value = $2,000, deprec. rate = $200/yr. | 2,000 | 2,000 | 2,000 |
| 7. Other buildings | | | |
| a. Machinery storage shed; current value = $3,000, deprec. rate = $300/yr. | $ 3,000 | $ 3,000 | $ 3,000 |
| b. House & garage; current value = $20,000 | $20,000 | $20,000 | $20,000 |
| 8. Land | | | |
| 150 A. continuous corn (mostly class I and some class II soil); 5 A. buildings, fence rows, etc.; 30 A. watered permanent pasture (not suitable for anything else); 20 A. timber, rocks, etc. (useless) | | | |
| Total current value = $120,000 | $120,000 | $120,000 | $120,000 |
| *Purchase Prices* | | | |
| 1. Total alt. I (with purchased base) | $367,000 | $426,000 | $408,000 |
| 2. Total alt. I (no base) | 327,000 | 376,000 | 358,000 |
| 3. Total alt. II (with purchased base) | 179,000 | 233,000 | 215,000 |
| 4. Total alt. II (no base) | 139,000 | 183,000 | 165,000 |

## APPENDIX 2

### EXERCISES

Successful and profitable dairying requires proficiency in many skills and techniques as well as knowledge of what to do and why it should be done. Proficiency in these skills and techniques can best be acquired by doing them under the supervision of a competent and experienced person. The following procedure may be useful for teachers to offer students the opportunity to acquire many needed skills and techniques.

1. Prepare a list of needed herdsmanship skills and techniques (this list may vary somewhat from area to area) and have the students indicate their degree of experience in each. Exercise Table 1 is an example of a list or questionnaire that might be used.

2. Divide the skills into related areas that can be conveniently taught with similar facilities, equipment, and cattle. The right column in Exercise Table 1 indicates a logical division for skills listed in that table.

3. Summarize the results of the questionnaire and divide the students with a similar degree of proficiency into small groups.

4. Arrange laboratories and schedules to offer the students the opportunity to acquire needed skills.

The time required to acquire proficiency varies with cattle, facilities, and equipment availability, individual student aptitude and experience, and the skill itself. The following schedule might be feasible based on the division given in Exercise Table 1.

**Exercise 1.** Divide the students into groups of four to five and arrange whenever calves are available.

**Exercise 2.** Assign each student the responsibility for heat checks (two to three times per day) for a herd for one to two days. Provide each with the herd breeding records so that he can identify which cows should or might be in heat. Also, provide forms for recording observed heats and time of breeding recommendations.

**Exercise 3.** This is difficult to schedule. These events often do occur during times when students are at the herd for other exercises.

**Exercise 4.** This should be scheduled as a short course or as a regularly scheduled course. This skill cannot be learned in one to two sessions.

589

**Exercise Table 1**  Dairy Skills and Techniques Analysis

| Skill or Technique | No experience | Some, but not proficient | Proficient | Lab exercise[a] |
|---|---|---|---|---|
| 1. Calf herdsmanship | | | | |
|   a. Calf ident.—ear tag | | | | 1 |
|         tattoo | | | | 1 |
|         freeze brand | | | | 1 |
|   b. Calf registration | | | | 1 |
|   c. Calf dehorning—caustic | | | | 1 |
|         electric | | | | 1 |
|         barnes | | | | 1 |
|   d. Extra teat removal | | | | 1 |
| 2. Reproductive management | | | | |
|   a. Heat detection | | | | 2 |
|   b. Determine time to breed | | | | 2 |
|   c. Aid cows at calving | | | | 3 |
|   d. Aid calf to breathe | | | | 3 |
|   e. Dip navel | | | | 3 |
|   f. Artificial insemination | | | | 4 |
| 3. Cow herdsmanship | | | | |
|   a. Barn clipping | | | | 5 |
|   b. Hoof trimming | | | | 5 |
|   c. Halter breaking—cow or heifer | | | | 6 |
|   d. Show clipping | | | | 6 |
|   e. Show a cow or heifer | | | | 6 |
|   f. Load and unload cattle | | | | 6 |
| 4. Health care | | | | |
|   a. Temperatures | | | | 7 |
|   b. Balling gun | | | | 7 |
|   c. Frick tube | | | | 7 |
|   d. Stomach tube and pump | | | | 7 |
|   e. Sample blood—tail | | | | 7 |
|         jugular | | | | 7 |
|   f. Intramuscular injection | | | | 7 |
|   g. Intravenous infusion | | | | 7 |
|   h. Intramammary infusion | | | | 7 |
|   i. Leucocyte test (CMT) | | | | 7 |
|   j. Sample milk for bacteriological test | | | | 7 |
| 5. Milking | | | | |
|   a. Milking system analysis | | | | 8 |
|   b. Milking system maintenance | | | | 8 |
|   c. Milking procedure analysis | | | | 8 |
|   d. Milking | | | | 9 |
| 6. Cattle marketing | | | | |
|   a. Prepare sale catalog pedigree | | | | 10 |
|   b. Prepare ad material | | | | 10 |
|   c. Prepare cattle for sale | | | | 10 |
|   d. Sell cattle | | | | 10 |
|   e. Buy cattle at auction | | | | 10 |

[a] The numbers in this column refer to specific exercises in the text.

**Exercise 5.** Divide the students into groups of four to five and arrange whenever cattle are available. Hoof trimming may be more conveniently handled in a separate lab or a short course.

**Exercise 6.** Divide the students into groups of four to five and arrange whenever cattle are available. This may be combined with the barn clipping exercise if hoof trimming is omitted. These experiences can also be achieved by participation in dairy fitting and showmanship contests held on many colleges or as part of the mock sale exercise (Exercise 10).

**Exercise 7.** This lab should be taught by a veterinarian. Divide the students into groups of 6 to 10 and arrange when cattle and a veterinarian are available.

**Exercise 8.** Provide the students with a milking machine and procedure analysis form that can be used to evaluate proper installation and maintenance of the milking system (Appendix Table G) and of the milking procedures (Chapter 12).

Demonstrate the use of an air flow meter to determine vacuum pump capacity, a vacuum gauge to determine vacuum level and fluctuation at the teat cup, how to calculate CFM needs of a system, etc., and assign each student the responsibility for a thorough analysis of a herd milking system and procedures.

**Exercise 9.** Work students into a herd milking operation on an individual basis.

**Exercise 10.** Conduct a mock sale with the students as consignors and buyers. An outline of the mock sale procedures used at V.P.I. & S.U. follows.

**Purpose.** To acquire experience and learn how to:
1. Prepare dairy cattle for a consignment auction sale.
2. Prepare catalog pedigrees.
3. Prepare ad materials.
4. Help sell consignments.
5. Purchase cattle at reasonable values.

**Materials**
1. One dairy animal per student. These should be both heifers and cows and may be grades or purebreds or a combination of both.
2. Materials that can be used to prepare pedigrees such as registration papers, DHI reports, classification reports, sire summaries, U.S.D.A. cow indices, etc. These should be made readily available to the students.
3. Facilities and equipment to prepare cattle (clip, wash, etc.) and display them before the sale.
4. A sale ring and auction box.
5. A knowledgeable dairy auctioneer, ringmen, and leaders.

## Procedures

1. Assign each student a heifer or cow two to three weeks before the sale date.
2. Inform the students of their responsibilities to: (a) prepare the animal for sale; (b) prepare the catalog pedigree, (c) prepare the ad for their consignment, and (d) help sell their consignment.
3. Inform the students of the location of their animal, location of equipment and facilities for cattle preparation and permissible work times, and location of other needed materials such as registration papers, DHI reports, etc.
4. Divide the students into bidding syndicates of two to four each and allot each syndicate a given sum.
5. Distribute copies of the PDCA Code of Ethical Sales Practices (Appendix L), the procedural details for the sale, and the grading criteria.

## Grading criteria

|  |  | *Points* |
|---|---|---|
| 1. | Condition of the animal (leading, fitting, etc.) | 0–100 |
| 2. | Catalog material (completeness, accuracy, etc.) | 0–100 |
| 3. | Merchandising (ad, identification in the barn, etc.) | 0–100 |
| 4. | Selling—the consigner (student) of any animal that is sold at $105 to $200 higher than its assigned value will receive 50 points and for $205 or higher, 100 points | 0–100 |
| 5. | Identification of the best and worst. Each student will select his choice for high and low animal in the sale and deliver this to the instructor not later than one hour before the sale.<br>Both correct = 100 points; one correct = 50 points; none correct = 0 points | 0–100 |
| 6. | Syndicate buying | |
| a. | Each animal purchased within $100 of its assigned value is worth 100 points | 0–? |
| b. | $105 to $150 more than its assigned value is worth 50 points | 0–? |
| c. | $105 to $150 less than its assigned value is worth 125 points | 0–? |
| d. | $155 or more under its assigned value is worth 150 points | 0–? |
| e. | $155 or more over its assigned value costs 1 point | 0–? |

       for each $1 over its assigned value to a maximum of
       250 points

If the sale average is:

f.   $101 or more higher than the average assigned value   0–?
     of the cattle, purchasers receive 1 point for every
     $200 not spent at the sale

g.   $101 or more lower than the average assigned value   0–?
     of the cattle, purchasers lose 1 point for every $200
     not spent at the sale

                                        Total   ? to +1200

**Grading Scale**

    A = 600 or more points
    B = 500 to 599 points
    C = 400 to 499 points
    D = 300 to 399 points
    E = 299 or less points

6.  Have knowledgeable persons (the auctioneer, dairyman, etc.) assign a price to each animal before the sale.

7.  Conduct the auction. The auctioneer is a key factor in the success of this exercise. An experienced, respected dairy auctioneer is best. Suggest the auctioneer review the sale with the students after it is over.

# APPENDIX 3

## TABLES

**Appendix Table A** Daily Nutrient Requirements of Dairy Cattle[a]

594

| Body weight (kg.) | Daily gain (g.) | Dry feed (kg.) | Protein | | Energy[b] | | | Ca (g.) | P (g.) | Carotene (mg.) | Vitamin A (1,000 I.U.) | Vitamin D (I.U.) |
|---|---|---|---|---|---|---|---|---|---|---|---|---|
| | | | Total (g.) | Digest. (g.) | NEm[b] (Mcal.) | NEgain (Mcal.) | TDN (kg.) | | | | | |
| **Growing heifers (large breeds)** | | | | | | | | | | | | |
| 40 | 200 | 0.5[c] | 110 | 100 | 0.9 | 0.4 | 0.5 | 2.2 | 1.7 | 4.2 | 1.7 | 265 |
| 45 | 300 | 0.6 | 135 | 120 | 1.1 | 0.5 | 0.6 | 3.2 | 2.5 | 4.8 | 1.9 | 300 |
| 55(5)[d] | 400 | 1.2 | 180 | 145 | 1.3 | 0.6 | 0.9 | 4.5 | 3.5 | 5.8 | 2.3 | 360 |
| 75(10) | 750 | 2.1 | 330 | 245 | 1.5 | 0.9 | 1.5 | 9.1 | 7.0 | 7.9 | 3.2 | 495 |
| 100(15) | 750 | 2.9 | 370 | 260 | 2.0 | 1.1 | 2.0 | 10.9 | 8.4 | 11 | 4 | 660 |
| 150(24) | 750 | 4.1 | 435 | 295 | 3.1 | 1.5 | 2.7 | 15 | 12 | 16 | 6 | 990 |
| 200(34) | 750 | 5.3 | 500 | 330 | 4.1 | 1.8 | 3.4 | 18 | 14 | 21 | 8 | 1320 |
| 250(43) | 750 | 6.5 | 570 | 365 | 4.8 | 2.2 | 4.0 | 21 | 16 | 26 | 10 | — |
| 300(53) | 750 | 7.5 | 640 | 395 | 5.6 | 2.5 | 4.5 | 24 | 18 | 32 | 13 | — |
| 350(62) | 750 | 8.4 | 715 | 430 | 6.2 | 2.8 | 4.9 | 25 | 19 | 37 | 15 | — |
| 400(72) | 750 | 9.3 | 800 | 465 | 6.9 | 3.1 | 5.2 | 26 | 20 | 42 | 17 | — |
| 450(82) | 700 | 9.5 | 885 | 495 | 7.5 | 3.1 | 5.3 | 27 | 21 | 48 | 19 | — |
| 500(93) | 600 | 9.5 | 935 | 505 | 8.1 | 2.9 | 5.3 | 27 | 21 | 53 | 21 | — |
| 550(107) | 400 | 8.9 | 915 | 475 | 8.7 | 2.0 | 5.0 | 26 | 20 | 58 | 23 | — |
| 600(133) | 150 | 8.6 | 810 | 405 | 9.3 | 0.7 | 4.3 | 24 | 18 | 64 | 26 | — |
| **Growing heifers (small breeds)** | | | | | | | | | | | | |
| 20 | 100 | 0.3[c] | 65 | 60 | 0.6 | 0.2 | 0.3 | 1.1 | 0.8 | 2.1 | 0.8 | 130 |
| 25 | 150 | 0.4 | 90 | 80 | 0.8 | 0.3 | 0.4 | 1.5 | 1.1 | 2.6 | 1.0 | 165 |
| 35(5)[d] | 300 | 0.8 | 135 | 110 | 0.9 | 0.5 | 0.6 | 3.2 | 2.5 | 3.7 | 1.5 | 230 |
| 50(10) | 500 | 1.2 | 215 | 160 | 1.0 | 0.9 | 0.9 | 4.9 | 3.8 | 5.3 | 2.1 | 330 |
| 75(17) | 550 | 1.7 | 275 | 190 | 1.5 | 1.0 | 1.2 | 7 | 5.4 | 7.9 | 3.2 | 495 |
| 100(23) | 550 | 2.4 | 320 | 210 | 2.1 | 1.1 | 1.6 | 9 | 7 | 11 | 4 | 660 |
| 150(36) | 550 | 3.6 | 390 | 245 | 3.7 | 1.3 | 2.3 | 12 | 9 | 16 | 6 | 990 |
| 200(49) | 550 | 4.8 | 465 | 280 | 4.1 | 1.6 | 2.9 | 15 | 11 | 21 | 8 | 1320 |
| 250(62) | 550 | 6.1 | 550 | 320 | 4.8 | 1.9 | 3.5 | 17 | 13 | 26 | 10 | — |
| 300(76) | 500 | 6.8 | 590 | 330 | 5.6 | 2.0 | 3.8 | 19 | 14 | 32 | 13 | — |
| 350(93) | 350 | 6.6 | 585 | 315 | 6.2 | 1.5 | 3.7 | 19 | 14 | 37 | 15 | — |
| 400(121) | 150 | 6.4 | 555 | 290 | 6.9 | 0.7 | 3.6 | 19 | 14 | 42 | 17 | — |
| 450(192) | 50 | 6.1 | 580 | 290 | 7.5 | 0.5 | 3.4 | 19 | 14 | 48 | 19 | — |

**Growing bulls (large breeds)**

| | | | | | | | | | | | | |
|---|---|---|---|---|---|---|---|---|---|---|---|---|
| 40 | 200 | 0.5[c] | 110 | 100 | 0.9 | 0.4 | 0.5 | 2.2 | 1.7 | 4.2 | 1.7 | 265 |
| 55(5)[d] | 400 | 1.2 | 180 | 145 | 1.3 | 0.6 | 0.9 | 4.5 | 3.5 | 5.8 | 2.3 | 360 |
| 100(13) | 1000 | 3.2 | 455 | 320 | 2.1 | 1.3 | 2.2 | 13 | 10 | 11.0 | 4.0 | 660 |
| 200(27) | 1000 | 5.9 | 595 | 390 | 4.5 | 2.2 | 3.8 | 21 | 16 | 21 | 8 | 1320 |
| 300(41) | 1000 | 8.7 | 745 | 465 | 7.2 | 3.0 | 5.2 | 27 | 20 | 32 | 13 | — |
| 400(56) | 1000 | 11.8 | 930 | 540 | 9.0 | 3.8 | 6.6 | 30 | 23 | 42 | 17 | — |
| 500(70) | 900 | 13.0 | 1110 | 610 | 10.6 | 4.0 | 7.3 | 30 | 23 | 53 | 21 | — |
| 600(88) | 700 | 13.8 | 1190 | 630 | 12.1 | 3.5 | 7.7 | 30 | 23 | 64 | 26 | — |
| 700(112) | 500 | 13.4 | 1235 | 630 | 13.6 | 2.8 | 7.5 | 30 | 23 | 74 | 30 | — |
| 800 | 250 | 12.7 | 1165 | 570 | 15.1 | 1.4 | 7.1 | 30 | 23 | 85 | 34 | — |

**Growing bulls (small breeds)**

| | | | | | | | | | | | | |
|---|---|---|---|---|---|---|---|---|---|---|---|---|
| 20 | 150 | 0.4[c] | 90 | 80 | 0.6 | 0.3 | 0.4 | 1.5 | 1.1 | 2.6 | 1.0 | 165 |
| 50(8)[d] | 650 | 1.4 | 265 | 200 | 1.0 | 1.1 | 1.0 | 6.5 | 5.0 | 5.3 | 2.1 | 330 |
| 100(18) | 750 | 2.8 | 390 | 255 | 2.1 | 1.6 | 1.9 | 11 | 8 | 11 | 4 | 660 |
| 200(37) | 750 | 5.7 | 530 | 330 | 4.5 | 2.3 | 3.4 | 18 | 14 | 21 | 8 | 1320 |
| 300(56) | 750 | 8.2 | 680 | 395 | 7.2 | 3.1 | 4.6 | 23 | 17 | 32 | 13 | — |
| 400(76) | 700 | 10.2 | 820 | 450 | 8.9 | 3.6 | 5.7 | 25 | 19 | 42 | 17 | — |
| 500(106) | 400 | 10.0 | 885 | 455 | 10.6 | 2.3 | 5.6 | 26 | 20 | 53 | 21 | — |
| 600 | 100 | 9.8 | 800 | 385 | 12.1 | 0.6 | 5.5 | 24 | 18 | 64 | 26 | — |

**Veal calves**

| | | | | | | | | | | | | |
|---|---|---|---|---|---|---|---|---|---|---|---|---|
| 35 | 500 | 0.7[c] | 155 | 130 | 1.0 | 0.8 | 0.7 | 3.0 | 2.3 | 3.7 | 1.5 | 230 |
| 40 | 800 | 1.1 | 240 | 205 | 1.5 | 1.4 | 1.1 | 4.8 | 3.7 | 5.3 | 2.1 | 330 |
| 75 | 1000 | 1.4 | 310 | 260 | 1.9 | 1.8 | 1.4 | 7.9 | 5.9 | 7.9 | 3.2 | 495 |
| 100 | 1150 | 1.7 | 375 | 320 | 2.3 | 2.2 | 1.7 | 11.1 | 8.0 | 11.0 | 4.0 | 660 |
| 150 | 1300 | 2.4 | 485 | 410 | 3.0 | 3.0 | 2.4 | 16.0 | 11.0 | 16.0 | 6.0 | 990 |

**Maintenance of mature breeding bulls**

| | | | | | | | | | | | | |
|---|---|---|---|---|---|---|---|---|---|---|---|---|
| 500 | — | 8.3 | 640 | 300 | 9.5 | — | 4.6 | 20 | 15 | 53 | 21 | — |
| 600 | — | 9.6 | 735 | 345 | 10.8 | — | 5.4 | 22 | 17 | 64 | 26 | — |
| 700 | — | 10.9 | 830 | 390 | 12.3 | — | 6.1 | 25 | 19 | 74 | 30 | — |
| 800 | — | 12.0 | 915 | 430 | 13.9 | — | 6.7 | 27 | 21 | 85 | 34 | — |
| 900 | — | 13.1 | 1000 | 470 | 15.2 | — | 7.3 | 30 | 23 | 95 | 38 | — |
| 1000 | — | 14.1 | 1075 | 505 | 16.9 | — | 7.9 | 32 | 25 | 106 | 42 | — |
| 1100 | — | 15.1 | 1160 | 545 | 18.2 | — | 8.4 | 35 | 27 | 117 | 47 | — |
| 1200 | — | 16.1 | 1235 | 580 | 19.5 | — | 9.0 | 38 | 29 | 127 | 51 | — |
| 1300 | — | 17.1 | 1310 | 615 | 20.7 | — | 9.6 | 40 | 31 | 138 | 55 | — |
| 1400 | — | 18.1 | 1380 | 650 | 21.9 | — | 10.1 | 43 | 33 | 148 | 59 | — |

[a] Adapted from Table 1, *Nutrient Requirements of Dairy Cattle*, 4th rev. ed., 1971. Reproduced with the permission of the National Academy of Sciences. [b] $NE_m$ = Net energy for maintenance; $NE_{gain}$ = net energy for gain. [c] Based on milk replacer. [d] Weeks of age.

**Appendix Table B**  Daily Nutrient Requirements of Lactating Dairy Cattle[a]

| Body weight (kg.) | Dry feed (kg.) | Protein Total (g.) | Protein Digestible (g.) | Energy NE lactating cows (Mcal.)[b] | Energy TDN (kg.) | Ca (g.) | P (g.) | Carotene (mg.) | Vitamin A (1,000 I.U.) |
|---|---|---|---|---|---|---|---|---|---|
| Maintenance of mature lactating cows[c] | | | | | | | | | |
| 350 | 5.0 | 468 | 220 | 6.9 | 2.8 | 14 | 11 | 37 | 15 |
| 400 | 5.5 | 521 | 245 | 7.6 | 3.1 | 17 | 13 | 42 | 17 |
| 450 | 6.0 | 585 | 275 | 8.3 | 3.4 | 18 | 14 | 48 | 19 |
| 500 | 6.5 | 638 | 300 | 9.0 | 3.7 | 20 | 15 | 53 | 21 |
| 550 | 7.0 | 691 | 325 | 9.6 | 4.0 | 21 | 16 | 58 | 23 |
| 600 | 7.5 | 734 | 345 | 10.3 | 4.2 | 22 | 17 | 64 | 26 |
| 650 | 8.0 | 776 | 365 | 10.9 | 4.5 | 23 | 18 | 69 | 28 |
| 700 | 8.5 | 830 | 390 | 11.6 | 4.8 | 25 | 19 | 74 | 30 |
| 750 | 9.0 | 872 | 410 | 12.2 | 5.0 | 26 | 20 | 79 | 32 |
| 800 | 9.5 | 915 | 430 | 12.8 | 5.3 | 27 | 21 | 85 | 34 |
| Maintenance and pregnancy (last 2 months of gestation) | | | | | | | | | |
| 350 | 6.4 | 570 | 315 | 8.7 | 3.6 | 21 | 16 | 67 | 27 |
| 400 | 7.2 | 650 | 355 | 9.7 | 4.0 | 23 | 18 | 76 | 30 |
| 450 | 7.9 | 730 | 400 | 10.7 | 4.4 | 26 | 20 | 86 | 34 |
| 500 | 8.6 | 780 | 430 | 11.6 | 4.8 | 29 | 22 | 95 | 38 |
| 550 | 9.3 | 850 | 465 | 12.6 | 5.2 | 31 | 24 | 105 | 42 |
| 600 | 10.0 | 910 | 500 | 13.5 | 5.6 | 34 | 26 | 114 | 46 |
| 650 | 10.6 | 960 | 530 | 14.4 | 6.0 | 36 | 28 | 124 | 50 |
| 700 | 11.3 | 1000 | 555 | 15.3 | 6.3 | 39 | 30 | 133 | 53 |
| 750 | 12.0 | 1080 | 595 | 16.2 | 6.7 | 42 | 32 | 143 | 57 |
| 800 | 12.6 | 1150 | 630 | 17.0 | 7.1 | 44 | 34 | 152 | 61 |

**Milk production (nutrients required per kg. of milk)[d]**

| % fat | | | | | | |
|---|---|---|---|---|---|---|
| 2.5 | 66 | 42 | 0.59 | 0.255 | 2.4 | 1.7 |
| 3.0 | 70 | 45 | 0.64 | 0.280 | 2.5 | 1.8 |
| 3.5 | 74 | 48 | 0.69 | 0.305 | 2.6 | 1.9 |
| 4.0 | 78 | 51 | 0.74 | 0.330 | 2.7 | 2.0 |
| 4.5 | 82 | 54 | 0.78 | 0.355 | 2.8 | 2.1 |
| 5.0 | 86 | 56 | 0.83 | 0.380 | 2.9 | 2.2 |
| 5.5 | 90 | 58 | 0.88 | 0.405 | 3.0 | 2.3 |
| 6.0 | 94 | 60 | 0.93 | 0.430 | 3.1 | 2.4 |

[a] Adapted from Table 2, *Nutrient Requirements of Dairy Cattle*, 4th rev. ed., 1971. Reproduced with the permission of the National Academy of Sciences.

[b] The energy requirements for maintenance, reproduction, and milk production of lactating cows are expressed in terms of $NE_{lactating\,cows}$.

[c] Maintenance of lactating cows $= 0.085$ Mcal. $NE_{lactating\,cows}/kg.^{3/4}$. To allow for growth, add 20% to the maintenance allowance during the first lactation and 10% during the second lactation.

[d] The energy requirement is presented as the actual amount required with no adjustment to compensate for any reduction in feed value at high levels of feed intake. To account for depressions in digestibility, which occur at high planes of nutrition with certain types of rations, such as corn silage, coarse-textured grains, or forages with high cell-wall content (e.g., Bermuda grass, sorghum, etc.), an increase of 3% feed should be allowed for each 10 kg. of milk produced above 20 kg./day.

**Appendix Table C** Nutrient Content of Rations for Dairy Heifers and Bulls[a]

| Nutrients | Concentration in dry matter | | | | | | | |
|---|---|---|---|---|---|---|---|---|
| | Calf milk replacer[b] | | Calf starter | | Heifer grower ration | | Mature bull ration | |
| | Min. | Max. | Min. | Max. | Min. | Max. | Min. | Max. |
| Protein, % | 22.0 | | 16.0 | | 10.0 | | 7.7 | |
| Digestible, % | 20.0 | | 12.0 | | 6.2 | | 3.6 | |
| Energy, Mcal./kg. | | | | | | | | |
| Digestible (DE) | 4.2 | | 3.2 | | 2.9 | | 2.5 | |
| Metabolizable (ME) | 3.4 | | 2.6 | | 2.4 | | 2.0 | |
| NE$_m$ | 1.7 | | 0.8 | | 0.8 | | 1.2 | |
| NE$_{gain}$ | 0.8 | | 0.7 | | 0.4 | | | |
| TDN, % | 95.0 | | 72.0 | | 66.0 | | 56.0 | |
| Ether, extract,% | 10.0 | | 2.5 | | 2.0 | | 2.0 | |
| Crude fiber, % | 0 | 3.0 | | 15.0 | 15.0 | | 15.0 | |
| Calcium, % | 0.55 | | 0.41 | | 0.34 | | 0.24 | |
| Phosphorus, % | 0.42 | | 0.32 | | 0.26 | | 0.18 | |
| Magnesium, % | 0.06 | | 0.07 | | 0.08 | | 0.08 | |
| Potassium,% | 0.70 | | 0.70 | | 0.70 | | 0.70 | |
| Sodium, % | 0.10 | | 0.10 | | 0.10 | | 0.10 | |

598

| | IV | | III | | II | | I | |
|---|---|---|---|---|---|---|---|---|
| Sodium chloride, % | 0.25 | | 0.25 | | 0.25 | | 0.25 | |
| Sulfur, % | 0.20 | | 0.20 | | 0.20 | | 0.20 | |
| Iron, ppm. | 100.0 | | 100.0 | | 100.0 | | 100.0 | |
| Cobalt, ppm. | 0.1 | 10 | 0.1 | 10 | 0.1 | 10 | 0.1 | 10 |
| Copper, ppm. | 10.0 | 100 | 10.0 | 100 | 10.0 | 100 | 10 | 100 |
| Manganese, ppm. | 20.0 | | 20.0 | | 20.0 | | 20 | |
| Zinc, ppm. | 40.0 | 500 | 40.0 | 500 | 40.0 | 500 | 40 | 1000 |
| Iodine, ppm. | 0.1 | | 0.1 | | 0.1 | | 0.1 | |
| Molybdenum, ppm. | | 6 | | 6 | | 6 | | 6 |
| Fluorine, ppm. | | 40 | | 30 | | 30 | | 40 |
| Selenium, ppm. | 0.1 | 5 | 0.1 | 5 | 0.1 | 5 | 0.1 | 5 |
| Carotene, ppm. | 9.5 | | 4.2 | | 4.0 | | 8.0 | |
| Vit. A equiv., I.U./kg. | 3800 | | 1600 | | 1500 | | 3200 | |
| Vit. D, I.U./kg. | 600 | | 250 | | 250 | | 300 | |
| Vit. E, mg./kg. | 300 | | | | | | | |

[a] From Table 3, *Nutrient Requirements of Dairy Cattle*, 4th rev. ed., 1971. Reproduced with the permission of the National Academy of Sciences.

[b] The following minimum quantities of B-complex vitamins are suggested for milk replacers: niacin, 2.6 mg.; pantothenic acid, 13 mg.; riboflavin, 6.5 mg.; pyridoxine, 6.5 mg.; thiamine, 6.5 mg.; folic acid, 0.5 mg.; biotin, 0.1 mg.; vitamin $B_{12}$, 0.07 mg.; choline, 2.6 g. per kg. It appears that adequate amounts of these vitamins are furnished when calves have functional rumens (usually at 6 weeks of age) by a combination of rumen synthesis and natural feedstuffs.

**Appendix Table D**  Nutrient Content of Rations for Dry and Lactating Dairy Cows[a]

| | Concentration in dry matter | | | | | | | |
| | Dry cow ration | | Lactating cow | | | | | |
| | | | <20 kg. | | 20–30 kg. | | >30 kg. | |
| Nutrients | Min. | Max. | Min. | Max. | Min. | Max. | Min. | Max. |
|---|---|---|---|---|---|---|---|---|
| Protein, % | 8.5 | | 14.0 | | 15.0 | | 16.0 | |
| Digestible, % | 5.1 | | 10.5 | | 11.4 | | 12.3 | |
| Energy, Mcal./kg. | | | | | | | | |
| Digestible (DE) | 2.3 | | 2.7 | | 2.9 | | 3.1 | |
| Metabolizable (ME) | 1.9 | | 2.1 | | 2.3 | | 2.5 | |
| $NE_m$ | 1.1 | | | | | | | |
| $NE_{1,act}$ | | | 1.4 | | 1.6 | | 1.8 | |
| TDN, % | 53.0 | | 60.0 | | 65.0 | | 70.0 | |
| Ether extract, % | 2.0 | | 2.0 | | 2.0 | | 2.0 | |
| Crude fiber, % | 15.0 | | 13.0 | | 13.0 | | 13.0 | |
| Calcium, % | 0.34 | | 0.43 | | 0.47 | | 0.53 | |
| Phosphorus, % | 0.26 | | 0.33 | | 0.35 | | 0.39 | |
| Magnesium, % | 0.08 | | 0.10 | | 0.10 | | 0.10 | |
| Potassium, % | 0.70 | | 0.70 | | 0.70 | | 0.70 | |
| Sodium, % | 0.10 | | 0.18 | | 0.18 | | 0.18 | |
| Sodium chloride, % | 0.25 | | 0.45 | | 0.45 | | 0.45 | |
| Sulfur, % | 0.20 | | 0.20 | | 0.20 | | 0.20 | |
| Iron, ppm. | 100.0 | | 100.0 | | 100.0 | | 100.0 | |
| Cobalt, ppm. | 0.1 | 10 | 0.1 | 10 | 0.1 | 10 | 0.1 | 10 |
| Copper, ppm. | 10.0 | 100 | 10.0 | 100 | 10.0 | 100 | 10.0 | 100 |
| Manganese, ppm. | 20.0 | | 20.0 | | 20.0 | | 20.0 | |
| Zinc, ppm. | 40.0 | 1000 | 40.0 | 1000 | 40.0 | 1000 | 40.0 | 1000 |
| Iodine, ppm. | 0.6 | | 0.6 | | 0.6 | | 0.6 | |
| Molybdenum, ppm. | | 6 | | 6 | | 6 | | 6 |
| Fluorine, ppm. | | 40 | | 40 | | 40 | | 40 |
| Selenium, ppm. | 0.1 | 5 | 0.1 | 5 | 0.1 | 5 | 0.1 | 5 |
| Carotene, ppm. | 8.0 | | 8.0 | | 8.0 | | 8.0 | |
| Vit. A equiv., I.U./kg. | 3200 | | 3200 | | 3200 | | 3200 | |
| Vit. D, I.U. | 300 | | 300 | | 300 | | 300 | |

[a] From Table 3, *Nutrient Requirements of Dairy Cattle*, 4th rev. ed., 1971. Reproduced with the permission of the National Academy of Sciences.

**Appendix Table E**  Nutrient Composition of Some Commonly Used Dairy Feeds (Dry Matter Basis[a])

| Feed | DM (%) | Growing cattle NE$_m$ Mcal./kg. | Growing cattle NE$_{gain}$ Mcal./kg. | Lactating cows NE$_{lact.}$ Mcal./kg. | TDN (%) | Crude protein (%) | Digest. protein (%) | Crude fiber (%) | Ca (%) | P (%) | K (%) | Carotene mg./kg. | Vit. D$_2$ I.U./kg. |
|---|---|---|---|---|---|---|---|---|---|---|---|---|---|
| **Forages** | | | | | | | | | | | | | |
| Alfalfa, immature hay | 89.1 | 1.36 | 0.76 | 1.46 | 63 | 21.5 | 15.0 | 28.5 | 1.25 | 0.23 | — | — | — |
| Alfalfa, early (1/10) bloom, hay | 90.0 | 1.35 | 0.49 | 1.25 | 57 | 18.4 | 12.7 | 29.8 | 1.25 | 0.23 | 2.08 | 127.2 | 1991.0 |
| Alfalfa, early (1/10) bloom, haylage | 55.0 | 1.10 | 0.35 | 1.07 | 52 | 17.9 | 10.7 | 32.4 | 1.61 | 0.38 | — | — | — |
| Alfalfa, early (1/10) wilted silage | 36.2 | 1.31 | 0.69 | 1.28 | 58 | 17.8 | 11.9 | 28.8 | 1.74 | 0.31 | 1.46 | 97.2 | — |
| Alfalfa, mid (avg. ½) bloom, hay | 89.2 | 1.20 | 0.52 | 1.21 | 56 | 17.1 | 12.1 | 30.9 | 1.35 | 0.22 | 0.55 | 33.3 | — |
| Alfalfa, full (⅔ or more) bloom, hay | 87.7 | 1.22 | 0.55 | 1.25 | 57 | 15.9 | 11.4 | 33.9 | 1.28 | 0.20 | 2.03 | 37.0 | — |
| Alfalfa, pasture | 27.2 | 1.32 | 0.71 | 1.39 | 61 | 19.3 | 15.0 | 27.4 | 1.72 | 0.31 | 2.03 | 198.9 | 0.2 |
| Alfalfa, brome, hay | 82.5 | 1.19 | 0.50 | 1.19 | 55 | 16.2 | 10.9 | 33.7 | 1.03 | 0.30 | 1.85 | 26.0 | — |
| Alfalfa, orchardgrass, wilted silage | 40.0 | 1.15 | 0.44 | 1.14 | 54 | 17.2 | 10.0 | 31.4 | — | — | — | — | — |
| Bermudagrass, coastal, hay | 91.5 | 1.14 | 0.42 | 1.13 | 51 | 9.5 | 5.1 | 30.5 | 0.46 | 0.18 | — | — | — |
| Bluegrass (ky.) immature, pasture | 30.5 | 1.59 | 1.02 | 1.77 | 72 | 17.3 | 12.6 | 25.1 | 0.56 | 0.47 | 2.28 | 383.0 | — |
| Brome, hay | 89.7 | 1.15 | 0.44 | 1.14 | 54 | 11.8 | 5.0 | 32.0 | — | — | — | — | — |
| Clover, alsike, hay | 87.9 | 1.29 | 0.66 | 1.36 | 60 | 14.7 | 9.3 | 29.4 | 1.31 | 0.25 | 1.70 | 187.0 | — |
| Clover, crimson, hay | 87.4 | 1.29 | 0.66 | 1.36 | 60 | 16.9 | 11.8 | 32.2 | 1.42 | 0.18 | 1.54 | — | — |
| Clover, red, hay | 87.7 | 1.26 | 0.62 | 1.31 | 59 | 14.9 | 8.9 | 30.1 | 1.61 | 0.22 | 1.76 | 36.8 | 1912.5 |
| Clover, red, early bloom, pasture | 19.6 | 1.56 | 0.99 | 1.70 | 70 | 21.1 | 15.2 | 19.0 | 2.26 | 0.38 | 2.49 | — | — |
| Corn silage, mature, well eared, max. 30%DM | 27.9 | 1.56 | 0.99 | 1.70 | 70 | 8.4 | 4.9 | 26.3 | 0.28 | 0.21 | 0.95 | — | 119.0 |
| Corn silage, mature, well eared, max. 30–50% DM | 40.0 | 1.56 | 0.99 | 1.70 | 70 | 8.1 | 4.7 | 24.4 | 0.27 | 0.20 | 1.05 | — | — |
| Cowpea, hay | 90.5 | 1.36 | 0.76 | 1.46 | 63 | 18.4 | 12.9 | 27.3 | 1.34 | 0.32 | 1.99 | — | — |
| Fescue, meadow, hay | 88.5 | 1.33 | 0.72 | 1.41 | 62 | 10.5 | 6.0 | 31.2 | 0.50 | 0.36 | 1.87 | — | — |
| Lespedeza, prebloom | 92.1 | 1.26 | 0.62 | 1.31 | 59 | 17.8 | 8.3 | 23.7 | 1.14 | 0.26 | — | — | — |
| Lespedeza, early bloom, hay | 93.4 | 1.05 | 0.23 | 0.97 | 49 | 15.5 | 7.0 | 29.6 | 1.23 | 0.25 | 1.00 | — | — |

601

# Appendix Table E  Continued

| Feed | DM (%) | Growing cattle NE_m Mcal./kg. | NE_gain Mcal./kg. | Lactating cows NE_lact. Mcal./kg. | TDN (%) | Crude protein (%) | Digest. protein (%) | Crude fiber (%) | Ca (%) | P (%) | K (%) | Carotene mg./kg. | Vit. D₂ I.U./kg. |
|---|---|---|---|---|---|---|---|---|---|---|---|---|---|
| Oats, hay | 88.2 | 1.31 | 0.70 | 1.39 | 61 | 9.2 | 4.4 | 31.0 | 0.26 | 0.24 | 0.97 | 101.0 | — |
| Oats, silage | 31.7 | 1.27 | 0.64 | 1.31 | 59 | 9.7 | 5.5 | 31.6 | 0.37 | 0.30 | 3.41 | 119.5 | — |
| Orchardgrass, hay | 88.3 | 1.22 | 0.55 | 1.25 | 57 | 9.7 | 5.8 | 34.0 | 0.45 | 0.37 | 2.10 | 33.5 | — |
| Orchardgrass, immature, pasture | 23.8 | 1.46 | 0.88 | 1.59 | 67 | 18.4 | 13.5 | 23.6 | 0.58 | 0.55 | 3.38 | 337.4 | — |
| Orchardgrass, early bloom, pasture | 27.5 | 1.60 | 1.03 | 1.55 | 66 | 13.1 | 9.0 | 28.8 | 0.25 | 0.39 | — | — | — |
| Sorghum, grain variety, silage | 29.4 | 1.22 | 0.57 | 1.24 | 57 | 7.3 | 2.0 | 26.3 | 0.25 | 0.18 | — | — | — |
| Soybean, hay | 89.2 | 1.11 | 0.36 | 1.07 | 52 | 16.3 | 10.0 | 32.1 | 1.29 | 0.23 | 0.97 | 35.7 | 709.8 |
| Timothy, prebloom, hay | 88.6 | 1.36 | 0.76 | 1.46 | 63 | 12.3 | 6.6 | 32.9 | 0.66 | 0.34 | — | — | — |
| Timothy, midbloom, hay | 88.4 | 1.24 | 0.59 | 1.28 | 58 | 8.5 | 4.6 | 33.5 | 0.41 | 0.19 | — | 53.4 | — |
| Timothy, prebloom, pasture | 28.3 | 1.47 | 0.90 | 1.59 | 67 | 9.3 | 5.5 | 33.5 | 0.50 | 0.35 | 2.40 | — | — |
| Timothy, silage | 37.5 | 1.26 | 0.62 | 1.31 | 59 | 10.2 | 5.7 | 33.9 | 0.55 | 0.29 | 1.69 | 78.9 | — |
| Trefoil, birdsfoot, pasture | 20.0 | 1.68 | 1.10 | 1.87 | 75 | 28.0 | 23.0 | 13.0 | 2.20 | 0.25 | 2.30 | — | — |
| Wheat, pasture | 21.5 | 1.64 | 1.07 | 1.79 | 73 | 28.6 | 22.2 | 17.4 | 0.42 | 0.40 | 3.50 | 520.2 | — |
| **Concentrates-energy feeds** | | | | | | | | | | | | | |
| Barley, grain | 89.0 | 2.13 | 1.40 | 2.14 | 83 | 13.0 | 9.8 | 5.6 | 0.09 | 0.47 | 0.63 | — | — |
| Beet pulp, dried | 91.0 | 1.60 | 1.03 | 1.77 | 72 | 10.0 | 4.5 | 20.9 | 0.75 | 0.11 | 0.23 | — | 606.2 |
| Citrus pulp, dried WO fines | 90.0 | 1.97 | 1.32 | 1.94 | 77 | 7.3 | 3.9 | 14.4 | 2.18 | 0.13 | 0.69 | — | — |
| Corn, yellow dent #2 | 89.0 | 2.28 | 1.48 | 2.42 | 91 | 10.0 | 7.5 | 2.2 | 0.02 | 0.35 | — | 2.0 | 24.7 |
| Corn on cob meal | 87.0 | 2.04 | 1.36 | 2.21 | 85 | 9.3 | 4.6 | 9.2 | 0.05 | 0.31 | 0.61 | — | — |
| Hominy feed—min. 5% fat | 90.6 | 2.45 | 1.55 | 2.56 | 95 | 11.8 | 7.9 | 5.5 | 0.06 | 0.58 | 0.74 | 10.1 | — |
| Molasses, cane | 75.0 | 2.27 | 1.48 | 2.42 | 91 | 4.3 | 2.4 | — | 1.19 | 0.11 | 3.17 | — | — |
| Molasses, beet | 77.0 | 1.73 | 1.14 | 1.90 | 76 | 8.7 | 5.0 | — | 0.21 | 0.04 | 6.20 | — | — |
| Oats, grain | 89.0 | 1.73 | 1.14 | 1.90 | 76 | 13.2 | 9.9 | 12.4 | 0.11 | 0.39 | 0.42 | — | — |
| Sorghum, grain | 89.0 | 1.96 | 1.31 | 2.14 | 83 | 12.5 | 7.1 | 2.2 | 0.05 | 0.35 | 0.38 | — | — |
| Wheat, grain | 89.0 | 2.15 | 1.42 | 2.32 | 88 | 14.3 | 11.2 | 3.4 | 0.06 | 0.41 | 0.58 | — | — |

**Concentrates—protein supplements**

| Feed | | | | | | | | | | | | |
|---|---|---|---|---|---|---|---|---|---|---|---|---|
| Brewers dried grains | 92.0 | 1.42 | 0.83 | 1.55 | 66 | 28.1 | 16.3 | 0.29 | 0.54 | 0.09 | — | — |
| Corn distillers, dried grains | 92.0 | 1.99 | 1.33 | 2.18 | 84 | 29.5 | 23.1 | 13.0 | 0.10 | 0.40 | 0.10 | — | — |
| Corn distillers dried solubles | 93.0 | 2.15 | 1.42 | 2.32 | 88 | 28.9 | 22.6 | 4.3 | 0.38 | 1.47 | 1.87 | 0.8 | — |
| Corn gluten feed | 90.0 | 1.93 | 1.29 | 2.11 | 82 | 28.1 | 24.2 | 8.9 | 0.51 | 0.86 | 0.67 | — | — |
| Cottonseed meal, solvent, min. 41% P | 91.5 | 1.69 | 1.11 | 1.87 | 75 | 44.8 | 36.3 | 13.1 | 0.17 | 1.31 | 1.53 | — | — |
| Cottonseed meal, prepress solv., min. 50% P | 92.5 | 1.70 | 1.11 | 1.87 | 75 | 54.0 | 43.7 | 9.2 | 0.17 | 1.09 | 1.36 | — | — |
| Linseed meal—solvent | 91.0 | 1.90 | 1.27 | 1.91 | 81 | 38.6 | 34.1 | 9.9 | 0.48 | 0.98 | 1.36 | 0.2 | — |
| Peanut meal—solvent, min. 45% P | 92.0 | 1.76 | 1.16 | 1.94 | 77 | 51.5 | 46.4 | 14.1 | 0.22 | 0.71 | — | — | — |
| Soybean meal—mech. extracted | 90.0 | 2.06 | 1.37 | 2.21 | 85 | 48.7 | 41.4 | 6.7 | 0.30 | 0.70 | 1.90 | — | — |
| Soybean meal—solv., extracted | 89.0 | 1.93 | 1.29 | 2.07 | 81 | 51.5 | 43.8 | 6.7 | 0.36 | 0.75 | 2.21 | — | — |
| Wheat bran | 89.0 | 1.53 | 0.96 | 1.70 | 70 | 18.0 | 14.0 | 11.2 | 0.16 | 1.32 | 1.39 | — | — |
| Yeast, torula, dried | 93.0 | 1.86 | 1.24 | 2.05 | 80 | 51.9 | 47.2 | 2.2 | 0.61 | 1.81 | 2.02 | — | — |

**Calcium and phosphorus supplements**

| Feed | | | | | | | | | | | | |
|---|---|---|---|---|---|---|---|---|---|---|---|---|
| Bone meal, steamed | 95.0 | — | — | — | 16 | 12.7 | 8.6 | 2.1 | 30.51 | 32.0 | 14.31 | — | — |
| Defluorinated phosphate rock | 99.8 | — | — | — | — | — | — | — | 33.07 | 18.04 | 0.09 | — | — |
| Dicalcium phosphate | 96.0 | — | — | — | — | — | — | — | 23.13 | 18.65 | — | — | — |
| Limestone, ground—min. 33% Ca | 100.0 | — | — | — | — | — | — | — | 33.84 | 0.02 | — | — | — |
| Monosodium phosphate | 87.0 | — | — | — | — | — | — | — | — | 25.80 | — | — | — |
| Oyster shell flour—min. 33% Ca | 100.0 | — | — | — | — | 1.0 | — | — | 38.05 | 0.07 | 0.10 | — | — |

[a] From Table 4, *Nutrient Requirements of Dairy Cattle*, 4th rev. ed., 1971. Reproduced with the permission of the National Academy of Sciences.

Composition is based on a dry-matter basis. To convert to an-as fed basis, multiply the nutrient content by the DM percentage. Example: To determine the as-fed protein content of high-moisture corn (70% DM), multiply the DM protein content of corn, yellow dent #2 (10%) by 70% = 7% protein. The $NE_{lact.}$ of the high-moisture corn on an-as fed basis would be: (2.42 Mcal./kg. × .70) = 1.69 Mcal./kg.

**Appendix Table F**  Approximate Dry Matter Capacity of Tower Silos[a]

| Depth of settled silage (feet) | Silo diameter (feet) | | | | | | | | | | |
|---|---|---|---|---|---|---|---|---|---|---|---|
| | 10' | 12' | 14' | 16' | 18' | 20' | 22' | 24' | 26' | 28' | 30' |
| Tons of dry matter | | | | | | | | | | | |
| 20' | 8 | 12 | 16 | 21 | 27 | 33 | 40 | 47 | 56 | 65 | 74 |
| 22 | 9 | 14 | 19 | 24 | 30 | 38 | 48 | 54 | 64 | 74 | 85 |
| 24 | 11 | 15 | 21 | 27 | 34 | 43 | 52 | 61 | 72 | 83 | 96 |
| 26 | 12 | 17 | 23 | 30 | 38 | 48 | 58 | 68 | 81 | 94 | 107 |
| 28 | 13 | 19 | 26 | 35 | 44 | 53 | 64 | 76 | 90 | 104 | 119 |
| 30 | 15 | 21 | 29 | 38 | 47 | 59 | 71 | 84 | 99 | 115 | 132 |
| 32' | 16 | 23 | 32 | 41 | 52 | 65 | 78 | 93 | 109 | 127 | 145 |
| 34 | 18 | 25 | 34 | 45 | 57 | 70 | 85 | 101 | 119 | 137 | 158 |
| 36 | 19 | 28 | 37 | 48 | 62 | 76 | 92 | 109 | 129 | 150 | 172 |
| 38 | 21 | 30 | 41 | 53 | 67 | 82 | 100 | 118 | 139 | 161 | 185 |
| 40 | 22 | 32 | 44 | 57 | 72 | 89 | 107 | 127 | 150 | 173 | 199 |
| 42' | | 34 | 47 | 61 | 77 | 95 | 115 | 137 | 161 | 186 | 214 |
| 44 | | 37 | 50 | 65 | 82 | 102 | 123 | 146 | 172 | 200 | 229 |
| 46 | | 39 | 53 | 69 | 88 | 108 | 131 | 155 | 183 | 212 | 244 |
| 48 | | 42 | 56 | 74 | 93 | 115 | 140 | 166 | 195 | 226 | 260 |
| 50 | | 44 | 60 | 78 | 99 | 122 | 148 | 175 | 206 | 239 | 274 |
| 52' | | | 64 | 83 | 105 | 129 | 157 | 186 | 219 | 254 | 291 |
| 54 | | | 67 | 88 | 111 | 137 | 165 | 197 | 231 | 267 | 306 |
| 56 | | | 71 | 93 | 117 | 144 | 174 | 207 | 243 | 282 | 324 |
| 58 | | | 74 | 98 | 123 | 151 | 183 | 218 | 261 | 297 | 339 |
| 60 | | | 78 | 102 | 129 | 159 | 192 | 228 | 273 | 309 | 357 |
| 62' | | | | | 135 | 167 | 201 | 239 | 287 | 324 | 374 |
| 64 | | | | | 142 | 174 | 210 | 250 | 301 | 339 | 391 |
| 66 | | | | | 149 | 182 | 219 | 260 | 314 | 354 | 407 |
| 68 | | | | | 155 | 190 | 228 | 271 | 328 | 369 | 424 |
| 70 | | | | | 162 | 198 | 237 | 282 | 342 | 384 | 441 |
| 72' | | | | | | | | 293 | 356 | 400 | 458 |
| 74 | | | | | | | | 305 | 371 | 415 | 476 |
| 76 | | | | | | | | 316 | 385 | 431 | 493 |
| 78 | | | | | | | | 328 | 400 | 446 | 511 |
| 80 | | | | | | | | 339 | 414 | 462 | 528 |

[a] Reprinted with permission from *Dairy Housing and Equipment Handbook,* Midwest Plan Service, Ames, Ia 50010, 1971 and 1976.

To estimate tons of silage of various moisture contents multiply tons of dry matter

$$\times \frac{100}{\text{estimated \% DM in silage}} = \text{tons actual silage.}$$

**Appendix Table F-1** Estimating Silage Capacity of Horizontal Silos: Horizontal Silo Dimensions, Cross-sectional Area of Horizontal Silo, and Weight of Silage in 4-inch Slice and Per Lineal Foot (as-fed basis)[a]

| Side slope per feet of depth (in.) | Depth ft. | Bottom width ft. | Top width Ft. | Top width In. | Cross-sectional area sq. ft. | Weight of silage 4-in. slice lb. | Weight of silage 1-ft. slice lb. |
|---|---|---|---|---|---|---|---|
| | 4 | 6 | 7 | 0 | 26.0 | 346 | 1,040 |
| | 4 | 10 | 11 | 0 | 42.0 | 560 | 1,680 |
| | 6 | 8 | 9 | 6 | 52.5 | 700 | 2,100 |
| 1½ | 6 | 12 | 13 | 6 | 76.5 | 1,015 | 3,060 |
| | 8 | 10 | 12 | 0 | 88.0 | 1,173 | 3,520 |
| | 8 | 14 | 16 | 0 | 120.0 | 1,600 | 4,800 |
| | 10 | 12 | 14 | 6 | 132.5 | 1,767 | 5,300 |
| | 10 | 16 | 18 | 6 | 172.5 | 2,300 | 6,900 |
| | 4 | 6 | 8 | 0 | 28.0 | 373 | 1,120 |
| | 4 | 10 | 12 | 0 | 44.0 | 587 | 1,760 |
| | 6 | 8 | 11 | 0 | 57.0 | 760 | 2,280 |
| 3 | 6 | 12 | 15 | 0 | 81.0 | 1,080 | 3,240 |
| | 8 | 10 | 14 | 0 | 96.0 | 1,280 | 3,840 |
| | 8 | 14 | 18 | 0 | 128.0 | 1,767 | 5,120 |
| | 10 | 12 | 17 | 0 | 145.0 | 1,933 | 5,800 |
| | 10 | 16 | 21 | 0 | 185.0 | 2,467 | 7,400 |
| | 4 | 6 | 8 | 8 | 29.0 | 387 | 1,160 |
| | 4 | 10 | 12 | 8 | 45.2 | 603 | 1,808 |
| | 6 | 7 | 11 | 0 | 54.0 | 720 | 2,160 |
| 4 | 6 | 12 | 16 | 0 | 84.0 | 1,120 | 3,360 |
| | 8 | 10 | 15 | 4 | 102.7 | 1,369 | 4,108 |
| | 8 | 14 | 19 | 4 | 134.7 | 1,796 | 5,388 |
| | 10 | 12 | 18 | 8 | 153.3 | 2,044 | 6,132 |
| | 10 | 16 | 22 | 8 | 193.3 | 2,577 | 7,732 |

[a] For silage weighing 40 lb./cu. ft. The rate of feeding is assumed to be 4 in. lengthwise of the silo per day. From *Farm Silos* by J. R. McCalmont, 1960. USDA Miscellaneous Publication No. 810, 1960.

Calculate the cubic feet capacity of the structure by multiplying the average depth of settled silage × the average width of the silo × the length of the silo. Multiply this figure by 12 lb. (weight of average silage DM/cu. ft.) and divide by 2,000 to get the DM capacity in tons.

Example: Silage depth = 10 ft., silage length = 110 ft., silo width = 28 ft. at the bottom and 32 ft. at the top (avg. 30 ft.)

10 × 110 × 30 = 33,000 cu. ft. × 12 lb./cu. ft. = 396,000 lb. ÷ 2,000 = 198T. of

silage DM capacity. If the silage contained 35% DM 198 ÷ .35 = 566 T. of wet silage capacity.

**Appendix Table F-2**    Space Required for Storing Various Feeds[a]

| Material | Weight per cubic foot in pounds | Cubic feet per ton |
|---|---|---|
| Hay, loose in shallow mows | 4 | 512 |
| Hay, loose in deep mows | 5 | 400 |
| Hay, baled loose | 10 | 200 |
| Hay, baled tight | 25 | 80 |
| Hay, chopped long cut | 8 | 250 |
| Hay, chopped short cut | 12 | 167 |
| Straw, loose | 4 | 512 |
| Straw, baled | 12 | 167 |
| Silage, shallow | 30 | 67 |
| Silage, deep | 50 | 40 |
| Barley, 48 lb./bu. | 39 | 51 |
| Corn, ear, 70 lb./bu. (legal wt.) | 28 | 72 |
| Corn, shelled, 56 lb./bu. | 45 | 44 |
| Corn, cracked or corn meal, 50 lb./bu. | 40 | 50 |
| Corn-and-cob meal, 45 lb./bu. | 36 | 56 |
| Oats, 32 lb./bu. | 26 | 77 |
| Oats, ground, 22 lb./bu. | 18 | 111 |
| Oats, middlings, 48 lb./bu. | 39 | 51 |
| Rye, 56 lb./bu. | 45 | 44 |
| Wheat, 60 lb./bu. | 48 | 42 |
| Soybeans, 62 lb./bu. | 50 | 40 |
| Most concentrates | 45 | 44 |

[a] Adapted from *Hoard's Dairyman*, 1951.


**Appendix Table G**    Milking Machine Preventive Maintenance Guide (O: to be performed by operator; S: to be performed by serviceman)[a]

| | System | | |
|---|---|---|---|
| | Bucket | Pipeline | Parlor |
| **A. Each milking:** | | | |
|   1. Check all rubber goods for breaks, tears, or water in shells, replace when necessary. | O | O | O |
|   2. Make sure pulsators are operating (check operation of inflation with thumb or finger) air bleeder vents open, clean. | O | O | O |
|   3. Check vacuum level. | O | O | O |
|   4. Empty trap, check belt tension and oil supply in vacuum pump. | O | O | O |
|   5. Install clean filters in milk-filtering equipment. | O | O | O |

| | System | | |
|---|---|---|---|
| | Bucket | Pipeline | Parlor |
| 6. Make sure milk releaser or pumping equipment is operating and that line joints are tight. | | O | O |
| 7. Make sure cleaned-in-place (CIP) equipment is properly detached or capped off. | | O | O |
| 8. Prepare udder wash and sanitizers at correct concentration. | O | O | O |
| 9. Check receiver jar probes for cleanliness. | | O | O |
| 10. Check slot in sensing chamber of automatic take-offs for foreign material. | | | O |
| B. 50-hr. service (approx. 1 week): | | | |
| 1. Inspect and change inflations. | O | O | O |
| 2. Check pulsator filters and clean or replace. | O | O | O |
| 3. Check stall cock for leaks and electrical connections for tightness. | O | O | O |
| 4. Clean vacuum controller and moisture drain valves. | O | O | O |
| 5. Clean and brush all air tubes. | O | O | O |
| 6. Fill oil reservoir on vacuum pump and check tension on vacuum pump belts. | O | O | O |
| 7. Disassemble and check probes in sensing chamber of automatic take-offs for cleanliness, check probe wire connection for tightness. | | | O |
| C. 250-hr. service (approx. 6 weeks), in addition to 50-hr. check points: | | | |
| 1. Milking units: | | | |
| a. Check condition of milker unit and cleanliness. | O | O | O |
| b. Change inflations. | O | O | O |
| c. Check condition of air tubes, vacuum hoses, and milk hoses. | O | O | O |
| d. Check for air leaks. | | O | O |
| 2. Pulsation system: | | | |
| a. Inspect operation and condition of pulsator control. | O | O | O |
| b. Clean air ports and screens of pulsators. | O | O | O |
| 3. Vacuum lines: | | | |
| a. Check vacuum gauge setting against teat gauge. | O | O | O |
| b. Lubricate stall cocks and check for leaks. | O | O | O |
| c. Vacuum line checked for leaks and drainage. | O | O | O |
| 4. Vacuum regulators: | | | |
| a. Disassemble and clean. | O | O | O |
| b. Replace air filters. | O | O | O |
| c. Calibrate vacuum level for pulsator and milk lines. | S | S | S |
| 5. Vacuum pumps: | | | |
| a. Inspect condition of oil. | O | O | O |
| b. Adjust belt tension, check pulley alignment. | O | O | O |
| c. Change air filter in vacuum tank. | O | O | O |
| d. Vacuum pump maintenance as recommended by operator's manual. | O | O | O |
| e. Check air flow of pump and lines. | S | S | S |
| 6. Check operation of automatic washing equipment and inlet screens. | | O | O |

607

| | System | | |
|---|---|---|---|
| | Bucket | Pipeline | Parlor |

7.  Milker stalls and prep stalls.:

    a.  Inspect and service stalls as necessary.                                    O

    b.  Lubricate and check hangers for tightness.                          O

8.  Milker supports:

    a.  Lubricate slide bar and yoke assembly.                              O

    b.  Check alignment for smooth operation.                             O

9.  Crowd gate:

    a.  Cycle gate to observe proper movement and operation.                O

    b.  Lubricate gear bearings and steel support cable.            O

D.  500-hr. service (approx. 3 mo.), in addition to 250-hr. check points:

  1.  Sanitation equipment:

| Item | Bucket | Pipeline | Parlor |
|---|---|---|---|
| D.1.a. Check automatic cycling for cleanliness, operation, flow rate and water temperature. | | S | S |
| D.1.b. Check in-parlor mainfold for proper operation and condition of cover. | | S | S |
| D.1.c. Check diverter operation. | | S | S |
| D.1.d. Check tank drains for proper operation and seal. | | S | S |
| 2. Cooling system: | | | |
| a. Check starting of condensing unit. | S | S | S |
| b. Inspect milk tank interior for cleanliness. | S | S | S |
| c. Check system for refrigerant leaks and charge. | S | S | S |
| d. Install test-gauge manifold to check pressures. | S | S | S |
| e. Check and adjust fan controls as required. | S | S | S |
| f. Check low-pressure control and adjust for proper pump down. | S | S | S |
| g. Check all automatic washer connections, pump seal, and water inlet screens. | S | S | S |
| h. Inspect sprayball and clean all openings. | S | S | S |
| 3. Automatic take-off milkers: | | | |
| a. Test power supply. | | | S |
| b. Set sensing point. | | | S |
| c. Check 5-sec. delay of each sensor card. | | | S |
| d. Check 70-sec. delay of each lodgic card. | | | S |
| e. Check operation of each cylinder. | | | S |
| f. Lubricate air cylinder and insulate electrical quick-connects in base. Clean contacts. | | | S |
| E. 1,250-hr. service (approx. 7–8 mo.), in addition to 250-hr. check points: | | | |
| 1. Check all gaskets, flappers, "O" rings, and caps which come in contact with milk. | S | S | S |
| 2. Pulsation system: | | | |
| a. Clean all electric pulsator selectors and activators. Check solenoids and coils. Clean plungers. | S | S | S |
| b. Rebuild pneumatic pulsators. | S | S | S |
| 3. Clean vacuum lines. | S | S | S |
| 4. Check electrical equipment for loose or frayed wires. Lubricate all bearings. | S | S | S |
| 5. Vacuum pumps: | | | |
| a. Change oil in reclaimer and clean. | S | S | S |
| b. Replace oil filter in reclaimer. | S | S | S |

| | System | | |
|---|---|---|---|
| | **Bucket** | **Pipeline** | **Parlor** |
| 6. Vacuum regulators—change cushion control fluid. | S | S | S |
| 7. Cooling system: | | | |
| a. Check crankcase heater. | S | S | S |
| b. Check thermometer and temperature control calibration. | S | S | S |
| 8. Check condition of milk valves and repair. | | S | S |
| 9. Check milk pump for leaks and timer operation. | | S | S |
| 10. Milker stalls and prep-stalls: | | | |
| a. Service and clean spray nozzles. | | | S |
| b. Repair or replace bent or worn parts. | | | S |
| 11. Milker supports: | | | |
| a. Replace torsion spring. | | | S |
| b. Check adjustments of locking assembly. | | | S |
| 12. Crowd gates: | | | |
| a. Check condition of power cable. | | | S |
| b. Check condition and alignment of wheels in track. | | | S |
| c. Check oil level in gear box. | | | S |
| d. Check drive couplings for wear. | | | S |
| F.  2,500-hr. service: | | | |
| 1. Complete 250-, 500-, and 1,250-hr. service check. | S | S | S |
| 2. Change all solenoid coils, plungers, hoses, diaphragms, caps gaskets, flappers, and rubber vacuum connectors. | S | S | S |
| 3. Check electric timer controls, switches, motors, and parts. Grease all bearings. | S | S | S |
| 4. Check milk pump seal, rubber spring, and clearances. Change gaskets. | S | S | S |
| 5. Check milking units for vacuum fluctuations. | S | S | S |
| 6. Clean vacuum lines on pulsator line and between receiver and vacuum pump. | | S | S |
| 7. Replace brass valve and sleeve in vacuum regulators. | S | S | S |
| 8. Replace motor drive belts on vacuum pump. | O | O | O |
| 9. Lubricate piston seal of secondary drain in tank. | S | S | S |
| 10. Automatic take-offs—disassemble air cylinder, replace springs, filter, seals, and lubricate. | | | S |
| 11. Crowd gates—readjust lift cables on power-lift models. | | | S |
| G.  5,000-hr. service: | | | |
| 1. Complete 250-, 500-, 1,250-, and 2,500-hr. service check | S | S | S |
| 2. Recondition all motors, pumps, selectors, timers, starters, milker supports, stalls, milker units, automatic washing equipment, milk pumps, power or control panels, solenoids, switches, bulk tank washers, sprayers, etc. | S | S | S |
| 3. Replace all rubber coils, hoses, gaskets, "O" rings, springs, and plungers. Clean entire pipeline milking system. | S | S | S |
| 4. Check hangers and slope of vacuum and milk lines. | S | S | S |

**609**

**Milking Machine Checks and Disorders**

| Problem | Possible causes | Correction |
|---|---|---|
| Inadequate vacuum or vacuum fluctuation | Pump too small | Replace (reserve should be 50% of needed capacity). |
| | Plugged line | Clean with lye once a month. |
| | Sticky vacuum regulator | Clean and oil, adjust if necessary. |
| | Worn belt | Replace. |
| | Worn pump | Repair or replace. |
| | Air leaks | Tighten couplings, repair or replace gaskets or stallcocks. |
| | Broken vacuum line | Replace. |
| | Vacuum or milk lines too small | Replace. |
| | Pump running too slow | Have serviceman check voltage and pulley sizes. |
| | Too many milking units | Eliminate units or get larger system. |
| High vacuum | Sticky or dirty vacuum regulator | Clean and oil regularly. |
| | Too much weight on regulator | Remove and adjust weights. |
| Pulsators too fast or too slow | Poor adjustment | Correct to manufacturer's recommended rate. |
| | High or low vacuum | Adjust to manufacturer's recommended vacuum. |
| | Dirty pulsators | Clean at least once a month. |
| | Worn pulsators | Repair. |
| Teat cup inflations | Ballooned, cracked or blistered | Replace entire set of four. |
| | Rough | Clean thoroughly. |
| Teat cups drop off | Pipeline couplings may leak, valves not properly seated, or poor air tube or hose | Tighten couplings or replace gaskets, valves or hose. |
| | Plugged air-bleeder vents | Clean daily. |
| | Inadequate vacuum | See above. |
| | Too many milking units | |
| | Milk line too high | See above. |
| Slow milking | Inflations in poor condition | Replace. |
| | Air tubes contain holes | Replace. |
| | Plugged air-bleeder vent | Clean daily. |
| | Inadequate vacuum | See above. |
| | Poor pulsation | Clean and service pulsator. |
| Pipeline not washing | Water may not be hot enough | Check hot water heater and water pressure. |
| | Washing detergent solution too weak | Add more detergent to washing cycle, use recommended amount. |
| | Wrong type of cleaner | Use total program with one manufacturer's products and use them as recommended. |

[a] Jones, G. M., *Managed Milking Guidelines,* V.P.I. & S.U. Extension Publication 633, 1975.

## APPENDIX H

### EXAMPLE, DAIRY FARM LABOR CONTRACT

"The following contract between the employer _____ _____ and the employee _____ is mutually agreed upon:

That the employee shall receive a base salary of _____ per month paid in equal amounts on the 16th day and the last day of each month, and

That a raise of _____ per month will be given after the first 6 months of work and a minimum annual raise of _____ per month will be given thereafter, and

That a modern house in good condition known as the farm tenant house shall be provided for the use of the employee. Major repairs and improvements (roofing, plumbing, outside painting, etc.) shall be provided by the employer while minor repairs and improvements (inside painting, etc.) shall be provided by the employee. The employee shall be responsible for maintaining the general condition of the house and grounds, and

That the employee shall have use of the garden at the tenant house, and

That the employee shall receive compensation for the electric bill up to a maximum of _____ per month, and

That the employee shall be given one dairy bull calf annually and receive space and feed to raise that animal for beef, and

That the starting time will be 6:00 a.m. each day, the quitting time will be after the evening milking at approximately 6:30 p.m. each day and that one hour will be allowed for breakfast and two hours for lunch unless mutually agreed on by both parties, and

That extra hours worked by the employee during busy seasons (silo filling, planting, etc.) shall be compensated for by equivalent time off during slack seasons, and

That every other weekend starting from after the Friday p.m. milking and extending until 6:00 a.m. Monday will be given as free time off, and

That a seven-day period the first year, a 10-day period the second year, and a 14-day period annually thereafter will be given as a paid vacation. This may be taken as individual days off or all at once during the period of June through August or during another time of the year if agreed on by both parties, and

That holidays are considered as time off except for morning and evening milking, feeding and cleaning chores, and when falling on a working weekend, and

That six days per year will be allowed as paid sick leave; however, if two or less of these are used the employee will receive a bonus of _____ on Decenber 20 of each year, and

That the employee shall be expected to milk, feed, clean barns, do general farm work, and other jobs assigned by the employer to the best of his ability, and

That either the employee or the employer may initiate renegotiation of this contract annually hereafter or may terminate this contract by giving 30 days written notice to the other.

These above statements are considered binding by employer and employee, and are therefore set down as a legal document."

Signed _____        Signed _____
　　　　　　Employer　　　　　　　　　　　　　　　　Employee
Date _____        Date _____

## EXAMPLE, DAIRY ENTERPRISE
## PARTNERSHIP CONTRACT

THIS PARTNERSHIP CONTRACT, made and entered into this _____ day of _____, in the year _____, by and between John Doe, party of the first part, and Richard Doe, party of the second part.

### :WITNESSETH:

THAT FOR AND IN CONSIDERATION of the sum hereinafter mentioned, the parties hereto do by this Agreement enter into a Partnership for the purpose of owning and operating a dairy and general farming enterprise:

(1)  The party of the second part convenants and agrees to pay to the party of the first part the sum of _____ Dollars, at the rate of _____ Dollars per year, plus interest of _____ per centum on the unpaid balance, payments to be made on the 1st day of each year hereafter until principal and interest are paid in full; and the party of the first part does hereby sell, assign and transfer unto the party of the second part, a one-half (½) undivided interest in and to all milk base, cattle, farm machinery, and equipment owned by him. Also, the party of the first part agrees to rent to the partnership the farm and farm buildings now owned by him for the annual sum of _____.

(2)  The parties hereto to be equal partners, sharing equally in all profits and losses; an accounting to be made by the parties hereto on the first day of each year during the duration of this Agreement.

(3)  John Doe to be the Managing Partner.

(4)  The parties hereto to set and establish a monthly drawing account for themselves; the amounts to be agreed upon.

(5)  In the event either of the parties hereto desires to sell or otherwise dispose of his interest in the assets of the partnership, then the partner who desires to sell, or otherwise dispose of his interest in the partnership, shall give the other partner first refusal to purchase same. Also, in the event of the death of either partner, during the existence of this partnership, he does hereby bind his heirs, executor, or administrator to give the said surviving partner the first option and refusal to buy the deceased partner's interest in the partnership assets.

613

(6) It is further agreed by and between the parties hereto that if any farm building or buildings are built on the farm now owned by the party of the first part or to be owned by either of the parties hereto, that the cost of said building or buildings is to be paid for jointly be the partners hereto. And it is further agreed that in the event of the sale or transfer of any land on which the parties hereto as partners have constructed any buildings at the time of such sale or transfer, each partner would be entitled to and would be reimbursed for one-half (½) of the undepreciated value of the said farm building.

(7) In the event either party becomes physically or mentally disabled to the extent he/she cannot or is unable to perform the duties and obligations under this Agreement, then the surviving partner shall be the Managing Partner and shall do what he may deem necessary to carry on the partnership business; and the incapacitated partner shall continue to be able to receive a monthly payment for a period of not to exceed five (5) years from date of this Contract.

(8) Both parties hereto convenant and agree they will devote full time to the partnership business and will do what is necessary to promote the interest of both parties of this Agreement.

(9) Both parties convenant and agree they will pay their individual debts and obligations as they become due and will not suffer or permit any of the assets, equipment, or property of the partnership to be encumbered, pledged, levied upon, or sold for any individual indebtedness or other obligation for which either party hereto may be liable or responsible, either now or at anytime during the existence of this partnership.

WITNESS the following signatures and seals.

_____(SEAL)
John Doe

_____(SEAL)
Richard Doe

STATE OF VIRGINIA )
                  )   To-wit:
COUNTY OF _____ )

     I, _____, a Notary Public of and for the County and State aforesaid, do hereby certify that John Doe and Richard Doe, whose names are signed to the foregoing Partnership Contract, bearing date on the _____ day of _____, 19\_\_\_\_, have this day personally appeared before me in my County and State aforesaid and acknowledged the same.

     Given under my hand this _____ day of _____, 19\_\_\_\_.

     My commission expires _____.

                             ―――――――――――
                              Notary Public

**Appendix Table J** Gestation Table for Dairy Cattle

| Service date | Date due to calve | | |
|---|---|---|---|
| | **Ayr., Hol., Jer.** (278 days) | **Guernsey** (283 days) | **Brown Swiss** (288 days) |
| Jan. 1 | Oct. 6 | Oct. 11 | Oct. 16 |
| Jan. 15 | Oct. 20 | Oct. 25 | Oct. 30 |
| Feb. 1 | Nov. 6 | Nov. 11 | Nov. 16 |
| Feb. 15 | Nov. 20 | Nov. 25 | Nov. 30 |
| Mar. 1 | Dec. 4 | Dec. 9 | Dec. 14 |
| Mar. 15 | Dec. 18 | Dec. 23 | Dec. 28 |
| Apr. 1 | Jan. 4 | Jan. 9 | Jan. 14 |
| Apr. 15 | Jan. 18 | Jan. 23 | Jan. 28 |
| May 1 | Feb. 3 | Feb. 8 | Feb. 13 |
| May 15 | Feb. 17 | Feb. 22 | Feb. 27 |
| June 1 | Mar. 6 | Mar. 11 | Mar. 16 |
| June 15 | Mar. 20 | Mar. 25 | Mar. 30 |
| July 1 | Apr. 5 | Apr. 10 | Apr. 15 |
| July 15 | Apr. 19 | Apr. 24 | Apr. 29 |
| Aug. 1 | May 6 | May 11 | May 16 |
| Aug. 15 | May 20 | May 25 | May 30 |
| Sept. 1 | June 6 | June 11 | June 16 |
| Sept. 15 | June 20 | June 25 | June 30 |
| Oct. 1 | July 6 | July 11 | July 16 |
| Oct. 15 | July 20 | July 25 | July 30 |
| Nov. 1 | Aug. 6 | Aug. 11 | Aug. 16 |
| Nov. 15 | Aug. 20 | Aug. 25 | Aug. 30 |
| Dec. 1 | Sept. 5 | Sept. 10 | Sept. 15 |
| Dec. 15 | Sept. 19 | Sept. 24 | Sept. 29 |

## APPENDIX K

**PDCA SHOW RING CODE OF
ETHICS(Adopted by PDCA in March 1970;
Revised March 1971)**

The showing of registered dairy cattle is an important part of the promotion, merchandising and breeding program of many breeders. Additionally, it is an important part of the program of The Purebred Dairy Cattle Association to stimulate and sustain interest in breeding registered dairy cattle. This relates to both spectators and exhibitors. In this connection, the PDCA believes that it is in the best interests of the breeders of registered dairy cattle to maintain a reputation of integrity and to present a wholesome and progressive image of their cattle in the show ring. It recognizes that there are certain practices in the proper care and management of dairy cattle, which are necessary in the course of moving dairy cattle to and between shows that are advisable to keep them in a sound, healthy condition so that they might be presented in the show ring in a natural, normal appearance and condition. Conversely, it recognizes certain practices in the cataloging, handling and presentation of cattle in the show ring which are unacceptable.

The following practices or procedures are considered unacceptable and defined as being unethical in the showing of registered dairy cattle:

1. Misrepresenting the age and/or milking status of the animal for the class in which it is shown.

2. Balancing the udder by any means other than by leaving naturally produced milk in any or all quarters.

3. Setting the teats with a mechanical contrivance or with the use of a chemical preparation.

4. Treating or massaging any part of the animal's body, particularly the udder, internally or externally with an irritant, counter-irritant, or other substance to temporarily improve conformation or produce unnatural animation.

5. Minimizing the effects of crampiness by feeding or injecting drugs, depressants or applying packs or using any artificial contrivance or therapeutic treatment excepting normal exercise.

6. Blocking the nerves to the foot to prevent limping by injecting drugs.

7. Striking the animal to cause swelling in a depressed area.

617

8. Surgery of any kind performed to change the natural contour or appearance of the animal's body, hide or hair. Not included is the removal of warts, teats and horns, clipping and dressing of hair and trimming of hooves.

9. Insertion of foreign material under the skin.

10. Changing the color of hair at any point, spot or area on the animal's body.

11. The use of alcoholic beverages in the feed or administered as a drench.

12. Administration of a drug of any kind or description internally or externally prior to entering the show ring, except for treating a recognized disease or injury and for tranquilizing bulls that may otherwise be dangerous or females in heat. For the purpose of this Code, the term "drug" shall mean any substance, the sale, possession or use of which is controlled by license under federal state or local laws of regulations and any substance commonly used by the medical or veterinary professions which affect the circulatory, respiratory or central nervous system of a cow.

13. Criticizing or interfereing with the judge, show management or other exhibitors while in the show ring or other conduct detrimental to the breed or show.

In keeping with the basic philosophy of The Purebred Dairy Cattle Association, ethics are an individual responsibility of the owner of each animal shown. Violations of this code are subject to the disciplinary provisions of the appropriate dairy breed association.

**APPENDIX TABLE L**

**PDCA CODE OF ETHICAL SALES
PRACTICES (Excerpts from the PDCA code,
Revised February 1969)**

The Purebred Dairy Cattle Association recommends that the following sales practices and procedures be approved by the five dairy cattle registry associations, and a condensed version be published in the sale catalogs of dairy cattle auction sales. This Code sets forth the terms and conditions of sales, the responsibilities of sellers, sale managers and buyers, defines misrepresentations and unethical practices, recommends a uniform method of reporting sales averages and provides that disciplinary action be taken by breed registry associations against violators.

   I.  *Definitions of Auction Sales*

       In the selling of registered dairy cattle, a true statement of facts shall be made relative to pertinent information available on each animal to be sold. Advertisements and sale catalogs shall accurately and completely describe the nature of the sale, such as:

       (a)  *Dispersal Sale:*  Indicates a complete sell-out of all sound and salable cattle owned by the seller.

       (b)  *Partial Dispersal Sale, Reduction Sale or Production Sale:*  Indicates that only a part of the herd will be offered for sale.

       (c)  *Consignment Sale:*  Indicates that animals from two or more breeders will be offered for sale. The sponsorship of a consignment sale shall be clearly set forth in the advertising material and in the catalog, and the name and address of each consigning seller shall be accurately printed in the catalog.

  II.  *Terms and Conditions of Sale*

       Every animal cataloged for a public auction sale is pledged to absolute sale if there be two or more bidders, except that an animal which has become sick or unfit to offer for sale may be withdrawn by the seller or sale management. Guarantees, if any, shall be set forth in the catalog and if there are none, that fact shall be also stated.

       *Responsibility:*  The seller subscribes to this Code and assumes full responsibility for his cattle. Except as herein provided, the seller shall not be responsible for any damages beyond the selling price of the animal. The sales management acts as agent of the seller and shall only to the extent provided by law, share the seller's responsibility, but it shall endeavor to protect the interests of the buyer as well.

       *Errors:*  If errors are noted in the catalog or advertising, announcement of such errors shall be made from the auction stand just prior to the sale of the animal

**619**

and such announcements shall take precedence over the printed matter in the catalog or advertising.

*Health:* A statement as to the condition of the health of a herd from which animals are to sell shall be given in the catalog and announced from the auction stand prior to the sale.

*Registry Certificates:* Each animal which is cataloged, offered and sold as registered shall be recorded in the herd book of the breed registry association. Any animal for which a certificate of registry is not available or which has not been issued by the breed registry association must be represented as an unregistered animal at the time the animal is offered for sale.

*Transfer:* The sales management shall be responsible that every animal be transferred within 30 days to the purchaser at the seller's expense, such transfer to be recorded on the registry certificate by the breed registry association. An animal may not be transferred back to the seller within one year except at the discretion of the breed registry concerned. The transfer of ownership by the seller of animals advertised for the sale subsequent to the announcement of a sale, but prior to the sale, may subject the seller and/or the sales management to charges of misrepresentation.

*Guarantees:* Each seller warrants clear title to the animal and the right to sell the animal. Unless otherwise noted in the catalog, or announced from the auction stand prior to the animal being struck off, each animal is offered for sale as sound. If any animal is not found to be sound as represented, the buyer shall report this fact within one hour of the close of the sale to the seller or the sales management. In sales other than dispersal sales, animals are guaranteed to be breeders in accordance with the following:

(a) Bulls are sold as being able to serve and settle after reaching 14 months of age. The buyer is expected to provide reasonable care and feeding. Should the bull prove to be a non-breeder, the buyer shall notify the seller within four months after the bull reaches 14 months of age. The seller shall then have six months to prove the bull to be a breeder. Older bulls are warranted to be breeders unless otherwise announced from the auction stand. If semen is retained when a bull is sold, the amount and circumstances shall be stated in the catalog and announced from the auction stand.

(b) Females pregnant when sold or which have freshened normally within 60 days prior to the date of the sale are by that fact considered breeders. Should any other female not pregnant when sold fail to become pregnant within six months of sale date, or, if less than 15 months old when sold by the time she reaches the age of 21 months, after having been bred to a bull known to be a breeder and after having been treated by a licensed veterinarian, the matter shall be reported in writing to the seller who shall then have the privilege of six months time in which to prove the animal a breeder before refunding the purchase price.

*Bids:* Buyers not present may send their bids or instructions to buy, with necessary funds, to a person known to be in attendance at the sale who will use such information and funds as if he were buying for himself. The auctioneer shall accept bids from prospective buyers and the person or company of persons offering the highest bid shall be the buyer. In case of dispute between bidders, the animal shall again be put up for advance bids and if there be no advance bid, the animal shall be sold to the person from whom the auctioneer accepted the final bid.

*By-bidding is Prohibited:* By-bidding is the bidding, or the contrivance to bid, on an animal by the owner of that animal, or his agent.

*Risk:* Animals are held at purchaser's risk and expense as soon as they have been struck off to the purchaser. However, care and feed will be furnished by the seller or sales management for 24 hours free of charge.

*Sale Terms:* The terms of payment for animals purchased shall be cash unless other arrangements have been previously made with the seller and sales management. Fictitious selling, or any understanding or arrangement between the seller and any buyer before the auction of an animal which provides for the return or re-sale within one year of the sale shall be presumptive evidence of fictitious selling.

III. *Penalties*

Violation of this code shall be considered as an unethical sales practice and shall subject the violator to such penalties as may be imposed by the applicable breed registry association.

IV. *Publication of Sales Prices*

In order to establish uniformity in the methods of reporting prices received for registered dairy cattle sold at public auction, the following shall apply:

(a) The sale price of an animal shall be considered to be the amount of money for which the BUYER is legally obligated and required to pay the SELLER, exclusive of any other conditions of sale.

(b) Record breaking and unusually high sale prices shall be adequately and completely qualified in all published reports.

(c) If an "average sale price" is published, it shall be computed by dividing the gross bid sale price by the number of animals sold. But calves under three months of age may be listed as "half lots" and may be counted (though sold separately from their dams) as though part of the dam's selling price, and as though dam and calf were one animal.

V. *Registry Breed Association Statement*

Each registry breed association shall prepare for publication and distribution in substantial conformity with this Code such shorter statement of the provisions thereof as may be considered vital for assuring the integrity of public sales and fairly protecting the interests of both buyers and sellers there at, but shall not be limited to these provisions.

**Appendix Table M** Age-Correction Factors for Cows Milked in Region I (The Northeastern States—States Included Vary with Breed—Milking Shorthorns Nationwide) (ARS-NE40-1974)

| | Ayrshire | | | | Brown Swiss | | | | Guernsey | | | |
|---|---|---|---|---|---|---|---|---|---|---|---|---|
| | January[a] | | July | | January | | July | | January | | July | |
| Age (Mo.) | Milk | Fat | Milk | Fat | Milk | Fat | Milk | Fat | Milk | Fat | Milk | Fat |
| 22 | 1.25 | 1.22 | 1.36 | 1.33 | 1.47 | 1.43 | 1.66 | 1.62 | 1.22 | 1.22 | 1.34 | 1.32 |
| 24 | 1.21 | 1.18 | 1.31 | 1.28 | 1.42 | 1.38 | 1.60 | 1.56 | 1.19 | 1.18 | 1.30 | 1.28 |
| 28 | 1.16 | 1.13 | 1.25 | 1.22 | 1.34 | 1.29 | 1.50 | 1.46 | 1.13 | 1.12 | 1.24 | 1.21 |
| 32 | 1.13 | 1.10 | 1.21 | 1.18 | 1.27 | 1.23 | 1.42 | 1.38 | 1.10 | 1.08 | 1.19 | 1.16 |
| 36 | 1.10 | 1.08 | 1.19 | 1.17 | 1.20 | 1.17 | 1.34 | 1.31 | 1.07 | 1.06 | 1.17 | 1.14 |
| 42 | 1.05 | 1.04 | 1.16 | 1.15 | 1.11 | 1.09 | 1.26 | 1.23 | 1.04 | 1.02 | 1.14 | 1.11 |
| 48 | 1.02 | 1.01 | 1.14 | 1.13 | 1.06 | 1.04 | 1.20 | 1.18 | 1.00 | 0.99 | 1.12 | 1.10 |
| 60 | 0.98 | 0.97 | 1.09 | 1.08 | 1.00 | 0.98 | 1.13 | 1.11 | 0.97 | 0.96 | 1.08 | 1.07 |
| 72 | 0.97 | 0.96 | 1.08 | 1.08 | 0.97 | 0.96 | 1.08 | 1.09 | 0.96 | 0.96 | 1.08 | 1.08 |
| 90 | 0.96 | 0.96 | 1.07 | 1.08 | 0.95 | 0.95 | 1.08 | 1.08 | 0.95 | 0.97 | 1.08 | 1.10 |
| 108 | 0.97 | 0.99 | 1.09 | 1.11 | 0.95 | 0.95 | 1.08 | 1.08 | 0.96 | 0.99 | 1.10 | 1.12 |

| | Jersey | | | | Holstein | | | | Milking Shorthorn | | | |
|---|---|---|---|---|---|---|---|---|---|---|---|---|
| | January | | July | | January | | July | | January | | July | |
| Age (Mo.) | Milk | Fat | Milk | Fat | Milk | Fat | Milk | Fat | Milk | Fat | Milk | Fat |
| 22 | 1.31 | 1.29 | 1.42 | 1.39 | 1.37 | 1.36 | 1.44 | 1.42 | 1.37 | 1.33 | 1.49 | 1.45 |
| 24 | 1.26 | 1.25 | 1.37 | 1.34 | 1.32 | 1.31 | 1.39 | 1.37 | 1.30 | 1.26 | 1.41 | 1.37 |
| 28 | 1.19 | 1.17 | 1.29 | 1.26 | 1.25 | 1.24 | 1.31 | 1.30 | 1.22 | 1.18 | 1.31 | 1.28 |
| 32 | 1.14 | 1.11 | 1.24 | 1.21 | 1.19 | 1.18 | 1.25 | 1.25 | 1.19 | 1.15 | 1.28 | 1.25 |
| 36 | 1.10 | 1.08 | 1.20 | 1.18 | 1.13 | 1.13 | 1.22 | 1.21 | 1.18 | 1.15 | 1.27 | 1.24 |
| 42 | 1.05 | 1.03 | 1.16 | 1.14 | 1.07 | 1.07 | 1.16 | 1.16 | 1.15 | 1.12 | 1.25 | 1.24 |
| 48 | 1.01 | 1.00 | 1.14 | 1.12 | 1.03 | 1.03 | 1.13 | 1.12 | 1.10 | 1.08 | 1.22 | 1.21 |
| 60 | 0.98 | 0.97 | 1.09 | 1.08 | 0.98 | 0.98 | 1.08 | 1.08 | 1.03 | 1.02 | 1.16 | 1.16 |
| 72 | 0.96 | 0.96 | 1.07 | 1.07 | 0.96 | 0.96 | 1.06 | 1.07 | 0.99 | 0.99 | 1.12 | 1.13 |
| 90 | 0.96 | 0.97 | 1.06 | 1.08 | 0.97 | 0.97 | 1.07 | 1.08 | 0.96 | 0.96 | 1.08 | 1.09 |
| 108 | 0.96 | 0.98 | 1.07 | 1.09 | 0.99 | 1.00 | 1.09 | 1.11 | 0.96 | 0.97 | 1.08 | 1.10 |

[a] Month of calving.

**Appendix Table M-1**   Factors for Adjusting Age-Corrected Records to a Twice-a-Day Milking Basis (All Breeds) (ARS-52-1, 1955)

| No.of days milked | 3 times to twice a day | | | 4 times to twice a day | | |
|---|---|---|---|---|---|---|
| | 2–3 yr. of age | 3–4 yr. of age | 4 yr. and over | 2–3 yr. of age | 3–4 yr. of age | 4 yr. and over |
| 26–35 | 0.98 | 0.98 | 0.98 | 0.96 | 0.97 | 0.97 |
| 46–55 | 0.97 | 0.97 | 0.97 | 0.94 | 0.95 | 0.96 |
| 66–75 | 0.95 | 0.96 | 0.96 | 0.92 | 0.93 | 0.94 |
| 86–95 | 0.94 | 0.95 | 0.96 | 0.90 | 0.91 | 0.93 |
| 106–115 | 0.93 | 0.94 | 0.95 | 0.88 | 0.90 | 0.91 |
| 126–135 | 0.92 | 0.93 | 0.94 | 0.87 | 0.88 | 0.90 |
| 146–155 | 0.91 | 0.92 | 0.93 | 0.85 | 0.87 | 0.88 |
| 166–175 | 0.90 | 0.91 | 0.92 | 0.83 | 0.85 | 0.87 |
| 186–195 | 0.89 | 0.90 | 0.91 | 0.82 | 0.84 | 0.86 |
| 206–215 | 0.88 | 0.89 | 0.90 | 0.80 | 0.83 | 0.85 |
| 226–235 | 0.87 | 0.88 | 0.90 | 0.79 | 0.81 | 0.83 |
| 246–255 | 0.86 | 0.88 | 0.89 | 0.77 | 0.80 | 0.82 |
| 266–275 | 0.85 | 0.87 | 0.88 | 0.76 | 0.79 | 0.81 |
| 286–295 | 0.84 | 0.86 | 0.87 | 0.74 | 0.78 | 0.80 |
| 295–305 | 0.83 | 0.85 | 0.87 | 0.74 | 0.77 | 0.79 |

# Appendix Table M-2 Factors for Projecting Incomplete Records to 305 Days (ARS-44-239, 1972)

## For cows calving at less than 36 months of age

| Days in milk | Ayrshire Milk | Ayrshire Fat | Brown Swiss Milk | Brown Swiss Fat | Guernsey Milk | Guernsey Fat | Jersey Milk | Jersey Fat | Holstein Milk | Holstein Fat | Other breeds Milk | Other breeds Fat |
|---|---|---|---|---|---|---|---|---|---|---|---|---|
| 30  | 8.15 | 8.30 | 8.28 | 8.37 | 7.89 | 8.51 | 7.65 | 8.22 | 8.32 | 7.99 | 8.25 | 8.05 |
| 60  | 4.07 | 4.20 | 4.18 | 4.29 | 4.00 | 4.31 | 3.89 | 4.17 | 4.16 | 4.10 | 4.13 | 4.13 |
| 90  | 2.75 | 2.93 | 2.85 | 2.94 | 2.74 | 2.93 | 2.68 | 2.85 | 2.82 | 2.82 | 2.81 | 2.84 |
| 120 | 2.11 | 2.18 | 2.19 | 2.25 | 2.12 | 2.25 | 2.09 | 2.19 | 2.16 | 2.18 | 2.15 | 2.19 |
| 150 | 1.73 | 1.79 | 1.79 | 1.84 | 1.75 | 1.83 | 1.73 | 1.79 | 1.77 | 1.79 | 1.76 | 1.79 |
| 180 | 1.48 | 1.52 | 1.53 | 1.56 | 1.50 | 1.56 | 1.48 | 1.53 | 1.51 | 1.52 | 1.50 | 1.53 |
| 210 | 1.31 | 1.33 | 1.33 | 1.36 | 1.32 | 1.36 | 1.31 | 1.34 | 1.32 | 1.34 | 1.32 | 1.34 |
| 240 | 1.18 | 1.20 | 1.20 | 1.21 | 1.19 | 1.21 | 1.19 | 1.20 | 1.19 | 1.20 | 1.19 | 1.20 |
| 270 | 1.08 | 1.09 | 1.09 | 1.10 | 1.09 | 1.10 | 1.08 | 1.09 | 1.08 | 1.09 | 1.08 | 1.09 |
| 300 | 1.01 | 1.01 | 1.01 | 1.01 | 1.01 | 1.01 | 1.01 | 1.01 | 1.01 | 1.01 | 1.01 | 1.01 |

## For cows calving at 36 months of age or older

| Days in milk | Ayrshire Milk | Ayrshire Fat | Brown Swiss Milk | Brown Swiss Fat | Guernsey Milk | Guernsey Fat | Jersey Milk | Jersey Fat | Holstein Milk | Holstein Fat | Other breeds Milk | Other breeds Fat |
|---|---|---|---|---|---|---|---|---|---|---|---|---|
| 30  | 7.06 | 6.78 | 7.57 | 7.36 | 7.06 | 7.25 | 7.14 | 7.27 | 7.42 | 6.89 | 7.37 | 6.96 |
| 60  | 3.55 | 3.51 | 3.84 | 3.82 | 3.61 | 3.74 | 3.63 | 3.73 | 3.74 | 3.60 | 3.72 | 3.63 |
| 90  | 2.43 | 2.44 | 2.63 | 2.65 | 2.49 | 2.58 | 2.50 | 2.58 | 2.56 | 2.52 | 2.55 | 2.53 |
| 120 | 1.89 | 1.91 | 2.04 | 2.06 | 1.94 | 2.01 | 1.96 | 2.01 | 1.98 | 1.97 | 1.97 | 1.98 |
| 150 | 1.58 | 1.60 | 1.68 | 1.70 | 1.61 | 1.66 | 1.63 | 1.67 | 1.64 | 1.64 | 1.64 | 1.65 |
| 180 | 1.37 | 1.39 | 1.44 | 1.46 | 1.40 | 1.44 | 1.41 | 1.44 | 1.41 | 1.42 | 1.41 | 1.42 |
| 210 | 1.23 | 1.24 | 1.28 | 1.29 | 1.25 | 1.27 | 1.26 | 1.28 | 1.26 | 1.27 | 1.26 | 1.27 |
| 240 | 1.13 | 1.14 | 1.16 | 1.17 | 1.14 | 1.16 | 1.15 | 1.16 | 1.14 | 1.15 | 1.14 | 1.15 |
| 270 | 1.06 | 1.06 | 1.07 | 1.08 | 1.07 | 1.07 | 1.07 | 1.07 | 1.06 | 1.07 | 1.06 | 1.07 |
| 300 | 1.01 | 1.01 | 1.01 | 1.01 | 1.01 | 1.01 | 1.01 | 1.01 | 1.01 | 1.01 | 1.01 | 1.01 |

**Appendix Table N**  Metric Conversion Tables

1. **Weight**
   1 oz. = 28.35 g. = .0283 kg.
   1 lb. = 453.59 g. = .4536 kg.
   100 lb. = 45.3592 kg.
   2,000 lb. (1 T.) = 907.1847 kg.
   1 g. = .0353 oz.
   1,000 g. = 1.0 kg. = 2.2046 lb.
   1 g. = 1,000 mg. = .0353 oz. = .0022 lb.
   1 kg. = 1,000 g. = 35.2740 oz. = 2.2046 lb.
   1 metric ton = 1,000 kg. = 1.1023 T.

2. **Energy**
   1 c. = heat to raise 1 ml. HOH 1°C.
   1,000 c. = 1C.
   1,000 C = 1 Mcal. (therm.)

3. **Capacity**
   1 l. = 1.057 qt. (liquid)
   1 l. = 0.264 gal. (liquid)
   44 l. = 100 lb. milk
   1 qt. = 0.946 l. (liquid)
   1 gal. = 3.785 l. (liquid)
   1 bu. = 35.239 l. (dry measure)

4. **Length**
   1 mm. = 0.0394 in.
   1 cm. = 0.394 in.
   1 m. = 39.370 in.
   1 m. = 3.281 ft.
   1 m. = 1.094 yd.
   1 km. = 0.621 mi.
   1 in. = 2.54 cm.
   1 ft. = 30.48 cm.
   1 yd. = 91.44 cm.
   1 yd. = 0.914 cm.
   1 mi. = 1.61 km.
   55 mi. = 88.51 km.

5. **Area**
   1 sq. cm. = 0.155 sq. in.
   1 sq. m. = 10.764 sq. ft.
   1 sq. m. = 1.196 sq. yd.
   1 sq. ft. = 0.093 sq. m.
   1 sq. yd. = 0.836 sq. m.
   1 A = 0.405 hectares
   1 sq. mi. = 2.590 sq. km.
   1 sq. mi. = 258.999 hectares

6. **Volume**
   1 cu. m. = 35.31 cu. ft.
   1 cu. m. = 1.31 cu. yd.
   1 cu. in. = 16.387 cu. cm.
   1 cu. ft. = 0.028 cu. m.
   1 cu. yd. = 0.765 cu. m.

7. **Temperature**

| °F | °C |
|---|---|
| 212 | 100 |
| 101.5 | 38.6 |
| 98.6 | 37.0 |
| 70 | 21 |
| 40 | 4 |
| 32 | 0 |
| −110 | −79 |
| −320 | −196 |

625

# Index

627